Marine Biodiversity

Marine Biodiversity

Edited by Simon Oakenfold

SYRAWOOD
PUBLISHING HOUSE
New York

Published by Syrawood Publishing House,
750 Third Avenue, 9th Floor,
New York, NY 10017, USA
www.syrawoodpublishinghouse.com

Marine Biodiversity
Edited by Simon Oakenfold

International Standard Book Number: 978-1-68286-418-0 (Hardback)

Cataloging-in-publication Data

Marine biodiversity / edited by Simon Oakenfold.
p. cm.
Includes bibliographical references and index.
ISBN 978-1-68286-418-0
1. Marine biodiversity. 2. Marine biodiversity conservation. 3. Marine habitats. 4. Marine ecology. I. Oakenfold, Simon.
QH91.8.B6 M37 2017
577.7--dc23

Printed in China.

TABLE OF CONTENTS

PREFACE

The richness of the species and the abundance of the world's oceans and seas are referred to as marine biodiversity. It is important to study and preserve the marine biodiversity because like other resources of earth, marine resources too are facing the threat of extinction. For all readers who are interested in marine biodiversity, the case studies included in the book will serve as an excellent guide to develop a comprehensive understanding. This book is a compilation of chapters that discuss the most vital concepts and emerging trends in this field. Different approaches, evaluations, methodologies and advanced studies on marine biodiversity have been included in this book. This book includes contributions of experts and scientists which will provide innovative insights into this field.

The information contained in this book is the result of intensive hard work done by researchers in this field. All due efforts have been made to make this book serve as a complete guiding source for students and researchers. The topics in this book have been comprehensively explained to help readers understand the growing trends in the field.

I would like to thank the entire group of writers who made sincere efforts in this book and my family who supported me in my efforts of working on this book. I take this opportunity to thank all those who have been a guiding force throughout my life.

Editor

1

Habitat Use by Fishes in Coral Reefs, Seagrass Beds and Mangrove Habitats in the Philippines

Kentaro Honda[1]*, Yohei Nakamura[2], Masahiro Nakaoka[1], Wilfredo H. Uy[3], Miguel D. Fortes[4]

1 Akkeshi Marine Station, Field Science Center for Northern Biosphere, Hokkaido University, Aikappu, Akkeshi, Hokkaido, Japan, **2** Graduate School of Kuroshio Science, Kochi University, Nankoku, Kochi, Japan, **3** Institute of Fisheries Research and Development, Mindanao State University at Naawan, Misamis Oriental, The Philippines, **4** The Marine Science Institute, CS, University of the Philippines-Diliman; Diliman, Quezon City, The Philippines

Abstract

Understanding the interconnectivity of organisms among different habitats is a key requirement for generating effective management plans in coastal ecosystems, particularly when determining component habitat structures in marine protected areas. To elucidate the patterns of habitat use by fishes among coral, seagrass, and mangrove habitats, and between natural and transplanted mangroves, visual censuses were conducted semiannually at two sites in the Philippines during September and March 2010–2012. In total, 265 species and 15,930 individuals were recorded. Species richness and abundance of fishes were significantly higher in coral reefs (234 species, 12,306 individuals) than in seagrass (38 species, 1,198 individuals) and mangrove (47 species, 2,426 individuals) habitats. Similarity tests revealed a highly significant difference among the three habitats. Fishes exhibited two different strategies for habitat use, inhabiting either a single (85.6% of recorded species) or several habitats (14.4%). Some fish that utilized multiple habitats, such as *Lutjanus monostigma* and *Parupeneus barberinus*, showed possible ontogenetic habitat shifts from mangroves and/or seagrass habitats to coral reefs. Moreover, over 20% of commercial fish species used multiple habitats, highlighting the importance of including different habitat types within marine protected areas to achieve efficient and effective resource management. Neither species richness nor abundance of fishes significantly differed between natural and transplanted mangroves. In addition, 14 fish species were recorded in a 20-year-old transplanted mangrove area, and over 90% of these species used multiple habitats, further demonstrating the key role of transplanted mangroves as a reef fish habitat in this region.

Editor: Brian Gratwicke, Smithsonian's National Zoological Park, United States of America

Funding: This study was supported by Japan Science and Technology Agency (http://www.jst.go.jp/EN/index.html)/Japan International Cooperation Agency (http://www.jica.go.jp/english/index.html), Science and Technology Research Partnership for Sustainable Development (http://www.jst.go.jp/global/english/index.html), for the project Coastal Ecosystem Conservation and Adaptive Management under Local and Global Environmental Impacts in the Philippines (CECAM, http://www.cecam-project.net/). The funders had no role in study design, data collection and analysis, decision to publish, or preparation of the manuscript.

Competing Interests: The authors have declared that no competing interests exist.

* E-mail: fbmods@fish.hokudai.ac.jp

Introduction

In the tropics, seagrass beds and mangroves are formed in the shallow reef flat zone and the near coastline/estuarine region, respectively. Habitat-specific fish species inhabit either of these habitats, whereas some coral reef fishes, such as Lutjanidae, Haemulidae, Lethrinidae, Scaridae, Siganidae, and several other families, utilize these habitats as their nursery grounds [1–13]. In addition, several fish species show diel movements among these habitats for feeding or shelter [14,15]. Local connectivity by fishes and its importance among coral, seagrass, and mangrove ecosystems have received a great deal of attention in recent years [16–19]. Previous studies have indicated that the intensity and characteristics of connectivity by reef fishes widely fluctuate depending on regional differences and/or geographical conditions (see [18]), while the intensity also weakens depending on the distance between habitats [20–22]. Although several studies have been conducted in various regions under different conditions, few have been performed in Southeast Asian countries relative to other regions such as the Caribbean and Australia (see [18]). Furthermore, only a few studies have evaluated differences in the intensity or effectiveness of connectivity by reef fishes between non-estuarine and transplanted mangroves in the Indo-Pacific [11,21–24].

The rates of environmental deterioration at various scales and unabated overfishing continue to increase worldwide, resulting in reductions of fishery resources. Open access to fishing grounds [25] has become a popular trend, leaving resources at the brink of collapse [26,27]. While fish resources face threats of the loss of both biodiversity and stock replenishment, the degradation of coral reef ecosystems has also become a huge social problem (e.g., [28–30]). The establishment of marine protected areas (MPAs) is seen as an effective tool to protect coastal habitats and to enhance nearshore fisheries, especially in tropical regions (e.g., [31–33]). When establishing a MPA to protect fishery resources, the life history and habitat use of target species must first to be clearly determined. If target fishes exhibit ontogenetic habitat shifts (i.e., habitat changes with growth stages) or if fishes move daily among different habitats for feeding or shelter, all habitats being used may be equally important, regardless of scale and type; therefore, each of these habitats must be included in the establishment of MPA boundaries. Even when establishing a MPA for biodiversity conservation, the inclusion of multiple habitats within a MPA could further enhance its effectiveness, as such a MPA may be able to protect not only habitat-specific species, but also those that inhabit multiple habitats. For these reasons, understanding the interconnectivity of reef fishes among different habitats is a key requirement for making effective management plans in coastal ecosystems, particularly for determining the component habitat structures in a MPA.

Most coastal areas in the Philippines are located in the Coral Triangle, an area known for the highest biodiversity of coral worldwide [34,35]. However, habitat loss along the Philippine coasts has remarkably increased in recent years [36–38]. In particular, more than half of natural mangroves had disappeared by 1994, mainly due to the establishment of fish ponds [36,39]. Furthermore, seagrass beds are drastically decreasing, even though these habitats serve as equally important fishing grounds as coral reefs for various commercially important species such as Lethrinidae or Siganidae [40–42]. Since the 1930s–1950s, mangrove replantation projects have been implemented and have emphasized the participation of local communities in the Visayas region [43]. In recent years, however, sustaining the health of fishery resources has become difficult in the Philippines, particularly due to poor environmental governance and lack of effective, coherent monitoring programs. This situation eventually led to increased overfishing and other forms of environmental deterioration, which consequently became the topmost increasing concern regarding the proper management of fishery resources and MPAs [44–46]. From 1967 to the present, nearly 1,000 MPAs have been established in the Philippines [46]. Upon careful review, most of them focus on coral reefs, and those that incorporate multiple habitats number relatively few. In addition, the implications of connectivity among different habitats have been underexplored and are poorly understood [47,48]. In the Philippines, several studies have evaluated the effectiveness of MPA management in terms of fishery regulations (e.g., [49,50]). Furthermore, the effectiveness of fishery resource conservation was verified by a series of studies in the Sumilon and Apo islands (e.g., [51–53]). These studies documented the effects of MPAs on the enhancement of fisheries components; however, they only focused on one type of ecosystem (i.e., coral reefs) and disregarded important effects of the multiple habitats used by some commercially important fish species. If the present study can determine the

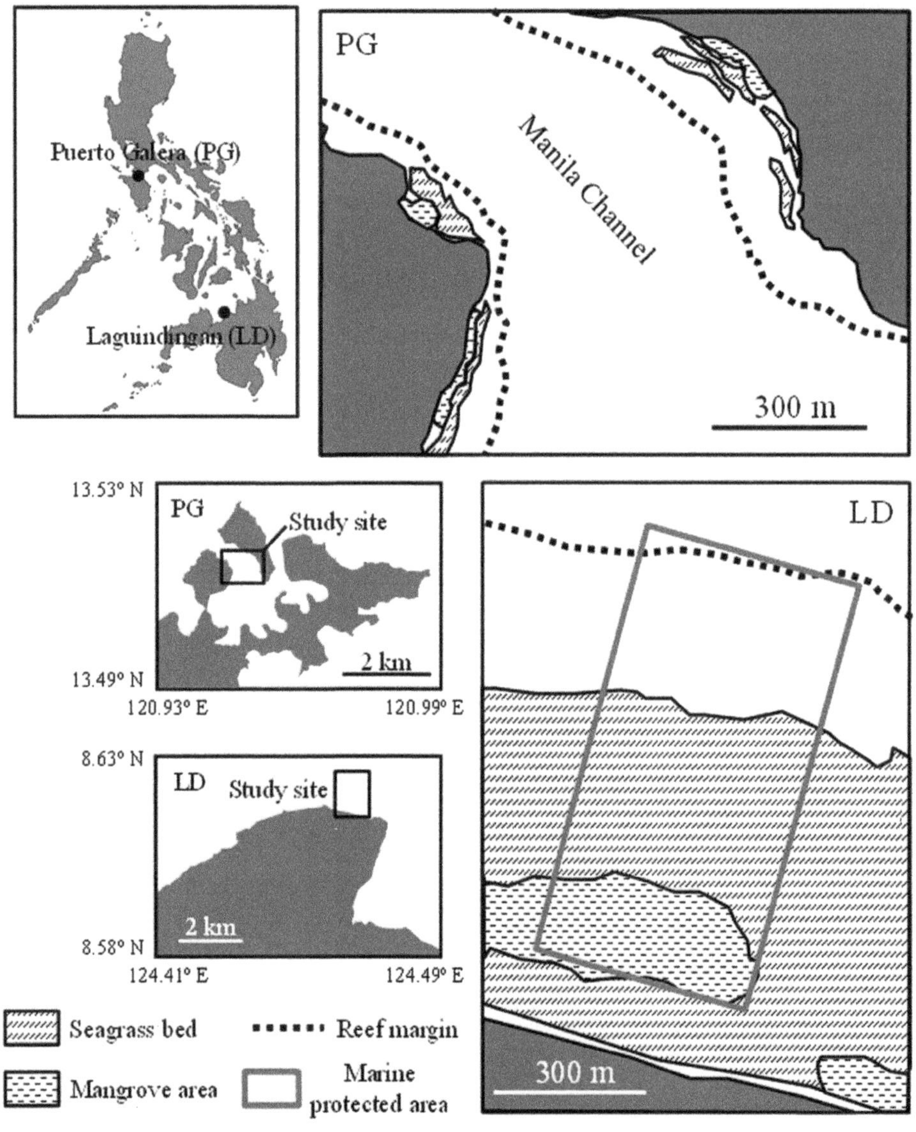

Figure 1. The two study sites in the Philippines.

specific features of each habitat, their connectivity, and corresponding importance, then our findings would be valuable for fishery resource conservation and management in regions of the Philippines. Furthermore, we also compared the transplanted mangroves to other types of habitats; such comparisons have been rare in previous studies.

The present study was designed to address differences in the pattern of habitat use by fishes, with a focus on commercial fishery species, among coral, seagrass, and mangrove habitats in the Philippines and whether transplanted and natural mangroves are used as common habitats for adult fishes and/or as potential nursery habitats for juveniles [54]. Based on our results, we further discuss the importance of including multiple habitats within MPAs.

Materials and Methods

Study Design

Field surveys were conducted semiannually for 2 years (2010 and 2011) during months representing the rainy season (September) and the dry season (March) at Puerto Galera (PG; 13°30′ N, 120°57′ E) off of northern Mindoro Island, and at Laguindingan (LD; 8°37′ N, 124°28′ E) off of northern Mindanao Island, the Philippines (Figure 1). The study site at PG was situated in a fringing reef with the reef flat zone located along both the western and eastern side of Manila Channel within Puerto Galera Bay. The study site at LD was located in a fringing reef where the reef flat zone faces the open sea. The MPAs of PG (entire study site) and LD were established in 2006 and 2002, respectively, both with a strict no-take-zone policy (Figure 1). Coral reefs at both sites are composed of hermatypic corals (e.g., tabular and branching *Acropora*; living coral coverage, >80%), which are more abundant near the reef margins. Dominant seagrass species at PG were *Thalassia hemprichii* (15.9% of cover), *Halodule pinifolia* (15.0%), and *Cymodocea rotundata* (12.0%), whereas *T. hemprichii* (63.6%) and *Enhalus acoroides* (4.3%) were dominant at LD. The mean (± SD) canopy heights at PG and LD were 6.4±5.0 cm ($n = 148$ quadrats) and 11.4±3.4 cm ($n = 156$ quadrats), respectively. *Rhizophora apiculata* and *Sonneratia* sp. were the dominant mangrove species at PG, whereas only *R. apiculata* was present at LD. Mangrove

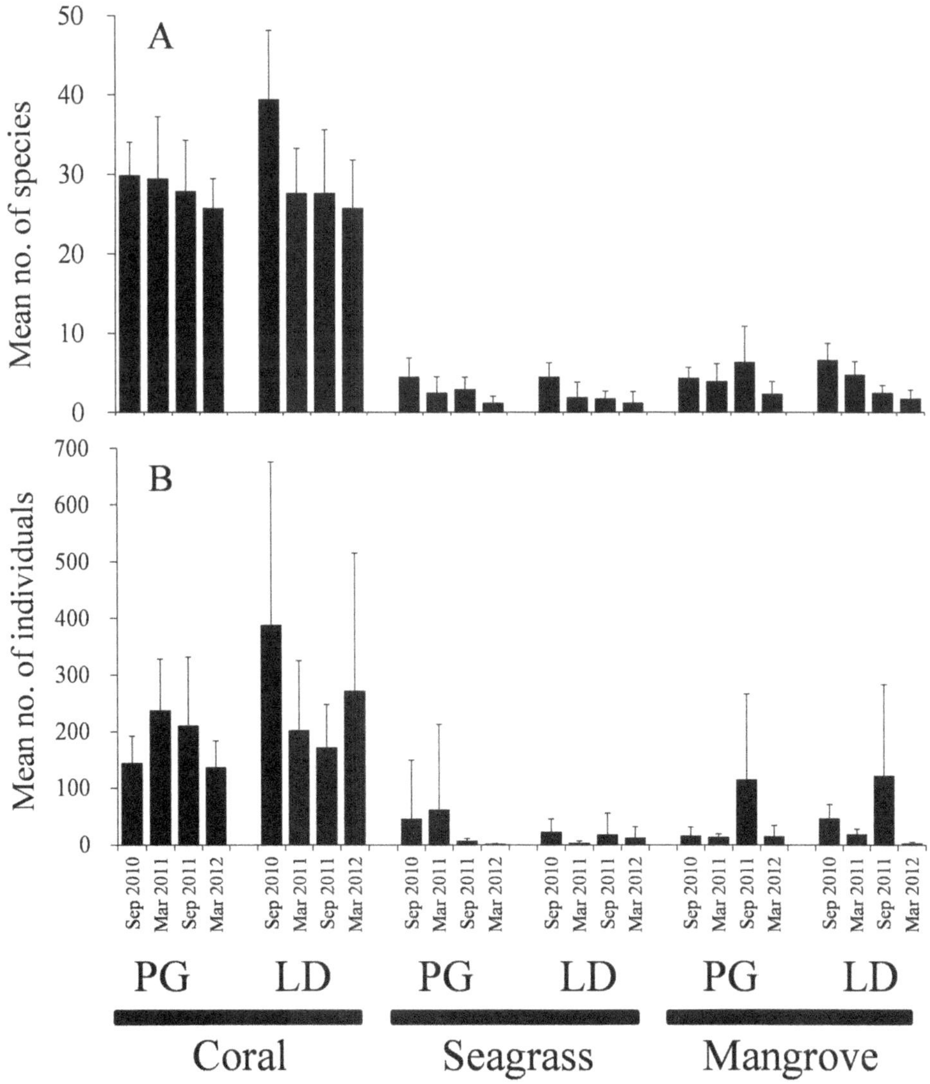

Figure 2. Mean number of fish species (A) and individuals (B) per transect (1 ×20 m) during sampling months in each habitat at each study site. Error bars are standard deviations of $n = 7$ transects. PG and LD represent Puerto Galera and Laguindingan, respectively.

areas at both sites were composed of clear-water non-estuarine mangrove (Figure 1). The mangroves at LD have been planted along seagrass beds near the shoreline since 1992, and they presently form a band of young and mature trees that protect the coastal communities from strong winds (Honda, personal communication). Fish distribution patterns among coral reefs, seagrass beds and mangrove areas were assessed during each season using an underwater visual transect survey method. In each habitat, seven 1×20-m (20 m^2) belt transects were established haphazardly using a scaled rope (see also [8,9]). Transects were separated from one another by at least 5 m, In PG, three and four line transects were established in each habitat along the western and eastern side, respectively, of Manila Channel. All fish visual censuses (FVCs) were conducted in daytime between 08:00 and 16:00 h, and fishes were identified to the lowest taxonomic level whenever possible. Individual fish size (total length) was also recorded underwater using a ruler attached to the recording slate. In coral reefs, FVC were conducted via SCUBA or snorkeling depending on the water depth (2.0–8.0 m). In seagrass beds (0.5–1.0 m deep at low tide, 1.5–2.0 m at high tide) and mangrove areas (0.5–1.0 m at low tide, 1.0–1.5 m at high tide), only snorkeling was used and FVCs were conducted when the depth ranged from 1.0 to 1.5 m to avoid tidal effects [55]. Visibility within the water at any transect generally exceeded 7 m. Sea surface water temperature at PG and LD was 30.0°C and 30.4°C in September and 27.8°C and 28.8°C in March, respectively. Salinity at both sites was about 34‰, and no estuaries were present near either site. All methods utilized in the present study were conducted under the permit requirements of the municipal government of PG and the Barangay Tubajon in LD.

Data Analysis

At both sites, data collected from each habitat type were analyzed for species composition and density. Because assumptions of homogeneity of variance could not be met by some data even after transformations, nonparametric Steel–Dwass tests were used to determine whether species and individual numbers of fishes differed among habitats for each site, month, and year (see also [9]). Moreover, these variables were compared between PG and LD mangrove areas in each sampling month using Mann–Whitney *U*-tests. Family composition of species and individuals in each habitat at each site was also estimated.

The similarity of fish assemblages among habitats was examined using data from the seven transects within each habitat for every month. The Chao index [56] was used for this analysis and results were visualized using nonmetric multidimensional scaling (NMDS). Similarity tests among four variables (year, month, site and habitat) were conducted using nonparametric multivariate analysis of variance (NPMANOVA; $\alpha = 0.05$). All statistical analyses were performed using the "vegan" package of R ver. 2.14.1 (R Development Core Team).

Counts of each fish species that occurred in two or more habitats were analyzed to verify the presence of fish that use multiple habitats. To avoid incidental detection, instances of only one individual-of a species recorded in a habitat, along with unidentified species, were excluded from analysis. Moreover, based on the commercial fishery species listed in Fishbase [57], the number of commercial fish species utilizing a particular habitat type or a combination of habitats was determined. Using these data, the habitat or combination of habitas favored by a large number of commercial species was evaluated. Here, commercial species included species listed as "highly commercial" or "commercial" in Fishbase, while other categories, such as "minor commercial," "subsistence fisheries," "of no interest," and "no information," were not regarded as commercial species. Here, *Pomacentrus lepidogenys*, which was categorized as "highly commercial" in Fishbase, was considered a noncommercial species together with other pomacentrids, because it is highly unlikely that this species was of high fishery importance in the Philippines. Moreover, for cases in which fish exhibited possible ontogenetic habitat shifts, the size distribution pattern in each habitat was visualized. Fish species were considered to undergo possible ontogenetic habitat shifts based on individual counts or mean length. If the individual count of a fish species reached 10 or more within juvenile habitats (seagrass and/or mangrove) and five or more in coral reefs, then this fish species was considered to undergo a potential ontogenetic habitat shift. In addition, if the mean total length of a fish species from the coral reef was significantly longer than that in the juvenile habitat (Mann–Whitney *U*-test, $\alpha = 0.05$), then such a fish species may also exhibit an ontogenetic habitat shift. Species belonging to Atherinidae and Gobidae families were excluded from all analyses because they are pelagic and small cryptic fishes, respectively.

Results

Fish Assemblage Structure

In total, 15,930 individuals, belonging to 265 species in 45 families were recorded (Table S1). In coral reefs, 12,305 individuals comprising 234 species in 37 families were recorded. In contrast, fewer fish were recorded in seagrass beds (1,198 individuals belonging to 38 species in 18 families) and mangrove areas (2,426 individuals belonging to 47 species in 24 families). The mean numbers of species and individuals per transect in coral areas at both PG and LD sites were significantly higher than those in seagrass and mangrove habitats ($P<0.05$), with four exceptions for the number of individuals (seagrass beds in September 2010 and March 2011 at PG and mangrove areas in September 2011 in both PG and LD; Figure 2). Seagrass and mangrove habitats did not significantly differ in terms of either the number of species or individuals ($P>0.05$), although the numbers of both species and individuals at LD in March 2011 differed significantly between seagrass and mangrove habitats. Neither the number of species nor individuals ($P>0.05$) significantly differed between PG and LD mangroves, except during September 2011, when the number of species in mangroves was higher at PG than that at LD.

The three most dominant families in terms of the number of species in PG coral reefs were Pomacentridae, Labridae, and Chaetodontidae (Figure 3a). Dominant families in LD were Labridae and Pomacentridae, and the combined abundance of these families accounted for more than 40% at both PG and LD. For fish in seagrass beds at PG, Labridae accounted for about 30% of all species, followed by Muraenidae, Syngnathidae, Nemipteridae, and Scaridae, in that order. Labridae and Apogonidae together accounted for more than half of the fish species in LD seagrass beds. In mangrove areas, the three most dominant families were Nemipteridae, Pomacentridae and Labridae at PG and Nemipteridae, Siganidae and Lutjanidae at LD. Pomacentridae (represented by *Chromis ternatensis*, *Acanthochromis polyacanthus*, and *Pomacentrus moluccensis*) was the most dominant family in terms of the number of individuals in coral reefs at both PG (69.5%) and LD (58.6%), followed by Seranidae (represented by *Pseudanthias huchti*; Figure 3b). Fish family composition in seagrass beds differed between PG and LD. Only two species, *Plotosus lineatus* (Plotosidae) and *Siganus spinus* (Siganidae), accounted for over 80% of all fish species at PG. In contrast, at LD, Apogonidae (represented by *Apogon ceramensis*), Labridae (represented by *Halichoeres argus* and *Halichoeres scapularis*), and Siganidae (represented by *S. spinus*) were

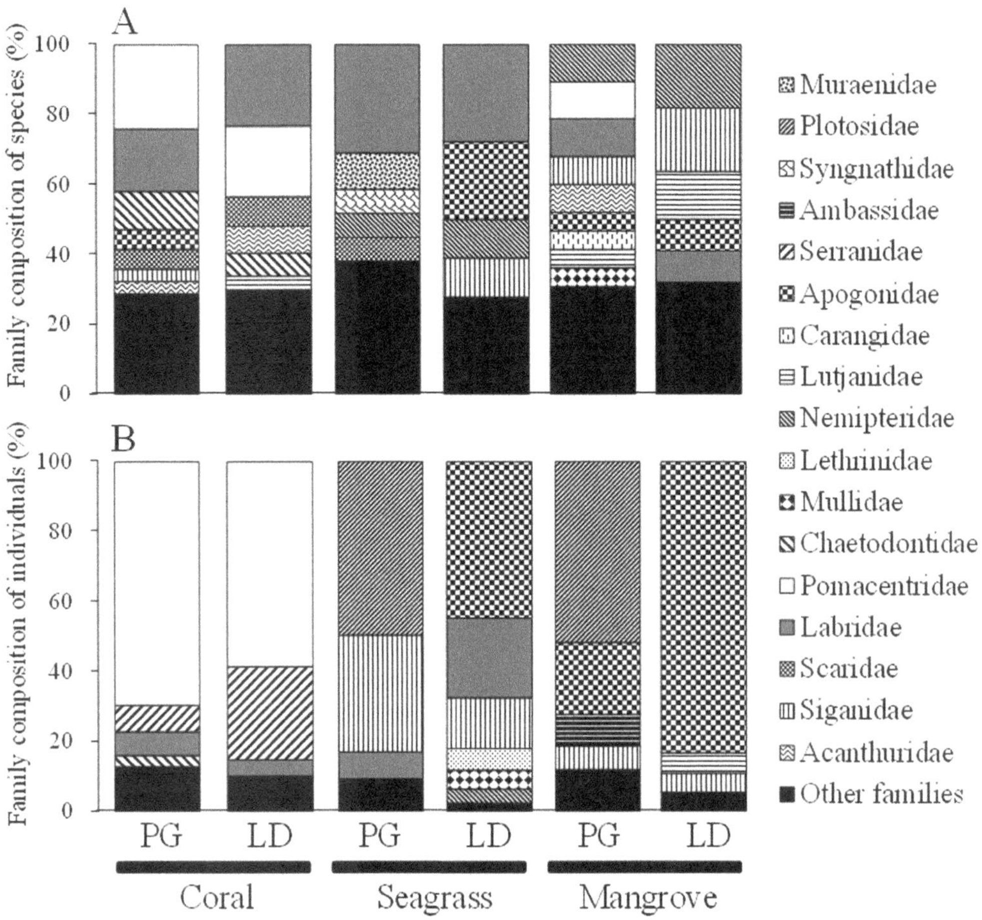

Figure 3. Relative family composition of fish species (A) and number of individuals (B) in the different habitats at each study site.

the three most dominant families, together comprising 80% of fish individuals. In mangrove areas, Plotosidae (represented only by *P.lineatus*) and Apogonidae (represented by *Sphaeramia orbicularis* and *A. ceramensis*) together accounted for about 70% of fish individuals at PG. Apogonidae also represented by *S. orbicularis* and *A. ceramensis*) accounted for more than 80% of fish individuals at LD.

At the species level, 199 species were unique to coral reefs, whereas nine and 15 species were unique to seagrass beds and mangrove areas, respectively (Table S1). At the family level, 14 families were only recorded in coral reefs: Holocentridae, Aulostomidae, Centriscidae, Pseudochromidae, Priacanthidae, Caesionidae, Chaetodontidae, Kyphosidae, Blennidae, Gobiesocidae, Zanclidae, Balistidae, Ostraciidae, and Diodontidae. Five families were only recorded in the mangrove areas: Mugilidae, Hemiramphidae, Belonidae, Ambassidae, and Terpontidae. No families were unique to seagrass habitats.

Similarity indices revealed that fish communities could be divided into three large groups (coral, seagrass, and mangrove habitat types) regardless of sampling month and site (Figure 4). Results of similarity tests using NPMANOVA revealed a highly significant difference among habitats ($F = 11.28$, $P < 0.001$). Other variables, such as sampling period (year, month), did not significantly affect patterns of fish structure ($F < 2.0$, $P > 0.05$), although a marginal difference was observed between sites ($F = 1.92$, $P = 0.08$).

Fishes Utilized Multiple Habitats

In total, 29 fish species accounting for 14.4% of species abundance were recorded in multiple habitats (Table 1). Six species were recorded in both coral and seagrass habitats and three of these belonged to Labridae (Table S1). Nine species were recorded in both coral and mangrove habitats: three species belonged to Pomacentridae, while Lutjanidae and Siganidae were represented by two species each. Six species were observed in both seagrass and mangrove habitats, and each belonged to different families. Eight species were recorded in all three habitats, and Labridae and Siganidae were represented by two species each.

For commercial species, the total number of species in coral reefs was 27 (62.8%), which was relatively greater than the numbers found in seagrass beds (two species, 4.7%) and mangrove areas (four species, 9.3%; Table 1). Ten commercial species (23.3%) were recorded in multiple habitats. Sixteen commercial species used multiple habitats or exclusively used either seagrass beds or mangrove areas, accounted for 37.2% of commercial species (i.e., "all except coral reef only" group in Table 1).

Fourteen fish species were recorded in the transplanted mangrove area at LD, 13 of which also utilized coral and/or seagrass habitats (Table 1). Even though minimal differences were found between PG and LD in terms of the numbers of total species, commercial species, and multiple habitat users, the number of all fish species that were mangrove users, including those with commercial value, was twice higher at PG than at LD (Table 1). In addition, PG and LD differed greatly in the number

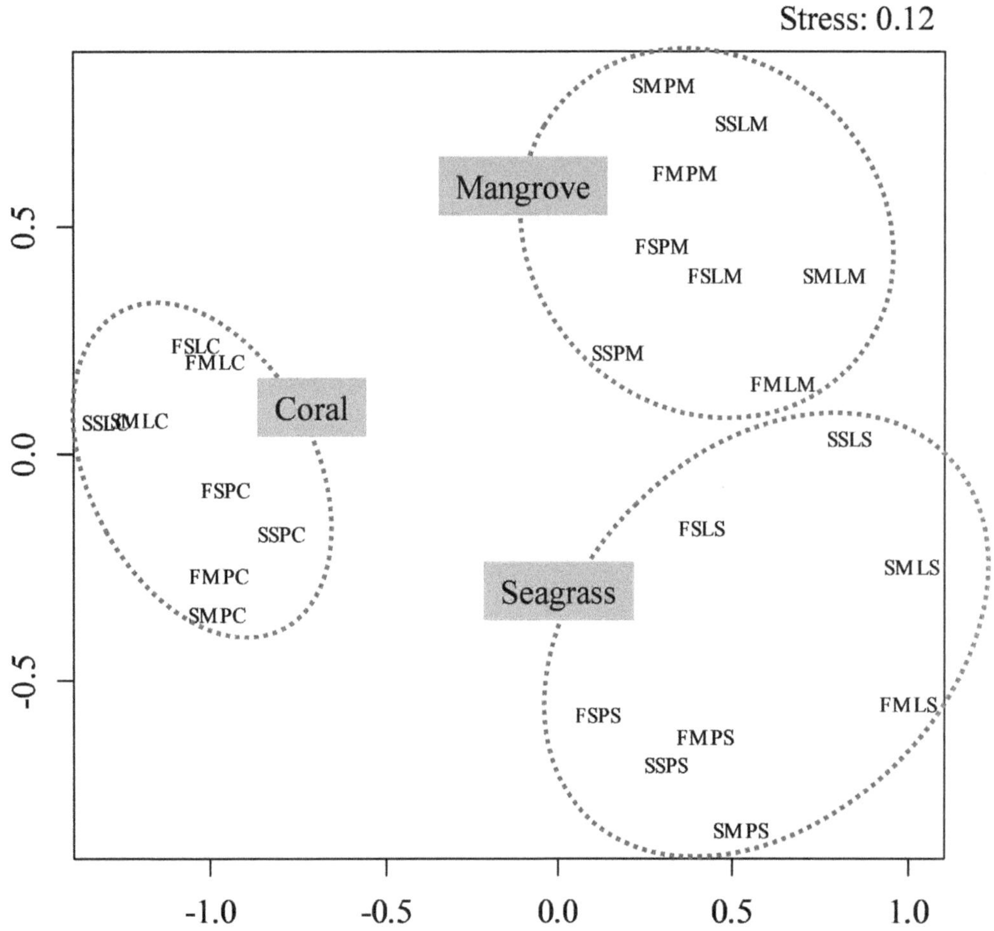

Figure 4. Nonmetric multidimensional scaling (NMDS) of data from all seven transects in each habitat at each site in each month of each year. Each data set is represented by four letters in the order from Year (First or Second), Month (September or March), Site (Puerto Galera or Laguindingan), and Habitat (Coral reef, Seagrass bed, or Mangrove area).

Table 1. Number of fish species at all sampled sites for each habitat and each combination of different habitats.

Habitat number	Observed habitat	All species			Commercial species		
		Puerto Galera	Laguindingan	Both sites	Puerto Galera	Laguindingan	Both sites
1	Coral reef only	110 (74.3)	103 (83.1)	157 (78.1)	17 (54.8)	20 (71.4)	27 (62.8)
2	Seagrass bed only	5 (3.4)	4 (3.2)	6 (3.0)	2 (6.5)	0 (0.0)	2 (4.7)
3	Mangrove area only	14 (9.5)	1 (0.8)	9 (4.5)	6 (19.4)	0 (0.0)	4 (9.3)
4	Both coral reef & seagrass bed	5 (3.4)	3 (2.4)	6 (3.0)	0 (0.0)	2 (7.1)	1 (2.3)
5	Both coral reef & mangrove area	7 (4.7)	5 (4.0)	9 (4.5)	3 (9.7)	3 (10.7)	3 (7.0)
6	Both seagrass bed & mangrove area	3 (2.0)	7 (5.6)	6 (3.0)	2 (6.5)	2 (7.1)	2 (4.7)
7	All three habitats	4 (2.7)	1 (0.8)	8 (4.0)	1 (3.2)	1 (3.6)	4 (9.3)
4–7	Multiple habitats	19 (12.8)	16 (12.9)	29 (14.4)	6 (19.4)	8 (28.6)	10 (23.3)
3, 5–7	Mangrove area	28 (18.9)	14 (11.3)	32 (15.9)	12 (38.7)	6 (21.4)	13 (30.2)
2–7	All except coral reef only	38 (25.7)	21 (16.9)	44 (21.9)	14 (45.2)	8 (28.6)	16 (37.2)
1–7	All	148 (100)	124 (100)	201 (100)	31 (100)	28 (100)	43 (100)

Percentages of totals are shown in parentheses. Instances for which only one individual was recorded in a habitat for each species were excluded (i.e., numbers differ from those in Table S1). All unidentified species were also excluded.

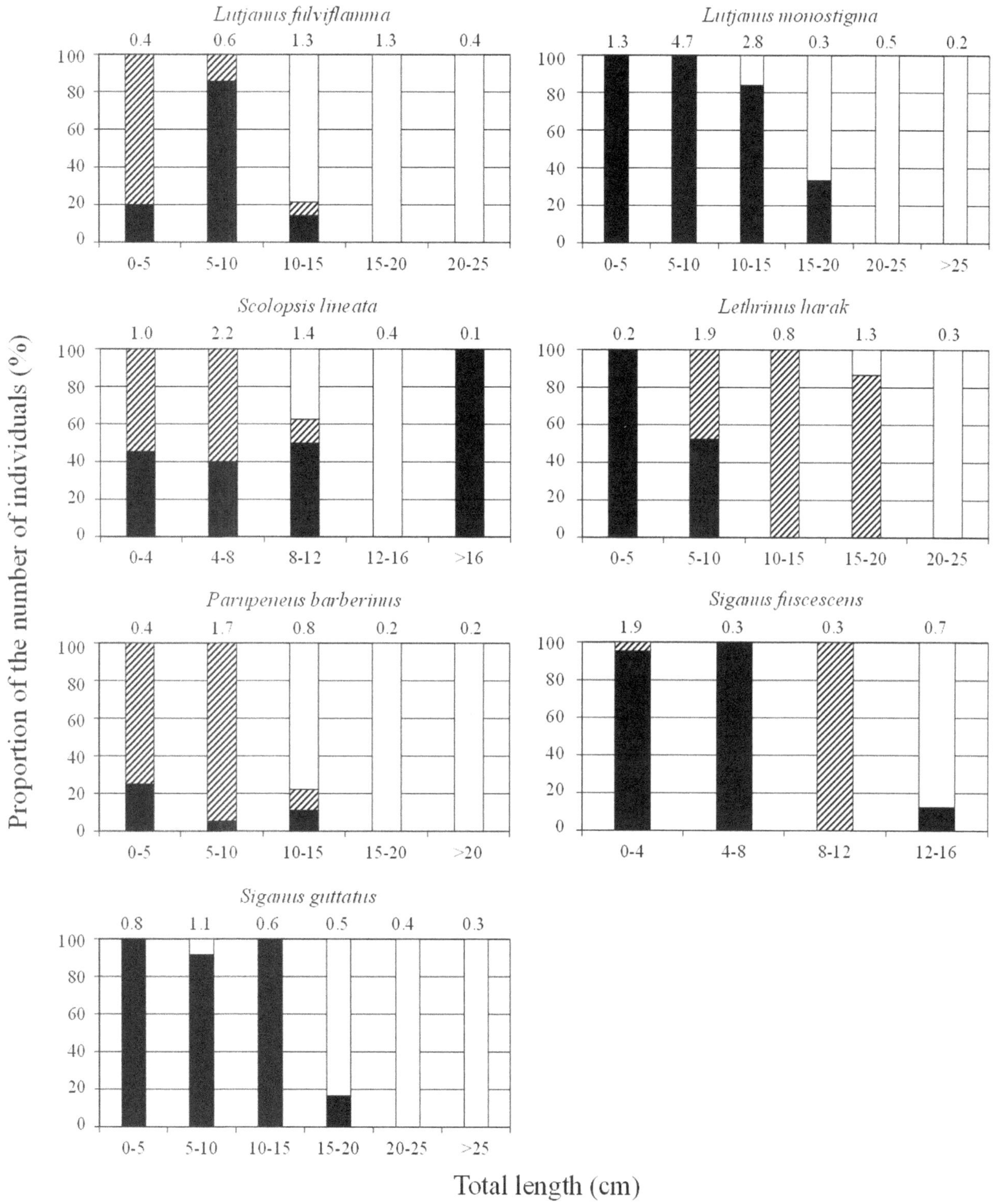

Figure 5. Relative abundance of the seven fish species in coral reefs (white), seagrass beds (hatched), and mangrove areas (black) across habitats according to class sizes using pooled data from Puerto Galera and Laguindingan. See Table S1 for size distribution of fishes in each habitat for every site. Mean individual numbers per 1,000 m^2 during the period are given above each size-class column.

of species observed in only mangrove habitat (Table 1): *S. orbicularis* was the only species recorded as a mangrove only user in LD, whereas 14 species were exclusively recorded in the mangrove areas at PG.

In the present study, seven species (Lutjanus fulviflamma, Lutjanus monostigma, Scolopsis lineata, Lethrinus harak, Parupeneus barberinus, Siganus fuscescens, and Siganus guttatus) exhibited possible ontogenetic habitat shifts from seagrass beds and/or mangrove areas to coral reefs (Figure 5). Of these seven species, multiple adult-sized individuals of only L. harak (n = 8) occurred in seagrass beds.

Discussion

The present study revealed that fish assemblage structure varied significantly among coral, seagrass, and mangrove habitats at the study sites, although differences between seagrass and mangrove habitats in terms of species richness and abundance were not significant (Figures 2, 4). The 199 fish species recorded only in coral reefs was accounted for approximately 75% of all fish species recorded, whereas only nine and 15 fish species exclusively utilized in seagrass and mangrove habitats, respectively (Table S1). Although the majority of fish species was found in coral reefs, the other habitats also yielded several unique species. The fact that five families of fish were found only in mangrove habitats emphasizes the need to conserve multiple habitats even without considering connectivity. Different habitats exhibit different environmental conditions and fish assemblage structures; thus, managing all of these habitats can serve as a very effective method of conserving coastal biodiversity.

Even though each habitat exhibited different fish compositions, many fish species were found to use multiple habitats, and these fish could be categorized into two groups. The first group includes those species that generally did not change habitat preference after settlement although they did inhabit more than one habitat (also see [19]). The second group includes fishes that use seagrass and/or mangrove habitats as their feeding or shelter grounds at the adult stage or as nursery grounds in the juvenile stage. Adult-sized *Lutjanus griseus* and mullids are examples of fish that migrate daily into seagrass/mangrove habitats for feeding, similar to the observations of Nakamura and Tsuchiya [8] and Luo et al. [15]. In this study, several adult-sized *L. harak* were observed not only in coral reefs but also in seagrass beds. Although we could not determine if these individuals migrated between habitats or if each habitat harbored its own population, adult individuals of coral fishes have often been observed in seagrass beds and/or mangrove areas in previous studies (e.g., [11,24]). Adult *P. barberinus* and *S. guttatus* were sometimes observed in seagrass beds at the study sites, although these fish were not recorded in the transect survey. Feeding behavior of *L. harak* and *P. barberinus* was also observed in seagrass beds (Honda, personal observation). Moreover, seven species were confirmed to use seagrass and/or mangrove habitats during the juvenile stage. Even though fish species commonly exhibit ontogenetic habitat shifts (see [18,19]), it is also important to recognize that the juvenile fish of some species were found only in seagrass beds and/or mangroves (e.g., [2,9]). Such habitat fidelity is considered to reduce the likelihood of fish flexibility or opportunism in habitat use.

More than 37% of the commercial fish recorded in this study utilized seagrass and/or mangrove habitats or one of these habitats in combination with coral reefs, and more than 34% of the fish that utilized multiple habitats in this study were commercial species (Table 1). In addition, six of seven species that exhibited possible ontogenetic habitat shifts (the exception being *S. lineata*) were commercial species. Based on previous reports, many of the species that exhibit ontogenetic habitat shifts, as well as those species that migrate among different habitats in their adult stage, have fishery value (e.g., [2,12,15]). Thus, the inclusion of adjacent seagrass beds and mangrove areas connected to coral reefs in the same MPA offers several important benefits: increased carbon dioxide fixation or sequestration by seagrasses and mangroves [58], buffering against disasters such as high waves [59], enhanced conservation of biodiversity of organisms, and increased sustainability of fishery resources [21,22,42,60].

Regrettably, seagrass beds and mangrove forests are disappearing worldwide [61–63]; consequently, the species richness and biomass of fishes and invertebrates decrease with such habitat losses [16,64]. In the present study, 14 fish species that utilize mangroves were recorded in the transplanted mangrove area, and most of these species were multiple habitat users (Table 1), indicating that both natural and transplanted mangroves play an important role as habitat for some reef fishes. Moreover, this finding suggests that transplanting mangroves can be useful in terms of fishery resource conservation and recovery. The total number of fish species considered as mangrove users at LD was half in the value observed at PG, even though the two sites did not greatly differ in the number of multiple habitat users (Table 1). Almost two decades have passed since mangroves were transplanted at LD; however, much more time may be needed for the colonization of fish species that are dependent on mangroves for their recovery and population replenishment, as only a few natural mangroves exist nearby as sources of populations of fish species. The fact that *S. orbicularis* was the only fish species unique to mangrove areas at LD supports this hypothesis.

Blaber and Milton [65] and Thollot [66] reported that fish species diversity in clear-water mangroves was lower than that in estuarine mangroves. In most cases, estuarine mangroves are surrounded by more simply structured habitats, such as mud or sand habitats, whereas clear-water non-estuarine mangroves connected to coral reefs exhibit a more complex structure. Such differences in the complexity of the surrounding habitats may affect the structure of species diversity between clear-water non-estuarine mangroves and estuarine mangroves [24]. In addition, a few reports have indicated that clear-water non-estuarine mangroves in the Indo-Pacific serve as juvenile habitat for reef fishes [11]. Barnes et al. [24] compared fish assemblage structure between coral reefs and clear-water non-estuarine mangrove areas near Orpheus Island in the Great Barrier Reef, and they found that no specific reef fish species used clear-water non-estuarine mangroves as their juvenile habitats. Nevertheless, we found that seven reef fish species exhibited possible ontogenetic habitat shifts from clear-water non-estuarine mangroves to coral reefs. This finding strongly indicates that clear-water non-estuarine mangroves in the Indo-Pacific function as the juvenile habitat of some reef fishes, similar to mangroves in the Caribbean region (e.g., [1,67]) and estuarine mangroves in the Indo-Pacific [6,9,68].

Human population near coastal areas in the tropics are expected to double in the next 50–100 years, ultimately leading to increased fishing pressure and therefore accelerated biodiversity loss and depletion of fishery resources [69]. Even though several limited MPAs contribute to fishery resource conservation and recovery (e.g., [51,70]), coastal fisheries are not effectively managed in most coastal areas in Southeast Asian regions, including the Philippines [46]. This situation will presumably get worse in the future with increases in the human population. In the tropics, not only overfishing but also juvenile habitat loss, followed by decreases in the survival rate of juveniles, have strongly contributed to declines in fishery resources in recent years. The conservation and development of juvenile habitats would delay the exhaustion of such resources.

Supporting Information

Table S1 Total number of individuals and size range of 265 species in three habitats at two study sites during

the study period. PG, LD, and TL indicate Puerto Galera, Laguindingan, and total length, respectively.

Acknowledgments

We thank K. Nadaoka and E. Tsukamoto of the Tokyo Institute of Technology, Y. Hamamitsu and Y. Nagahama of the Japan International Coperation Agency, M. Watai of Kochi University, K.O. Bolisay and J. Leriorato of the University of the Philippines Diliman, and A.D.S. Pantallano and V.E. Leopardas of Mindanao State University at Naawan for their cooperation and assistance. We also thank anonymous reviewers for their helpful comments.

Author Contributions

Conceived and designed the experiments: YN MN WHU MDF. Performed the experiments: KH YN WHU. Analyzed the data: KH. Contributed reagents/materials/analysis tools: MN WHU. Wrote the paper: KH YN MN. Acquired budgets: MN WHU MDF.

References

1. Nagelkerken I, Dorenbosch M, Verberk WCEP, Cocheret de la Morinière E, van der Velde G (2000) Importance of shallow water biotopes of a Caribbean bay for juvenile coral reef fishes: patterns in biotope association, community structure and spatial distribution. Mar Ecol Prog Ser 202: 175–192.
2. Nagelkerken I, van der Velde G, Gorissen MW, Meijer GJ, van't Hof T, et al. (2000) Importance of mangroves, seagrass beds and the shallow coral reef as a nursery for important coral reef fishes, using a visual census technique. Estuar Coast Shelf Sci 5: 31–44.
3. Nagelkerken I, Roberts CM, van der Velde G, Dorenbosch M, van Riel MC, et al. (2002) How important are mangroves and seagrass beds for coral-reef fish? The nursery hypothesis tested on an island scale. Mar Ecol Prog Ser 244: 299–305.
4. Nakamura Y, Sano M (2004) Overlaps in habitat use of fishes between a seagrass bed and adjacent coral and sand areas at Amitori Bay, Iriomote Island, Japan: importance of the seagrass bed as juvenile habitat. Fish Sci 70: 788–803.
5. Dorenbosch M, Grol MGG, Christianen MJA, Nagelkerken I, van der Velde G (2005) Indo-Pacific seagrass beds and mangroves contribute to fish density and diversity on adjacent coral reefs. Mar Ecol Prog Ser 302: 63–76.
6. Nakamura Y, Horinouchi M, Shibuno T, Tanaka Y, Miyajima T, et al. (2008) Evidence of ontogenetic migration from mangroves to coral reefs by black-tail snapper *Lutjanus fulvus*: stable isotope approach. Mar Ecol Prog Ser 355: 257–266.
7. Nakamura Y, Hirota K, Shibuno T, Watanabe Y (2012) Variability in nursery function of tropical seagrass beds during fish ontogeny: timing of ontogenetic habitat shift. Mar Biol 159: 1305–1315.
8. Nakamura Y, Tsuchiya M (2008) Spatial and temporal patterns of seagrass habitat use by fishes at the Ryukyu Islands, Japan. Estuar Coast Shelf Sci 76: 345–356.
9. Shibuno T, Nakamura Y, Horinouchi M, Sano M (2008) Habitat use patterns of fishes across the mangrove-seagrass-coral reef seascape at Ishigaki Island, southern Japan. Ichthyol Res 55: 218–237.
10. Unsworth RKF, Salinas De Leon P, Garrard SL, Jompa J, Smith DJ, et al. (2008) High connectivity of Indo-Pacific seagrass fish assemblages with mangrove and coral reef habitats. Mar Ecol Prog Ser 353: 213–224.
11. Unsworth RKF, Garrard SL, Salinas De Leon P, Cullen LC, Smith DJ, et al. (2009) Structuring of Indo-Pacific fish assemblages along the mangrove-seagrass continuum. Aquat Biol 5: 85–95.
12. Verweij MC, Nagelkerken I, Hans I, Ruseler SM, Mason PRD (2008) Seagrass nurseries contribute to coral reef fish populations. Limnol Oceanogr 53: 1540–1547.
13. Kimirei IA, Nagelkerken I, Trommelen M, Blankers P, van Hoytema N, et al. (2013) What drives ontogenetic niche shifts of fishes in coral reef ecosystems? Ecosystems. doi:10.1007/s10021-013-9645-4.
14. Krumme U (2009) Diel and tidal movements by fish and decapods linking tropical coastal ecosystems. In: Nagelkerken I, editor. Ecological Connectivity among Tropical Coastal Ecosystems. Dordrecht, The Netherlands: Springer Science and Business Media. 271–324.
15. Luo J, Serafy JE, Sponaugle S, Teare PB, Kieckbusch D (2009) Diel and seasonal movement of gray snapper (*Lutjanus griseus*) among subtropical seagrass, mangrove and coral reef habitats. Mar Ecol Prog Ser 380: 255–269.
16. Mumby PJ, Edwards AJ, Arias-Gonzalez JE, Lindeman KC, Blackwell PG, et al. (2004) Mangroves enhance the biomass of coral reef fish communities in the Caribbean. Nature 427: 533–536.
17. Nagelkerken I, van der Velde G (2004) Relative importance of interlinked mangroves and seagrass beds as feeding habitats for juvenile reef fish on a Caribbean island. Mar Ecol Prog Ser 274: 153–159.
18. Nagelkerken I (2009) Evaluation of nursery function of mangroves and seagrass beds for tropical decapods and reef fishes: patterns and underlying mechanisms. In: Nagelkerken I, editor. Ecological Connectivity among Tropical Coastal Ecosystems. Dordrecht, The Netherlands: Springer Science and Business Media. 357–399.
19. Adams A, Dahlgren C, Kellison GT, Kendall MS, Layman CA, et al. (2006) Nursery function of tropical back-reef systems. Mar Ecol Prog Ser 318: 287–301.
20. Dorenbosch M, Verberk WCEP, Nagelkerken I, van der Velde G (2007) Influence of habitat configuration on connectivity between fish assemblages of Caribbean seagrass beds, mangroves and coral reefs. Mar Ecol Prog Ser 334: 103–116.
21. Olds AD, Connolly RM, Pitt KA, Maxwell PS (2012) Primacy of seascape connectivity effects in structuring reef fish assemblages. Mar Ecol Prog Ser 462: 191–203.
22. Olds AD, Connolly RM, Pitt KA, Maxwell PS (2012) Habitat connectivity improves reserve performance. Conserv Lett 5: 56–63.
23. Crona BI, Rönnbäck P (2007) Community structure and temporal variability of juvenile fish assemblages in natural and replanted mangroves, *Sonneratia alba* Sm., of Gazi Bay, Kenya. Est Coast Shelf Sci 74: 44–52.
24. Barnes L, Bellwood DR, Sheaves M, Tanner JK (2012) The use of clear-water non-estuarine mangroves by reef fishes on the Great Barrier Reef. Mar Biol 159: 211–220.
25. Gordon HS (1954) The economic theory of a common property resource: the fishery. J Polit Econ 62: 124–142.
26. Pauly D, Christensen V, Guenette S, Pitcher TJ, Sumaila UR, et al. (2002) Towards sustainability in world fisheries. Nature 418: 689–695.
27. Berkes F, Hughes TP, Steneck RS, Wilson JA, Bellwood DR, et al. (2006) Globalization, roving bandits and marine resources. Science 311: 1557–1558.
28. Bellwood DR, Hughes TP, Folke C, Nystrom M (2004) Confronting the coral reef crisis. Nature 429: 827–832.
29. Wilkinston CR, editor. (2004) Status of Coral Reefs of the World 2004. Cape Ferguson, Qld, Australia: Australian Institute of Marine Science. 557 p.
30. Kleypas JA, Yates KK (2009) Coral reefs and ocean acidification. Oceanography 22: 108–117.
31. Hannesson R (1998) Marine reserves: what should they accomplish? Mar Resour Econ 13: 159–170.
32. Roberts CM, McClean DJ, Veron JEN, Hawkins JP, Allen GR, et al. (2002) Marine biodiversity hotspots and conservation priorities for tropical reefs. Science 295: 1280–1284.
33. Edgar GJ, Russ GR, Babcock RC (2007) Marine protected areas. In: Connell SJ, Gillanders BM, editors. Marine Ecology. Melbourne, Australia: Oxford University Press.
34. Spalding M, Ravilious C, Green EP (2001) World Atlas of Coral Reefs. Berkeley: University of California Press.
35. Carpenter KE, Springer VG (2005) The center of the center of marine shore fish biodiversity: the Philippine Islands. Environ Biol Fish 72: 467–480.
36. Primavera JH (2000) Development and conservation of Philippine mangroves: institutional issues. Ecol Econ 35: 91–106.
37. Bruno JF, Selig ER (2007) Regional decline of coral cover in the Indo-Pacific: timing, extent, and subregional comparisons. PLoS ONE 2: e711.
38. Fortes MD (2008) Ecological changes in seagrass ecosystems in Southeast Asia. In: Mimura N, editor. Chapter 3, Asian-Pacific Coasts: States of Environments and Their Management: States of environment. Dordrecht, The Netherlands: Springer Science and Business Media. 131–136.
39. Primavera JH (1995) Mangroves and brackish water pond culture in the Philippines. Hydrobiologia 295: 303–309.
40. Kirkman H, Kirkman JA (2002) The management of seagrasses in Southeast Asia. Bull Mar Sci 71: 1379–1390.
41. Fortes MD (2004) Wetland conservation and management in the Philippines: where are we now? The case of seagrass and mangrove. In: Wong MH, editor. Wetlands Ecosystems in Asia. Amsterdam, The Netherlands: Elsevier. 233–262.
42. Unsworth RKF, Cullen LC (2010) Recognising the necessity for Indo-Pacific seagrass conservation. Conserv Lett 3: 63–73.
43. Primavera JH, Esteban JMA (2008) A review of mangrove rehabilitation in the Philippines: successes, failures and future prospects. Wetl Ecol Manage 16: 345–358.
44. Christie P, White A, Deguit E (2002) Starting point or solution? Community-based marine protected areas in the Philippines. J Environ Manage 66: 441–454.
45. Alcala AC, Russ GR (2006) No-take marine reserves and reef fisheries management in the Philippines: a new people power revolution. Ambio 35: 245–254.
46. Weeks R, Russ GR, Alcala AC, White AT (2009) Effectiveness of marine protected areas in the Philippines for biodiversity conservation. Conserv Biol 24: 531–540.
47. Dolar MLL, Lepiten LV (1991) A study on the mangrove, seagrass and coral reef fishes of North Bais Bay, Central Philippines. Proceedings of a workshop on Mangrove Fisheries and Connections. ASEAN-Australia Marine Science Project: Living Coastal Resources at Ipoh, Perak, University of Malaysia. 137–172.

48. Kochzius M (1999) Interrelation of ichthyofauna from a seagrass meadow and coral reef in the Philippines. In: Proceedings of the 5th Indo-Pacific Fisheries Conference, Nouméa, 1997. Moorea: Antenne Museum-Ephe. 517–535.
49. Russ GR, Alcala AC, Maypa AP, Calumpong P, White AT (2004) Marine reserve benefits local fisheries. Ecol Appl 14: 597–606.
50. Garces LR, Pido MD, Tupper MH, Silvestre GT (2012) Evaluating the management effectiveness of three marine protected areas in the Calamianes Islands, Palawan Province, Philippines: process, selected results and their implications for planning and management. Ocean Coast Manage. Available: http://dx.doi.org/10.1016/j.ocecoaman.2012.07.014. Accessed 2013 May 7.
51. Russ GR, Alcala AC (1996) Do marine reserves export adult fish biomass? Evidence from Apo Island, central Philippines. Mar Ecol Prog Ser 132: 1–9.
52. Russ GR, Alcala AC, Maypa AP (2003) Spillover from marine reserves: the case of *Naso vlamingii* at Apo Island, Philippines. Mar Ecol Prog Ser 264: 15–20.
53. Alcala AC, Russ GR, Maypa AP, Calumpong HP (2005) A long-term, spatially replicated experimental test of the effect of marine reserves on local fish yields. Can J Fish Aquat Sci 62: 98–108.
54. Beck MW, Heck KL, Able KW, Childers DL, Eggleston DB, et al. (2001) The identification, conservation, and management of estuarine and marine nurseries for fish and invertebrates. BioScience 51: 633–641.
55. Unsworth RKF, Bell JJ, Smith DJ (2007) Tidal fish connectivity of reef and seagrass habitats in the Indo-Pacific. J Mar Biol Assoc UK 87: 1287–1296.
56. Chao A, Chazdon RL, Colwell RK, Shen TJ (2005) A new statistical approach for assessing similarity of species composition with incidence and abundance data. Ecol Lett 8: 148–159.
57. Froese R, Pauly D (2012) FishBase. World Wide Web electronic publication. Available: www.fishbase.org, version (12/2012) Accessed 2012 Dec 14.
58. McLeod E, Chmura GL, Bouillon S, Salm R, Björk M, et al. (2011) A blueprint for blue carbon: toward an improved understanding of the role of vegetated coastal habitats in sequestering CO_2. Front Ecol Environ 9: 552–560.
59. Alongi DM (2008) Mangrove forests: resilience, protection from tsunamis, and responses to global climate change. Estuar Coast Shelf Sci 76: 1–13.
60. Gladstone W (2009) Conservation and management of tropical coastal ecosystems. In: Nagelkerken I, editor. Ecological Connectivity among Tropical Coastal Ecosystems. Dordrecht, The Netherlands: Springer Science and Business Media. 565–605.
61. Valiela I, Bowen JL, York JK (2001) Mangrove forests: one of the world's threatened major tropical environments. BioScience 51: 807–815.
62. Orth RJ, Carruthers TJB, Dennison WC, Duarte CM, Fouqurean JW, et al. (2006) A global crisis for seagrass ecosystems. BioScience 56: 987–996.
63. Waycott M, Duarte CM, Carruthers TJB, Orth RJ, Dennison WC, et al. (2009) Accelerating loss of seagrasses across the globe threatens coastal ecosystems. Proc Natl Acad Sci USA 106: 12377–12381.
64. Hughes JE, Deegan LA, Wyda JC, Weaver MJ, Wright A (2002) The effects of eelgrass habitat loss on estuarine fish communities of southern New England. Estuaries 25: 235–249.
65. Blaber SJM, Milton DA (1990) Species composition, community structure and zoogeography of fishes of mangrove estuaries in the Solomon Islands. Mar Biol 105: 259–267.
66. Thollot P (1992) Importance of mangroves for Pacific reef fish species, myth or reality? Proc 7th Int Coral Reef Symp 2: 934–941.
67. Dorenbosch M, van Riel MC, Nagelkerken I, van der Velde G (2004) The relationship of reef fish densities to the proximity of mangrove and seagrass nurseries. Estuar Coast Shelf Sci 60: 37–48.
68. Kimirei IA, Nagelkerken I, Griffioen B, Wagner C, Mgaya YD (2011) Ontogenetic habitat use by mangrove/seagrass-associated coral reef fishes shows flexibility in time and space. Estuar Coast Shelf Sci 92: 47–58.
69. Mora C, Aburto-Oropeza O, Ayala Bocos A, Ayotte PM, Banks S, et al. (2011) Global human footprint on the linkage between biodiversity and ecosystem functioning in reef fishes. PLoS Biology 9: e1000606. doi: 10.1371/journal.pbio.1000606.
70. Christie MR, Tissot BN, Albins MA, Beets JP, Jia Y, et al. (2010) Larval connectivity in an effective network of marine protected areas. PLoS ONE 5: e15715. doi:10.1371/journal.pone.0015715.

2

Does 'You Are What You Eat' Apply to Mangrove Grapsid Crabs?

Thi Hong Hanh Bui, Shing Yip Lee*

Australian Rivers Institute and School of Environment, Griffith University Gold Coast Campus, Southport, Qld, Australia

Abstract

In tropical mangroves, brachyuran crabs have been observed to consume high percentages of leaf litter production. However, questions concerning their ability to assimilate this low-quality food remain, as stable isotope analysis of C and N does not seem to support assimilation. Individuals of the common eastern Australian mangrove grapsid *Parasesarma erythodactyla* feeding on a mangrove leaf litter or mangrove+microphytobenthos diet developed a significantly higher hepatosomatic index than those with access to only sediment. Lipid biomarker analysis and feeding experiments using ^{13}C and ^{15}N-enriched mangrove leaf litter confirmed rapid assimilation of mangrove C and N by *P. erythodactyla*. Eight-week feeding experiments utilizing three food types (mangrove leaf litter, microphytobenthos and prawn muscle) established different food-specific trophic discrimination values ($\Delta\delta^{13}C$ and $\Delta\delta^{15}N$) that are significantly different from those commonly applied to mixing model calculations. The mean $\Delta\delta^{13}C_{(crab\text{-}mangrove)}$ of +5.45‰ was close to the mean and median literature values for grapsid-mangrove pairs in 29 past studies (+5.2±1.8‰ and +5.6‰, respectively), suggesting that this large discrimination may generally be characteristic of detritivorous grapsid crabs. Solutions from the IsoConc mixing model using our determined trophic discrimination values suggest significantly higher and dominant contributions of mangrove C to the diet than those based on the global mean trophic discrimination values. Our results reaffirm the physiological capacity for and important mediating role of grapsid crabs in processing low-quality mangrove C in tropical estuaries, and caution against the use of global trophic discrimination values in stable isotope analysis of food-web data, especially those involving detritivores. While recent studies have questioned the trophic significance of mangrove detritus in coastal food chains, the contribution of this productive carbon source needs to be re-assessed in the light of these data.

Editor: Craig A. Layman, North Carolina State University, United States of America

Funding: THHB is supported by an Australian Endeavour Postgraduate Award. The funders had no role in study design, data collection and analysis, decision to publish, or preparation of the manuscript.

Competing Interests: The authors have declared that no competing interests exist.

* E-mail: joe.lee@griffith.edu.au

Introduction

Despite their limited global areal extent of *ca.* 1.52×10^5 km^2 [1–2], tropical mangrove forests are amongst the most productive ecosystems on earth [3–4]. Early paradigms on estuarine carbon dynamics emphasized the 'outwelling' role of mangroves and saltmarshes in subsidising nearshore consumers with organic matter [5–7], but the spatial extent and the trophic significance of exported vascular plant detritus in nearshore food webs seem to be overstated [4,8]. The generally low nutritive value (e.g. high C/N ratio) and recalcitrant nature (e.g. high structural carbon content) of mangrove and saltmarsh plant detritus has prompted recent suggestions that this production may be consumed or mineralized minimally (eg. [9–10]), thus promoting storage. Other reports argued that even when this detritus is consumed, it may lead to trophic 'cul-de-sacs' [11].

The notion that mangrove detritus contributes little to nearshore consumer food chains is apparently supported by the stable isotope tracer analysis data. Numerous studies, using stable isotope analysis of carbon, nitrogen and occasionally sulphur, covering a range of mangrove-dominated systems, have yielded consumer signatures deemed too distant (e.g. >+5‰ for $\delta^{13}C$) to directly relate consumer biomass to assimilated mangrove detritus material (e.g. [12–14]; review in [15]). Alternatively, more ^{13}C-enriched producers such as the microphytobenthos (MPB), seagrass and phytoplankton have been suggested as the primary C sources. This notion also corroborates with the paradigm that algal C is more easily utilizable by animals compared to mangrove leaf litter (with a low N content but high concentration of secondary metabolites).

While the trophic dependence of nearshore consumers on mangrove detritus is debatable, there is much direct evidence of consumption of this relatively poor food source by grapsid crabs (Brachyura: Grapsidae) throughout the Indo-west-Pacific (IWP) [16–19], where these crabs are diverse and abundant [20]. Gut content analysis revealed that mangrove leaf litter was the dominant food item of the mangrove grapsids, though evidence of animal tissue and sediment consumption was also recorded [21–23]. With high consumption rates, significant percentages (>50%) of mangrove leaf litter production were retained and processed via this pathway, potentially 'short-circuiting' the food chain [24]. Initial processing by grapsid crabs may physically and biologically facilitate subsequent utilization of mangrove detritus by other consumers in nearshore food chains. The crabs act as shredders of fresh or aged leaf litter, greatly reducing the size but increasing the surface area to volume ratio of processed fragments, thus enhancing microbial colonisation and physical leaching of feeding deterrents [25–26]. The crabs therefore act as an efficient initial processor for low-quality mangrove leaf litter material, mediating its eventual utilization by other estuarine consumers [20]. This role is, however, dependent on the crabs' ability to effectively digest and assimilate the low-quality mangrove leaf litter diet.

Recently, the nutritional dependence of mangrove grapsids on mangrove leaf litter, and thus implicitly their role in acting as initial processors of mangrove organic production, has been questioned. Firstly, the physiological mechanism that allows profitable utilization of this poor food source is unknown. Secondly, consumption of mangrove leaf litter, often fresh and without microbial enrichment, by the grapsids is considered trophically non-viable and there needs to be substantial additional nutrient sources particularly for N [27–28]. Finally, stable isotope analysis of the leaf litter-crab link apparently fails to produce data that support the dominant contribution of mangrove C in the grapsids' diet. For example, recently Mazumder and Saintilan [9] claimed that mangrove (*Avicennia marina*) leaf litter could not be an important food source for the temperate grapsid crabs *Helograpsus haswellianus* and *Parasesarma erythodactyla* in Australia, as the $\delta^{13}C$ of the crabs were too high to indicate substantial utilization of this food item. There is a consistent large difference between the $\delta^{13}C$ of mangrove leaf litter and those of consumers potentially benefitting from the mangrove detritus-based food chain, e.g. [15,27,29]. It should be noted, however, that the interpretation of stable isotope data (particularly $\delta^{13}C$ and $\delta^{15}N$) are based on a few assumptions, such as a constant trophic discrimination factor between consumer and food. In most analyses, values of ~ +1‰ and ~+3‰ have been used for generating solutions from mixing models relating mangrove detritus and crab biomass $\delta^{13}C$ and $\delta^{15}N$ values, respectively. The applicability of these values generated from a wide range of consumer-food combinations to explaining specific trophic paths such as the grapsid crab-mangrove detritus link has not been tested. Significant deviations from these assumed values would have strong impacts on the interpretation of the stable isotope data, and thus the importance of the trophic links concerned.

In order to reconcile the observation that mangrove grapsids do consume large quantities of mangrove leaf litter, and the apparently contradictory stable isotope data related to this important link in mangrove food chains, we conducted a series of feeding experiments utilizing naturally abundant and isotopically enriched substrates to (1) investigate the ability of the mangrove grapsid *P. erythodactyla*, a common grapsid crab of the subtropical mangroves in eastern Australia, to directly utilize leaf litter of *Avicennia marina*; (2) empirically determine the trophic discrimination values of its potential food sources, namely mangrove leaves, MPB, and animal tissue; (3) estimate and compare diet compositions predicted from IsoConc mixing modelling based on the commonly assumed and our determined trophic discrimination values, and (4) discuss the implication of our results for the evaluation of the mangrove detritus food chain.

Methods

No ethics approval is required for this study according to current Australian law. The species used (*Parasesarma erythodactyla*) is not an endangered or a protected species. No specific permission is required for collection of grapsid crabs from the study location.

Collection of Crabs, Mangrove Leaf Litter and Sediment

Intermoult male individuals of *P. erythodactyla* of carapace width ranging from 11 to 20 mm were collected from an *A. marina*-dominated intertidal mangrove forest at Tallebudgera Creek, southeast Queensland, Australia. Crabs that were analyzed later as 'field' samples (hereafter referred as T0 sampling event) were rinsed with distilled water and frozen immediately upon arrival to the laboratory while those for the laboratory experiments were transferred to individual growth compartments ($l\times w\times h = 13\times 7\times 4$ cm, each containing 50 ml of sea water) and starved for two days for acclimation to laboratory condition and gut evacuation.

Freshly senescent mangrove leaves (yellow leaves that were easily detached from the trees), MPB, and sediment were collected from the same sites where the crabs were caught. Leaves were soaked in seawater for 24 h to remove feeding deterrents, e.g. tannin, before being offered to crabs. Sediments were collected by scraping the top 1 cm surface sediment, homogenized with a shovel before adding into experimental tanks as an organic substrate in experiment 1 or used for MBP isolation.

Extraction of MPB from sediment. MPB was isolated from sediment by density gradient centrifugation in colloidal silica. Sediment was spread to ca. 3 cm depth in plastic trays ($45\times 30\times 5$ cm), which were exposed to white fluorescent light for 16 h to mediate vertical migration of MBP to the surface of the sediment. The top ca. 0.5 cm sediment was then scrapped, suspended in seawater, and then sieved through a 63 μm to remove large detritus and nematodes. The filtrate was centrifuged at 4400 rpm for 5 min. Supernatant was poured off. Pellets were resuspended in left over supernatant, divided into 5 ml aliquots in individual centrifuge tubes, mixed with 40 ml of 30% Ludox colloidal silica (Sigma), and centrifuged at 4400 rpm for 5 min. The distinct layer of MPB, which was confirmed by microscopic examination, was collected, and washed with distilled water to remove Ludox before collected on pre-combusted glass filters. Each filter was loaded with the MPB suspension until it was clogged to maximize amount of MPB provided as a food source to the crabs in experiment 3. MPB containing filters were stored at −20°C until used. Aliquots of MPB were dried at 60°C for stable isotope analysis.

Preparation of ^{13}C and ^{15}N enriched mangrove leaves. Thirty *Avicennia* seedlings each with 4–6 leaves, were planted in two glass chambers ($h\times l\times w = 40\times 50\times 30$ cm) containing 10 cm deep sediment. Seedlings were grown at 24°C, under lighting from fluorescent tubes. The seedlings were labelled with ^{13}C and ^{15}N using methods modified from Bromand *et al.* [30] and Unsicker *et al.* [31]. Growth chambers were left open for two days, allowing water to evaporate from the top sediment before 1 ml of 61.2 mM $^{15}NH_4Cl$ (99 atom% ^{15}N, Cambridge Isotope Laboratories) was injected at each of the 15 injection points evenly distributed in each chamber. Injection was done at the depth of 2 cm from sediment surface using a 1 ml syringe. The top sediment was then re-wetted with distilled water. For ^{13}C labelling, a bottle containing 25 ml of 1 M $NaH^{13}CO_3$ (99 atom% ^{13}C, Cambridge Isotope Laboratories) was placed in each chamber before it was tightly sealed. One ml of 1 M HCl was added to the enriched bicarbonate bottle via a glass pipette passing through the top of the chamber every two days for 45 days to generate $^{13}CO_2$ *in situ*. A small fan (d = 8 cm) was turned on for 30 min after the addition of acid to facilitate even distribution of $^{13}CO_2$ within the growth chamber.

Experiment 1. Condition of Crabs

To evaluate relative contributions of organic matter from mangrove leaf litter and sediment to crab's diet, 18 crabs were randomly divided into three groups, each provided with either only mangrove leaf litter (the L treatment), only sediment (the S treatment), or a combination of both leaf litter and sediment (the L+S treatment). Each crab was allocated to an experimental tank ($l\times w\times h = 40\times 30\times 25$ cm) such that crab density was equivalent to ca. 9 individual per m^2. Tanks of the S and L+S treatments were filled with 3 cm of mangrove sediment and those of the L treatment were filled with 3 cm sand. Sand was collected from the foreshore of a local beach, cleared of organic debris by elutriation before adding into the tanks. It is assumed that the low organic matter oceanic sand would serve only as a substrate but not a significant source of carbon for the crabs (organic contents of the sediment and sand used in this experiment were 5.27±0.33% and 0.80±0.02%, respectively). Crabs were maintained at temperature of 24°C and a photoperiod of 16 h light : 8 h dark. A recirculation water supply system was set up for each tank such that seawater of salinity of 25 (PSU) was supplied to each tank according to a semi-diurnal tidal cycle. The reservoir water was replaced with fresh sea water at the end of every week.

Crabs in the L and L+S treatments were provided one *Avicennia* leaf per day while those in the S treatment were not. Uneaten feed was removed from the tanks the following morning. Survival of crabs in the three treatments was recorded every day. After 63 days, all crabs were collected, rinsed with distilled water, and stored at −20°C until dissection of muscle and hepatopancreas tissues from each individual. These tissues and the remaining parts of each crab were freeze-dried and weighed. Hepatosomatic index (HSI) of each crab was determined as the percentage of the dry weight of the hepatopancreas to the rest of the body [32]. Freeze-dried muscle tissues were used for fatty acid analysis.

Experiment 2. Leaf Litter Assimilation

Another batch of crabs was fed dual ^{13}C and ^{15}N enriched mangrove leaves for four weeks and temporal changes in stable isotope values of their muscle tissue were studied. Crabs were randomly assigned into the treatment group (32 crabs) and the control group (8 crabs), each were allocated to individual rearing compartments (13×7×4 cm, containing 50 ml of seawater). Crabs in the treatment and control group were provided one enriched and one non-enriched *A. marina* leaf every three days, respectively. Potential stable isotope enrichment in crab tissue due to the utilization of water containing leachate from enriched leaves was assessed by soaking one enriched leaf in each control compartment. This leaf was separated from the crab by two layers of plastic sheets, which had small holes at alternate positions at the bottom, allowing free movement of water across the sheets but preventing the crabs from direct access to the leaf. Enriched leaves in the control compartments were also replaced by fresh ones every three days. Crabs were maintained at temperature of 24°C and a photoperiod of 16 h light: 8 h dark. Water in the rearing compartments was changed once every week.

Eight crabs fed on the enriched-leaf diet were randomly sampled every week for four weeks (these sampling events hereafter are referred to as T1, T2, T3 and T4, respectively). At each sampling time, feed was removed from rearing compartments; crabs were then left to evacuate their gut for 24 h, rinsed with distilled water, and stored at −20°C until dissection. Control crabs were sampled similarly at the end of week 4. Crabs were dissected to collect muscle tissue, which was then dried at 60°C for 24 h. Eight 'field' crabs were also dissected to collect samples for T0 sampling event. To calculate assimilation efficiency, faeces were collected twice daily, dried at 60°C for 24 h. Faeces of individual crabs were pooled together at the end of the experiment. Dried muscle tissue and faeces from individual crabs were used for stable isotope analysis.

Experiment 3. Trophic Discrimination Values of Potential Food Sources

Carbon and nitrogen trophic discrimination values ($\Delta\delta^{13}C$ and $\Delta\delta^{15}N$, respectively) in *P. erythodactyla* for three common food sources (mangrove leaf litter, microphytobenthos and animal food) were determined by a third feeding experiment. Crabs were fed freshly senescent *A. marina* leaves, MPB or frozen prawn muscle (*Metapenaeus* spp.) for eight weeks. Previously work on similar animals (S.Y. Lee, unpublished data) suggests that eight weeks is usually sufficient for tissue turnover to result in isotopic equilibrium with the diet. The same feeding experiment was conducted twice (in 2011 and 2012). Twenty crabs were used in the first experiment, half of which was fed prawn while the other half was offered mangrove leaves. In the latter experiment, 40 crabs were randomly assigned into two groups of 20 individuals each, which had either MPB or mangrove leaf diet. Crabs were maintained in individual rearing compartments (13×7×4 cm, containing 50 ml of seawater). Feed was provided *ad libitum*. Uneaten glass filter containing MPB and leaves were replaced with fresh ones the next morning while unconsumed prawns were removed at the end of the day.

At the end of week 1, 3, 5 and 8 (hereafter referred to as T1, T3, T5 and T8, respectively), five crabs were samples from each diet in the MPB-feeding experiment while the numbers of crabs sampled in the prawn-feeding experiment were 1, 2, 2 and 5, respectively. These experimental crabs and five 'field' crabs (T0) of each feeding experiments were treated similarly with those sampled in experiment 2.

Stable isotope analysis. Carbon and nitrogen stable isotopic values of muscle tissue and faeces of crabs, *Avicennia* leaves, MPB, prawn used as food sources were determined using a Europa GSL (Sercon) elemental analyzer coupled to a Hydra 20–22 (Sercon) isotope ratio mass spectrometer in continuous flow mode. Samples are weighed in tin capsules (about 1 mg for animal sample and 4 mg for MPB and plant samples). PeeDee Belemnite and atmospheric air were used as standards for C and N, respectively. Stable isotope values are reported in δ-notation (‰), i.e. $\delta^{13}C$ or $\delta^{15}N = (R_{sample}/R_{standard} - 1) \times 1000$, where R is C^{13}/C^{12} and N^{15}/N^{14} ratios for carbon and nitrogen analyses, respectively. The analysis also provided %C and %N of the samples, from which assimilation efficiency (A) was determined by the following equation:

$$A = \frac{F - E}{(1 - E) \times F} \times 100 \quad [33]$$

where F = fraction of carbon or nitrogen in feed; and E = fraction of carbon or nitrogen in faeces.

Fatty acid analysis. Freeze-dried tissues were ground into powder using a mortar and pestle. Lipid extraction was performed by the one-step method of Abdulkadir and Tsuchiya [34]. In brief, each sample was mixed with 5 ml of hexane and 2 ml of 14% BF_3 in methanol in a 50 ml glass tube. The tube was heated under reflux on a hot plate at 100°C for 120 min and continuously stirred using a magnetic stirrer. The tube was left to cool to room temperature before 1 ml of hexane and 2 ml of distilled water were added. The tube was then vortexed for 1 min and then centrifuged at 2500 rpm for 3 min. The upper phase was transferred into a fresh tube using a Pasteur pipette. Fatty acid methyl esters (FAMEs) were separated from the extracted FA mix by thin layer chromatography following the method of Meziane and Tsuchiya [35].

FAMEs were then analyzed using a Varian CP-3800 gas chromatograph (GC) with an Omegawax 320 fussed-silica capillary column (Supelco) and flame ionisation detector. Helium was used as carrier gas. Column flow rate was set at 1.5 ml/min. Injector and detector temperatures were set at 240°C and 260°C, respectively. Temperature of the column oven was programmed at 60°C, held for 1 min, then increased to 150°C at the rate of 40°C/

min, held for 3 min, then increased to 210°C at 3°C/min, held for 17 min, then increased to 240°C at 5°C/min and held for 10 min. The resulting peaks were identified by comparing their retention times to those of standard FAMEs (Supelco).

Percentage contribution to total FA of FA markers of mangroves (18:2ω6, 18:3ω3, and the long-chain FA 26:0, 28:0, 30:0), branched FAs makers of bacteria (15:0 iso, 15:0 ant, 16:0 iso, 16:0 ant, 17:0 iso, 17:0 ant), and MPB (20:5ω3, a diatom marker) in the crab muscle tissues were determined. These FA markers were selected based on Meziane and Tsuchiya [35] and Hall *et al.* [36], who used FA analysis to investigate the transfer of mangrove organic matter in *P. erythodactyla*.

Data Analysis

One-way ANOVA ($\alpha = 0.05$) was used for the comparisons of HSI, the abundance of FA biomarkers, temporal changes in stable isotope values of crabs fed enriched mangrove leaves. Tests for normality (Shapiro-Wilk's test) and homogeneity of variance (Levene's test) were performed to check if assumptions of ANOVA were met before analysis. The Kruskal-Wallis H test was used instead of one-way ANOVA when these assumptions were violated. If significant ANOVA or Kruskal-Wallis H test results were obtained, statistical difference between specific treatments were determined by applying Tukey's Honestly Significant Difference (HSD) post-hoc test or Mann-Whitney U test, respectively.

FA profiles of crabs of the three treatments in the leaf litter utilization study were compared using multivariate analyses with PRIMER version 6 software. Exploratory multi-dimensional scaling (MDS) was conducted based on similarity data built using the Bray-Curtis similarity co-efficient. No transformations were used to avoid giving artificial weight to FAs that only have minor contributions to FA profiles. Statistical differences between treatments were determined using ANOSIM and similarities between them were estimated using the SIMPER function.

Determination of trophic discrimination values. Change in isotopic value of the consumers in response to diet shift can be modelled as a function of time as

$$\delta_t = \delta_f + \left(\delta_i - \delta_f\right) \times e^{-kt}$$

where δ_t is the isotopic value of the tissue at time t (in days); δ_f is the isotopic value when consumer reaches isotopic equilibrium with the new diet; δ_i denotes isotopic composition before the diet shift; and k is the turnover rate of the isotope of interest in the tested tissue [37]. This exponential model was fit to the C and N isotopic data of each feeding experiment to determine the isotopic composition of the muscle at equilibrium with the tested diet (δ_f) by least square method using SigmaPlot 10.0. Model fitting was performed with the δ_i term was fixed to the mean isotopic values of the T0 samples. Food specific trophic discrimination factors ($\Delta\delta^{13}C$ or $\Delta\delta^{15}N$) were then calculated as the difference between δ_f and the mean isotopic value of the food.

Assessing diet composition by the IsoConc mixing model. Using the trophic discrimination values determined from experiment 3, the stable isotope values and the C and N concentrations of the tested food, the contribution of these food sources to the crabs' diet was assessed using the IsoConc mixing model [38]. To compare the diet compositions predicted based on our determined isotopic discrimination values with those estimated based on the trophic fractionation factors that are commonly used in food web data analysis, IsoConc mixing model was also run using the mean trophic discrimination values reported for aquatic consumers of 0.5‰ for $\delta^{13}C$ and 2.9‰ for $\delta^{15}N$ [39] with the same data set for other parameters.

Results

Experiment 1. Condition of Crabs

All six crabs on the sole mangrove leaf diet (the L treatment) survived through the nine-week experiment. Number of survivors in the L+S treatment, in which crabs had access to both mangrove leave and organic matter from the sediment, and the S treatment, where sediment was the only food source, were five and four, respectively. The type of available food sources significantly affected the hepatosomatic index of the crabs (one-way ANOVA, $p = 0.001$). Post-hoc analysis showed that there was no significant difference between the L and L+S crabs (Tukey's HSD, $p = 0.116$). HSI of the S crabs was, however, significantly lower than those of both the L ($p<0.001$) and L+S treatments ($p = 0.009$) (Fig. 1A).

Fatty acid analysis showed that there were significant differences in the proportions of mangrove (one-way ANOVA, $p = 0.001$), bacterial (one-way ANOVA, $p = 0.002$), and MPB (Kruskal-Wallis H, $p = 0.014$) FA biomarkers to total FA in crab muscle tissue between treatments. Proportions of all three FA biomarkers were not significantly different in crabs which had access to mangrove leaves in the presence (the L+S treatment) or absence (the L treatment) of additional organic matter from the sediment (Tukey's HSD post-hoc test, $p = 0.58$ for mangrove, $p = 0.98$ for bacterial, and Mann-Whitney U test, $p = 0.46$ for MPB markers). Crabs in the S treatment had significantly lower percentage of mangrove FA biomarker but significantly higher proportions of MPB and bacterial biomarkers ($p<0.05$) than those offered mangrove leaves (Fig. 1B).

A similar trend was also obtained when the fatty acid profiles (i.e. the collections of all FA detected from the muscle tissue) of crabs in the three treatments were compared by ANOSIM analysis. There were no significant differences in the profiles of the L and the L+S crabs ($p = 0.08$). The FA profiles of crabs of the S treatment, however, were significantly different from those of the L and L+S treatments ($p \leq 0.001$). In addition, SIMPER analysis indicated that similarity between the S treatment and the other two treatments (<80% similarity) was lower than that between the L and L+S treatments (85% similarity). This confirms the observed separation of the S treatment from the L and L+S treatments on the MDS plot (Fig. 1C).

Experiment 2. Leaf Litter Assimilation

Muscle tissues of crabs fed the enriched mangrove leaves were clearly enriched in both ^{13}C and ^{15}N than those of the field crabs. After only one week on the enriched-leaf diet (T1), stable isotope values of the crabs were significantly higher than those of the field crabs (T0, Mann-Whitney U tests, $p = 0.006$ for $\delta^{13}C$ and $p = 0.009$ for $\delta^{15}N$). The T4 crabs were significantly enriched in both ^{13}C and ^{15}N than crabs of the T1 and T2 sampling events but not significantly different from the T3 crabs (Fig. 2), showing that the enrichment was slowing down by the end of the feeding experiment. Mean $\delta^{13}C$ and $\delta^{15}N$ of the field crabs and those in the control treatment, in which crabs were provided non-enriched leaves but exposed to water containing leachate from enriched ones, were not significantly different ($p>0.05$). Enrichment in crabs fed on the enriched leaves is, therefore, solely attributed to the digestion and assimilation of the mangrove leaves, with negligible contamination from the leachate. Crabs assimilated C and N from the mangrove leaves with an efficiency of 36.0 and 56.6%, respectively.

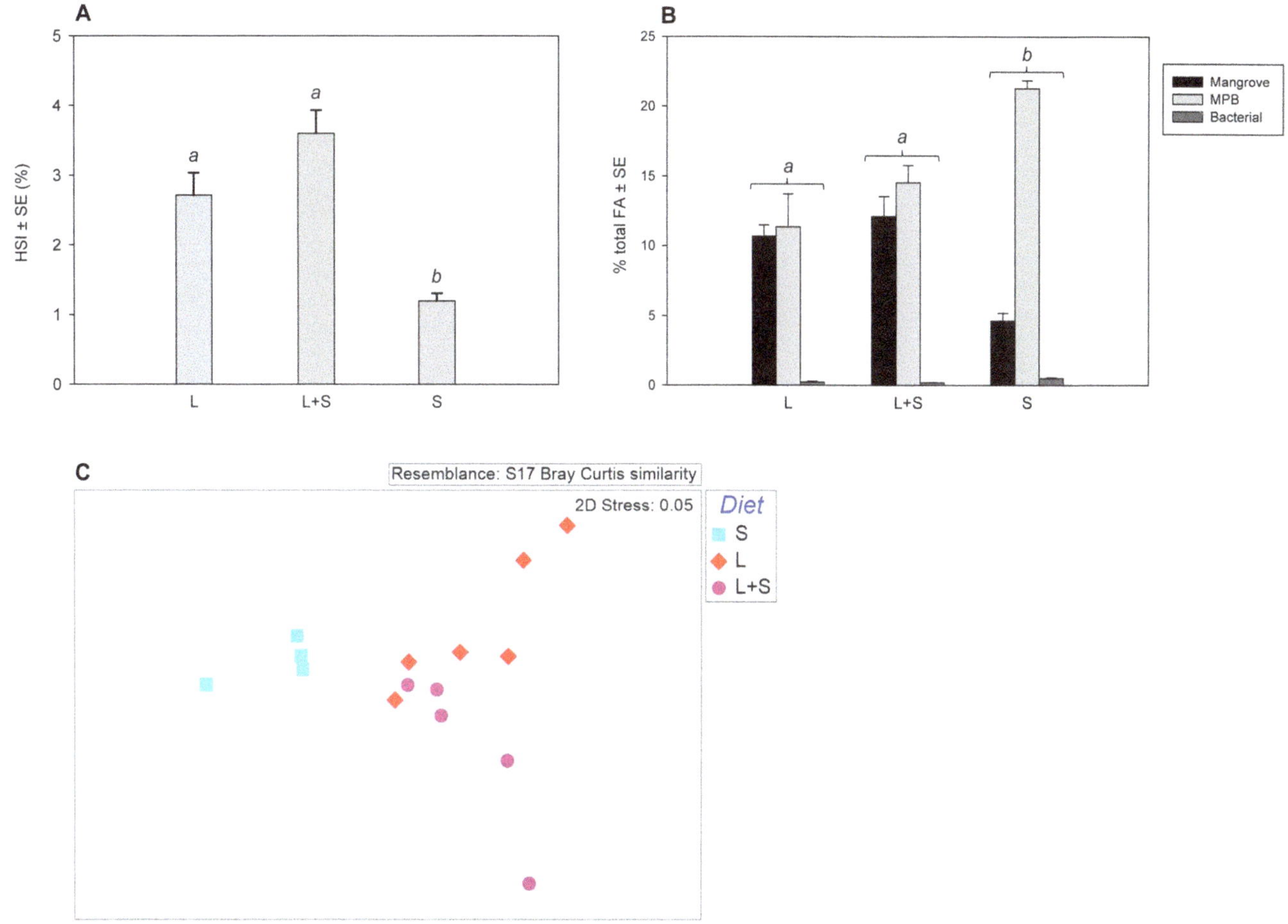

Figure 1. Influence of available organic matter on the fitness of *P. erythodactyla.* Crabs were offered mangrove leaves only (L, n = 6), mangrove leaves and sediment (L+S, n = 5) or sediment only (S, n = 4) for 63 days. Means HSI (A) and % to total FA in muscle tissue of FA biomarkers of the mangrove leaf litter, bacteria and MPB (B) were compared by one-way ANOVA or Kruskal-Wallis H non-parametric tests. Means marked with different letters are significantly different (Tukey's HSD or Mann-Whitney U, $p<0.05$). Similarity of FA profiles of crabs in the three treatments was compared by ANOSIM and graphically presented in a MDS plot (C).

Experiment 3. Trophic Discrimination Values of the Potential Food Sources

There are no obvious temporal changes in either $\delta^{13}C$ or $\delta^{15}N$ in crabs fed mangrove leaves and MPB (One-way ANOVA, $p>0.05$, Fig. 3). The exponential equation predicted isotopic compositions of the consumer in response to diet shift, hence, was not fitted to these data. C and N isotopic profiles of crabs on the prawn-diet, however, showed a good fit with the model ($p<0.001$, $r^2 = 0.89$ and 0.86, respectively). δ_f values were estimated to be -16.4‰ ($p = 0.037$) and +8.3‰ ($p<0.01$) for $\delta^{13}C$ and $\delta^{15}N$, respectively.

Table 1 presents the trophic discrimination values ($\Delta\delta^{13}C$ and $\Delta\delta^{15}N$) determined for each food item. $\Delta\delta^{13}C$ and $\Delta\delta^{15}N$ for the prawn diet were calculated from the differences between the respective δ_f values estimated from model fitting and the mean stable isotope values of the prawn muscle tissue. For the mangrove leaves and MPB, trophic discrimination factors were determined as the differences between the mean stable isotope values of the final sampling event and those of the food (this calculation will be justified in the discussion).

Assessing Diet Composition by the IsoConc Mixing Model

Results of IsoConc modelling using our trophic discrimination values (Table 1) show that mangrove leaf is the primary contributor to the diet of *P. erythodactyla* in terms of biomass (89%) and C (92%) but less important in term of N (48%, Fig. 4A). On the contrary, animal tissue makes up a minor proportion of the biomass (2%) and C intake (2%) but has a significant contribution to N (19%). Modelling based on the global mean trophic fractionation values reported for aquatic consumers ([39]), however, suggests that MPB but not the mangrove leaf would play a key role in the nutrition of this crab. In this scenario, mangrove litter was estimated to account for only 32, 41 and 6% of the biomass, C and N intakes, respectively (Fig. 4B).

Discussion

The Trophic Role of Mangrove Leaf Litter

Results of experiment 1 suggest that mangrove leaf litter was the 'staple' food of *P. erythodactyla* while the sediment organic matter sources have a minor contribution to its diet. Crabs on the mangrove litter only diet had comparable fitness and fatty acid profiles to those that had access to additional food sources

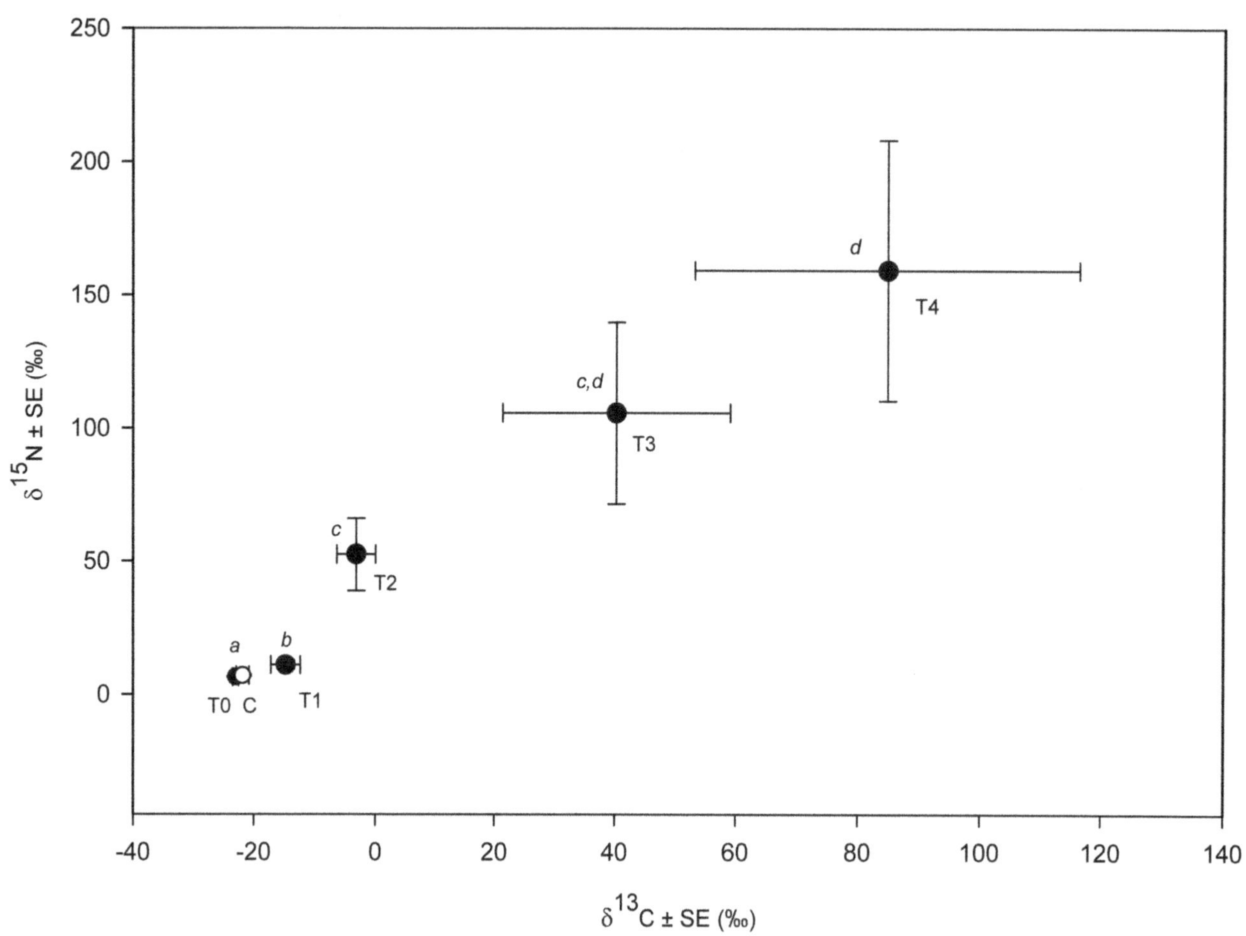

Figure 2. Temporal changes in $\delta^{13}C$ and $\delta^{15}N$ values of *P. erythodactyla* fed enriched mangrove leaves. Shown in the plot are the mean $\delta^{13}C$ and $\delta^{15}N$ values of crabs sampled at the start of the feeding experiment (T0, n = 7), at the end of every week (T1–T4, n = 8 for each sampling event), and those in the control (C, open circle, n = 8). Data points marked with different italic letters are significantly different (Mann-Whitney U tests, $p<0.05$).

available from the sediment, such as benthic bacteria, micro-algae or aged detritus. Feeding on the sediment alone, on the contrary, may reduce the fitness of the crabs as indicated by the significant lower HSI (a proxy for fitness) of crabs in this treatment (Fig. 1). We were able to maintain *P. erythodactyla* for more than six months on the sole mangrove leaf diet in the laboratory with a good number of them moulted during that time (unpublished data), suggesting that these crabs had access to sufficient food supply [40]. Bacteria and benthic microalgae in the sediment may provide some subsidy to the N deficit leaf diet, but the crabs are not likely to depend strongly on this source as has been suggested from some studies [9,28]. From the analysis of C and N balance in the grapsid crab *Neoepisesarma versicolour*, Thongtham and Kristensen [41] also reported that mangrove C was the major C source of this crab while sediment bacteria and MPB, due to their low availability, may have only a minor contribution to the crab's N requirement. In addition, leaf tissues were the dominant food items found in the gut of other mangrove grapsids such as *Chiromanthes onychophorum* [22], *N. versicolour*, *N. mederi* [23], *Aratus pisonii* [21].

Our feeding experiment using $\delta^{13}C$ and $\delta^{15}N$ enriched leaf (Fig. 2) provides direct evidence of the assimilation of mangrove organic matter in *P. erythodactyla.* In previous studies on the feeding ecology of mangrove grapsids, the contribution of mangrove leaf litter in their diets has been demonstrated from observations of the crabs' foraging activities on leaf litter [19,42], the presence of leaf tissues in the stomach contents [22–23], the growth and survival of crabs in long-term feeding experiments [43], and the transfer of terrestrial plant biomarkers such as long-chain fatty acids to the animals [36]. The lack of convincing direct evidence showing the assimilation of mangrove organic matter in these detritivores is obviously one of the reasons why the role of mangrove leaf litter in their diets has been questioned in some recent studies [9,28]. We also showed that *P. erythodactyla* assimilated C and N from *A. marina* at efficiencies (36% for C and 57% for N) comparable to those reported for *Neosarmatium smithi* fed on *Ceriops tagal* leaf detritus [44] and *N. versicolour* consuming brown (for % C assimilation) and green (for % N assimilation) leaves of *Rhizophora apiculata* [41].

The Change of Isotope Compositions in Experiment 3

When a new diet is introduced, isotopic compositions of the consumer will shift gradually toward those of the diet in time following an exponential function, i.e. $\delta_t = \delta_f + \left(\delta_i - \delta_f\right) \times e^{-kt}$ [37,45–46]. The isotope values of crabs on the prawn-diet showed a similar trend (Fig. 3) and fitted well to the proposed exponential model. Asymptotic values of the exponential curves (i.e. the δ_f term of the exponential function) were, therefore, considered as the

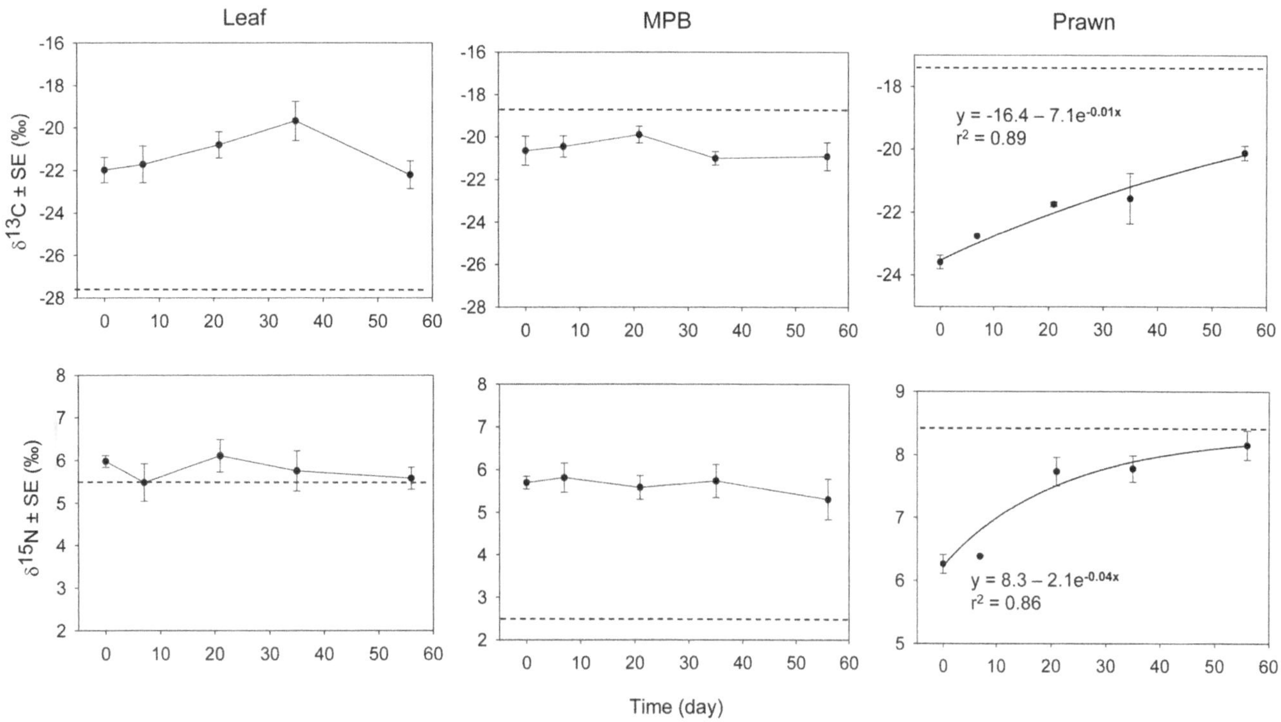

Figure 3. δ^{13}C and δ^{15}N profiles of crabs on the mangrove leaf, MPB, and prawn diet. Temporal changes in stable isotope values of the muscle tissues of crabs fed the potential dietary items are indicated by the solid lines. Mean δ^{13}C and δ^{15}N values of the tested food are presented as the broken horizontal lines. For the prawn-feeding data, the best fitting curves are plotted with the respective exponential functions and r^2 values.

expected isotope values of the crab muscle tissue when it reached equilibrium with the prawn food, which was subsequently used in the calculation of trophic discrimination values (Table 1). It must be noted that there was only a small difference between $\delta_f{}^{15}$N (+8.3‰) and the mean δ^{15}N (+8.2‰) of the samples collected at the final sampling event (T8), while the difference in δ^{13}C was larger, −16.4 vs. −20.1‰, respectively. This suggests that δ^{15}N of the muscle had almost approached equilibrium with the diet within the time frame of the experiment but C isotope composition may require a longer feeding time to reach isotopic equilibrium. A similar response resulting from diet shift was reported in the mysid *Mysis mixta*, whose δ^{15}N reached equilibrium with the *Artemia* used as feed about 8–9 weeks after the diet switch while δ^{13}C took longer than 12 weeks to attain isotopic equilibrium [47].

In contrast to the prawn-feeding treatment, isotopic compositions of crabs on the mangrove leaves and MPB diets stayed more or less constant during the experiment (Fig. 3). Herbon and Nordhaus (2013) also reported the lack of change in δ^{13}C in the sesarmid crabs *Episesarma singaporense* and *E. versicolor* over the 12 week long feeding experiment, in which *R. apiculata* senescent leave was provided as the sole dietary item. Stabilization of isotopic values in our case may due to: 1) no food ingestion and/or assimilation in the tested animals; 2) the time frame of the experiment was not long enough for an isotopic change following diet switch to occur; 3) the tested diet was similar to the natural diet of the animal; or 4) isotopic discrimination values of the tested diet were coincidentally similar to the difference in stable isotope values of the consumer from the diet.

In our feeding experiments, feed consumption in all experimental crabs was confirmed by the observation of feed removal and defecation. If food was ingested but not assimilated, the animal would have been on long term starvation. Reduction in food consumption and starvation often cause enrichment in δ^{15}N of the animals [48–51]. The increase in the ^{15}N/^{14}N ratio is probably due to the same mechanism that causes trophic discrimination of nitrogen: the animals metabolized their own proteins, continuously excreted ^{14}N without replenishment from the diet, and thus progressively become enriched in ^{15}N [52]. The

Table 1. Data used in the estimation of relative contributions of mangrove leaf, MPB and animal tissue in the diet of *P. erythodactyla* using the IsoConc mixing model.

Food	δ^{13}C (‰)	δ^{15}N (‰)	$\Delta\ \delta^{13}C_{(Crab\text{-}Food)}$#	$\Delta\ \delta^{15}N_{(Crab\text{-}Food)}$#	[C] (%)	[N] (%)
Mangrove	−27.7	+5.5	+5.5	+0.1	45.6	0.8
MPB	−19.0	+2.3	−1.9	+3.0	30.6	5.4
Prawn	−17.5	+8.3	+1.1	0.0	41.1	12.3

#$\Delta\ \delta^{13}C_{(Crab\text{-}Food)}$ and $\Delta\delta^{15}N_{(Crab\text{-}Food)}$ were determined by subtracting the mean C and N isotopic values of the food from the asymptotic values of the exponential curves for the prawn-diet data or the mean isotopic values of the last time point samples for the leaf and MPB diet data (Fig. 3, refer to text for justification).

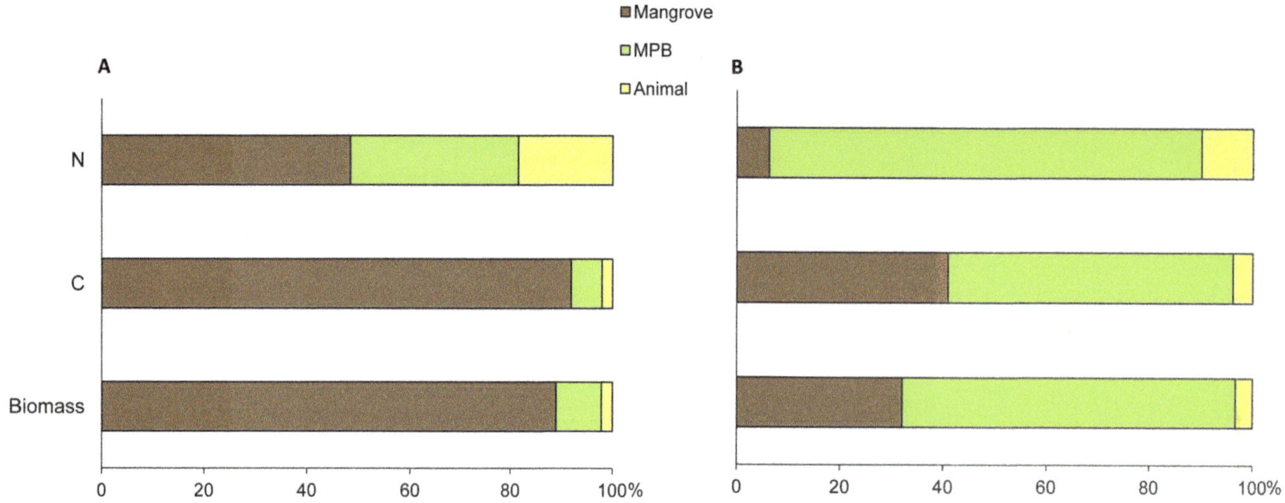

Figure 4. Estimation of the diet composition of *P. erythodactyla* by the IsoConc mixing model. Percentage contribution of the mangrove leaf, MPB and animal food sources in the crab's diet were estimated by the IsoConc mixing model using (A) our determined isotopic discrimination values (Table 1) and (B) the global means of trophic discrimination values for aquatic food web of 0.5‰ for C and 2.9‰ for N [39].

lack of change in δ^{15}N of the MPB and leaf feeding crabs during the feeding experiments suggests that these animals were not starved, thus ruling out the first explanation.

The results of the prawn-feeding treatment in experiment 3 and the enriched-leaf feeding experiment showed significant enrichment of both δ^{13}C and δ^{15}N after the first three- and one-week periods on the tested diets (Fig. 2 and 3). Eight weeks would therefore be long enough for any change in isotope composition in crabs' muscle tissue to be detected. Stabilization of stable isotope values throughout the experimental period, therefore, could be attributed to either of the last hypotheses. Under either scenario, difference between the mean stable isotope values of crabs of the last time point from values of the corresponding diet would be close to the true trophic discrimination values for the food of interest (Table 1).

Trophic Discrimination Values

Our determined trophic discrimination values (Table 1) are significantly different from the mean literature values (e.g.[39,53–54]). These reviews, however, also indicated that variability in $\Delta\delta^{13}$C and $\Delta\delta^{15}$N is significant, i.e. values for specific pairs of consumer-diet vary widely from the global mean values. For example, McCutchan *et al.* [53] showed that C and N discrimination values reported in literature may span the range of −2.7 to +5.5‰ and −2.4 to +9.2‰, respectively. Variation in trophic discrimination values are attributed to food quality (e.g. lipid and protein contents), tissue type analyzed, feeding mode (e.g. fluid-feeding vs. others), and habitat of consumers [55–57]. However, mechanisms for the discrepancies are obscure.

Our results suggest that the degree of trophic discrimination is specific to particular food-consumer pair, with a variation of >7‰ for $\Delta\delta^{13}$C and ~ 3‰ for $\Delta\delta^{15}$N between the three food types tested. Large variations in the C and N isotopic discrimination values of different food sources have also been reported in other animals but smaller differences are more common [39]. For examples, locust *Locusta migratoria* fed with corn and wheat had differences in $\Delta\delta^{13}$C and $\Delta\delta^{15}$N of these dietary items at 5.3‰ and 2.8‰, respectively [58]. In the grasshopper *Melanoplus sanguinipes*, $\Delta\delta^{15}$N of the corn seedling diet was 2.5‰ higher than that of the wheat seedling diet [59]. Indeed, trophic discrimination values may be influenced by food quality. Changes in nitrogen discrimination values in response to variations in protein quality and quantity [60] and C/N ratio [61] of diets have been reported. In addition, selective assimilation and differential routing of nutrient components (i.e. carbohydrates, proteins, lipids), which are expected to differ among the foods tested in our experiment, may also have contributed to the variation in trophic discrimination factors [52,62].

The difference in C^{13}/C^{12} ratio between the mangrove leaf and *P. erythodactyla* (5.5‰) is significantly higher than the commonly assumed trophic discrimination values used for analysing and interpreting grapsid crab food web studies. However, our $\Delta\delta^{13}C_{(crab\text{-}mangrove)}$ is close to C discrimination values recorded for *R. apiculata* in the grapsid crabs *E. singaporense* (5.1‰) and *E. versicolour* (4.1‰) [63]. Interestingly, our value also only slightly deviates from the mean (5.2) and median (5.6) of the $\Delta\delta^{13}$C values of the grapsid crab-mangrove leaf potential feeding link reported in the literature (Table 2). Caut *et al.* [55] reported a negative linear relationship between $\Delta\delta^{13}$C and food δ^{13}C, which is consistent with the large discrimination value we have recorded for the crab-mangrove leaf feeding relationship. The high $\Delta\delta^{13}$C may be due to selective assimilation of ^{13}C-enriched dietary components. Carbohydrates such as monosaccharides or cellulose are often more enriched than bulk leaf tissue by ~1 ‰ [64–66]. This difference could increase in extreme growing conditions, e.g. the leaves of *A. marina* growing in a hypersaline lagoon demonstrated a difference up to 5‰ [67]. In leaves of mangroves and other terrestrial plants, δ^{13}C values of individual amino acids may vary greatly; variations of *ca.* 20‰ or more have been recorded [68–69]. Selective assimilation of carbohydrates or the more enriched amino acids, therefore, would increase the discrimination value. The unusually high $\Delta\delta^{13}C_{crab\text{-}mangrove}$ values probably reflect some interesting and potentially unique features of the digestive physiology of the mangrove grapsid crabs, which invites further investigation. Nonetheless, these differences from the assumed values used in analyzing stable isotope data, whether through direct comparison or mixing model estimations, have significant implications for the results (see below).

Table 2. Difference in $\delta^{13}C$ values of the mangrove grapsid and their associated mangrove leaf litter from the literature.

Crab species	Mangrove species	Crab $\delta^{13}C$	Mangrove $\delta^{13}C$	$\Delta\ \delta^{13}C$	Reference
Australoplax tridentata	AM	−19.2	−27.9	8.7	[80]
	AM	−2.8	−28.8	6	[10]
Episesarma tetragonum	AO	−24.2	−28.6	4.4	[29]
	AO/BG/RA/RM	−25.2	−30.4	5.2	[81]
Episesarma versicolor	AO	−23.9	−28.6	4.7	[29]
	AM/AO/EA	−25.4	−28.6	3.2	[81]
Helice formosensis	KC	−21.5	−28.3	6.8	[82]
Helograpsus haswellianus	AM	−21.4	−27	5.6	[9]
Neoepisesarma versicolor	RA	−24.3	−28.5	4.2	[27]
	RM	−24.2	−29.3	5.1	[9]
Paragrapsus laevis	AM	−21	−27.8	6.8	[9]
Parasesarma asperum	AO	−25.5	−28.6	3.1	[29]
	AM/AO/EA	−23.8	−28.6	4.8	[81]
Parasesarma erythodactyla	AM	−22	−27.8	5.8	[9]
	AM	−22	−27.9	5.9	[80]
	AM	−23	−28.8	5.8	[10]
	AM	−20.7	−27.8	7.1	[83]
Parasesarma plicata	KC/AM/AC	−24	−26.2	2.2	[84]
Parasesarma plicatum	AM/AO/EA	−19.5	−28.6	9.1	[81]
Perisesarma bengalensis	AM/AO/EA	−25.4	−28.6	3.2	[81]
	RA/EA/BG	−25.7	−31.4	5.7	[81]
Perisesarma bidens	KC/AM/AC	−24.2	−26.2	2	[84]
Perisesarma dussumieri	AO/BG/RA/RM	−27.3	−30.4	3.1	[81]
	RA/EA/BG	−25.8	−31.4	5.6	[81]
Perisesarma guttatum	AM/CT/XG/RM	−23.3	−29.3	6	[81]
Perisesarma sp.	AM/CT/XG/RM	−22.1	−29.3	7.2	[81]
Psuedosesarma crassimanum	AO/BG/RA/RM	−24.3	−30.4	6.1	[81]
Sesarma spp.	AA	−24.0	−27.4	3.4	[14]
Mean		**−23.1**	**−28.3**	**5.2**	
SD		**2.6**	**2.2**	**1.8**	
Median		**−23.9**	**−28.6**	**5.6**	

AM – *Avicennia marina;* AO – *A. officinalis;* AC – *Aegiceras corniculatum;* BG – *Bruguiera gymnorhiza;* CT – *Ceriops tagal;* EA – *Excoecaria agallocha;* KC – *Kandelia candel;* RA – *Rhizophora apiculata;* RM – *Rhizophora mucronata;* XG – *Xylocarpus granatum.*

Implication of the Food-Specific Trophic Discrimination Values for the Mangrove-Grapsid Link

Mixing model calculations using our food-specific isotopic discrimination values suggest a significantly higher contribution of mangrove leaf litter to the crab's diet compared to results generated using the global mean values reported for aquatic consumers [39]. Increase in the contribution from mangrove leaf litter also means that contribution from the other common sources, such as the MPB, would decrease. The diet composition predicted using the experimentally determined values is in line with reports on the dominant proportion of leaf tissues in the gut contents of the mangrove grapsids. Microalgae and animal tissue remnants were also found, but at much lower fractions [21–23]. Our data question the proposed dominant dietary role of MPB based on modelling results obtained using the global mean trophic discrimination values. Firstly, there is no direct evidence on the consumption of large amount of MPB in mangrove grapsids. In addition, although MPB does offer some apparent nutritional advantages, such as a higher N content and presumably higher digestibility, the grapsid crabs' feeding appendages are not morphologically adapted to collecting microscopic food from the sediment. Mangrove grapsids have chelae with pointed 'finger-tips' that are more adapted to capturing, tearing and cutting large food items, rather than the spoon-tip feeding chelae found in deposit-feeding species such as the fiddler crabs (*Uca* spp.). Intertidal crabs that rely on microscopic food (e.g. MPB or the meiofauna) also possess specialised mouth parts or sediment processing behaviour, which are not present in the grapsids, to help efficiently extract food particles from the large volume of sediment that needs to be handled, e.g. 'floatation feeding' in soldier crab *Mictyris longicarpus* [70]. Further, the limited growth of microalgae on the poorly lighted mangrove forest floor [71–72] would hardly meet nutritional requirement of the highly abundant grapsid communities if it is their main food item. The strong competition for MPB from ocypodid crabs, such as *Heloecius cordiformis* on sub-tropical eastern Australian coast, which are also found at high density in more open mangroves, further reduces the availability of MPB to the grapsids. From the analysis of the N budget of the grapsid crab *N. versicolor* and the estimation of the nitrogen content of MPB in the sediment at the same site, Thongtham and Kristensen [41] showed that this crab must consume an unrealistically high amount of sediment to obtain sufficient N for its growth. N demand in the mangrove grapsids was more likely met by occasional consumption of animal tissue, through cannibalism, predation of other invertebrates or scavenging rather than regular consumption of MPB [21,27].

Our results not only confirm the significance of grapsid crabs in directly processing mangrove leaf organic matter, but also highlights the risk of applying 'global' trophic discrimination values to analyzing stable isotope food web data. The values commonly used are mean values derived from a large number of consumer-food combinations, which expectedly result in a wide distribution of actual discrimination values (see [39,53–54]). The mean values, while offering some statistical information on the discrimination values, are of little direct value in analyzing specific feeding relationships [56]. Values specific to potential consumer-food combinations need to be obtained before their application to mixing model calculations. While this may not be logistically feasible for all potential food items, a more practical approach is to obtain the values at least at the food category level (e.g. vascular plant vs. animal food) in animals that use diverse food sources.

Studies on C dynamics in estuaries have been dominated by the 'outwelling' paradigm for decades [5,7,73]. While earlier rates of export may have been over-estimated [8], recent attention on tropical mangroves has taken a dramatic turn in highlighting the role of mangroves as prime 'blue-carbon' storages [74–76]. In-situ consumption of fresh mangrove leaf litter by detritivores such as grapsid crabs and gastropods is a major fate of mangrove production, especially in the Indo-west-Pacific [20,77]. In poorly flushed mangroves with little tidal export, the sediment carbon accumulation rate depends on the balance between litter production and mineralization (in-situ consumption by macro-detritivores and microbial decomposition). Under-estimating detritus utilization resulting from inappropriate use of global trophic discrimination values in isotopic analyses would emphasize the storage role of mangroves while undervaluing the trophic contribution of wetland vascular plant production to coastal food chains. With recent reports on common occurrence of cellulases in estuarine invertebrates [78–79], these data together demand a re-examination of the general significance of the detritus-based food chain.

In conclusion, our study confirms the dominant role of mangrove leaf litter in the diet of grapsid crabs. Mechanisms by which these animals can benefit from a diet dominated by an apparently low-quality food item require further investigation. With a capacity to convert the low-quality mangrove C into biomass, the grapsid crabs would mediate the transfer of mangrove primary production to nearshore consumers. This link is particularly important in the Indo-west-Pacific mangroves, where the abundance and diversity of mangrove grapsids are maximal [20]. Other trophic links originating from vascular plant detritus may similarly have been undervalued because of the use of inappropriate trophic discrimination values in past isotopic studies. Our data suggest that the tide has not yet turned for mangrove trophic support to nearshore consumer communities.

Acknowledgments

We thank Brian Fry (Griffith University) and Wenqing Wang (Xiamen University, China) for useful discussions and advice on the enriched isotope experiment, and Rene Diocares (Griffith University) for expert technical help with stable isotope analysis.

Author Contributions

Conceived and designed the experiments: THHB SYL. Performed the experiments: THHB. Analyzed the data: THHB SYL. Contributed reagents/materials/analysis tools: THHB SYL. Wrote the paper: THHB SYL.

References

1. Spalding MD, Kainuma M, Collins L (2010) World Atlas of Mangroves. London: Earthscan. 319 p.
2. FAO (2007) The World's Mangroves: 1980–2005. Rome: FAO. 89 p.
3. Alongi DM (2009) The energetics of mangrove forests. Heidelberg: Springer. 216 p.
4. Bouillon S, Borges AV, Castaneda-Moya E, Diele K, Dittmar T, et al. (2008) Mangrove production and carbon sinks: A revision of global budget estimates. Global Biogeochem Cycles 22: 1–12.
5. Odum WE, Heald EJ (1975) The detritus-based food web of an estuarine mangrove community. In: Cronin LE, editor. Estuarine Research. New York: Academic Press. 265–286.
6. Odum EP (1980) The status of three ecosystem level hypotheses regarding salt marshes: tidal subsidy, outwelling and the detritus based food chain. In: Kennedy VS, editor. Estuarine Perspectives. New York: Academic Press. 485–496.
7. Teal JM (1962) Energy-Flow in Salt-Marsh Ecosystem of Georgia. Ecology 43: 614–624.
8. Lee SY (1995) Mangrove Outwelling - a Review. Hydrobiologia 295: 203–212.
9. Mazumder D, Saintilan N (2010) Mangrove Leaves are Not an Important Source of Dietary Carbon and Nitrogen for Crabs in Temperate Australian Mangroves. Wetlands 30: 375–380.
10. Oakes JM, Connolly RM, Revill AT (2010) Isotope enrichment in mangrove forests separates microphytobenthos and detritus as carbon sources for animals. Limnol Oceanogr 55: 393–402.
11. Bishop MJ, Cole MR, Taylor SL, Wilkie EM, Kelaher BP (2008) Size-specific predation by dominant consumers maintains a 'trophic cul-de-sac'. Mar Ecol Prog Ser 354: 75–83.
12. Chong VC, Low CB, Ichikawa T (2001) Contribution of mangrove detritus to juvenile prawn nutrition: a dual stable isotope study in a Malaysian mangrove forest. Mar Biol 138: 77–86.
13. Newell RIE, Marshall N, Sasekumar A, Chong VC (1995) Relative importance of benthic microalgae, phytoplankton, and mangroves as sources of nutrition for penaeid prawns and other coastal invertebrates from Malaysia. Mar Biol 123: 595–606.
14. Rodelli MR, Gearing JN, Gearing PJ, Marshall N, Sasekumar A (1984) Stable isotope ratio as a tracer of mangrove carbon in Malaysian ecosystems. Oecologia 61: 326–333.
15. Lee SY (2005) Exchange of organic matter and nutrients between mangroves and estuaries: myths, methodological issues and missing links. Int J Eco Environ Sci 31: 163–175.
16. Lee SY (1989) The importance of sesarminae crabs *Chiromanthes* spp. and inundation frequency on mangrove (*Kandelia candel* (L.) Druce) leaf litter turnover in a Hong-Kong tidal shrimp pond. J Exp Mar Biol Ecol 131: 23–43.
17. Robertson AI, Daniel PA (1989) The influence of crabs on litter processing in high intertidal mangrove forests in tropical Australia. Oecologia 78: 191–198.
18. Emmerson WD, Mcgwynne LE (1992) Feeding and assimilation of mangrove leaves by the crab *Sesarma meinerti* Deman in relation to leaf-litter production in Mgazana, a warm temperate Southern African mangrove swamp. J Exp Mar Biol Ecol 157: 41–53.
19. Camilleri JC (1992) Leaf-litter processing by invertebrates in a mangrove forest in Queensland. Mar Biol 114: 139–145.
20. Lee SY (2008) Mangrove macrobenthos: Assemblages, services, and linkages. J Sea Res 59: 16–29.
21. Lopez B, Conde JE (2013) Dietary variation in the crab *Aratus Pisonii* (H. Milne Edwards, 1837) (Decapoda, Sesarmidae) in a mangrove gradient in northwestern Venezuela Crustaceana 86: 1051–1069.
22. Malley DF (1978) Degradation of mangrove leaf litter by tropical sesarmid crab *Chiromanthes onychophorum*. Mar Biol 49: 377–386.
23. Poovachiranon S, Tantichodok P (1991) The role of sesarmid crabs in the mineralisation of leaf litter of *Rhizophora apiculata* in a mangrove, southern Thailand. Phuket Mar Biol Cent Res Bull 56: 63–74.
24. Sheaves M, Molony B (2000) Short-circuit in the mangrove food chain. Mar Ecol Prog Ser 199: 97–109.
25. Lee SY (1997) Potential trophic importance of the faecal material of the mangrove sesarmine crab *Sesarma messa*. Mar Ecol Prog Ser 159: 275–284.

26. Werry J, Lee SY (2005) Grapsid crabs mediate link between mangrove litter production and estuarine planktonic food chains. Mar Ecol Prog Ser 293: 165–176.
27. Kristensen DK, Kristensen E, Mangion P (2010) Food partitioning of leaf-eating mangrove crabs (*Sesarminae*): Experimental and stable isotope (^{13}C and ^{15}N) evidence. Estuar Coast Shelf Sci 87: 583–590.
28. Skov MW, Hartnoll RG (2002) Paradoxical selective feeding on a low-nutrient diet: why do mangrove crabs eat leaves? Oecologia 131: 1–7.
29. Bouillon S, Koedam N, Raman AV, Dehairs F (2002) Primary producers sustaining macro-invertebrate communities in intertidal mangrove forests. Oecologia 130: 441–448.
30. Bromand S, Whalen JK, Janzen HH, Schjoerring JK, Ellert BH (2001) A pulse-labelling method to generate C-13- enriched plant materials. Plant Soil 235: 253–257.
31. Unsicker SB, Renker C, Kahmen A, Spindler S, Buchmann N, et al. (2005) Testing the efficiency of three N-15-labeled nitrogen compounds for indirect labeling of grasshoppers via plants in the field. Entomol Exp Appl 116: 219–226.
32. Kennish R (1997) Seasonal patterns of food availability: Influences on the reproductive output and body condition of the herbivorous crab *Grapsus albolineatus*. Oecologia 109: 209–218.
33. Conover RJ (1966) Assimilation of organic matter by zooplankton. Limnol Oceanogr 11: 338–354.
34. Abdulkadir S, Tsuchiya M (2008) One-step method for quantitative and qualitative analysis of fatty acids in marine animal samples. J Exp Mar Biol Ecol 354: 1–8.
35. Meziane T, Tsuchiya M (2000) Fatty acids as tracers of organic matter in the sediment and food web of a mangrove/intertidal flat ecosystem, Okinawa, Japan. Mar Ecol-Prog Ser 200: 49–57.
36. Hall D, Lee SY, Meziane T (2006) Fatty acids as trophic tracers in an experimental estuarine food chain: Tracer transfer. J Exp Mar Biol Ecol 336: 42–53.
37. Tieszen LL, Boutton TW, Tesdahl KG, Slade NA (1983) Fractionation and turnover of stable carbon isotopes in animal tissues: Implications for δ^{13}C analysis of diet. Oecologia 57: 32–37.
38. Phillips D, Koch P (2002) Incorporating concentration dependence in stable isotope mixing models. Oecologia 130: 114–125.
39. Vander Zanden MJ, Rasmussen JB (2001) Variation in delta N-15 and delta C-13 trophic fractionation: Implications for aquatic food web studies. Limnol Oceanogr 46: 2061–2066.
40. Micheli F (1993) Effect of mangrove litter species and availability on survival, moulting, and reproduction of the mangrove crab *Sesarma messa*. J Exp Mar Biol Ecol 171: 149–163.
41. Thongtham N, Kristensen E (2005) Carbon and nitrogen balance of leaf-eating sesarmid crabs (*Neoepisesarma versicolor*) offered different food sources. Estuar Coast Shelf Sci 65: 213–222.
42. Camilleri J (1989) Leaf choice by crustaceans in a mangrove forest in Queensland. Mar Biol 102: 453–459.
43. Kwok PW, Lee SY (1995) The growth performances of two mangrove crabs, *Chiromanthes bidens* and *Parasesarma plicata* under different leaf litter diets. Hydrobiologia 295: 141–148.
44. Giddins RL, Lucas JS, Neilson MJ, Richards GN (1986) Feeding ecology of the mangrove crab *Neosarmatium smithi* (Crustacea : Decapoda : Sesarmidae) Mar Ecol Prog Ser 33: 147–155.
45. Guelinckx J, Maes J, Van Den Driessche P, Geysen B, Dehairs F, et al. (2007) Changes in delta C-13 and delta N-15 in different tissues of juvenile sand goby *Pomatoschistus minutus*: a laboratory diet-switch experiment. Mar Ecol Prog Ser 341: 205–215.
46. Buchheister A, Latour RJ (2010) Turnover and fractionation of carbon and nitrogen stable isotopes in tissues of a migratory coastal predator, summer flounder (*Paralichthys dentatus*). Can J Fish Aquat Sci 67: 445–461.
47. Gorokhova E, Hansson S (1999) An experimental study on variations in stable carbon and nitrogen isotope fractionation during growth of *Mysis mixta* and *Neomysis integer*. Can J Fish Aquat Sci 56: 2203–2210.
48. Hobson KA, Alisauskas RT, Clark RG (1993) Stable-Nitrogen isotope enrichment in avian tissues due to fasting and nutritional stress: Implications for isotopic analyses of siet. Condor 95: 388–394.
49. Olive PJW, Pinnegar JK, Polunin NVC, Richards G, Welch R (2003) Isotope trophic-step fractionation: a dynamic equilibrium model. J Anim Ecol 72: 608–617.
50. Scrimgeour CM, Gordon SC, Handley LL, Woodford JAT (1995) Trophic levels and anomalous δ^{15}N of iInsects on raspberry (*Rubus Idaeus* L.). Isotopes Environ Health Stud 31: 107–115.
51. Oelbermann K, Scheu S (2002) Stable isotope enrichment (δ^{15}N and δ^{13}C) in a generalist predator (*Pardosa lugubris*, Araneae: Lycosidae): Effects of prey quality. Oecologia 130: 337–344.
52. Gannes LZ, O'Brien DM, del Rio CM (1997) Stable isotopes in animal ecology: assumptions, caveats, and a call for laboratory experiments. Ecology 78: 1271–1276.
53. McCutchan JH, Lewis WM, Kendall C, McGrath CC (2003) Variation in trophic shift for stable isotope ratios of carbon, nitrogen, and sulfur. Oikos 102: 378–390.
54. Peterson BJ, Fry B (1987) Stable isotopes in ecosystem studies. Annu Rev Ecol Syst 18: 293–320.
55. Caut S, Angulo E, Courchamp F (2009) Variation in discrimination factors (δ^{15}N and δ^{13}C): the effect of diet isotopic values and applications for diet reconstruction. J Appl Ecol 46: 443–453.
56. Layman CA, Araujo MS, Boucek R, Hammerschlag-Peyer CM, Harrison E, et al. (2012) Applying stable isotopes to examine food-web structure: an overview of analytical tools. Biol Rev Camb Philos Soc 87: 545–562.
57. Malpica-Cruz L, Herzka SZ, Sosa-Nishizaki O, Lazo JP (2012) Tissue-specific isotope trophic discrimination factors and turnover rates in a marine elasmobranch: empirical and modeling results. Can J Fish Aquat Sci 69: 551–564.
58. Webb SC, Hedges REM, Simpson SJ (1998) Diet quality influences the δ^{13}C and δ^{15}N of locusts and their biochemical components. J Exp Biol 201: 2903–2911.
59. Deniro MJ, Epstein S (1981) Influence of diet on the distribution of nitrogen isotopes in animals. Geochim Cosmochim Acta 45: 341–351.
60. Florin ST, Felicetti LA, Robbins CT (2011) The biological basis for understanding and predicting dietary-induced variation in nitrogen and sulphur isotope ratio discrimination. Funct Ecol 25: 519–526.
61. Adams TS, Sterner RW (2000) The effect of dietary nitrogen content on trophic level N-15 enrichment. Limnol Oceanogr 45: 601–607.
62. DeNiro MJ, Epstein S (1978) Influence of diet on the distribution of carbon isotopes in animals. Geochim Cosmochim Acta 42: 495–506.
63. Herbon CM, Nordhaus I (2013) Experimental determination of stable carbon and nitrogen isotope fractionation between mangrove leaves and crabs. Mar Ecol Prog Ser 490: 91–105.
64. Dungait JAJ, Docherty G, Straker V, Evershed RP (2010) Seasonal variations in bulk tissue, fatty acid and monosaccharide δ^{13}C values of leaves from mesotrophic grassland plant communities under different grazing managements. Phytochemistry 71: 415–428.
65. Hyodo F, Kohzu A, Tayasu I (2010) Linking aboveground and belowground food webs through carbon and nitrogen stable isotope analyses. Ecol Res 25: 745–756.
66. Macko SA, Engel MH, Hartley G, Hatcher P, Helleur R, et al. (1991) Isotopic compositions of individual carbohydrates as indicators of early diagenesis of organic matter in peat. Chem Geol 93: 147–161.
67. Moers MEC, Jones DM, Eakin PA, Fallick AE, Griffiths H, et al. (1993) Carbohydrate diagenesis in hypersaline environments: application of GC-IRMS to the stable isotope analysis of derivatized saccharides from surficial and buried sediments. Org Geochem 20: 927–933.
68. Larsen T, Taylor DL, Leigh MB, O'Brien DM (2009) Stable isotope fingerprinting: a novel method for identifying plant, fungal, or bacterial origins of amino acids. Ecology 90: 3526–3535.
69. Larsen T, Wooller MJ, Fogel ML, O'Brien DM (2012) Can amino acid carbon isotope ratios distinguish primary producers in a mangrove ecosystem? Rapid Commun Mass Spectrom 26: 1541–1548.
70. Quinn RH (1980) Mechanisms for obtaining water for flotation feeding in the soldier crab, *Mictyris longicarpus* Latreille, 1806 (Decapoda, Mictyridae). J Exp Mar Biol Ecol 43: 49–60.
71. Alongi DM (1994) Zonation and seasonality of benthic primary production and community respiration in tropical mangrove forests. Oecologia 98: 320–327.
72. Lee SY (1990) Primary productivity and particulate organic-matter flow in an estuarine mangrove-wetland in Hong-Kong. Mar Biol 106: 453–463.
73. Odum EP (1980) The status of three ecosystem level hypotheses regarding salt marshes: tidal subsidy, outwelling and the detritus based food chain. In: Kennedy VS, editor. Estuarine Perspectives. New York: Academic Press. 485–496.
74. Alongi DM (2012) Carbon sequestration in mangrove forests. Carbon Manage 3: 313–322.
75. Donato DC, Kauffman JB, Murdiyarso D, Kurnianto S, Stidham M, et al. (2011) Mangroves among the most carbon-rich forests in the tropics. Nat Geosci 4: 293–297.
76. Mcleod E, Chmura GL, Bouillon S, Salm R, Bjork M, et al. (2011) A blueprint for blue carbon: toward an improved understanding of the role of vegetated coastal habitats in sequestering CO2. Front Ecol Environ 9: 552–560.
77. Fratini S, Vigiani V, Vannini M, Cannicci S (2004) *Terebralia palustris* (Gastropoda; Potamididae) in a Kenyan mangal: size structure, distribution and impact on the consumption of leaf litter. Mar Biol 144: 1173–1182.
78. Adachi K, Toriyama K, Azekura T, Morioka K, Tongnunui P, et al. (2012) Potent cellulase activity in the hepatopancreas of mangrove crabs. Fish Sci 78: 1309–1314.
79. Niiyama T, Toyohara H, Tanaka K (2012) Cellulase Activity in Blood Cockle (*Anadara granosa*) in the Matang Mangrove Forest Reserve, Malaysia. Jarq-Japan Agricultural Research Quarterly 46: 355–359.
80. Guest MA, Connolly RM (2004) Fine-scale movement and assimilation of carbon in saltmarsh and mangrove habitat by resident animals. Aquat Ecol 38: 599–609.
81. Bouillon S, Moens T, Overmeer I, Koedam N, Dehairs F (2004) Resource utilization patterns of epifauna from mangrove forests with contrasting inputs of local versus imported organic matter. Mar Ecol Prog Ser 278: 77–88.
82. Hsieh HL, Chen CP, Chen YG, Yang HH (2002) Diversity of benthic organic matter flows through polychaetes and crabs in a mangrove estuary: delta C-13 and delta S-34 signals. Mar Ecol Prog Ser 227: 145–155.

83. Nerot C, Meziane T, Provost-Govrich A, Rybarczyk H, Lee SY (2009) Role of grapsid crabs, *Parasesarma erythrodactyla*, in entry of mangrove leaves into an estuarine food web: a mesocosm study. Mar Biol 156: 2343–2352.

84. Lee SY (2000) Carbon dynamics of Deep Bay, eastern Pearl River estuary, China. II: Trophic relationship based on carbon- and nitrogen-stable isotopes. Mar Ecol Prog Ser 205: 1–10.

3

Estimates of Marine Debris Accumulation on Beaches Are Strongly Affected by the Temporal Scale of Sampling

Stephen D. A. Smith*, Ana Markic

National Marine Science Centre, Southern Cross University, Coffs Harbour, New South Wales, Australia

Abstract

Marine debris is a global issue with impacts on marine organisms, ecological processes, aesthetics and economies. Consequently, there is increasing interest in quantifying the scale of the problem. Accumulation rates of debris on beaches have been advocated as a useful proxy for at-sea debris loads. However, here we show that past studies may have vastly underestimated the quantity of available debris because sampling was too infrequent. Our study of debris on a small beach in eastern Australia indicates that estimated daily accumulation rates decrease rapidly with increasing intervals between surveys, and the quantity of available debris is underestimated by 50% after only 3 days and by an order of magnitude after 1 month. As few past studies report sampling frequencies of less than a month, estimates of the scale of the marine debris problem need to be critically re-examined and scaled-up accordingly. These results reinforce similar, recent work advocating daily sampling as a standard approach for accurate quantification of available debris in coastal habitats. We outline an alternative approach whereby site-specific accumulation models are generated to correct bias when daily sampling is impractical.

Editor: Brock Fenton, University of Western Ontario, Canada

Funding: The authors have no support or funding to report.

Competing Interests: The authors have declared that no competing interests exist.

* E-mail: steve.smith@scu.edu.au

Introduction

Marine debris is a key threatening process for marine organisms, with reports of fatal interactions becoming all too frequent [1–3]. While the discovery of vast concentrations of debris in ocean gyres over the past 2 decades [4–6], mostly comprising plastics, has increased awareness of the issue, the low cost and broad utility of plastic continues to drive growth in its production, with 265 Million tonnes produced in 2010 [7] and a forecast of 300 Million tonnes by 2020 [8]. Plastics have been recorded from some of the remotest beaches on the planet [9] and it is consequently highly likely that debris-free beaches have been consigned to history. Gaining accurate information on how much debris is in the marine environment is a critical step in targeted management, and assessments of accumulation rates on beaches are often used to provide such estimates for coastal environments [10–12].

There has been increasing recognition that accumulation studies may underestimate available debris and that the scale of this underestimation is dependent on the interval between accumulation studies [13,14]. In the majority of accumulation and trend assessment studies, sampling was conducted at a minimum frequency of monthly [15,16–18]. However, a few studies have employed bi-weekly intervals [19,20], a three-day interval [21] and daily intervals [13,14,22]. Unsurprisingly, the highest time-standardised accumulation rates result from daily surveys, but this timeframe is impractical for ongoing monitoring across numerous sites or for protracted periods.

Quite apart from the large range of methods applied to assessing marine debris densities on beaches [10], the lack of standardised approaches to accumulation studies makes it difficult to assess comparative debris loads at different sites (but note that substantial progress has recently been made through the development of recommended international protocols [23]). Further, translating debris loads on beaches into estimates of available debris in coastal waters is rendered almost impossible in most cases.

An understanding of overall debris dynamics is clearly needed to provide greater certainty about debris densities in coastal waters: one approach is to develop models based on empirical data [13]. Such models require metrics on the amount of debris arriving on a beach (loading rate [10]) as well as the relative importance of different removal pathways such as lateral drift [24], *in situ* burial [19,25], Aeolian transport, re-suspension and wash-out [13], and cleaning [26–28]. However, given the range of additional factors that can affect accumulation rates (e.g. extreme weather events, smaller-scale morphology of beaches, proximity to population centres, visitation rates and the socio-economic background of visitors, and other factors operating over various temporal scales) [13,14,20,24,29–34], it is unlikely that one model will fit all situations [13]. Nevertheless, the development of such models will facilitate progressive understanding of accumulation rates and, importantly, how these are correlated with the availability of debris in coastal habitats [14].

In this study, we take the first step in developing a model of marine debris dynamics in subtropical eastern Australia by assessing the effect of temporal scale on the estimated daily loading rates for a small depositional beach. We do not attempt to relate the patterns to the many, specific factors affecting debris accumulation rates – rather, we simply present the model as the product of these factors. We then explore the implications of our findings with respect to bias in accumulation studies with intervals of up to 6 months.

Methods

Our study was conducted on a small (350-m long) beach, Charlesworth Bay (30.26692 S, 153.13975 E), immediately north of Coffs Harbour, the main population centre on this stretch of the NSW mid-north coast (Fig. 1). Charlesworth Bay is protected from the dominant south-easterly swell by Diggers Headland and its associated reefs. As a result, this beach experiences some of the lowest wave energy of those in the Coffs Harbour region [35], and is considered to be depositional. Classified as a reflective beach, it has a steep beach face with wave heights usually ≤0.5 m [35,36]. Beach sediment consists of coarse sands and pebbles. Access to the beach is through a resort complex and visitation rates are substantially lower than for adjacent beaches with greater access (pers. obs.).

Sampling consisted of the removal of all items of macro-debris (≥5 mm) from the entire beach face. This involved searching from the waterline to the highest strandline, which was often within the vegetation line at the top of the beach. Each survey was completed over a single low-tide cycle. Only surface debris was removed and no attempt was made to exhume buried items unless they protruded through the beach surface.

Prior to commencing the accumulation study, we cleared the beach of all debris. Subsequently, over a 20-month period (July 2011 to March 2013), we conducted surveys of debris accumulation at intervals ranging from 1-165 days (Table 1). Given the considerable effort required to complete the surveys, some sampling periods were timed to coincide with larger community events (e.g. Clean Up Australia Day) or were carried out as a practical component in teaching (a 3rd year undergraduate unit on marine pollution). In these latter situations, all removal was very carefully supervised to ensure that sampling intensity was the same as at other times. Converting debris loads to estimated daily accumulation rates, we modelled loss of debris from the beach by plotting estimated daily accumulation rate against the period of accumulation, and fitting the most parsimonious regression.

No permits were required for this study and the field work did not involve protected or endangered species.

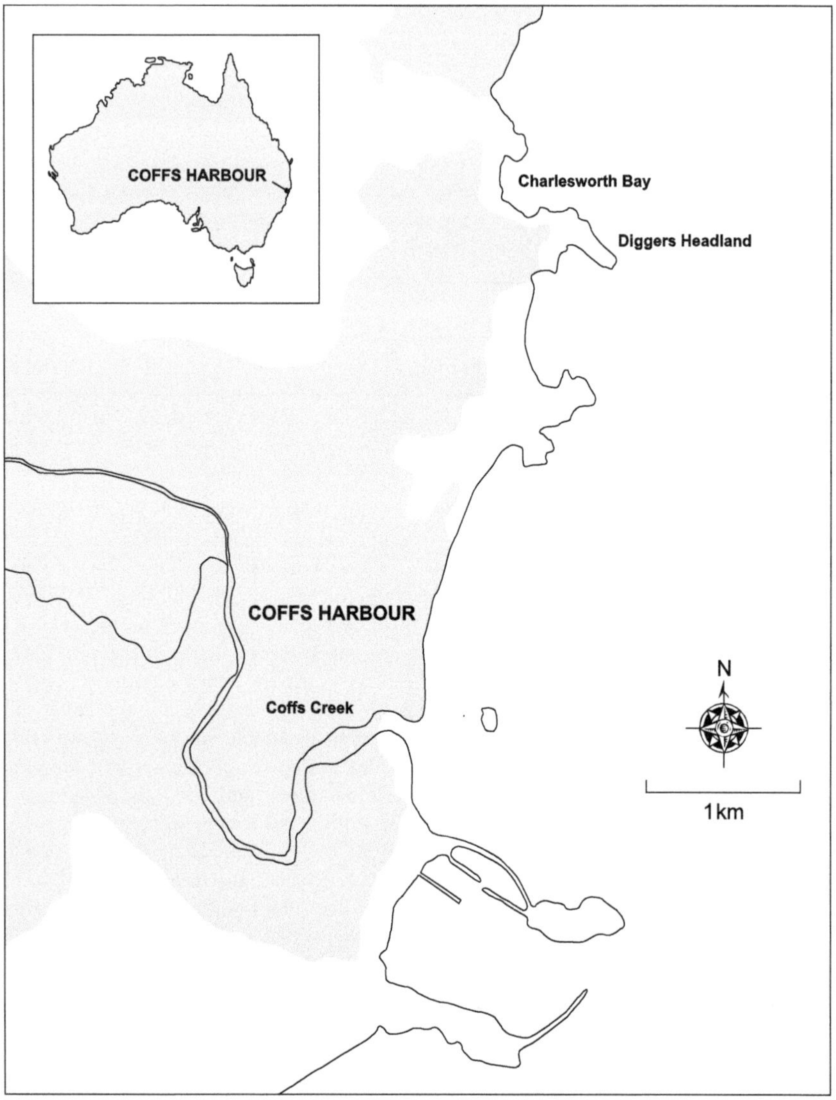

Figure 1. Map of the study area showing Charlesworth Bay and the extent of the urbanised area of Coffs Harbour (shaded).

Table 1. Intervals used for accumulation studies and the number of items found in in each interval category (rounded to the nearest integer) (n = the number of replicate surveys for that interval).

Interval (days)	n	Mean	Min.	Max.
1	7	772	540	928
4	1	1211	1211	1211
14	6	1506	825	2944
21	1	3762	3762	3762
28	2	4565	2080	7049
63	1	4571	4571	4571
84	1	1795	1795	1795
126	1	2360	2360	2360
165	1	5118	5118	5118

Results

The initial standing stock of debris at Charlesworth Bay was 4,044 items (0.24 items m^{-2}) and we collected a further 42,684 items over the duration of the study. Plastic items contributed a total of 91.4% of the total debris. Within the plastics category, fragments of plastic (26.4% of total debris load) and monofilament fishing line (25.0% of total debris load), mostly from recreational fishing activities, predominated. Other plastic items included plastic bags (9.5% of total debris), food wrappers (7.4%), food containers (3.9%) and foamed plastic (styrofoam) (2.3%). The balance of the debris comprised items made from cloth (2.7%), metal (2.0%), rubber (1.6%), paper (0.8%), processed wood (0.8%), glass (0.6%) and "other" (0.1% e.g. bricks and building materials). Of interest was the presence of 21 items that, based on date stamps (e.g. on plastic food containers), or information from the manufacturers, were 25–35 years old. These items primarily comprised plastic bottle tops, food containers and beer cans.

The line-of-best-fit ($r^2 = 0.872$, $P < 0.001$) for the plot of estimated daily accumulation rate against interval between samples was provided by a power curve (Fig. 2) with the following equation:

$$\textit{estimated daily accumulation rate}(y) = 749.81\ \textit{interval}\ (x)^{-0.67}$$

Thus, there was a rapid decline in estimated daily accumulation rate with increasing interval between sampling. Surveys conducted after an interval of only 3 days had daily accumulation rates <50% of the mean rate calculated from daily sampling, and apparent loads decreased by an order of magnitude by 1 month (~30 days).

As previous studies have suggested that variation in estimated loads is greatest at the shortest intervals [10], we calculated the coefficient of variation (CV) for intervals at which we had 2 or more observations (intervals of 1, 14 and 28 days). There was a clear trend of increasing variation with increasing interval: 1 day, CV = 21.0: 14 days, CV = 49.0; 28 days, CV = 94.4 (but this was based on only 2 replicates – Table 1).

Discussion

While this study was conducted on a single small beach, the results support those from other studies that have assessed accumulation over a range of temporal scales [10,13,14] - once deposited, debris is rapidly lost from the surface of beaches. Based on the few studies presenting comparable data, the rate-of-loss is surprisingly consistent regardless of geographic location. Thus, Eriksson et al. [14] found accumulation rates at Macquarie Island in the subantarctic to be underestimated by an order of magnitude after a period of one month. Similarly, Ryan et al. [10] reported that daily accumulation rates on beaches in South Africa were 100–600% greater than estimates based on weekly sampling: the modelled value in our study was ~360%. This concordance between studies is made more remarkable given the likelihood of very different values of parameters known to affect debris accumulation. The implications of these studies are profound - that most accumulation studies vastly under-estimate the abundance of marine debris in coastal habitats and thus the scale of the problem is much greater than initially thought.

Although this study had the sole objective of generating an accumulation-sampling interval model, it is nevertheless instructive to examine the likely pathways through which debris is lost from the system. Many of these have been documented from previous beach debris work and it is highly likely that there is considerable interaction between most mechanisms.

Tidal inundation is a primary mechanism not only for transport of debris onto a beach [14] but also for removing it from the beach [21]. Different studies have found correlations between debris loads and the strength and direction of wind [14,17,19], which also affects the distribution of debris at the scale of the beach, including its burial. Extreme weather events (storms) can have a major impact through intensification of wind and wave action and through run-off into adjacent waterways. Intense storms may also introduce debris from adjacent subtidal habitats and this is likely to have been a principal source of the fishing monofilament that comprised 25% of the total debris load in this study. Monofilament does not float and most of the 11,660 pieces found during this study were entangled around kelp or other types of detached sessile benthos. This reflects the fact that monofilament is by far the most common type of debris found on local reefs, with the majority resulting from recreational fishing activities [37].

Burial is thought to be a major sink for debris on many beaches [38], although this is mediated by the size of the debris items relative to beach grain size [19]. The importance of this loss mechanism is emphasised by Kusui and Noda [25] who found that the average weight ratio of buried to stranded debris was 0.65 on beaches in Russia and Japan. Mechanical degradation is likely to be an important mechanism on beaches where abrasion processes are high. However, unless plastic debris arriving on a beach is already made brittle by photodegradation [22], it is unlikely to be a major contributor to short-term loss. Finally, with increasing public awareness about marine debris, and concerns for the health and aesthetics of beaches and marine environments [26,39,40], removal by visitors may be an important loss mechanism at popular beaches [41]. Clearly, the scale of removal will vary from place to place, and is unlikely to have contributed greatly to the observations in this study (on a beach that has low visitation rates), and would have been absent at Macquarie Island [14].

Whilst we have primarily focused on parallels with other studies, there are many additional site- and region-specific factors that are likely to result in different findings if a study such as this was conducted elsewhere. For example, beach width, slope, small-scale topographical features, proximity to debris sources (e.g. waterways

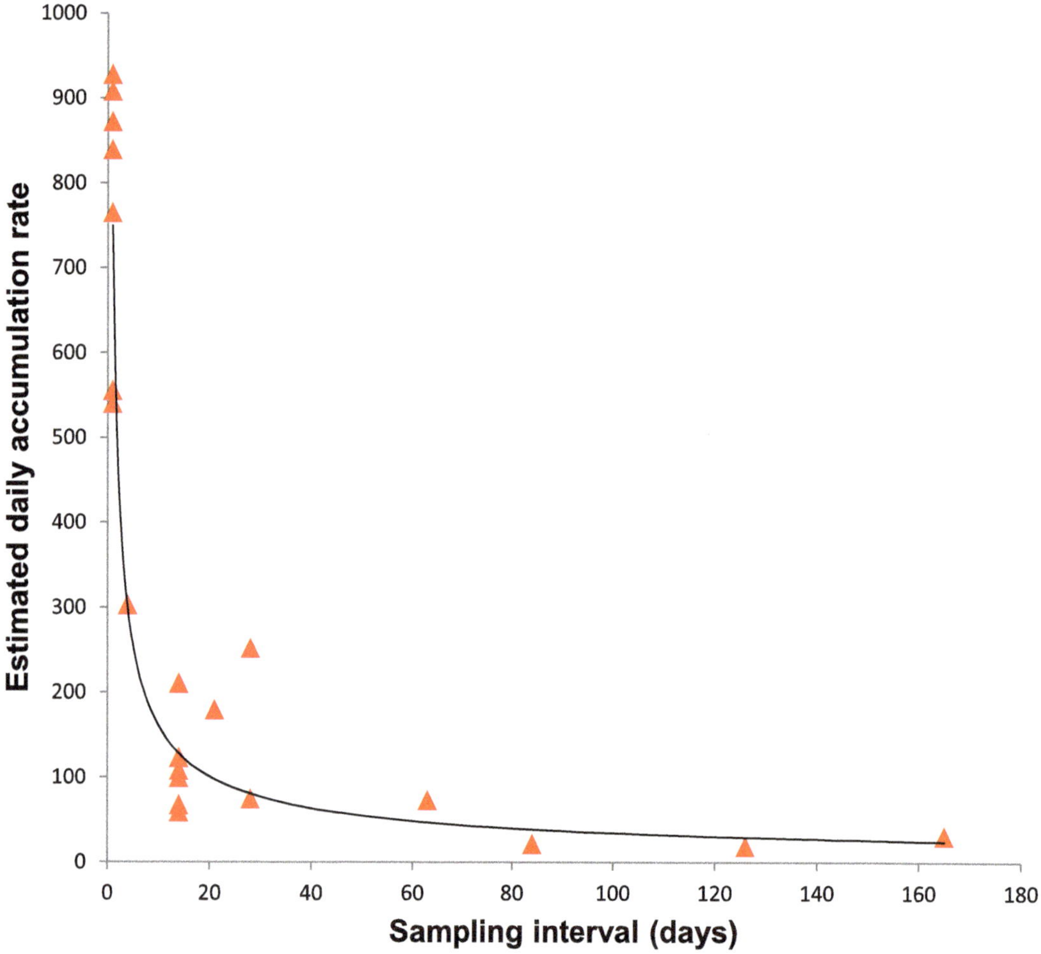

Figure 2. Plot of estimated daily accumulation rate of marine debris against interval between sample periods (days) for Charlesworth Bay. The regression is a power function with an r^2 of 0.872 and the following equation: *estimated daily accumulation (y)* $= 749.81\,interval(x)^{-0.67}$.

or urban areas), and usage rates, all contribute to rates of debris retention and loss [11,13,34]. In addition, broad oceanographic patterns [20,42,43] have been demonstrated to influence accumulation rates in different geographic regions. Very sheltered beaches in the tropics may have the added factor of biogenic habitat (mangrove vegetation) that can trap and concentrate debris [44–46].

The necessity of daily sampling to gain a realistic estimate of loading rate for specific beaches poses a number of substantial challenges. Firstly, the sheer effort required to clean even a small beach, such as Charlesworth Bay, is considerable and thus incurs high time-costs. It is thus impractical to do this over a long-term period unless a large pool of volunteers is available. Indeed, the utility of volunteers has been widely recognised and they are, increasingly, being successfully engaged to deal with burgeoning worldwide debris loads [23,41,47]. However, the novelty of removing debris from a beach is likely to wear off, even for the most committed volunteers. Eriksson et al. [14] recognised this problem and suggested that, for example, instead of 12 samples at monthly intervals, accumulation studies should consider 12 consecutive days to provide more realistic estimates. However, this approach carries the inherent assumption that the 12 sampling days will be representative of the mean pattern of accumulation for the beach. In our study, we found that the coefficient of variation was lowest for daily accumulation rates with a progressive increase for intervals of 14 and 28 days. While this trend is opposite to that recorded by Ryan et al. [10], it was based on relatively few observations at the longest interval. Clearly, better estimates of temporal variation in daily accumulation rates are required before clear protocols can be recommended.

We suggest that there may be an alternative approach to repeated sampling over consecutive days – sampling at a range of intervals and constructing site-specific accumulation models as presented here. This would allow correction of future assessments of loading rates for sampling conducted at a variety of intervals: new data points can also be used to further refine the model. This is appealing as it not only provides models at a local scale, but also allows for less rigid sampling agendas in a habitat that can be difficult to work. However, to avoid bias with this approach, it is important that sampling is not simply opportunistic, or conducted in response to specific events (e.g. good weather, availability of volunteers): program planning should include appropriate *a priori* randomisation of sampling periods over the duration of the survey.

Despite the efforts of a number of research teams in the past [13,19], it is clear that we still have a long way to go to generate realistic models for the dynamics of marine debris on beaches, let alone in less accessible habitats. This study helps to fill one of the gaps by providing a model for rates of accumulation and loss on an ocean beach. A key challenge remains not only to allocate the "lost" debris to the various possible pathways, but also to differentiate between, and quantify, the input sources which include: "new" items arriving by floating; items sourced from adjacent subtidal habitats; items delivered by wind and by runoff from adjacent terrestrial areas; and items retained within the system through a cyclical process of burial, exhumation and further transportation. The potentially extended temporal scale of the latter process is illustrated by the old items we found that had clearly been recently exhumed from adjacent beaches through coastal erosion.

This study provides strong evidence for rapid loss of debris from beaches following stranding, which has obvious implications for the interpretation of past accumulation studies. Thus, given that few studies have used sampling intervals of <1 month, and with the assumption that our model is more generally applicable, the scale of the marine debris problem in coastal waters may have been underestimated by at least an order of magnitude. This conclusion, which is supported by other recent studies [14], highlights the need for concomitant scaling up of measures to manage and mitigate the problem.

Acknowledgments

We thank the many volunteers who donated many hours of their time to assist with field and laboratory work for this project. In particular, we thank Matt Broadhurst, Shiori Naka, Paul Butcher, Jaqueline Thorner, staff and students from the 2012 Pollution of the Marine Environment course at Southern Cross University's National Marine Science Centre, and all the participants in the March 2013 Clean Up Australia Day event at Charlesworth Bay. Part of the data collected for this study contributed to a Masters thesis by the second author (Markic). Kathryn James produced Fig. 1. Martin Thiel, and an anonymous reviewer, made helpful comments on the manuscript.

Author Contributions

Conceived and designed the experiments: SDAS. Performed the experiments: AM SDAS. Analyzed the data: SDAS AM. Wrote the paper: SDAS AM.

References

1. Hong S, Lee J, Jang YC, Kim YJ, Kim HJ, et al. (2013) Impacts of marine debris on wild animals in the coastal area of Korea. Mar Pollut Bull 66: 117–124.
2. Moore E, Lyday S, Roletto J, Litle K, Parrish JK, et al. (2009) Entanglements of marine mammals and seabirds in central California and the north-west coast of the United States 2001–2005. Mar Pollut Bull 58: 1045–1051.
3. Gregory MR (2009) Environmental implications of plastic debris in marine settings-entanglement, ingestion, smothering, hangers-on, hitch-hiking and alien invasions. Philos Trans Roy Soc B Biol Sci 364: 2013–2025.
4. Moore CJ, Moore SL, Leecaster MK, Weisberg SB (2001) A comparison of plastic and plankton in the North Pacific Central Gyre. Mar Pollut Bull 42: 1297–1300.
5. Martinez E, Maamaatuaiahutapu K, Taillandier V (2009) Floating marine debris surface drift: Convergence and accumulation toward the South Pacific subtropical gyre. Mar Pollut Bull 58: 1347–1355.
6. Pichel WG, Churnside JH, Veenstra TS, Foley DG, Friedman KS, et al. (2007) Marine debris collects within the North Pacific Subtropical Convergence Zone. Mar Pollut Bull 54: 1207–1211.
7. Hammer J, Kraak MHS, Parsons JR (2012) Plastics in the Marine Environment: The Dark Side of a Modern Gift. Rev Environ Contam Toxicol 220: 1–44.
8. Thompson RC, Swan SH, Moore CJ, vom Saal FS (2009) Our plastic age. Philos Trans Roy Soc B Biol Sci 364: 1973–1976.
9. Convey P, Barnes DKA, Morton A (2002) Debris accumulation on oceanic island shores of the Scotia Arc, Antarctica. Polar Biol 25: 612–617.
10. Ryan PG, Moore CJ, van Franeker JA, Moloney CL (2009) Monitoring the abundance of plastic debris in the marine environment. Philos Trans Roy Soc B Biol Sci 364: 1999–2012.
11. Hinojosa IA, Thiel M (2009) Floating marine debris in fjords, gulfs and channels of southern Chile. Mar Pollut Bull 58: 341–350.
12. Ribic CA, Dixon TR, Vining I (1992) Marine debris survey manual. NOAA Technical Report NMFS 108. Springfield, VA: US Department of Commerce.
13. Bowman D, Manor-Samsonov N, Golik A (1998) Dynamics of litter pollution on Israeli Mediterranean beaches: a budgetary, litter flux approach. J Coastal Res 14: 418–432.
14. Eriksson C, Burton H, Fitch S, Schulz M, van den Hoff J (2013) Daily accumulation rates of marine debris on sub-Antarctic island beaches. Mar Pollut Bull 66: 199–208.
15. Martinez-Ribes L, Basterretxea G, Palmer M, Tintore J (2007) Origin and abundance of beach debris in the Balearic Islands. Sci Mar 71: 305–314.
16. Silva-Cavalcanti JS, De Araujo MCB, Costa MF (2009) Plastic litter on an urban beach - a case study in Brazil. Waste Manag Res 27: 93–97.
17. Thornton L, Jackson NL (1998) Spatial and temporal variations in debris accumulation and composition on an estuarine shoreline, Cliffwood Beach, New Jersey, USA. Mar Pollut Bull 36: 705–711.
18. Thiel M, Hinojosa IA, Miranda L, Pantoja JF, Rivadeneira MM, et al. (2013) Anthropogenic marine debris in the coastal environment: A multi-year comparison between coastal waters and local shores. Mar Pollut Bull 71: 307–316.
19. Williams AT, Tudor DT (2001) Temporal trends in litter dynamics at a pebble pocket beach. J Coastal Res 17: 137–145.
20. Morishige C, Donohue MJ, Flint E, Swenson C, Woolaway C (2007) Factors affecting marine debris deposition at French Frigate Shoals, Northwestern Hawaiian Islands Marine National Monument, 1990–2006. Mar Pollut Bull 54: 1162–1169.
21. Vauk GJM, Schrey E (1987) Litter pollution from ships in the German Bight. Mar Pollut Bull 18: 316–319.
22. Cooper DA, Corcoran PL (2010) Effects of mechanical and chemical processes on the degradation of plastic beach debris on the island of Kauai, Hawaii. Mar Pollut Bull 60: 650–654.
23. Cheshire AC, Adler E, Barbière J, Cohen Y, Evans S, et al. (2009) UNEP/IOC Guidelines on Survey and Monitoring of Marine Litter. UNEP Regional Seas Reports and Studies, No. 186; IOC Technical Series No. 83: Nairobi: UNEP. 120 p.
24. Garrity SD, Levings SC (1993) Marine debris along the Caribbean coast of Panama. Mar Pollut Bull 26: 317–324.
25. Kusui T, Noda M (2003) International survey on the distribution of stranded and buried litter on beaches along the Sea of Japan. Mar Pollut Bull 47: 175–179.
26. Ballance A, Ryan PG, Turpie JK (2000) How much is a clean beach worth? The impact of litter on beach users in the Cape Peninsula, South Africa. S Afr J Sci 96: 210–213.
27. Moore SL, Gregorio D, Carreon M, Weisberg SB, Leecaster MK (2001) Composition and Distribution of Beach Debris in Orange County, California. Mar Pollut Bull 42: 241–245.
28. Bravo M, de los Ángeles Gallardo M, Luna-Jorquera G, Núñez P, Vásquez N, et al. (2009) Anthropogenic debris on beaches in the SE Pacific (Chile): Results from a national survey supported by volunteers. Mar Pollut Bull 58: 1718–1726.
29. Frost A, Cullen M (1997) Marine debris on northern New South Wales beaches (Australia): Sources and the role of beach usage. Mar Pollut Bull 34: 348–352.
30. Willoughby NG (1986) Man-made litter on the shores of the Thousand Island Archipelago, Java. Mar Pollut Bull 17: 224–228.
31. Ariza E, Jimenez JA, Sarda R (2008) Seasonal evolution of beach waste and litter during the bathing season on the Catalan coast. Waste Manag 28: 2604–2613.
32. Madzena A, Lasiak T (1997) Spatial and temporal variations in beach litter on the Transkei coast of South Africa. Mar Pollut Bull 34: 900–907.
33. Santos IR, Friedrich AC, Wallner-Kersanach M, Fillmann G (2005) Influence of socio-economic characteristics of beach users on litter generation. Ocean Coast Manag 48: 742–752.
34. Eastman LB, Núñez P, Crettier B, Thiel M (2013) Identification of self-reported user behavior, education level, and preferences to reduce littering on beaches – A survey from the SE Pacific. Ocean Coast Manag 78: 18–24.
35. Rollason V, Goodwin I (2009) Coffs Harbour Coastal Processes Progress Report. BroadmeadowNSW: BMT-WBM. 148p .
36. Short AD (2003) Australia beach systems - the morphodynamics of wave through tide-dominated beach-dune systems. JCoastal Res 35: 7–20.
37. Smith SDA, Rule MJ, Harrison M, Dalton SJ (2008) Monitoring the sea change: Preliminary assessment of the conservation value of nearshore reefs, and existing impacts, in a high-growth, coastal region of subtropical eastern Australia. Mar Pollut Bull 56: 525–534.
38. Merrell TR (1980) Accumulation of plastic litter on beaches of Amchitka Island, Alaska. Mar Environ Res 3: 171–184.
39. Smith VK, Zhang X, Palmquist RB (1997) Marine debris, beach quality, and non-market values. Environ Resource Econ 10: 223–247.
40. Somerville SE, Miller KL, Mair JM (2003) Assessment of the aesthetic quality of a selection of beaches in the Firth of Forth, Scotland. Mar Pollut Bull 46: 1184–1190.
41. Bravo M, de los Angeles Gallardo M, Luna-Jorquera G, Nunez P, Vasquez N, et al. (2009) Anthropogenic debris on beaches in the SE Pacific (Chile): Results from a national survey supported by volunteers. Mar Pollut Bull 58: 1718–1726.
42. Ribic CA, Sheavly SB, Klavitter J (2012) Baseline for beached marine debris on Sand Island, Midway Atoll. Mar Pollut Bull 64: 1726–1729.

43. Santos IR, Friedrich AC, Ivar do Sul JA (2009) Marine debris contamination along undeveloped tropical beaches from northeast Brazil. Environ Monit Assess 148: 455–462.
44. Smith SDA (2012) Marine debris: A proximate threat to marine sustainability in Bootless Bay, Papua New Guinea. Mar Pollut Bull 64: 1880–1883.
45. Uneputty PA, Evans SM (1997) Accumulation of beach litter on islands of the Pulau Seribu Archipelago, Indonesia. Mar Pollut Bull 34: 652–655.
46. Debrot AO, Meesters HWG, Bron PS, de León R (2013) Marine debris in mangroves and on the seabed: Largely-neglected litter problems. Mar Pollut Bull 72: 1.
47. Martin JM (2013) Marine debris removal: One year of effort by the Georgia Sea Turtle-Center-Marine Debris Initiative. Mar Pollut Bull 74: 165–169.

4

Combining Telephone Surveys and Fishing Catches Self-Report: The French Sea Bass Recreational Fishery Assessment

Delphine Rocklin[1]*, Harold Levrel[2], Mickaël Drogou[3], Johanna Herfaut[2], Gérard Veron[3]

1 UMR M101, Aménagement des Usages, des Ressources et des Espaces marins et littoraux, Observatoire des Sciences de l'Univers – Institut Universitaire Européen de la Mer, Université de Brest, Brest, France, **2** UMR M101, Aménagement des Usages, des Ressources et des Espaces marins et littoraux, Unité d'Economie Maritime, Ifremer, Plouzané, France, **3** Département des Ressources Biologiques et Environnement, Unité Sciences et Techniques Halieutiques, Ifremer, Plouzané, France

Abstract

Fisheries statistics are known to be underestimated, since they are mainly based on information about commercial fisheries. However, various types of fishing activities exist and evaluating them is necessary for implementing effective management plans. This paper assesses the characteristics and catches of the French European sea bass recreational fishery along the Atlantic coasts, through the combination of large-scale telephone surveys and fishing diaries study. Our results demonstrated that half of the total catches (mainly small fish) were released at sea and that the mean length of a kept sea bass was 46.6 cm. We highlighted different patterns of fishing methods and type of gear used. Catches from boats were greater than from the shore, both in abundance and biomass, considering mean values per fishing trip as well as CPUE. Spearfishers caught the highest biomass of sea bass per fishing trip, but the fishing rod with lure was the most effective type of gear in terms of CPUE. Longlines had the highest CPUE value in abundance but not in biomass: they caught numerous but small sea bass. Handlines were less effective, catching few sea bass in both abundance and biomass. We estimated that the annual total recreational sea bass catches was 3,173 tonnes of which 2,345 tonnes were kept. Since the annual commercial catches landings were evaluated at 5,160 tonnes, recreational landings represent 30% of the total fishing catches on the Atlantic coasts of France. Using fishers' self-reports was a valuable way to obtain new information on data-poor fisheries. Our results underline the importance of evaluating recreational fishing as a part of the total amount of fisheries catches. More studies are critically needed to assess overall fish resources caught in order to develop effective fishery management tools.

Editor: Konstantinos I. Stergiou, Aristotle University of Thessaloniki, Greece

Funding: This study was carried out thanks to Data Collection Framework fundings of the European Union, the French "Direction des Pêches Maritimes et de l'Aquaculture" and IFREMER. The funders had no role in study design, data collection and analysis, decision to publish, or preparation of the manuscript.

Competing Interests: The authors have declared that no competing interests exist.

* E-mail: delphine.rocklin@gmail.com

Introduction

Marine waters support intensive fishing activities. The total worldwide marine fish production has been estimated as increasing steadily from 16.7 million tonnes in the mid-20th century to about 90 million tonnes at present [1]. However, these figures, mainly based on official fishery landings statistics, are incomplete [2]. They generally do not take into account discarded by-catch, illegally landed catches of the commercial fisheries and unregulated high-seas fisheries [3,4]. Moreover, they often ignore commercial small-scale artisanal fisheries catches, as well as non-commercial catches such as recreational, sport and subsistence fishing ones [5], which are still under-evaluated. But to manage marine resources responsibly, trustworthy data for all components of fishing activities are necessary.

Recreational fisheries recently appeared to be more important than was once thought, and their potential role in marine resources overexploitation has been pointed out [6,7]. The number of "recreational fisheries" citations in the research platform "Web of Knowledge" (www.webofknowledge.org) has indeed exponentially increased since 1990 (Fig. 1), demonstrating the quite recent interest in this subject and awareness of its relevance.

The earliest assessments regarded recreational fishing as of low importance. Indeed, in 1994, the National Research Council estimated that US recreational fisheries accounted for about 2% of total marine landings [8]. More recent studies, improving the assessment methodologies, indicated that this percentage ought to be revised upward. It has been estimated that recreational fisheries may account for a percentage of about 10% to 64% of the global fisheries production [6,9], depending on the considered areas and species [7]. Moreover, in some cases, recreational fishing catches can exceed that of commercial fishing [10]. Based on an extrapolation from the patterns of Canadian recreational fishing onto a global scale, Cooke and Cowx [6] estimated that the total recreational fishing catches may be as much as 47.1 billion fish, of which two-thirds are released, with a retained amount estimated at about 10.86 million metric tonnes. However, these estimates were made using highly general assumptions without taking into account each country's characteristics such as the environmental configuration (freshwater surface area, marine coastline), local policies and/or recreational fishing traditions.

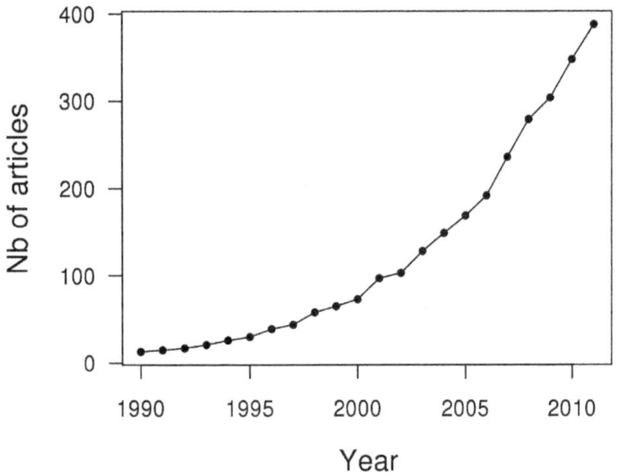

Figure 1. Number of scientific papers citing the term "recreational fisheries" from 1990 until now. Data derived from the research platform "Web of Knowledge" (www.webofknowledge.org).

Estimating the characteristics and catches of recreational fisheries thus remains a challenge. Unlike commercial fishing, carried out for profit, and subsistence fishing, carried out for food, recreational fishing is primarily a leisure activity [11] performed in both freshwater and marine environments. It is characterized by a large number of involved fishermen, by the great diversity of participants, the varied timing and frequency of fishing trips, the variability of used gears and access points and the lack of systematic recording of the catches [12,13]. It is spatially and temporally heterogeneous and there is no uniform method for valuing them [14]. Moreover, the licensing system is not systematic, leading to an overall miscalculation of the population of recreational fishers. All these features make the characteristics and catches of recreational fisheries hard to estimate.

Various methods have been developed for estimating recreational fishing [15]; these generally fall into two parts. The first part focuses on fishing effort: this can be calculated using registries (when a license system exists), aerial flights over geographically limited areas, or telephone surveys for large-scale studies. The second part focuses on evaluating catch rates and species composition: it can be calculated using on-site intercept surveys based on interviews with recreational fishers at the end of their fishing trips, or can draw on a group of fishers, called a panel, declaring their daily catches through diaries or web-based declaration. For example, the US Marine Recreational Information Program (MRIP) [16], formerly called MRFSS (Marine Recreational Fisheries Statistics Survey), was created in 1970 by the National Marine Fisheries Service (NMFS) of the NOAA. It is a large-scale information system which developed various monitoring methods regarding recreational fishing efforts and catches nationwide, through telephone surveys, access point surveys, web-based declaration and mail surveys [5,17]. Other methods have also been used, such as low-altitude flights for counting fishing boats and shore anglers, thus evaluating the fishing effort and areas, combined with on-site interviews for evaluating catches [18–23].

Numerous species, including some viewed as emblematic and/or threatened, are targeted by recreational fishers. Through the European Union Data Collection Framework (DCF), the European Commission has recently required the quantification of the recreational catches of some important European species targeted by both commercial and recreational fisheries: cod (*Gadus morhua*), bluefin tuna (*Thunnus thynnus*), eel (*Anguilla anguilla*), Atlantic salmon (*Salmo salar*) and European sea bass (*Dicentrarchus labrax*) (EC Regulations 199/2008 of 25 February 2008 and EC Decision 2008/949/EC of 6 November 2008). In France, an estimate of the recreational catches of the European sea bass (to be referred to here as sea bass) has been requested for the North Atlantic V-XIV ICES divisions [24]. By the late 1990s, it was already known that recreational fishing for sea bass was popular in England, Wales, the Channel Islands, Ireland, France (especially Brittany), Spain, Portugal and Italy [12]. This species is still highly attractive both on the French Atlantic coasts and in the Mediterranean, and is targeted both by commercial and recreational fishers [25–27]. In France, the sea bass is one of the most economically valuable species [12,28], although it is dependent on the considered métier and season, as well as on the condition of the fish and landed quantities [29]. However, the mean first sale price constantly increased between 1972 and 1984 [30]. Its commercial catches as well as its consumption also raised from 1998 to 2005, significantly contributing to the French market [31]. Nevertheless, although the commercial catches of sea bass in the French Atlantic are well known, no study evaluated the importance of its recreational part.

Our study aimed to enhance the still rare and sparse information on the French sea bass recreational fisheries. Increasing our knowledge of this subject is more than ever a crucial issue since overexploitation is threatening the availability of marine living resources, and managing coastal human activities and natural resources requires a well understanding of their interactions.

This study thus addressed three questions: How can we develop a robust large-scale monitoring system using fishing diaries? What are the main characteristics of the French recreational sea bass fishery? How important are recreational sea bass catches in France?

Materials and Methods

Ethics statement

All the participants of this study, only performed in mainland France, were called and interviewed by the BVA polling institute. The data collection and analyses, involving humans categorized by age, gender and other socially constructed groupings, were subjected to a declaration to the CNIL (Commission Nationale de l'Informatique et des Libertés – The French Data Protection Authority), in accordance with the decree n°78-17 of 6 January 1978 on Data Processing, Data Files and Individual Liberties, under the document number 1394854. The BVA polling institute obtained the 3 November 2009 from the CNIL the permanent authorization for performing such studies. The answer of all the called households, whether they consented or not to participate to the study, was recorded into a response file document: the variable "Acceptance" was filled up with "yes" or "no", and only the participants who provided their verbal consent were interviewed. During the call, the objectives and the proceedings of the study were initially described. Then, the first question was: "Do you agree to answer our questions?". If they agreed, the telephone interview was conducted. Finally, if they were sea bass recreational fishers, they were asked about their interest in participating as a volunteer for the panel. The whole study was based on the volunteer participation of sea bass recreational fishermen. The volunteers were contacted every three months to verify the smooth proceedings of their data collection and if they still wanted to continue the study. All the volunteers had the direct phone number of the scientist in charge of the study and were encouraged to contact him when necessary. All of the collected data was anonymously analyzed. At the end of the study, a thank you letter and a document summarizing the main results were sent by mail to each volunteer.

Sampling design

Estimating the number of marine recreational fishers in France is a challenge. This activity is free and no licensing system exists at this time. We focused this first study on the French littoral of the Bay of Biscay, English Channel and North Sea (VIIIa, VIIIb, VIIe, VIIh and VIId ICES areas). In accordance with previous work on recreational fishery surveys [32–34], we adopted a dual method combining two large-scale telephone surveys with a fishing diary survey, where the fishing diaries were filled by the recreational fishers themselves.

For the two telephone surveys, the phone numbers were randomly selected in an existing database constituted by the landline and cell phone numbers of French telephone subscribers appearing in the national telephone directory. The sampling design used to contact each household was based on the principle that it should offer to each of them the same and non-negative probability to be contacted. For not inducing bias in the representativeness of the sample, not only the households easily reachable (those often present at their house) were interviewed. Thus, a strict and rigorous call process was used, following these assumptions: (1) the call numbers appeared in a random order (no pool); (2) the calls were done with the same insistence; and (3) the call hours permitted to contact the whole population. Each phone number was therefore called at different hours and days until it was reached or was definitely abandoned after 12 unsuccessful calls.

A first telephone survey, sea bass-specific, was conducted in June and November 2009 in the French coastal departments of the considered study. 172,054 telephone numbers were exploited to obtain a representative sample of 15,091 interviewed households (9%) (Fig. 2). The 91% non-interviewed households corresponded, among others, to an occupied number (37%), a clear refusal to answer the interview (29%), or the non-response to the call (10%). At least one sea bass recreational fisher (here by definition a fisher targeting the sea bass and who has caught at least one sea bass during the last 12 months) was present in 535 (3.5%) of the 15,091 households successfully interrogated. These sea bass fishers were asked if they would agree to continue the interview and 467 of them did so (87.3%). The interview permitted us to collect socio-professional information on recreational fishers targeting sea bass during their fishing trips, as well as general information on their fishing activity during the previous year. The main questions were: What was the main type of gear you used? Did you fish from the shore or from a boat? In which area(s) did you go fishing? When did you go fishing? Approximately how many sea bass and other species did you catch during the last 12 months? Did you release any of these catches? This survey made it possible to identify and precisely describe the main characteristics and practices of recreational sea bass fishers.

At the end of this telephone survey, sea bass fishers were asked to join a panel, that is a group of recreational fishers who agreed to voluntary fill in a fishing diary to report their catches information, during one year. 256 (54.8%) of the 467 sea bass fishers agreed to join the panel. We sent the fishing diaries to the volunteers (one every three months or every 20 fishing trips), as well as a species identification guide describing the main characteristics of the commonly fished species, a spring balance and a measuring tape.

For each fishing trip, various items of information were to be recorded: date, main gear used (we made the hypothesis that they only used one type of gear during a fishing trip), whether they fished from a boat or from the shore, the travel duration (from home to the fishing site, and travel duration by boat if used), the fishing site (town and fishing sector, based on precise sectors on an attached map), the port of departure, the fishing duration, the description of the sea bass catches (weight, length, whether kept or not), and the description of the other catches (species common name, weight and number of both kept and released individuals). The panel methodology offers the possibility of obtaining precise information about released catches, which can generally not be inspected during on-site surveys [15], and about night fishing. 190 (74.2%) of the 256 fishers constituting the initial panel returned at least 1 fishing diary and 40 fishers (15.6%) returned the whole-year set of diaries, providing a seasonal picture of the fishery.

However, the sea bass-specific telephone survey combined with the panel study was inadequate for assessing the size of the recreational sea bass fishers population in whole France and the total French catches, since it only focused on the coastal departments. For extrapolating the results at a national scale, a pilot telephone survey, carried out in 2006 nationwide, was used. This second telephone survey aimed at providing the first global estimate of the French mainland recreational fishing (all species considered, including shellfish) by evaluating the total number of recreational fishers, the main target species and the related practices [35]. Two criteria of the nationwide telephone survey were considered for extrapolating the data of the sea bass-specific telephone survey: (1) having fished at least one time the last year in a coastal department included in the study area and (2) citing the sea bass among the three main targeted and fished species. Thus, using both the structure of the fishers' population living mainland (except in the coastal departments) and the respective weight of the fishers living in the coastal departments and in the rest of France, a crossing coefficient between fishers living in the coastal departments and fishers whom not was evaluated. By implementing this crossing coefficient, an estimation of the number of sea bass fishers living in the non-coastal departments was obtained (Table 1).

Weighting the data

To ensure the panel sample to be representative of the coastal and non-coastal French recreational sea bass fishers population, and to obtain reliable results at a national scale, the panel data was corrected using a set of weighting correction factors [35]. The calculation of the weighting factors was done applying the CALMAR program (http://www.insee.fr/fr/methodes/default.asp?page=outils/calmar/accueil_calmar.htm), used since 1990 by the French National Institute of Statistics and Economic Studies (INSEE) [36,37].

The data was corrected by weighting the individuals, here the fishers. This was done using the values of some variables of adjustment, which were known both for the sample and the population. In our case, the considered variables used for calculating the weighting factors were the fishing frequency, the main fishing gear and the fishing mode (Table 1). Since the considered variables are category-based and since we knew the number of fishers belonging to each of these categories' modalities in the whole French population (2009 and 2006 telephone surveys, Table 1), the "ranking ratio method", also named "iterative proportional fitting", was used [38]. A weighting factor was attributed to each fisher for every month, and permitted to extrapolate the panel data to the whole French sea bass recreational fishers population.

Data analysis

15,085 households were interviewed for the global telephone survey and 15,091 for the sea bass-specific one. A total of 134

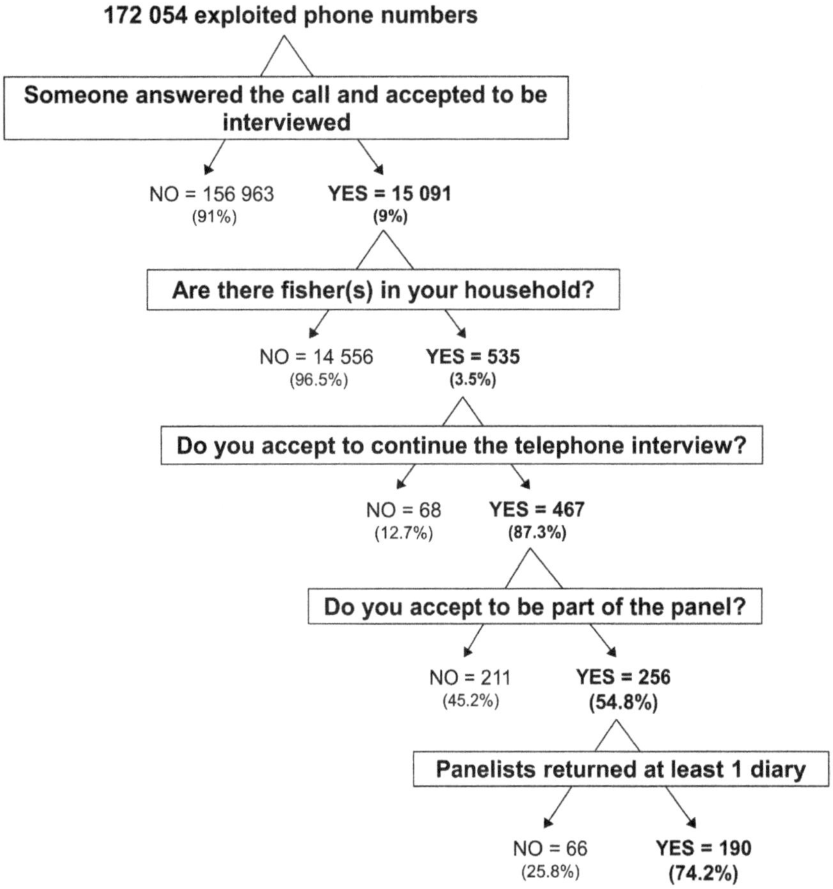

Figure 2. The sea bass-specific survey: followed sampling design.

(global survey) and 535 (sea bass-specific survey) sea bass fishing households were identified from these interviews. In 2009–2010, a total of 256 fishers agreed to be part of the panel. A total of 1,190 fishing trips and 1,383 sea bass catches were recorded from the 190 fishers who returned at least one fishing diary.

Table 1. Variables used for calculating the weighting factors and evaluated number of sea bass fishers in each stratum (coastal, non-coastal and whole France population).

Fishing frequency	Main fishing gear	Fishing mode	Coastal sea bass fishers population (2009 survey)	Non-coastal sea bass fishers population (estimation)	French sea bass fishers population (estimation)
Occasional	All	Shore	22,634	15,401	38,036
	All	Boat	21,450	7,006	28,455
Regular	Spearfishing	Shore	5,509	1,336	11,583
	Spearfishing	Boat	4,738		
	Lines	Shore	47,438	34,399	81,837
	Lines	Boat	44,523	58,023	102,546
Highly regular	Spearfishing	Shore	4,972	4,079	14,428
	Spearfishing	Boat	5,377		
	Lines	Shore	29,606	13,811	43,416
	Lines	Boat	39,004	10,682	49,687
Total fishers population			225,252	144,737	369,989

Occasional: fewer than 3 fishing trips/year targeting sea bass; Regular: between 3 and 15 fishing trips/year targeting sea bass; Highly regular: more than 15 fishing trips/year targeting sea bass. Population: number of coastal sea bass fishers in France, based on the national telephone interview.

Fishing effort was equated with fishing duration, in hours. The CPUE (Catch Per Unit of Effort) was expressed as the total catch (in biomass [in g.] and/or abundance [in ind.]) per hour of fishing (g h^{-1} or ind h^{-1}). The uncertainties of the catches estimations were assessed while analyzing the panel catches information per month, and calculating the precision of the information through the calculation of the mean, standard error and corresponding uncertainty of the sea bass catches information of the panel dataset. We found that over the year, the uncertainty could be evaluated at 51%.

The data analyses were performed using the R software [39]; all the results were standardized with the weighting factors using the "rgrs" – social sciences R-package.

Robustness of the dataset

In order to evaluate the robustness of the panel declarations, the length-weight (L-W) relationship of the 1,383 sea bass caught by the volunteer fishers of the panel, measured by themselves and recorded in the diaries, was compared to the L-W relationship of 911 sea bass collected and measured by scientists between 2007 and 2011 during the EVHOE scientific campaigns conducted in the Bay of Biscay, English Channel and North Sea (Fisheries Information System Resource, IFREMER Brest) [40]. The panel catches covered a wide range of sea bass lengths: the smallest sea bass caught by the panel was 6 cm long, whereas it was 26 cm for the scientist dataset, and the largest sea bass caught by the volunteers was 82 cm long compared to 85 cm in the scientific dataset. The mean observed sea bass length was 38 cm in the panel dataset and 53 cm in the scientific set. The biggest declared sea bass caught by volunteers was of 6 kg while the biggest sea bass of the scientific survey was of 6.3 kg.

The a and b parameters of the length-weight relationships were estimated using the "nls" non-linear least-squares function in R. The estimated L-W parameters of the panel dataset were a = 0.048536 and b = 2.615324 and those of the scientific dataset were a = 0.029122 and b = 2.734432. Although the "sm.ancova" statistical test of equality gave a p-value = 0, the regression of the panel dataset (bold solid line) and of the scientific one (dotted line) appeared extremely close (Fig. 3).

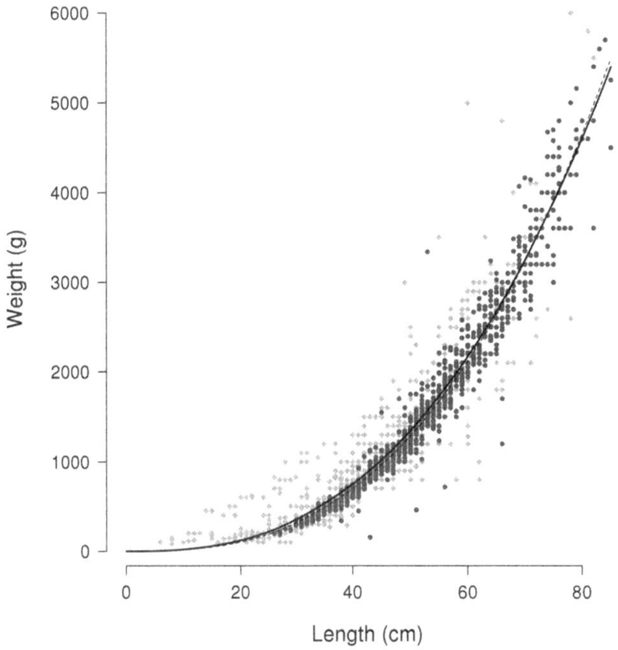

Figure 3. Length-weight (L-W) relationship of the European sea bass, based on the panel of volunteers and scientific catches. Dark grey circles: scientists' catches; Light grey squares: panel's catches; Solid line: L-W relationship based on scientists' dataset; Spotted line: L-W relationship based on panel's dataset.

Results

Fishing gears

Various types of gear were used for targeting the sea bass (Fig. 4). We recorded 7 of these: handlines, longlines, nets, spearguns, rod with bait, rod with lure, and fly-fishing rod. However, since only one member of the panel used this last type of fishing gear, it was not considered as representative of recreational fly-fishers as a whole, and was not considered.

The type of gear used depended on the fishing mode, meaning if the fisher fished from the shore or from a boat. The rod with bait was the mainly used type of gear from the shore, represented in 40.5% of the shore fishing trips while it was declared in only 8.6% of the boat fishing trips. However, the rod with lure was used in 52% of the boat fishing trips but only in 28.3% of the shore fishing events.

Longlines and spearguns were used by a small number of fishers. Both were used slightly more often from the shore (5.2% and 4.2% of the fishing trips, respectively) than from boats (2.2% and 3.3% of the fishing trips, respectively). Handlines were used as often from the shore as from boats, in about 20% of the reported fishing trips. Lastly, nets were mainly used from boats, representing 14% of the fishing trips but were only used in 1% of the fishing events displayed from the shore.

Catch-and-release

The minimum legal catch size of the European sea bass in the Atlantic was 36 cm at the time of this study. The panelists' records showed that the mean length of caught sea bass was 38 cm, corresponding to a weight of 825.2 g. The mean length of the kept sea bass was 46.6 cm (1,230 g) while it was 29 cm (407.5 g) for the released ones.

When looking at the minimum legal catch size, it appeared that 87.9% of the undersized sea bass were released at sea, while 12.1% of them were kept (Fig. 5). However, 23.4% of the legal-size sea bass were also released at sea. Although 36 cm was, at the time of the study, the minimum legal size for the sea bass catches in the French Atlantic, recreational fishing federations generally recommended a minimum catch size of 42 cm, accounting for the reproductive characteristics of Atlantic sea bass stocks. It is to note that since the 26th of October 2012, the minimum legal size for the Atlantic sea bass recreational catches is 42 cm. Considering the caught sea bass having a size between the former legal one of 36 cm and the former recommended one of 42 cm, we observed that half of them were released at sea (48.6%).

Catches characteristics per fishing trip

Analyzing the panel dataset, we found that a mean of 1.15 sea bass (or 953 g) were caught during a fishing trip; by mean, half of these catches were kept by the recreational fishers, and the other half released at sea.

However, the mean amount of sea bass caught from the shore and from boats differed significantly both in biomass and abundance (Fig. 6): from boats, a mean of 1,221.1 g of sea bass (1.4 ind.) were caught per fishing trip, while 650.9 g (0.87 ind.) were caught from the shore. The biomass of released fish was much lower than that of kept fish, both when fished from boats (256.6 g released, i.e. 21% of the caught biomass) and from the

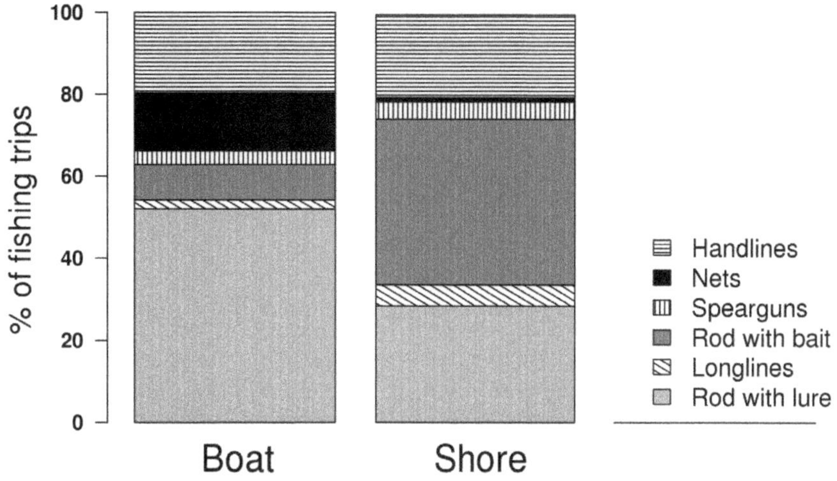

Figure 4. Percentage of fishing gear used for catching sea bass, from the shore and from boats.

shore (205.1 g released, i.e. 31.5% of the caught biomass). However, with respect to abundance, in each case about half of the caught sea bass were released at sea, that is 0.65 (46% of the catches) of those caught from boats and 0.46 (53% of the catches) of those caught from the shore.

Recorded catches per type of gear also highlighted some particular patterns (Fig. 7). Spearfishers caught the highest biomass of sea bass per fishing trip (1,572.35 g) whereas the mean abundance was one of the lowest (1 ind.). All the spearfishing catches were kept. The highest abundance of sea bass caught per fishing trip were reported by fishers using rod with lure and longlines, each type reporting a mean value of 1.6 sea bass caught per fishing trip; however, the mean biomass was higher when using rod with lure (1,287.9 g) than longlines (991.8 g). The released amount of sea bass was 381.6 g (29.6%) for the rod with lure and 166.1 g (16.7%) for the longlines. With respect to abundance, fishers using the rod with lure released a mean of 0.87 sea bass per trip (54.4% of the catches) while those using longlines released a mean of 0.67 sea bass per trip (41.9% of the catches). Fishing with nets produced a fairly small number of sea bass catches per trip (0.71 ind.), but all or almost all of them were kept (87%). This resulted in a mean caught biomass of 1,037.1 g and a released one of 64.8 g (i.e. 6.25% of the catches). Handlines produced the lowest catches per fishing trip, both in biomass and abundance: 390.2 g were caught and 36.15 g released (9.3% of the catches), corresponding respectively to 0.45 individuals caught per trip among which 0.12 were released (26.7% of the catches).

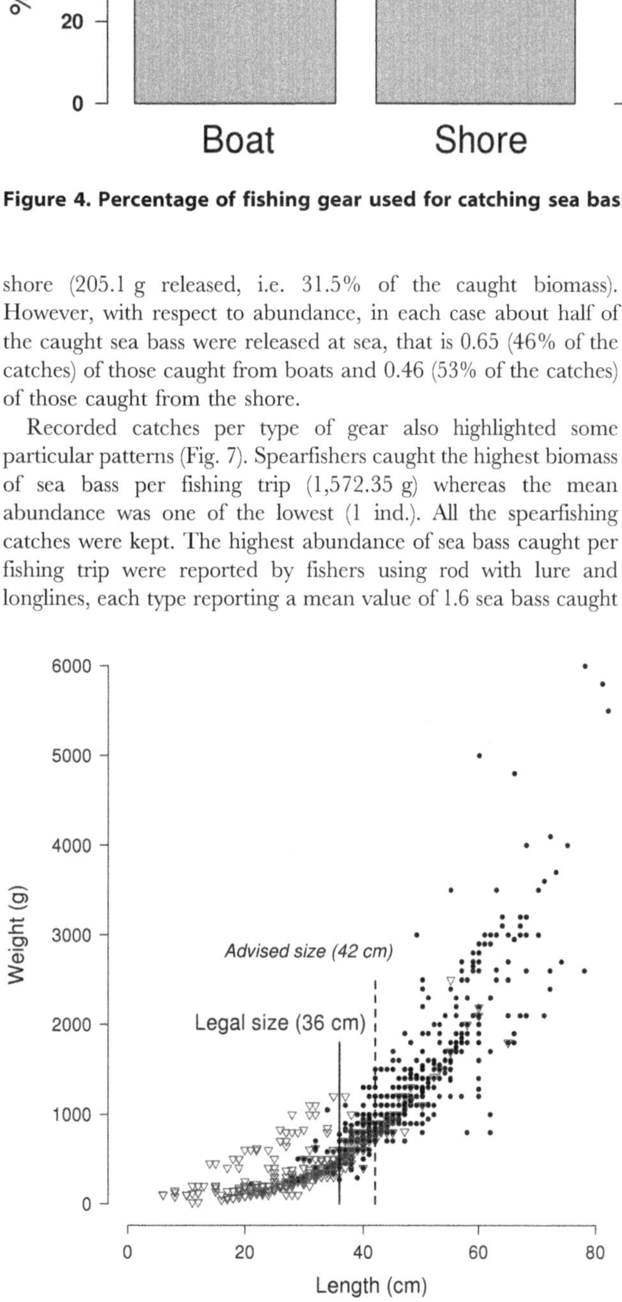

Figure 5. European sea bass kept and released by recreational fishers. Solid circle: kept sea bass; Filled triangle point down: released sea bass; Legal size: minimum legal catch length; Recommended size: length generally recommended by French recreational fishery federations.

Catch Per Unit of Effort (CPUE)

The mean observed recreational fishing duration, all types of gear included, was 3h20 (Table 2). However, the mean and maximum durations varied depending on the type of gear. The two passive gears displayed the greatest mean fishing duration: nets were generally set for 5h30 (with a maximum of 24 h), and longlines were set for a mean of 4 h per fishing operation (max. 12 h). Rods with bait were reportedly used for a mean of 3h20 (max. 9 h), spearguns for 3h10m (max. 7 h), rod with lure for 3 h (max. 12 h) and handlines for 2h57 (max. 12 h).

In calculating the CPUE, we observed that a mean of 321.5 g h^{-1} of sea bass (corresponding to 0.41 ind h^{-1}) were caught, all types of gear considered together. However, CPUE from boats were higher both in abundance and in biomass than CPUE from the shore (Fig. 8): 375.4 g h^{-1} and 0.46 ind h^{-1} were caught from boats compared to 259.2 g h^{-1} and 0.36 ind h^{-1} from the shore.

We also observed differences when viewing the fishing gears separately (Fig. 9). With respect to biomass, rod with lure was the most effective gear, allowing to catch a mean of 437 g h^{-1} of sea bass. Handlines were least effective, with 130 g h^{-1} caught. With respect to abundance, longlines caught the greatest number of sea bass per hour of fishing, with 0.64 ind h^{-1}, while handlines caught the lowest, with 0.16 ind h^{-1}. Differences in CPUE by type of gear, comparing both biomass and abundance data, brought out some gear-dependent fishing patterns: both nets and spearguns displayed high values of CPUE in biomass (266.51 g h^{-1} and 373.8 g h^{-1}, respectively) but low values of CPUE in abundance

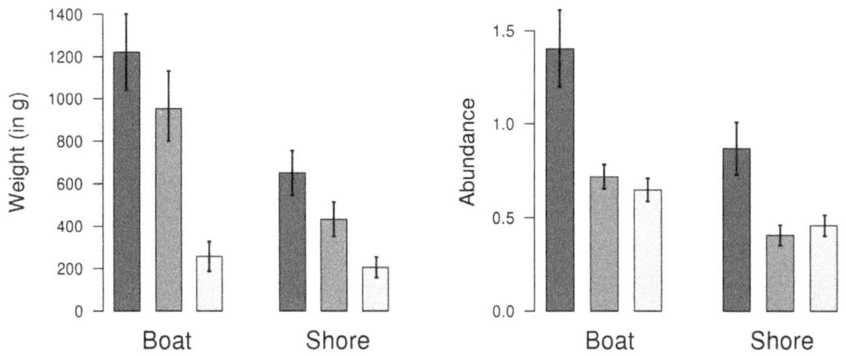

Figure 6. Mean weight and abundance of sea bass caught by recreational fishers from boats and from the shore. Dark grey: caught sea bass; Medium grey: kept sea bass; Light grey: released sea bass; Black bars: confidence interval ($\alpha = 0.95$).

(0.17 ind h^{-1} and 0.26 ind h^{-1}, respectively), indicating that mainly large individuals were caught, having a mean length of 53 cm and 50.6 cm respectively. In contrast, longlines had the highest CPUE in respect to abundance, with a mean of 0.64 ind h^{-1}, but were in third place with respect to biomass, with a mean value of 320.3 g h^{-1}, meaning that longlines mainly caught numerous but small sea bass, having an actual mean length of 36.2 cm.

Lastly, calculating the CPUE per fishing mode and type of gear brought out some additional differences (Fig. 10). For the majority of gear types, catches were observed to be greater when fishing from boats than from the shore, except in the case of longlines and handlines. However, the observed differences for these two gear types principally affected abundance and not biomass: in these cases, sea bass caught from the shore were smaller than those caught from boats. For spearfishing, differences were obvious both in biomass and abundance: there were more catches from boats than from the shore. The same pattern was also observed for nets: sea bass caught from boats were more numerous and the mean biomass was higher than when caught from the shore. However, these results must be viewed with caution since the dataset available for nets used onshore was relatively small ($n = 7$). The catches by rod with lure followed a different pattern: with respect to both abundance and biomass, while the mean CPUE in biomass was greater from boats than from the shore, this difference was even greater with respect to CPUE in abundance, indicating that more small individuals were caught from boats than from the shore. Lastly, CPUE observed when using a rod with bait did not differed when fishing from boats or from the shore.

Total catches

After adjusting the panel data with the weighting factors issued from the two telephone surveys, the total number of recreational sea bass fishers in the Bay of Biscay, Channel and North Sea regions was estimated at 370,000 and the overall recreational sea bass fishing effort was estimated at 2,177,378 fishing trips per year in these areas. In terms of catches, it was estimated that 3,935,024 sea bass (±51%; min: 1,928,162 and max: 5,941,886) were caught by recreational fishers, of which 1,948,888 were kept and the others released at sea. In terms of biomass, this corresponded to a total annual catch of 3,173 tonnes (±51%; min: 1,554.8 t and max: 4,791.2 t), of which 2,345 tonnes were kept (74%). Coastal fishers caught two-thirds of this total (2,079 t, of which 1,546 t were kept) and fishers living inland caught the other third (1,094 t, of which 799 t were kept).

Total annual catches varied significantly depending on the type of gear used (Fig. 11). Rod with lure, rod with bait and handlines were the most often used gears in the reported fishing trips (representing 40.8%, 20.1% and 17% of the total reported trips,

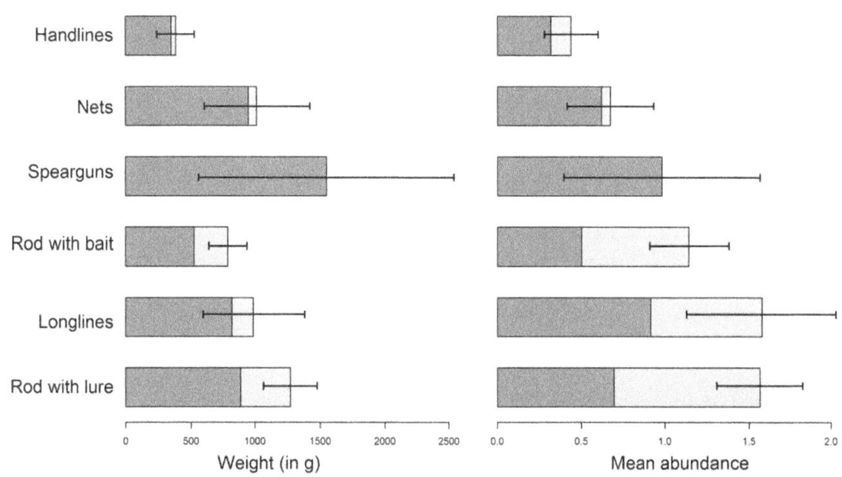

Figure 7. Mean weight and abundance of sea bass caught per gear type and fishing trip. Medium grey: kept sea bass; Light grey: released sea bass; Black bars: confidence interval ($\alpha = 0.95$).

Table 2. Mean, minimum and maximum observed fishing duration per type of gear, and number of fishing operations declared using the considered gear.

Gear	Mean	Min.	Max.	Nb of samples
Net	5h30	1 h	24 h	88
Longline	4 h	1 h	12 h	47
Fishing rod with bait	3h20	0h15	9 h	239
Fishing rod with lure	3 h	0h20	12 h	486
Handline	2h57	0h10	12 h	202
Speargun	3h10	0h15	7 h	49

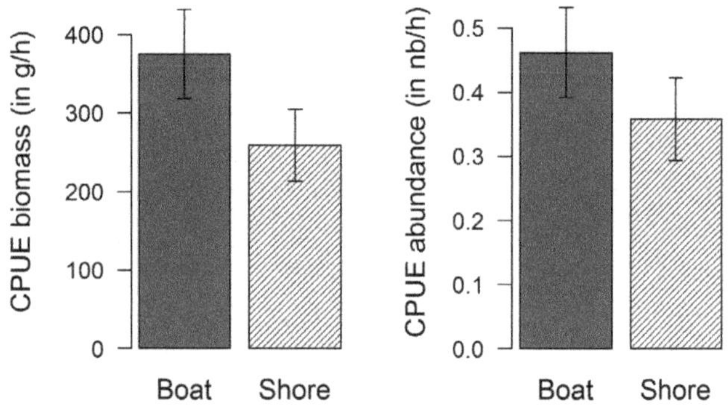

Figure 8. Mean CPUE of caught sea bass, from boats and from the shore. Dark grey: from boats; Shaded lines: from the shore; Black bars: confidence interval ($\alpha = 0.95$).

respectively). Fishers using rod with lure caught the largest proportion of sea bass annually, which was estimated at 1,683 tonnes (2.16 10^6 ind.). Fishers using rod with bait caught 674.5 tonnes of sea bass (0.96 10^6 ind.) in 2010. Fishers using other types of gear caught a smaller amount of sea bass, less than 300 tonnes per considered gear. Fishers using handlines, nets, spearguns and longlines caught the lowest amount of sea bass, that is 243.8 t (278.2 10^3 ind.), 228.3 t (156.7 10^3 ind.), 171.9 t (109.4 10^3 ind.) and 101.1 t (179.2 10^3 ind.), respectively. The total released biomass was observed to be around a third of the catches when using rod with lure and rod with bait (31% and 34% of the catches biomass, respectively). Fishers using other types of gear released a smaller percentage of the catches at sea: 8.4%, 10.5% and 17% of the sea bass biomass caught by nets, handlines and longlines, respectively. Since spearfishing generally results in the immediate death of the fish, no catch-and-release of sea bass was associated with this type of gear. The quantity of released fish was the highest for rod with lure (1.2 10^6 ind., i.e. 56% of the catches) and rod with bait (530 10^3 ind., i.e. 55% of the catches).

Discussion and Conclusions

Even though various recreational fishing studies have been carried out in European and non-European countries (see [7,18,23,24,32,33,35,41–48]), the lack of recreational and other non-commercial fishing data worldwide is still problematic [5,49]. Thus recently, many countries in Europe, supported by the Data Collection Framework, have developed recreational fishery evaluations [24,33]. Our study is the first to precisely describe and assess marine recreational fishing and catches of the European sea bass in France.

The nationwide telephone survey permitted to estimated at 2.5 million the number of marine recreational fishers in France (targeting all species including shellfish), representing 4% of the French population [35]. This can be compared to other countries records: in the Netherlands, it has also been estimated that 4% of the population are recreational fishers, fishing in marine coastal waters [50]; in Germany, including both inland freshwater and coastal marine waters, 4.7% of the population were considered as active anglers in 2002 [51], while the 2010 value, based on the sold fishing licenses, accounted 1,503,043 anglers, representing 1.8% of the German population [52]. In 2009, it was estimated that 137,000 anglers were fishing along the German Baltic Sea coast, representing only 0.17% of the population [33]. In contrast, in Australia the recreational fishing population has been estimated at 20% of the total population (80% of them fishing in marine and estuarine waters) [53]. Such a large number may be explained by the fact that 83% of the Australian population live within 50 km of the coastline. In summary, the proportion of the population involved in recreational fishing obviously varies substantially by country. Cooke and Cowx [6] have estimated that 11.5% of the world's population could be engaged in recreational fishing, but noted that this percentage can vary significantly according to the considered country.

The sea bass recreational fishery

We found that the mean weight of a sea bass caught from a boat (1.22 kg of sea bass caught and 0.96 kg kept per trip) was twice that when caught from the shore (0.65 kg caught and 0.45 kg kept). There was also a large difference in abundance: a mean of 1.4 sea bass was caught from boats compared to 0.87 from the shore. These differences between boat and shore catches have already been observed in other recreational fishing studies in the Atlantic and the Mediterranean, and also in freshwater areas. Pickett and Pawson [12] estimated that 0.94 kg of sea bass were retained by marine recreational fishers in the UK when fishing from boats and 0.8 kg when fishing from the shore. They also showed that a higher proportion of sea bass caught from the shore was released compared to that caught from boats, suggesting that shore recreational fishers usually catch smaller bass. In the Mediterranean, the same pattern has also been observed, since recreational fishing catches from boats were higher than ones from the shore [45].

The mean catch length varied according to the gear type. Our study showed that on the French Atlantic coasts, sea bass were targeted by recreational fishers using a wide variety of gears. This was also observed in recreational fisheries of other European countries like in Denmark, UK, Germany and Sweden [24]. We noticed that the type of gear used depended on the fishing mode: fishers fishing from a boat mainly used the rod with lure while those fishing from the shore mainly used the rod with bait. These observations could be explained owing to the fact that on one hand, bait used for catching sea bass, like sand eels or shrimp, can be easily found by shore anglers on the fishing site itself. On the other hand, there are various ways to catch sea bass with lure from a boat, like among others the troll, the spinning or the vertical fishing, whereas from the coast, the main used technique is the spinning. Thus, the possibilities to use lures from a boat are broader than from the coast.

These differences can help to explain the observed CPUE trends. The smallest sea bass were caught using longlines, which were used more often from the shore than from boats, which can signify that small sea bass are more numerous along the shore. In fact, additional information on sea bass ecological niches indicated

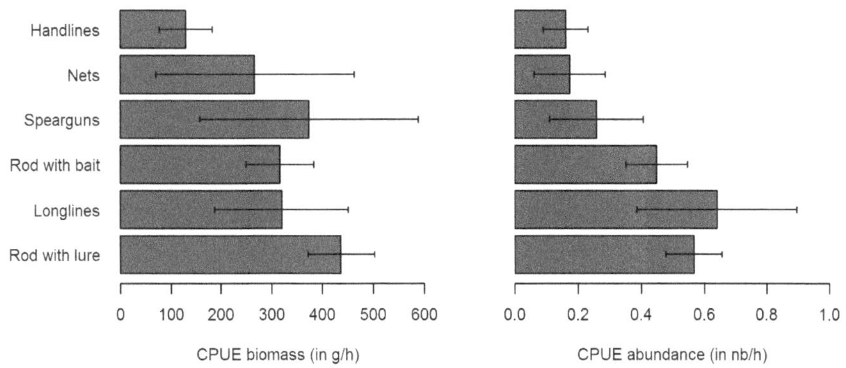

Figure 9. Mean CPUE of caught sea bass, per gear type. Black bars: confidence interval (α = 0.95).

that young fish mainly live near the shore and tend to move to deeper waters as they get older [12,54]. However, longlines have a great number of hooks, increasing the likelihood of catching several fish at once. That is why the CPUE of longlines used from shore was higher than that from boats, indicating larger catches mainly composed of small individuals.

The largest sea bass were caught by recreational fishers using nets. This gear was mainly used from boats, meaning that the fish could be pursued in deep water where the largest sea bass typically live. Nets may also be used primarily by fishers having a greater experience in fishing for sea bass, who may know more about where and how to catch larger fish.

Spearfishing is generally viewed as a highly selective fishing method [45]. Indeed, we observed in this study that the mean catch length was 50 cm, much larger than the minimum legal size, 36 cm, and the recommended size of 42 cm. With respect to CPUE relative to biomass, rod with lure and spearguns were the most effective gears, but spearfishing induced low CPUE relative to abundance, indicating that the targeted sea bass were mainly large ones. Spearfishers generally target fish considering their emblematic value, their taste, their ease of catching and their length [55]. Since in spearfishing there is no possibility of catch-and-release, and that all caught fish are kept, spearfishers generally target the largest ones, both to comply with the minimum legal catch length and for their trophy value.

Catch-and-release

Catch-and-release is known to be a common practice in recreational fisheries [10,56–61], as many people fish for fun and not mainly for their own consumption. Our study showed that this practice is prevalent in recreational sea bass fishing in France, since half of the caught sea bass were released. An equivalent release proportion of caught and released fish from recreational fishing has also been observed in various European countries [62] as well as in the US, with an estimate of 57% of released fish [10]. However, it is important to note that the released proportion of the catches was always greater in abundance than in biomass, showing that released sea bass catches were mainly composed of small fish.

Recreational fishers who keep the biggest fish while releasing the smallest ones can be seen as behaving responsibly, since catch-and-release is generally viewed as a conservation approach [61,63–65]. However, although it decreases direct mortality from recreational fishing, this must be set off against the potential lethal consequences of catch-and-release [66]. Indeed, various outcomes of catch-and-release have been demonstrated to have direct and/or indirect effects not only on the survival of the caught fish but also on the population structure [67]. Deep hooking, depending on factors such as type of bait, hook size and gear configuration can negatively impact the fish's survival [68,69]. The stress endured by the fish during the catch-and-release events can induce physiological and behavioral alterations (like increased vulnerability to

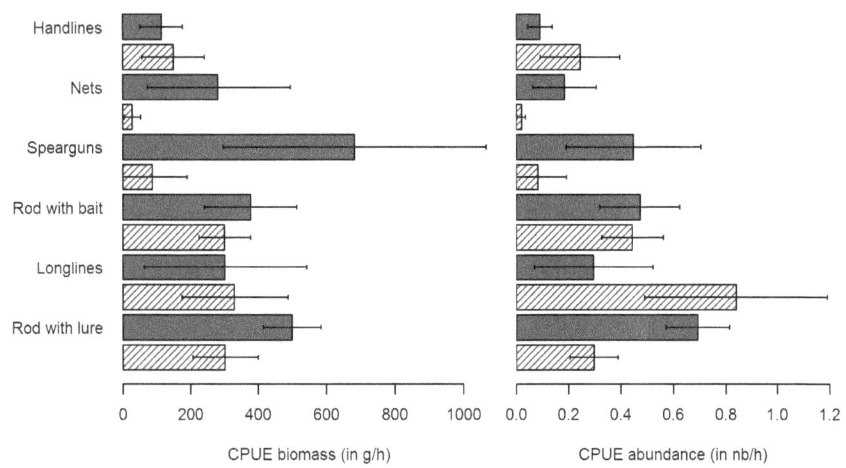

Figure 10. Mean CPUE of caught sea bass, per gear type and fishing mode. Dark grey: from boats; Shaded lines: from the shore; Black bars: confidence interval (α = 0.95).

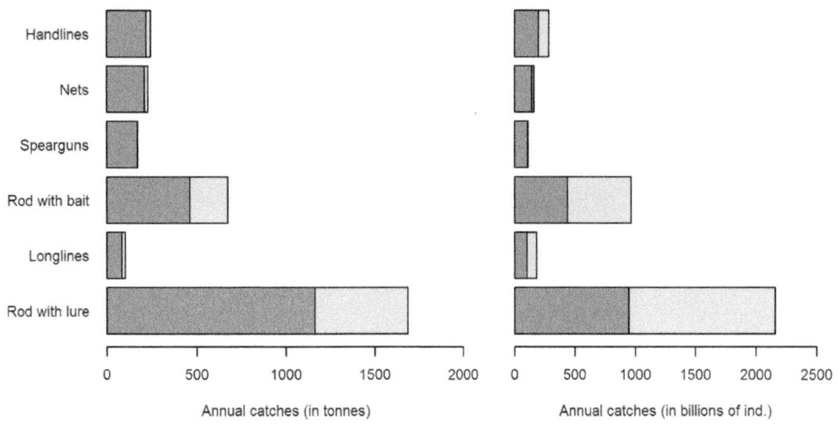

Figure 11. Annual recreational catch estimates of European sea bass in France (Atlantic coasts). Medium grey: kept amount; Light grey: released amount.

predation) which can lead to the death of the fish [67]. Fish mortality after catch-and-release is an essential element to evaluate for adapting fishing regulations; however, this is highly dependent on variables such as the considered species, the gear used [10] or the seawater temperature [70].

Commercial and recreational sea bass fisheries

The European sea bass is an important species, both for recreational and commercial fisheries. In the UK, it was yet targeted by anglers since the early 19th century, whom regarded this species as a gamefish, but was already known as a valuable food fish in the Mediterranean, particularly in France, Spain and Italy [12]. However, the interest of commercial fisheries is more recent, beginning to specifically target sea bass in the late 1970's [12,25]. Commercial sea bass fishing is divided into inshore fishing with small boats using a great variety of gears and displaying little activity during the winter, and offshore fishing, where mid-water pair-trawlers target pre-spawning and spawning sea bass during the spawning season between November and April [25,71].

We have estimated that recreational sea bass fishers of the Bay of Biscay, the Channel and the North Sea caught a total of 3,174 tonnes of sea bass and landed 2,345.5 tonnes per year, in other words a mean of 8.6 kg of sea bass caught and 6.3 kg kept per fisher per year. In the same areas and at the same time, landings by French commercial fishers have been estimated at 5,160 tonnes [71]. Thus, the total sea bass landings on the western coasts of France amounted to 7,505 tonnes, with recreational fishing accounting for around 30% of total landings.

Other studies have also shown that recreational fishing can represent a significant part of total landings [6], according to the considered species and area. In some cases, recreational fisheries are known to catch more fish than their commercial counterparts [4], like for the high-quality dolphin-fish (*Coryphaena hippurus*), targeted in the United States [72], and the lingcod (*Ophiodon elongatus*), where recreational landings account for 60% of total catches [15]. It has also been observed in the US that recreational landings of the Atlantic striped bass (*Morone saxatilis*), an emblematic species for recreational fishers, exceed those of commercial landings, and account for 74% of total landings [10]. In the Mediterranean, a global study estimated that leisure catches represent 10% of total landings [9]. In more specific areas, higher values can be reached, as observed in Cape Creus (Spain), where recreational fishing, while submitted to specific regulations and restrictions, is estimated to account for 30% of the commercial artisanal fishery production in the marine protected area [27], and in Mallorca, where recreational fishing production accounts for 27.5% of the commercial one [45].

Sampling design and data quality

The sampling design we used for the telephone surveys did not allow perfect coverage of the whole target population of marine sea bass recreational fishers. Indeed, for the two telephone surveys, a database of registered phone subscribers was used. However, this database did not record all the existing French phone numbers. It is known that some landline telephone numbers were registered in a red list and were thus not available in the used national telephone directory. This part of the population, whom decided to hide their phone number, has been yet studied by the BVA polling institute, and it has been evaluated that they well represent the rest of the subscribers (BVA, personal communication). Then, this missing part of the population should not induce bias in the results extrapolation. For the mobile-only group (people having only a cell phone and not a landline phone number), the subscribers could also decide not to be registered in the telephone directory. This represents ca. 15% of these subscribers; however, for these ones, we did not dispose of any information concerning their similarity to the rest of the population and at this point, we were not able to evaluate its associated bias.

The panel was constituted by fishers randomly called but whom deliberately decided to fill in fishing diaries. Generally, fishers who agreed to give their data are known to be the most involved in fishing and the most avid ones [52,73–77]. In such case, if it is not possible to evaluate this bias, the obtained total results would be biased upward. To counteract this phenomenon, we associated all fishers to 10 different strata depending on 3 categories of variables, available at both the panel and the population level (evaluated during the telephone surveys): the main gear used, the fishing mode and the fishing frequency. Increasing the number of variables and/or modalities would permit the reduction of the bias, but this would be possible only with a greater number of panelists, to have samples big enough to properly represent each stratum. The weighting factor attributed to each panelist fisher permitted to counteract the over representation of some stratum on the panel, regarding the whole fishers population [78]. Since the panelists were recruited during the second telephone interview, covering only the coastal departments, we did not dispose of fishing diaries filled in by non-coastal fishers, generally most occasional ones, fishing during holidays. Thus, a correction of the panelist records was done by the BVA polling institute whom evaluated the representativeness of the non-coastal fishers regarding the coastal ones.

Non-response as well as respondent refusal generate unknown bias in the sample estimates and the results extrapolation [79]. The non-response problematic can be induced by a limited coverage of a study or by the voluntary non-response of the interviewed persons. The non-response issue was thus present at various steps of our study, including the telephone surveys and the fishing diaries one. We highlighted various levels of non-response: (1) no phone answer (absence or no phone), representing a total of 59% of the exploited phone numbers; (2) persons who refused to answer, here 29% of the called numbers; (3) panelists who did not return all the fishing diaries (79% of the panelists whom returned at least one fishing diary), (4) panelists not declaring some fishing trips and (5) panelists not declaring all of the catches. In our study, 172,054 calls were done for obtaining 15,091 complete interviews (9%). Thus, 91% of the calls were not effective, due to various reasons (absence, refusal, occupied, fax/modem number, already contacted...). The persons who refused to answer could then represent a strong source of bias. The main problematic of the non-response is to evaluate if the non-respondents differ from the respondents [80]. Concerning the no-phone answer, it is particularly complicated to know if the non-respondents (here 59% of the generated phone numbers) were similar to the respondents. However, for limiting the bias associated with this type of non-response, the polling institute recalled all the generated phone numbers up to 12 times, at different days and hours. This sampling design permitted to assume that each household had the same probability to be contacted by the polling institute, and consequently that the respondent group was similar to the non-respondent one. However, 29% of the contacted households directly refused to answer the telephone interview, and any complementary survey was available to verify if this non-respondent group was similar to the respondent one. It was thus impossible to evaluate the bias induced by these non-respondents.

At the next step, we observed that 12.7% of the households with a sea bass recreational fisher refused to continue the phone interview, inducing a response rate of 87.3% at this step of the study. This rate is generally considered good enough to offer a good representation of the target population, inducing low risks of bias [80,81]. However, since some socio-demographic variables of the recreational fisher refusing to continue the interview were known, like the gender, the age and the activity, it was therefore possible to identify the category of the non-respondents and to compare it to the respondent ones for correcting the data.

Not all the panelists returned the fishing diaries for the whole year. Indeed, 190 of them returned at least one, and only 40 of them returned all the diaries. For limiting the bias induced by the fishers non-response during the fishing diary survey and to compensate for incomplete data, the applied weighting correction factor was fisher- and month-specific. The weighted component was not the individual but the fishing trips, which were monthly-representative. Thus it was not considered as problematic if some recreational fishers decided to stop the panel participation after returning the first, second or third fishing diary.

For the fishing diary survey, there was a risk that the fishers did not declare all their fishing trips neither all their catches. They could for example not declare an unsuccessful fishing trip, where any fish was caught, which could lead to an overestimation of the total catches, while missing the "zero-data". They could also be tempted to hide a high rate of catches and to not declare some of them, such as the kept undersized fish, which in this case would lead to a final underestimation of the total catches. For evaluating the importance of such bias, it has been proposed to 47 panelists to be contacted every week during 3 months to fill in the fishing diary with them. The objective was to compare the fishing diary information obtained with the fishers who filled in the diary alone and those when the fishers filled in the diary with help and intensive reminder. A group of 14 panelists (30%) agreed to be regularly followed by the polling institute. The socio-demographic and avidity profiles of the panelists who agreed to participate were the same as those of the non-participant (33 recreational fishers). Even if the success of returned diaries was higher for the followed participants, the mean number of trips recorded was not very different (5.2 fishing trips/week for the followed fishers and 4.9 fishing trips/week for the non-followed ones), neither was the mean number of sea bass caught per fishing trip (1.98 for the followed fishers vs. 1.89 for the non-followed fishers). Even if more diaries were returned by the fishers contacted weekly, the low frequency of contact did not seem to be a great source of response bias. Whereas the number of followed fishers was low, these results gave a preliminary information concerning the reliability of the data voluntarily provided by the fishers themselves.

Limitations

This study presents the first estimates for the recreational sea bass fishery on the Atlantic French coasts. However, although it gives much important information, other data are still needed to evaluate the entire French recreational sea bass fishery. This study did not account for the Mediterranean production, since it was designed to monitor sea bass catches requested by the DCF, which was limited to the Bay of Biscay, English Channel and North Sea. It also lacked reliable data concerning fly-fishing: only one fly-fisher agreed to be part of the panel and described only two fishing trips, which could not be considered as representative of this type of fishing activity. Thus, the annual amount of sea bass caught by fly-fishers, here estimated to be 5.2 tonnes, is still to be determined. Moreover, sport fishing, in the sense of official fishing competitions, as well as commercial angling charters were not taken into account in this study, since we did not dispose of any indicative estimates concerning their dimension. Finally, as already noted, catch estimates did not take into account the sea bass catch-and-release mortality [15], which is still unknown.

We used CPUE accounting for the fishing duration as the unit of effort, and compared various gears CPUE, while not accounting for the number of hooks or the length of the nets. Indeed, we aimed to compare the catches between different categories of fishers and not to compare the effectiveness of the gears themselves. However, comparing the CPUE of different gears, using for example the number of hooks as the unit of effort, would enhance the knowledge of the sea bass fishing techniques and characteristics. Since we did not dispose of such information in this study, it was not possible to evaluate these characteristics.

Methodologically, we defined a sea bass fisher as a fisher having caught at least one sea bass during the year. In other studies the only intention of catching sea bass was taken into account, and such fishers were also treated as sea bass fishers [12]. Our approach may thus have led us to underestimate the real number of recreational sea bass fishers.

Volunteer investment in scientific research can have some limitations that are important to point out. The panel survey may present some risks of bias by fishers who provided catch information. We have observed a tendency to round off, with many fish reported at exactly 40, 50, or 60 cm. This behavior has already been observed in other studies, and it seems that rounding errors are common [15]. This may be due to the fact that measuring a living fish is not easy before its release. However, we have seen that the length-weight relationship calculated on the panel dataset was very close to that of scientific estimates. The panel data can thus be viewed as of high quality despite these rounding errors.

Levrel et al. [82] have proposed three types of parameters which might influence the quality of volunteer-derived information: the availability of survey guidelines, the validation of datasets, and the ability to coordinate a network that includes different communities of practice. With respect to the first parameter, the polling institute directly contacted the volunteer fishers by telephone to explain the study design. The most important points were noted in the fishing diary supplied to each fisher, and a guide for fish identification was also sent out. Secondly, the validation of the dataset was managed by both the polling institute and the research team, while verifying all the parameters of the fishing trips and the catches amount, checking for extreme values or typing errors. Thirdly, recreational fishers were contacted by the polling institute every three months to get informed on the progress of the study, to answer their questions and to motivate the panelists to continue. The fishers also had direct phone access to the scientist in charge of the study, and were encouraged to phone at any time if they faced any problems.

Despite the possible limitations of a volunteer-based large-scale study, such a participative methodology has many advantages: it can help to secure the participation and cooperation of recreational fishers in management decisions [83]; it is cost-effective [82]; working in collaboration with recreational fishers is a good way to obtain data otherwise hard to collect and to foster the participation of fishers in recreational fisheries management [15,27]; and using volunteer self-declarations also makes it possible to collect information on discards [83], which is more important than ever but has been until now a gap in many catch estimates [3,84,85].

Implications for management

Our study gave new information and highlighted the importance of the sea bass recreational marine fishery in the French littoral of the Bay of Biscay, English Channel and North Sea. We have underlined that this concerns 370,000 fishers, that around 4 million of sea bass were caught, and half of them were kept, representing 30% of the total catches including commercial landings. Such values are not anecdotal, and the question of the impacts of these catches on the stocks, not taken into account until this time, should be considered.

Some regulations for recreational fisheries in general, which hence concerns the sea bass, are yet implemented: The ministry of Ecology, Sustainable Development and Energy, as part of the "Grenelle de la mer" (2009–2012), which was a public process for reflection and negotiation between the State, the elected representatives, the economic and professional stakeholders concerned by the sea and the civil society, has developed some documents and actions concerning marine recreational fishing. On July 7th 2010, a charter of commitments and objectives for ecologically responsible marine recreational fisheries (http://www.developpement-durable.gouv.fr/IMG/pdf/CHARTE_peche_maritime_de_loisir_eco-responsable_signee_-2.pdf) was signed. On the 1st of July 2012, the Ministry opened a website for the online declaration of marine recreational catches (http://www.developpement-durable.gouv.fr/Declarez-pechez.html), but such declaration is still not obligatory. Moreover, a decree was established on May 17th 2011, requiring all recreational fishers to mark the catches they keep by cutting the lower part of the caudal fin (http://www.legifrance.gouv.fr/jopdf/common/jo_pdf.jsp?numJO = 0&dateJO = 20110527&numTexte = 45&pageDebut = 09187&pageFin = 09188), in order to discourage the illegal sale of recreational catches. Moreover, the minimum catch length for the sea bass caught by recreational fishers fishing in the French Atlantic waters was raised from 36 cm to 42 cm on October 26th 2012, with the support of the recreational fishing federations.

Further management measures, concerning both commercial and recreational fishing of the sea bass, may be created later. Indeed, the ICES working group on Assessment of New MoU (Memorandum of Understanding) Species (WGNEW), created in 2005, aims to collect information on total international landings and research vessels survey data to evaluate stocks abundance trends, among which the sea bass stocks. The results of this working group are provided to the others ICES working groups WGHMM (Working Group on the Assessment of Southern Shelf Stocks of Hake, Monk and Megrim, for the Bay of Biscay Ecoregion) and WGCSE (Working Group for the Celtic Seas Ecoregion), in charge of modelling the stock evolution and proposing further recommendations if necessary. The 2013 ICES advices for the European sea bass are now available for the Bay of Biscay (http://www.ices.dk/sites/pub/Publication%20Reports/Advice/2013/2013/Bss-8ab.pdf) and for the English Channel and North Sea (http://www.ices.dk/sites/pub/Publication%20Reports/Advice/2013/2013/bss-47.pdf). For both regions, according to the ICES approach to data limited stock (DLS), a decrease of respectively 20% and 36% of the commercial landings is recommended. However, whereas the results of our study were presented and considered in these advices, recreational catches were still not taken into account by the ICES committee for the annual evaluations and recommendations, mainly owing to a lack of historical dataset.

Closer to the publication of these advices, the European Union has asked that each Member State recommend various management propositions for the European sea bass. All the possibilities were explored, concerning both commercial and recreational fisheries, but at this time, any official documents are available.

However, since we have developed another large-scale study on the European sea bass (among other species) recreational fishing in 2011–2012, including also the Mediterranean Sea, we hope that the new results, giving additional information and providing a second value of recreational fishing, will contribute to improve the management of the sea bass commercial and recreational fishery.

Acknowledgments

We would like to thank all the people who agreed to participate in telephone interviews as well as all the panelists who volunteered to take part in this study, thus providing a new and important overall view of the recreational fisheries sector. We would also like to thank Fabienne Boria and Celia Bensafi of the BVA polling institute, and all the interviewers from BVA, who conducted more than 30,000 telephone interviews for this study and recorded all the data. We also want to thank the SIH information system of IFREMER. Finally, we want to thank the reviewers for taking the time to review this manuscript and to give us suggestions and advice which permitted us to greatly enhance the quality of this paper.

Author Contributions

Conceived and designed the experiments: HL JH GV MD. Performed the experiments: HL DR JH. Analyzed the data: DR. Contributed reagents/materials/analysis tools: DR MD JH GV HL. Wrote the paper: DR MD HL.

References

1. FAO (2008) The state of world fisheries and aquaculture. Rome: FAO Fisheries and aquaculture department. 196 p.
2. Chuenpagdee R, Liguori L, Palomares MLD, Pauly D (2006) Bottom-up, global estimates of small-scale marine fisheries catches. Vancouver: Fisheries Centre, University of British Columbia. 112 p.
3. Pauly D (2006) Major trends in small-scale marine fisheries, with emphasis on developing countries, and some implications for the social sciences. Maritime Studies 4: 7–22.
4. Zeller D, Booth S, Pauly D (2007) Historic catch reconstruction for coral reef - and bottom - fisheries in the State of Hawaii, 1950 - 2002. VancouverBC, Canada: University of British Columbia. 24 p.
5. Zeller D, Darcy M, Booth S, Lowe MK, Martell S (2008) What about recreational catch? Potential impact on stock assessment for Hawaii's bottomfish fisheries. Fish Res 91: 88–97.
6. Cooke SJ, Cowx IG (2004) The role of recreational fishing in global fish crises. BioScience 54: 857–859.
7. Coleman FC, Figueira WF, Ueland JS, Crowder LB (2004) The impact of United States recreational fisheries on marine fish populations. Science 305: 1958–1960.
8. NRC (1999) Sustaining marine fisheries. Washington DC: National Research Council. 166 p.
9. EU (2004) Fishing in Europe. Mediterranean: guaranteeing sustainable fisheries. Brussels: European Commission. 12 p.
10. Bartholomew A, Bohnsack JA (2005) A review of catch-and-release angling mortality with implications for no-take reserves. Rev Fish Biol Fish 15: 129–154.
11. Pitcher TJ, Hollingworth CE (2002) Fishing for fun: Where's the catch? In: Pitcher TJ, Hollingworth CE, editors. Recreational fisheries: Ecological, economic and social evaluation. Oxford, UK: Blackwell Science. pp. 1–16.
12. Pickett GD, Pawson MG (1994) Sea Bass: Biology, exploitation and conservation. LondonUK: Chapman & Hall. 337 p.
13. Cardona-Pons F, Morales-Nin B, Sutton SG (2010) Scientists and recreational fishers: Communication manners and its efficiency. Fish Res 106: 575–578.
14. Kearney RE (2002) Recreational fishing: Value is in the eye of the beholder. In: Pitcher TJ, Hollingworth CE, editors. Recreational fisheries: Ecological, economic and social evaluation. Oxford, UK: Blackwell Science. pp. 17–33.
15. NRC (2006) Review of the recreational fisheries survey methods. Washington DC: National Research Council. 187 p.
16. NOAA (2011) Marine recreational information program. Implementation plan: 2011/2012 update.
17. Wilberg MJ (2009) Estimation of recreational bag limit noncompliance using contact creel survey data. Fish Res 99: 239–243.
18. Hartill BW, Watson TG, Bian R (2011) Refining and applying a maximum-count aerial-access survey design to estimate the harvest taken from New Zealand's largest recreational fishery. North American Journal of Fisheries Management 31: 1197–1210.
19. Euston ET, Haney SA, Hattala KA, Kahnle AW (2006) Overview of the Hudson River recreational fisheries, with an emphasis on striped bass. Hudson River Fishes and their Environment. American Fisheries Society Symposium. pp. 295–315.
20. Fraidenburg ME, Bargmann GG (1982) Estimating boat-based fishing effort in a marine recreational fishery. North American Journal of Fisheries Management 2: 351–358.
21. Pollock KH, Hoenig JM, Jones CM, Robson DS, Greene CJ (1997) Catch rate estimation for roving and access point surveys. North American Journal of Fisheries Management 17: 11–19.
22. Normandeau Associates (2007) Assessment of spring 2005 Hudson River recreational fisheries. Normandeau Associates Inc. 179 p.
23. Veiga P, Ribeiro J, Goncalves JMS, Erzini K (2010) Quantifying recreational shore angling catch and harvest in southern Portugal (north-east Atlantic Ocean): Implications for conservation and integrated fisheries management. J Fish Biol 76: 2216–2237.
24. ICES (2011) Report of the Planning Group on Recreational Fisheries Surveys (PGRFS). EsporlesSpain: ICES CM 2011/ACOM:23. 111 p.
25. Colman JE, Pawson MG, Holmen J, Haugen TO (2008) European sea bass in the North Sea. In: Aas O, Arlinghaus R, Ditton RB, Policansky D, Jr HLS, editors. Global challenge in recreational fisheries. Oxford, UK: Blackwell Publishing. pp. 111–129.
26. Gomez S, Lloret J, Demestre M, Riera V (2006) The decline of the artisanal fisheries in Mediterranean coastal areas: The case of Cap de Creus (Cape Creus). Coast Manage 34: 217–232.
27. Lloret J, Zaragoza N, Caballero D, Riera V (2008) Biological and socioeconomic implications of recreational boat fishing for the management of fishery resources in the marine reserve of Cap de Creus (NW Mediterranean). Fish Res 91: 252–259.
28. Steinmetz F, Thebaud O, Blanchard F, Le Floch P, Bihel J (2008) A bio-economic analysis of long term changes in the production of French fishing fleets operating in the Bay of Biscay. Aquat Liv Resour 21: 317–327.
29. Drogou M, Biseau A, Berthou P, De Pontual H, Habasque J, et al. (2011) Synthèse des informations disponibles sur le bar: Flottilles, captures, marché. Réflexions autour de mesures de gestion. Ifremer, Département Ressources Biologiques et Environnement. 148 p.
30. Bertignac M (1987) L'exploitation du bar (*Dicentrarchus labrax*) dans le Monbras (Bretagne sud) Rennes: Ecole Nationale Supérieure d'Agronomie.
31. Le Floc'h P, Poulard J-C, Thebaud O, Blanchard F, Bihel J, et al. (2008) Analyzing the market position of fish species subject to the impact of long-term changes: A case study of French fisheries in the Bay of Biscay. Aquat Liv Resour 21: 307–316.
32. Henry GW, Lyle JM (2003) The national recreational and indigenous fishing survey. Australian Government Department of Agriculture, Fisheries and Forestry. 190 p.
33. ICES (2010) Report of the Planning Group on Recreational Fisheries Surveys (PGRFS). BergenNorway: ICES CM 2010/ACOM:34. 168 p.
34. Lyle JM, Coleman APM, West L, Campbell D, Henry GW (2002) New large-scale survey methods for evaluating sport fisheries. In: Pitcher TJ, Hollingworth CE, editors. Recreational fisheries: Ecological, economic and social evaluation. Oxford, UK: Blackwell Science. pp. 207–226.
35. Herfaut J, Levrel H, Thébaud O, Véron G (2013) The nationwide assessment of marine recreational fishing: A French example. Ocean & Coastal Management 78: 121–131. doi:10.1016/j.ocecoaman.2013.02.026.
36. Le Guennec J, Sautory O (2002) CALMAR 2: Une nouvelle version de la macro CALMAR de redressement d'échantillon par calage. Insee-Méthodes: Actes des Journées de Méthodologie Statistique. 6 p.
37. Deville J-C, Särndal C-E, Sautory O (1993) Generalized raking procedures in survey sampling. Journal of the American Statistical Association 88: 1013–1020. doi:10.1080/01621459.1993.10476369.
38. Sautory O (1993) La macro CALMAR: Redressement d'un échantillon par calage sur marges. INSEE - Direction des statistiques démographiques et sociales. 51 p.
39. R_Development_Core_Team (2011) R: A language and environment for statistical computing. Vienna, Austria: ISBN 3-900051-07-0. Available: http://www.R-project.org/.
40. Chen Y, Jackson DA, Harvey HH (1992) A comparison of von Bertalanffy and polynomial functions in modelling fish growth data. Can J Fish Aquat Sci 49: 1228–1235.
41. Pitcher TJ, Hollingworth C (2002) Recreational Fisheries: Ecological, Economic and Social Evaluation.
42. Aas O, Arlinghaus R, Ditton RB, Policansky D, Schramm HL (2008) Global challenges in recreational fisheries. OxfordUK: Blackwell Publishing. 364 p.
43. Greiner R, Gregg D (2010) Considering recreational catch and harvest in fisheries management at the bio-regional scale. Fisheries Management and Ecology 17: 336–345.
44. Ihde TF, Wilberg MJ, Loewensteiner DA, Secor DH, Miller TJ (2011) The increasing importance of marine recreational fishing in the US: Challenges for management. Fish Res 108: 268–276.
45. Morales-Nin B, Moranta J, Garcia C, Tugores MP, Grau AM, et al. (2005) The recreational fishery off Majorca Island (western Mediterranean): Some implications for coastal resource management. ICES J Mar Sci 62: 727–739.
46. NMFS (2012) Marine recreational fisheries statistical survey real time data queries.
47. Rangel MO, Erzini K (2007) An assessment of catches and harvest of recreational shore angling in the north of Portugal. Fisheries Management and Ecology 14: 343–352.
48. Unal V, Acarli D, Gordoa A (2010) Characteristics of Marine Recreational Fishing in the Canakkale Strait (Turkey). Mediterranean Marine Science 11: 315–330.
49. Bray K (2000) A global review of illegal, unreported and unregulated (IUU) fishing. Document AUS:IUU/2000/6: FAO, Fisheries Department. 53 p.
50. Aarts TWPM (2008) Recreational angling in the Netherlands: Participation, trends and management. In: Aas O, Arlinghaus R, Ditton RB, Policansky D, Jr HLS, editors. Global challenge in recreational fisheries. Oxford, UK: Blackwell Publishing. pp. 39–42.
51. Arlinghaus R (2008) The social and economic significance of recreational fishing in Germany. In: Aas O, Arlinghaus R, Ditton RB, Policansky D, Jr HLS, editors. Global challenge in recreational fisheries. Oxford, UK: Blackwell Publishing. pp. 25–30.
52. Strehlow HV, Schultz N, Zimmermann C, Hammer C (2012) Cod catches taken by the German recreational fishery in the western Baltic Sea, 2005–2010: implications for stock assessment and management. ICES J Mar Sci 69: 1769–1780. doi:10.1093/icesjms/fss152.
53. Reid DD (2008) Overview of recreational fishing in Australia. In: Aas O, Arlinghaus R, Ditton RB, Policansky D, Jr HLS, editors. Global challenge in recreational fisheries. Oxford, UK: Blackwell Publishing. pp. 13–18.
54. Pawson MG, Pickett GD, Leballeur J, Brown M, Fritsch M (2007) Migrations, fishery interactions, and management units of sea bass (*Dicentrarchus labrax*) in Northwest Europe. ICES J Mar Sci 64: 332–345.
55. Rocklin D, Tomasini J-A, Culioli J-M, Pelletier D, Mouillot D (2011) Spearfishing Regulation Benefits Artisanal Fisheries: The ReGS Indicator and Its Application to a Multiple-Use Mediterranean Marine Protected Area. PLoS ONE 6.
56. Douglas J, Brown P, Hunt T, Rogers M, Allen M (2010) Evaluating relative impacts of recreational fishing harvest and discard mortality on Murray cod (*Maccullochella peelii peelii*). Fish Res 106: 18–21.

57. Dowling CE, Hall KC, Broadhurst MK (2010) Immediate fate of angled-and-released Australian bass *Macquaria novemaculeata*. Hydrobiologia 641: 145–157.
58. Gutowsky LFG, Harrison PM, Landsman SJ, Power M, Cooke SJ (2011) Injury and immediate mortality associated with recreational troll capture of bull trout (*Salvelinus confluentus*) in a reservoir in the Kootenay-Rocky Mountain region of British Columbia. Fish Res 109: 379–383.
59. Klefoth T, Kobler A, Arlinghaus R (2011) Behavioural and fitness consequences of direct and indirect non-lethal disturbances in a catch-and-release northern pike (*Esox lucius*) fishery. Knowledge and Management of Aquatic Ecosystems. Available: //WOS: 000299310800005.
60. Arlinghaus R, Cooke SJ, Lyman J, Policansky D, Schwab A, et al. (2007) Understanding the complexity of catch and release in recreational fishing: An integrative synthesis of global knowledge from historical, ethical, social, and biological perspectives. Reviews in Fisheries Science 15: 75–167.
61. Wallmo K, Gentner B (2008) Catch-and-release fishing: A comparison of intended and actual behavior of marine anglers. North American Journal of Fisheries Management 28: 1459–1471.
62. Ferter K, Weltersbach MS, Strehlow HV, Vølstad JH, Alos J, et al. (2013) Unexpectedly high catch-and-release rates in European marine recreational fisheries: implications for science and management. ICES Journal of Marine Science. doi:10.1093/icesjms/fst104.
63. Hall KC, Broadhurst MK, Butcher PA, Rowland SJ (2009) Effects of angling on post-release mortality, gonadal development and somatic condition of Australian bass *Macquaria novemaculeata*. J Fish Biol 75: 2737–2755.
64. Johnston FD, Arlinghaus R, Stelfox J, Post JR (2011) Decline in angler use despite increased catch rates: Anglers' response to the implementation of a total catch-and-release regulation. Fish Res 110: 189–197.
65. Landsman SJ, Wachelka HJ, Suski CD, Cooke SJ (2011) Evaluation of the physiology, behaviour, and survival of adult muskellunge (*Esox masquinongy*) captured and released by specialized anglers. Fish Res 110: 377–386.
66. Donaldson MR, Arlinghaus R, Hanson KC, Cooke SJ (2008) Enhancing catch-and-release science with biotelemetry. Fish Fish 9: 79–105.
67. Cooke SJ, Schramm HL (2007) Catch-and-release science and its application to conservation and management of recreational fisheries. Fisheries Management and Ecology 14: 73–79. doi:10.1111/j.1365-2400.2007.00527.x.
68. Alos J, Arlinghaus R, Palmer M, March D, Alvarez I (2009) The influence of type of natural bait on fish catches and hooking location in a mixed-species marine recreational fishery, with implications for management. Fish Res 97: 270–277.
69. Alos J, Palmer M, Maria Grau A, Deudero S (2008) Effects of hook size and barbless hooks on hooking injury, catch per unit effort, and fish size in a mixed-species recreational fishery in the western Mediterranean Sea. ICES J Mar Sci 65: 899–905.
70. Gale MK, Hinch SG, Donaldson MR (2011) The role of temperature in the capture and release of fish. Fish Fish: 1–33.
71. ICES (2010) Report of the working group on assessment of new MoU species (WGNEW). Denmark: ICES CM 2010/ACOM:21. 185 p.
72. Rudd MA, Folmer H, VanKooten GC (2002) Economic evaluation of recreational fishery policies. In: Pitcher TJ, Hollingworth CE, editors. Recreational fisheries: Ecological, economic and social evaluation. Oxford, UK: Blackwell Science. pp. 34–52.
73. Fisher MR (1996) Estimating the effect of nonresponse bias on angler surveys. transactions of the american fisheries society 125: 118–126. doi:10.1577/1548-8659(1996)125<0118:ETEONB>2.3.CO;2.
74. Thomson CJ (1991) Effects of the avidity bias survey estimates of fishing effort and economic value. Creel and angler surveys in fisheries management. D. Guthrie, et al., Vol. 12. pp. 356–366.
75. Griffiths SP, Zischke MT, Tonks ML, Pepperell JG, Tickell S (2013) Efficacy of novel sampling approaches for surveying specialised recreational fisheries. Rev Fish Biol Fisheries 23: 395–413. doi:10.1007/s11160-012-9299-x.
76. Frijlink K, Lyle JM (2010) A socio-economic assessment of the Tasmanian recreational rock lobster fishery. University of TasmaniaHobart: Tasmanian Aquaculture and Fisheries Institute. 67 p.
77. Andrews R, Salz RJ, Sminkey T (2012) Changes to sampling and estimation designs, including descriptions of completed and ongoing MRIP Projects, that could impact MRFSS/MRIP calibration efforts.
78. Dorow M, Arlinghaus R (2011) A telephone-diary-mail approach to survey recreational fisheries on large geographic scales, with a note on annual landings estimates by anglers in northern Germany. The Angler in the Environment: Social, economic, biological, and ethical dimensions. Beard TD, Arlinghaus R, Sutton SG. pp. 319–344.
79. O'Neil MJ (1979) Estimating the nonresponse bias due to refusals in telephone surveys. Public Opin Q 43: 218–232. doi:10.1086/268513.
80. Salz RJ (2011) Non-response bias study for the large pelagics telephone survey.
81. Groves RM (2006) Nonresponse rates and nonresponse bias in household surveys. Public Opin Q 70: 646–675. doi:10.1093/poq/nfl033.
82. Levrel H, Fontaine B, Henry P-Y, Jiguet F, Julliard R, et al. (2010) Balancing state and volunteer investment in biodiversity monitoring for the implementation of CBD indicators: A French example. Ecol Econ 69: 1580–1586.
83. Gerdeaux D, Janjua MY (2009) Contribution of obligatory and voluntary fisheries statistics to the knowledge of whitefish population in Lake Annecy (France). Fish Res 96: 6–10.
84. Singh R, Weninger Q (2009) Bioeconomies of scope and the discard problem in multiple-species fisheries. J Environ Econ Manag 58: 72–92.
85. Pitcher TJ, Watson R, Forrest R, Por Valtysson H, Guénette S (2002) Estimating illegal and unreported catches from marine ecosystems: A basis for change. Fish Fish 3: 317–339.

5

Mineralogy and Geochemistry of the Main Glauconite Bed in the Middle Eocene of Texas: Paleoenvironmental Implications for the Verdine Facies

Sherie C. Harding[1]*, Barbara P. Nash[1], Erich U. Petersen[1], A. A. Ekdale[1], Christopher D. Bradbury[1], M. Darby Dyar[2]

1 Department of Geology and Geophysics, University of Utah, Salt Lake City, Utah, United States of America, **2** Department of Astronomy, Mount Holyoke College, South Hadley, Massachusetts, United States of America

Abstract

The Main Glauconite Bed (MGB) is a pelleted greensand located at Stone City Bluff on the south bank of the Brazos River in Burleson County, Texas. It was deposited during the Middle Eocene regional transgression on the Texas Gulf Coastal Plain. Stratigraphically it lies in the upper Stone City Member, Crockett Formation, Claiborne Group. Its mineralogy and geochemistry were examined in detail, and verdine facies minerals, predominantly odinite, were identified. Few glauconitic minerals were found in the green pelleted sediments of the MGB. Without detailed mineralogical work, glaucony facies minerals and verdine facies minerals are easily mistaken for one another. Their distinction has value in assessing paleoenvironments. In this study, several analytical techniques were employed to assess the mineralogy. X-ray diffraction of oriented and un-oriented clay samples indicated a clay mixture dominated by 7 and 14Å diffraction peaks. Unit cell calculations from XRD data for MGB pellets match the odinite-1M data base. Electron microprobe analyses (EMPA) from the average of 31 data points from clay pellets accompanied with Mössbauer analyses were used to calculate the structural formula which is that of odinite: $Fe^{3+}_{0.89}\ Mg_{0.45}\ Al_{0.67}\ Fe^{2+}_{0.30}\ Ti_{0.01}\ Mn_{0.01})\ _{\Sigma=2.33}\ (Si_{1.77}\ Al_{0.23})\ O_{5.00}\ (OH)_{4.00}$. QEMSCAN (Quantitative Evaluation of Minerals by Scanning Electron Microscopy) data provided mineral maps of quantitative proportions of the constituent clays. The verdine facies is a clay mineral facies associated with shallow marine shelf and lagoonal environments at tropical latitudes with iron influx from nearby runoff. Its depositional environment is well documented in modern nearshore locations. Recognition of verdine facies clays as the dominant constituent of the MGB clay pellets, rather than glaucony facies clays, allows for a more precise assessment of paleoenvironmental conditions.

Editor: Lee A. Newsom, The Pennsylvania State University, United States of America

Funding: This project was funded by research grants to AA Ekdale, PI, from the ACS Petroleum Research Fund (PRF grant 47560-AC-8) and Division of Earth Sciences of the National Science Foundation (NSF grant EAR-1052661) and by student grants to SC Harding from the Gulf Coast Association of Geological Societies and the Department of Geology and Geophysics of the University of Utah. The funders had no role in study design, data collection and analysis, decision to publish, or preparation of the manuscript.

Competing Interests: The authors have declared that no competing interests exist.

* E-mail: sherie.harding@utah.edu

Introduction

Green clay-rich sediment is recognized throughout the geologic record, and the paleoenvironmental implications of green clay facies are not always understood clearly. This study reports the clay mineralogy and geochemistry of the Main Glauconite Bed (MGB), a shelly and sandy mudstone that has an olive-gray hue and is bioturbated, pelleted, and trace fossiliferous, with diverse body fossils. These characteristics, especially the abundant ovoidal, small clay pellets, probably led to its being named the Main Glauconite Bed. However, very few glauconitic minerals are present in the MGB.

The study area, Stone City Bluff, is located on the Gulf of Mexico Coastal Plain in east-central Texas (Figure 1). A detailed geologic map (Figure 2) shows the underlying Wilcox Group, formations of the Claiborne Group, and overlying formations of the Jackson Group. The Crockett Formation is also known as Cook Mountain Formation. The Middle Eocene, stratigraphic section (Figure 3) includes the Stone City Member, Crockett Formation. The upper part of the section outcrops along the south bank of the Brazos River in Burleson County, where the MGB is a prominent 1.7 m thick greensand unit. The MGB in outcrop at Stone City Bluff is pictured in Figure 4. Sample locations in the central zone and concretionary burrows at the top of the MGB are shown (Figure 5).

The mineralogy and geochemistry of the central zone and top layer of the MGB are different. The central part of the MGB contains heterogeneous green marine clay that is intensely bioturbated with a significant fraction of fecal pellets. The top part of the MGB is lined with concretions of filled burrows that extend down into the central MGB. These concretions are composed of siderite and apatite that formed as cementing agents. The green minerals in the central zone of the MGB warrant close examination, as do the contrasting minerals at the top of the MGB, because they reflect a change in the depositional environment.

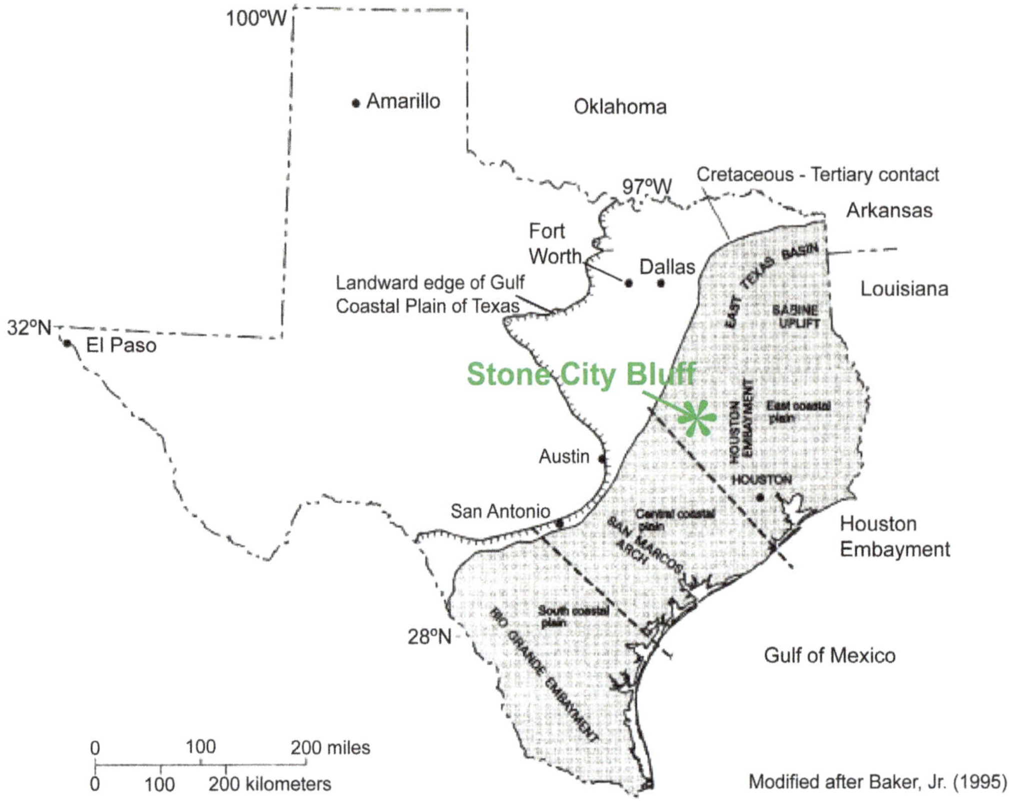

Figure 1. Location of the Stone City Bluff study area on the Texas Gulf Coastal Plain [59].

Previous Work on the MGB

The MGB and associated stratigraphy have been studied extensively by other workers who have offered various mineralogical and paleoenvironmental interpretations [1–9]. Thornton [10] reported, in a master's thesis, that the green marine clay of the MGB had the characteristics of odinite. Huggett et al. [11] described the geochemistry of authigenic minerals in the Claiborne Group, including samples from Stone City Bluff, and concluded that the composition was dark green serpentine-rich mixed layer clay. This characterization of the clay mineralogy and geochemistry of the MGB clay pellets, as well as its ichnology (evidence from trace fossils), have not received detailed treatment until now. It contributes to a coordinated approach to understanding the complex processes during accumulation of the MGB.

Previous interpretations of the depositional environment of the Stone City Member suggest that it was dynamic and complex. Davidoff and Yancey [12] documented eustatic sea-level changes and identified major sequence boundaries and maximum flooding surfaces in the Middle Eocene and some parasequences that may be capped by an exposure surface. The regional sequence stratigraphy testifies to numerous sea-level fluctuations during the early Tertiary (Figure 6). Stanton and Nelson [13] and Zuschin and Stanton [9] reconstructed the paleocommunity of the MGB, and they identified a time-averaged, parautochthonous, highly diverse fossil assemblage in the central bed along with a complex sedimentologic and taphonomic history due to reworking by winnowing and bioturbation. Stanton and Warme [3] examined the trace fossils at Stone City Bluff, and they suggested an interfingering of off-delta transgression of the sea into interdeltaic bays and sounds, marked by a gradual change from restricted and perhaps brackish delta-margin environments and biota to a normal marine environment and biota. Berg [14] interpreted the Stone City section as restricted marine or lagoonal environment. Yancey [6] reported on depositional trends of the Stone City transgressive-systems-tract, and he suggested that it is a transgressive depositional system in an area characterized by high sedimentation rate during progressive marine deepening. He proposed that the Stone City Bluff section is capped by a maximum flooding surface of regional extent. Previous interpretations of paleoenvironment are varied. The mineralogic and geochemical findings in this study offer further information and complement findings in previous studies.

Background on Green Clay Minerals

Green clay minerals. Published research on green minerals in sedimentary rocks reveals a lack of consistency in use of mineralogic terms. Terms such as "glauconite," "greensands," "glaucony," or "glauconitic minerals," are used frequently in the literature, and confusion persists. Glauconite, strictly speaking, is a series name for dioctahedral, interlayer-deficient micas with compositions of $K_{0.8}R^{3+}_{1.33}R^{2+}_{0.67}Al_{0.13}(Si_{3.87})O_{10}(OH)_2$ [15]. The term also has been employed as a common field term for small, rounded, green to black pellets and grains. Field identification often was confined to morphological appearance and color. In this report, as suggested by Odin [16], the term "glauconitic minerals" is used for the minerals that are rich in Fe^{3+} with a potassium content greater than 2 wt.% K_2O in the interlayer. In the case of the MGB, the few glauconitic minerals detected were usually confined to sparse, isolated grains.

Numerous studies have attempted to characterize clay mineralogy in marine sedimentary environments [11,16–24]. Odin's treatise on *Green Marine Clays* inventories green minerals in the

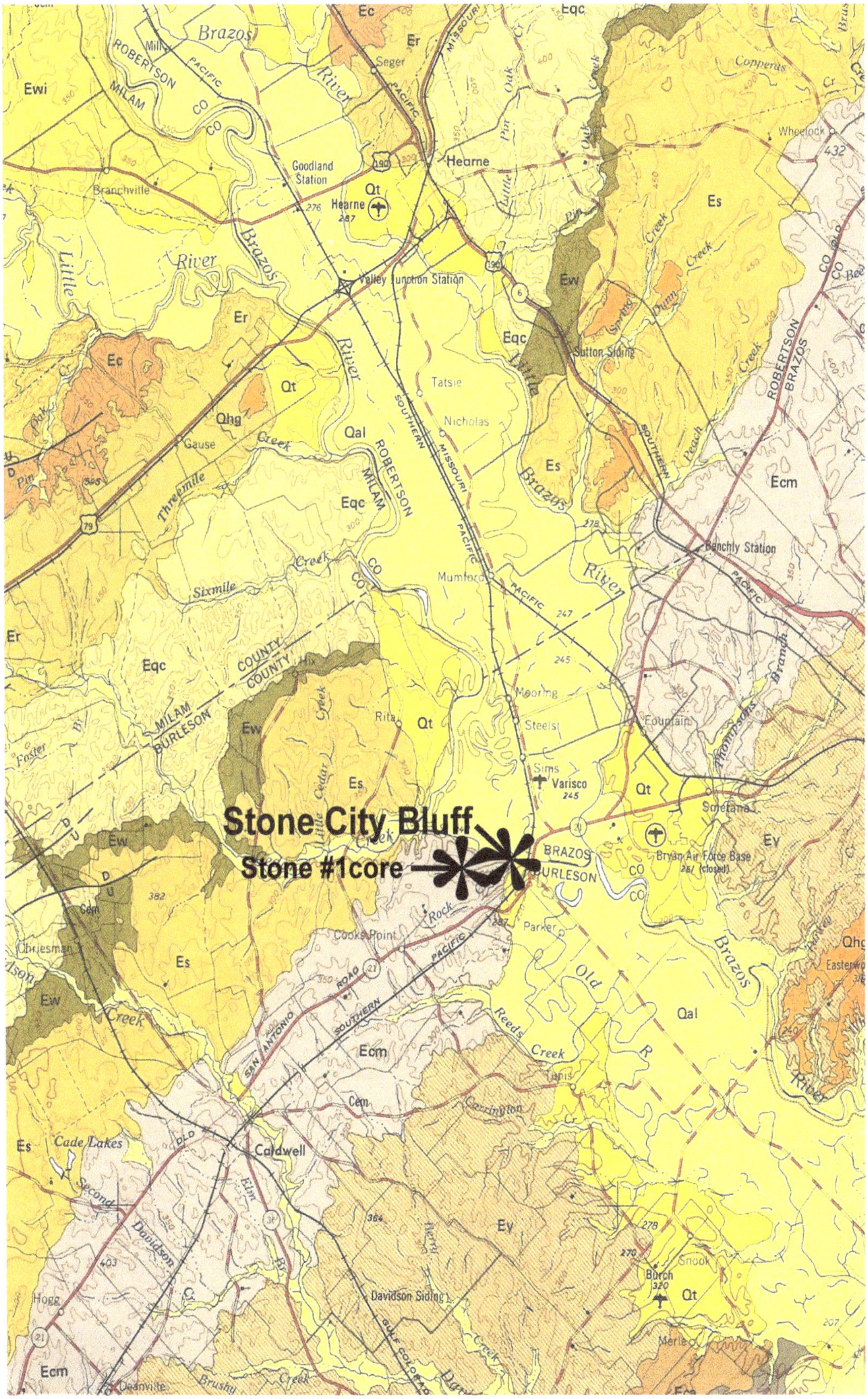

Figure 2. Geologic Atlas of Texas, Austin Sheet, Burleson and Brazos Counties [60]. Stone City Bluff and Stone #1core located. Wilcox Group (Ewi); Crockett Fm., also known as the Cook Mountain Fm. (Ecm); and Caddell Fm. of the Jackson Group (Eca).

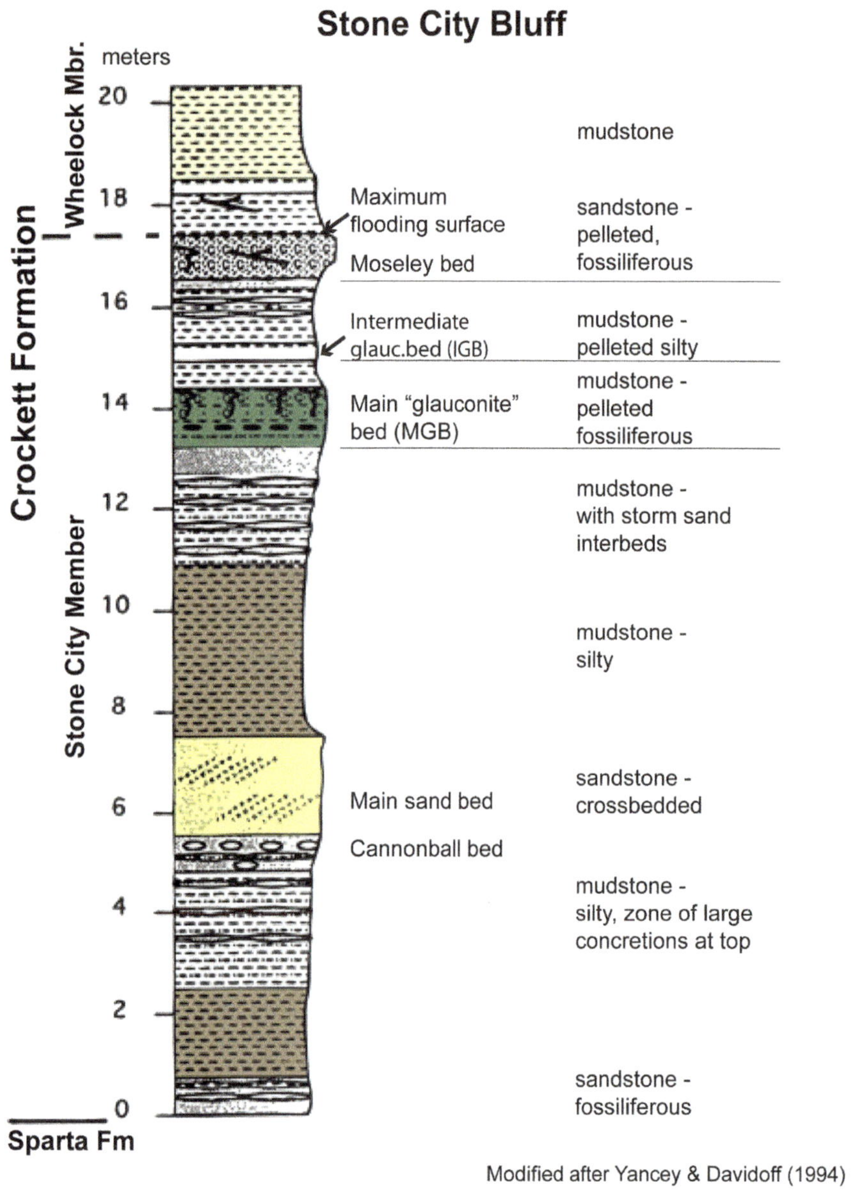

Figure 3. Middle Eocene stratigraphic section includes Stone City Member, Crockett Formation, Claiborne Group (modified after Yancey [6]).

modern marine environment and separates green marine clays into four facies: oolitic ironstone facies, verdine facies, glaucony facies and celadonite-bearing facies, based on examples on today's ocean floors. Odin [16] demonstrated that clay minerals on the sea floor progressively transform by slow cation exchange. The term "glaucony" is introduced to designate the fully marine facies characterized by a green pigment made of 2:1 type clay minerals that include glauconitic minerals, which form today at water depths around 50 m to as much as 1000 m on the sea floor.

The verdine facies is represented by a variety of chlorite-like clays, with an alternative mineralogy and geological significance. The verdine facies is identified in modern, shallow, tropical environments, in proximity to a terrigenous clastic source with iron influx. It is characteristic of bottom sediments of nearly normal salinity, but at a shallower depth range of 15 to 60 m. Mineralogically, verdine facies clays are uniquely different from those in the glaucony facies. Glauconitic minerals are absent from the verdine facies, while 7 and 14Å clays are present. Odin identified new clay minerals for the verdine facies, initially termed phyllite-V and phyllite-C clays, and later named odinite and iron smectite [16,24].

Authigenesis. Mineral authigenesis in sea-floor sediments can occur as glauconitization or verdinization, processes by which a sea-floor substrate is progressively modified to glauconitic minerals or to verdine minerals, respectively [16]. These result from interaction between open seawater and sea-floor sediment, and they are influenced by latitude, sea-floor temperature and bathymetric setting on the continental shelf.

Glauconitization results in a continuum of minerals from K-poor, disordered glauconitic smectite to K-rich, ordered glauconitic mica. Stages of development are defined by increasing potassium oxide content where 2 wt.%–4 wt.% K_2O is nascent stage, 4 wt.% -6 wt.% K_2O is slightly evolved (slightly mature), 6 wt.%–8 wt.% K_2O is evolved (mature), and greater than 8 wt.%

Figure 4. Stone City Bluff outcrop photos, the MGB is green and nearly vertical, A) looking west, upstream (the subject in this photograph has given written informed consent, as outlined in the PLOS consent form, to publication of their photograph), B) looking east, downstream.

K_2O is highly evolved (highly mature) glauconitic minerals [16,21,25–29]. The level of maturity is a reflection of residence time on the sea floor in sediment-starved conditions. Potassium, at 0.4 parts per thousand (ppt), is one of six major dissolved constituents in normal seawater [30]. It is available for uptake during glauconitization. Evolution to the highly mature stage takes 10^5 to 10^6 years in recent material [16]. Optimal conditions for glauconitization include a sea-floor temperature of 10–15°C, and unrestricted seawater circulation, allowing for winnowing accompanied by a slow sediment accumulation rate. Glauconitization commonly occurs in a substrate rich in fecal pellets and/or foraminiferal tests, which provide micro-reducing environments favorable for glauconitization. Oxidizing conditions may prevail in the surrounding substrate due to winnowing currents. Glauconitic

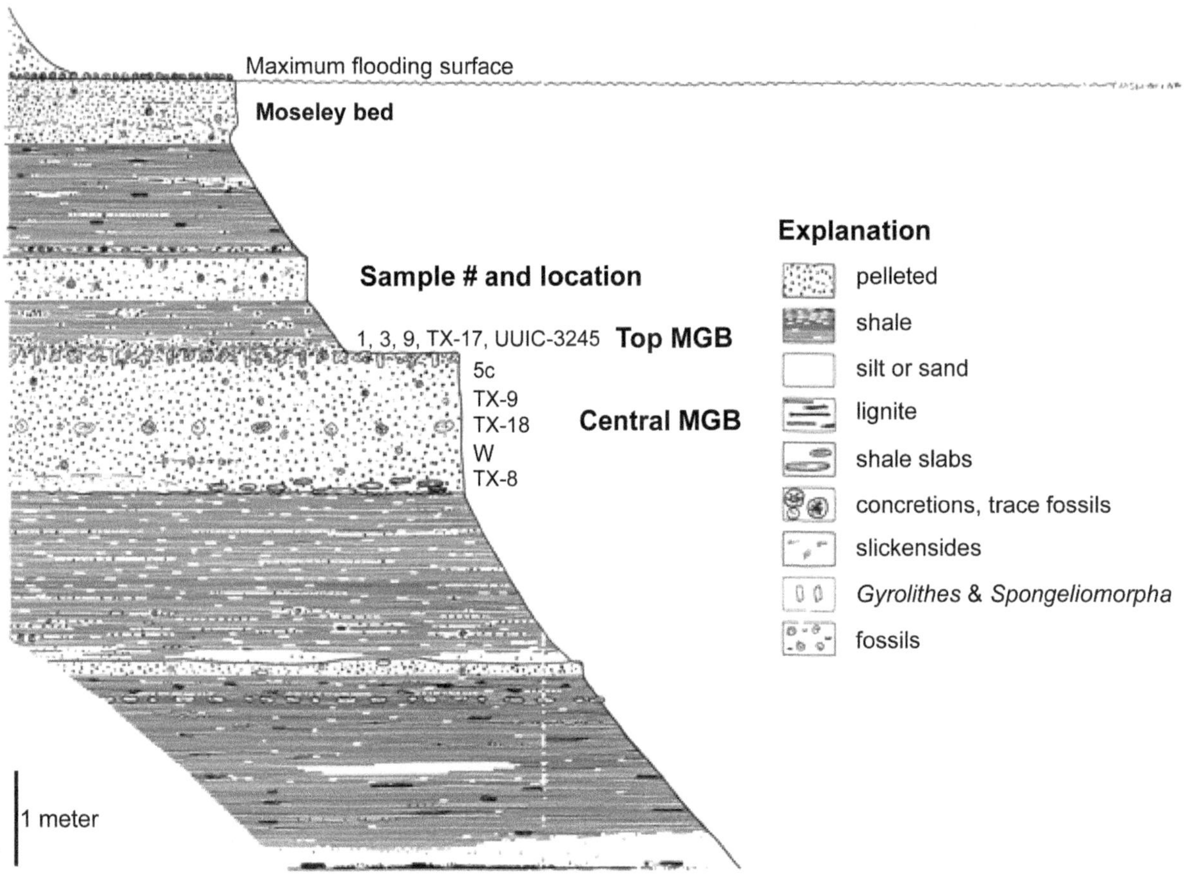

Figure 5. Upper units of the Stone City Member, sample number and location (modified after Stenzel [2]).

minerals commonly are associated with body fossils and trace fossils indicating sufficient oxygen to support benthic life. The glauconitization process may be halted at any stage of the maturity continuum if the environment becomes unsuitable because of a change in sea level or burial depth. A high rate of detrital influx will inhibit or entirely prevent glauconitization. Once formed, glauconitic minerals are highly stable and resistant to dissolution in the marine environment, which suggests that they can be reworked or transported with little degradation in a changing environment [16,21,25–29].

Verdinization, on the other hand, requires a shorter residence time on the sea floor [16]. Verdine facies minerals form rather quickly, probably in thousands of years, because of increased sea-floor temperature (~25°C). They tend to occur in shallower water under normal salinity and basic pH (7.5–8.5). Observed water depth where verdine minerals are forming today is between 15 and 60 m, and locally in 5 m depths, in common association with fecal pellets. Circulating currents are required in proximity to continental water input and abundant Si, Mg, and Fe in tropical to subtropical, warm water environments (Figure 7) [16,24,32]. Verdinization progresses from light green to dark green clay minerals. Verdine facies clay minerals are comprised primarily of two clay mineral species characterized by 7.2 and 14.5Å x-ray diffraction peaks [16].

The predominant verdine mineral is odinite, which is an iron and magnesium 1:1, 7Å clay. It is classified by some authors in the kaolin-serpentine group and the dioctahedral kaolin sub-group of phyllosilicates [15] or by other authors in the serpentine group [33]. Odinite was first described by Bailey [24] as serpentine-like, intermediate between dioctahedral and trioctahedral. It has total iron of about 75% Fe^{3+} and 25% Fe^{2+}, and a nominal structural formula of $(Fe^{3+},Mg,Al,Fe^{2+},Ti,Mn)_5(Si,Al)_4O_{10}(OH)_8$ [15]. Odinite is commonly identified in modern environments and is considered a possible analog to the oolitic ironstone facies in ancient rocks [16,24,32]. Although few possible occurrences of odinite from the geologic record have been reported in the literature odinite, or its alteration products, may be more common than is generally recognized.

Verdinization and glauconitization are processes that may be influenced by biogenic activity. Sediment-starved shelf environments often are characterized by high biological productivity. Biogenic activity, such as recycling of seafloor sediments by deposit-feeding and suspension-feeding organisms, can be prevalent in green marine clay environments at shelf depths. The effect of biological activity on clay mineral authigenesis, mineral growth and/or silicate weathering, is significant [23]. The MGB is a highly bioturbated, pelleted and fossiliferous unit. During deposition, biogenic activity on the seafloor was intense [2]. The existing mineralogy and geochemistry of the MGB may have been affected by animal-sediment interaction.

Diagenesis and weathering. Diagenesis refers to the physical and chemical reactions occurring in sediment after burial. Glauconitic minerals may be modified after burial, often showing an Al increase concomitant with Fe decrease. During weathering, glauconitic minerals may be altered in color and chemistry. In one study, Pestitschek et al. [34] described degradation over time

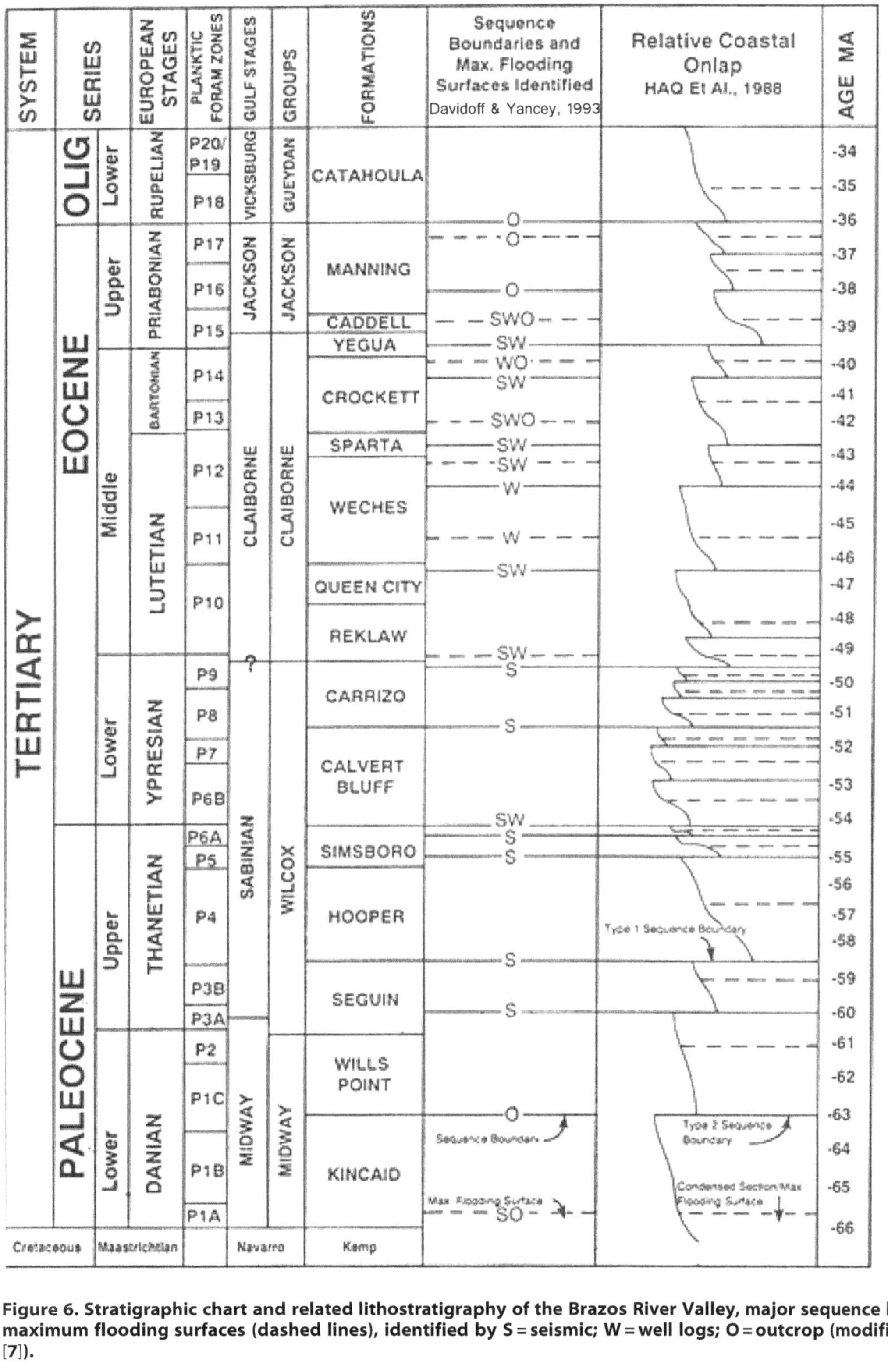

Figure 6. Stratigraphic chart and related lithostratigraphy of the Brazos River Valley, major sequence boundaries (solid lines) and maximum flooding surfaces (dashed lines), identified by S = seismic; W = well logs; O = outcrop (modified after Davidoff & Yancey [7]).

between fresh and weathered glauconite. They compared chemical results of Upper Cretaceous glauconitic minerals on the surface with core samples in the subsurface and demonstrated a loss of both Fe and K during weathering. They suggested that weathering reverses the glauconitization process, as glauconitic minerals degrade to smectite. They found that fresh samples were dark green, whereas the weathered samples were olive green to yellowish brownish green [34].

The MGB exhibits a variety of pellet colors, which may vary within a given pellet. This may be a function of different stages of diagenesis. Also, the possibility of weathering is suspected since the MGB is subaerially exposed. When mineralogical results of surface

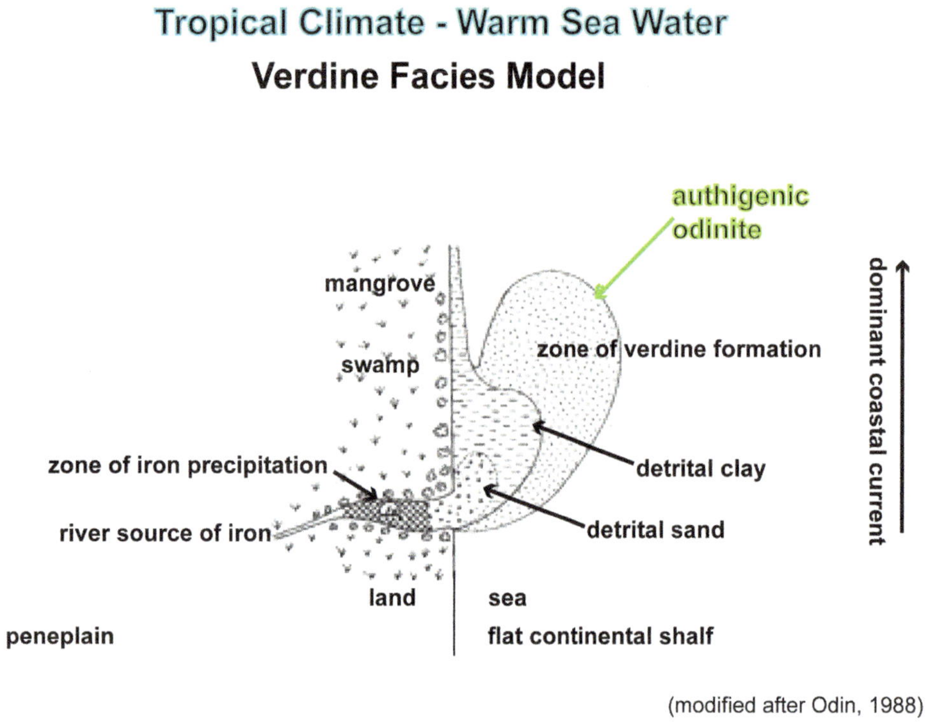

Figure 7. Diagrammatic verdine facies model depicting the idealized paleoenvironment at a tropical river mouth (modified after Odin [16]).

and core samples from the MGB were compared, there was no evidence of degradation due to weathering. The MGB showed similar mineral composition in both core and in outcrop samples. Therefore, it may be reasonably assumed that weathering is not an important influence on the MGB pellets, and that the lack of glauconitic minerals in the MGB is not due to weathering. Glauconitic minerals are not known to alter to odinite [16].

The fate of odinite, a verdine facies mineral, during diagenesis and weathering is not well known. According to Ku and Walter [32], odinite forms very quickly and apparently is metastable, because most of the reported occurrences of odinite are less than 20,000 years old. When odinite is exposed to fully oxygenated seawater or atmospheric oxygen, it changes color from grayish green to yellowish red, and it has been known to oxidize to goethite. Several workers have suggested the verdine facies as a modern precursor for ancient oolitic ironstones and grain-coating chlorite [16,32]. Huggett et al. [11] suggest that serpentine-rich mixed-layer clays of the Claiborne Group may be diagenetically altered odinite, and that serpentine is intermediate between odinite and berthierine. Ryan and Hillier [35] suggest that odinite, or evolved verdine minerals, may be precursor to authigenic chlorite in the Jurassic Sundance Formation of Wyoming. Whether the MGB is comprised of authigenic verdine facies minerals or altered authigenic verdine minerals, the paleoenvironmental implications are essentially the same.

Methods

Sampling

The Stone City Bluff outcrop of the Crockett Formation is readily accessible for sampling. Concretionary burrows at top of the MGB can be removed easily due to the friable nature of the sediment. The upper section of Stone City Bluff, particularly the MGB, was sampled for this study. Twenty-seven samples were collected at 60 cm intervals from two measured sections. Concretions, representative of filled burrows, were collected from the top of the MGB and from tumbled blocks. Eight samples were retrieved from a drill core (Stone #1 core at Texas A & M University). These represent unweathered samples of the Stone City Member. Additional samples, including concretionary burrows and pelleted sediment, were made available by JE Warme (Colorado School of Mines) [3]. Polished thin sections were prepared, and standard analytical procedures were used as described below.

The sections were measured and samples were collected via the public access site at the highway bridge over the Brazos River. No specific permissions were required. The field studies did not involve any endangered or protected species. The project did not involve any non-human primates or any other vertebrae animals.

Mineral Identification

Petrography of each thin section was described using a polarizing light microscope, and a variety of analytical techniques were employed to ascertain mineral composition and texture. X-ray diffraction (XRD) was carried out on oriented and unoriented samples to determine clay mineralogy. Electron microprobe (EMPA) analyses on selected pellets and matrix determined chemical composition and assisted with clay mineral identification. Mössbauer analyses were done on clay matrices and pellets to constrain mineralogy and determine iron oxidation state, which has bearing on EMPA data interpretation and significance for paleoenvironmental interpretation. QEMSCAN (Quantitative Evaluation of Minerals by Scanning Electron Microscopy) data were obtained to provide mineral maps and a textural view that complements, enhances, validates and quantifies the petrographic observations.

X-ray diffraction analysis. Clay minerals were examined from five central MGB samples (TX-8, W, TX-18, TX-9 & 5c) by X-ray diffraction (XRD) analyses. The 2 μm clay fraction was extracted by crushing, dispersion, and two-stage centrifuge. Oriented clay mounts were prepared and analyzed under conditions of air dried, ethylene glycolated, and heated to 375°C, then to 500°C for one hour. These were analyzed at 2° per minute from zero to 30° 2 θ, and examined in composite diffractograms. Randomly oriented powders were prepared by gently hand crushing bulk samples from TX-8 & TX-9 comprised of pellets and clay matrix from the central MGB. Pellets in samples 5c and TX-MGB were concentrated using a Frantz Magnetic Separator. TX-MGB is comprised of pellets from TX-8, W & TX-9. These were gently hand crushed and side loaded into an aluminum holder for analysis of unoriented powder. They were then analyzed at 2° per minute from zero to 66°. Eleven (11) peaks were indexed in the refinement, and quartz served as an internal standard. Selected samples were examined at high resolution, 0.5° per minute from 56° to 66° 2θ.

Electron microprobe analysis. Thirty clay pellets from both the central MGB and the concretionary burrow fill at the top of the MGB, and 14 non-pellet, cement areas were analyzed with a Cameca SX-50 electron microprobe equipped with four wavelength-dispersive spectrometers. Clay pellets were selected for analysis based on color in thin section and false color in QEMSCAN images. In an attempt to locate a potassium X-ray signal, some spot analyses focused on QEMSCAN-identified glauconitic grains, although very sparse. The more numerous clay pellets showed a low potassium signal. One to four spots were analyzed on each pellet for a total of 44 clay analyses. Analytical conditions were 15 keV accelerating voltage, 20 nA beam current, and a defocussed beam of 10–20 μm in diameter. A suite of natural minerals was employed as standards, and X-ray intensities were reduced using a phi-rho-z algorithm [36].

Mössbauer spectroscopy. Mössbauer analyses were used to constrain mineralogy and establish the iron oxidation state of pellets and matrix. The slightly magnetic pellets were separated from matrix in three central MGB samples (W, TX-9 & 5c) using a Frantz Magnetic Separator. Concentrated pellets and pellet free matrix were analyzed.

Sample mounts were prepared by gently mixing 30–40 mg of powdered sample with sugar to reduce preferred orientation. The mixtures were placed in a sample holder confined by Kapton tape. Mössbauer spectra were acquired at 295K using a source of ~40 mCi ^{57}Co in Rh on a WEB Research Co. model WT302 spectrometer. For each sample, the fraction of the baseline due to the Compton scattering of 122 keV gammas by electrons inside the detector was determined by measuring the count rate with and without a 14.4-keV stop filter (~2 mm of Al foil) in the gamma beam. Compton-corrected absorption was calculated for each individual spectrum using the formulation $A/(1-b)$, where b is the Compton fraction and A is the uncorrected absorption. Run times were 6–48 hours for each spectrum, and baseline counts were ~6–14 million after the Compton correction, as needed to obtain reasonable counting. Data were collected in 1024 channels and corrected for nonlinearity via interpolation to a linear velocity scale, which is defined by the spectrum of the 25 μm Fe foil used for calibration. Data then were folded before fitting.

Data were modeled using an in-house program from the University of Ghent, Belgium, called DIST_3E (an implementation of software described in Wivel and Mørup [37]), which uses model-independent quadrupole splitting distributions for which the subspectra are constituted by Lorentzian shaped lines. This program does not presume any particular shape of the distribution, in contrast to other distribution programs (e.g., Recoil). All Mössbauer data are posted for public use at http://www.mtholyoke.edu/courses/mdyar/database/.

In all fits, isomer shift (IS) and quadrupole splitting (QS) of the doublets were allowed to vary, and widths of both peaks in each pair were coupled to vary in unison (i.e. one width for each doublet, but every doublet independent). In a few cases it was necessary to constrain peak widths to lie above a certain value to obtain reasonable parameters, but most spectra were fit with only the minimal constraints described above.

Error bars for Mössbauer measurements are discussed at length by Dyar [31] and Dyar et al. [38] for fits to well-resolved spectra studied here: ±0.02 mm/s for IS and QS and ±3% absolute on areas.

QEMSCAN analysis. QEMSCAN is an automated mineralogy solution method capable of investigating fine details of samples of interest. Using both energy-dispersive X-ray (EDX) spectra and backscattered-electron imagery (BEI) obtained via an x-y raster scan pattern [39], QEMSCAN measures mineralogic variability across a sample's surface and can be used to quantify modal abundances of mineral species. Prior to each scan, instrument calibration is performed using the procedures outlined by Ayling [40].

Minerals are identified using a reference Species Identification Protocol (SIP), a hierarchical mineral database which determines what elements best fit a measured spectrum [41,42]. For this investigation, the Oil and Gas (O&G) v. 3.3 SIP is used, with an accelerating voltage of 20 Kev and specimen current of ~5 nA, and data are collected using iDiscover 5.2 beta software. A secondary SIP, Log5, also was used; however, it lacks detail in clay typing and therefore is less suitable for identification of verdine clays of interest.

Three representative polished and carbon-coated thin sections from the MGB (TX-18, TX-9 & TX-17) were analyzed using the O&G SIP. Total scan area was 3 by 3 mm, comprised of 9 blocks with sufficient overlap for easy stitching to produce a composite digital image. The resulting field images provide a 2-D view of the mineralogy of the sample illustrated in a color pattern from which quantitative modal mineral abundances can be extracted. Each analysis contains six false color digital images. The first image shows all the minerals detected, coded by various colors. The remaining five images characterize individual minerals textures and abundances. Pellets, grains and matrix are easily differentiated. Grain orientation, pellet size, shape, and abundance also are clearly apparent. QEMSCAN images were used to assist in selection of regions for analysis by electron microprobe.

Results

Petrography

Sediment. In a sedimentologic study of the central MGB, Zuschin and Stanton [9] recognized three microstratigraphic subunits: 1) homogeneous green minerals that they termed glauconitic siltstone, 2) bioturbated bioclastic, pelleted silty sandstone, 3) shell concentrations. Thin sections from the central MGB show bioclasts and pellets in a green clay matrix with quartz fragments. Area per cent, as determined by QEMSCAN, reveals sediment composed of clay pellets, as much as 50%, in a clay matrix. Quartz grains comprise up to 20%, and they are angular and poorly sorted. Shell fragments are a minor constituent, 2 to 4%. At top of the MGB, concretionary burrows contain varying amounts of clay pellets from 3 to 20%, with few quartz grains and shell fragments, less than 3%. Larger heterogeneous pellets (described below), are evident in the concretionary burrow fill.

Pellets. Two pellet types are recognized on the basis of size and composition: smaller clay pellets (cp) and larger heterogeneous pellets (hp). The clay pellets are green to black, tightly compacted ovoids that are easily observable with a hand lens. They are abundant in the central MGB (Figure 8). They also occur in varying amounts as burrow fill inside the concretions at top of the MGB (Figure 9). More obscure are the heterogeneous pellets that are found in both the central MGB and in burrows at the top of the MGB. In the concretionary burrow fill, the heterogeneous pellets were preferentially altered to apatite over siderite, which highlights their visibility, while the clay pellets remained unaltered. The two pellet types are displayed in photomicrographs of samples 3, UUIC-3245, and 9 top of the MGB with adjacent QEMSCAN false color images (Figure 9).

Abundance of clay pellets varies in the central MGB according to the three microstratigraphic subunits described above where they may comprise as much as 50 area %. At the top of the MGB within the concretionary burrows, clay pellets are scattered with some quartz and shell fragments in the matrix of siderite and apatite cement. The smallest clay pellets measure 0.2 mm length and 0.09 mm width, while the largest pellets average 1 mm length and 0.6 mm width. All are ovoid shape with some elongate ovoids. Clay pellets appear well indurated with sharp margins and smooth glossy surfaces. These pellets are thought to be of fecal origin based on their shape and the oxidation state of iron, which is chemically more reducing in the pellets compared to clay matrix. Very few pellets occur as internal filling of foraminifera tests. Angular quartz fragments occur primarily in the clay matrix and are not incorporated in the clay pellets. There is no evidence of internal canals within the clay pellets, as might be expected in decapod crustacean fecal pellets, such as *Palaxius* or *Favreina*. Although the most prominent burrows preserved in the MGB were excavated by decapod crustaceans, the fecal pellets in the sediment were produced by some other, presumably soft-bodied and unpreserved, animals.

The larger, heterogeneous pellets are not obvious in outcrop but are evident in thin section and in QEMSCAN images. Pellet shape in two-dimensional view is oval to circular, depending on thin section orientation. Pellets have sharp boundaries and consistent ovoid shape. Their size is consistently larger (average length 1 mm and width 0.6 mm) than the above-described green to black clay pellets. We see no evidence of diagenetic enlargement. These pellets are evident in the central bed and in the concretionary burrow fill at top of the MGB. In the central bed, microprobe and QEMSCAN analyses indicate that the heterogeneous pellet composition is dominated by clay minerals and tiny quartz fragments. At the top of the MGB, these pellets were altered to apatite and a lesser amount of siderite. Diagenetic processes did not alter or mask the original pellet shapes, and there is no evidence of internal canals. These larger, heterogeneous pellets indicate an alternate pellet producer. The pellet variety represents biological diversity in the MGB.

Color. The central MGB appears olive gray in outcrop. In thin section, the verdine clays appear green to brown. Stenzel [1] described the MGB as follows: "Olive-gray weathering to grayish-red, massive to poorly bedded, weathering to brownish-black, and grayish-red, sideritic calcareous concretions loosely spaced in the middle and crowded near the top to form a continuous clay ironstone layer, the bed forms a slight bench and a vertical face in the bluff." Upon close examination, the color of the smaller clay pellets varies from several hues of green to brown and black. Munsell colors of the pellets include light olive brown (5Y 5/6), moderate olive brown (5Y 4/4), olive gray (5Y 3/2), moderate brown (5YR 3/4), brownish black (5YR 2/1), and olive black (5Y 2/1) [43]. Clay pellets commonly appear homogeneous, while larger pellets are mottled. Rims are sometimes lighter in color than centers. Color often is linked with authigenic mineral maturity or weathering, because the color of clay changes during diagenesis [16,32,34,44].

Mineral Composition

Clay. XRD, EMPA and Mössbauer spectroscopy are common methods used to identify clay minerals of the verdine facies [16,24,32]. Each was employed in this study, and the results were compared with published data. Since a 7Å reflection characterizes a number of clay types including odinite, kaolinite, serpentine, vermiculite, berthierine, chlorite or mixtures of clays, resolution of which clay types are the most likely required detailed evaluations.

Odinite, the dominant verdine clay type, was indicated based on its characteristics as first described by Bailey [24]. It is a dioctahedral-trioctahedral Fe^{3+}-rich 1:1 layer, 7.2Å clay mineral. It was published with the following structural formula which is based on sample #699: $(Fe^{3+}_{0.78}, Mg_{0.77}\ Al_{0.56}\ Fe^{2+}_{0.28}\ Ti_{0.02}\ Mn_{0.02})_{\Sigma=2.42}\ (Si_{1.79}\ Al_{0.21})\ O_{5.00}\ (OH)_{4.00}$ [24]. This formula includes a correction for the amount of SiO_2 contamination (stated as ~3%, but actually 2.51%), ignores measured water (using stoichiometric water instead), ignores measured alkali elements (Na, Ca, K), assumes an Fe^{3+}/Fe^{2+} ratio measured on a different sample, and is normalized to 14 positive charges in the octahedral+tetrahedral sites. Recalculating the 699 analysis using the same total measured iron but fixing the Fe^{3+}/Fe^{2+} ratio at 3.0 ($X_{Fe} = 0.75$), yields a nearly identical formula.

$(Fe^{3+}_{0.79}\ Mg_{0.77}\ Al_{0.55}\ Fe^{2+}_{0.26}\ Ti_{0.02}\ Mn_{0.01})\ _{\Sigma=2.42}\ (Si_{1.79}\ Al_{0.21})\ O_{5.00}\ (OH)_{4.00}$.

X-ray diffraction analyses of both oriented and unoriented clay mounts consistently exhibited strong reflections at 7.2Å and 3.58Å. The oriented 2 µm clay, in a pellet-matrix mix from the central MGB, exhibited said reflections (Figure 10 and Figures S1 & S2). Another strong reflection occurs at 14Å in air-dried results. It is confirmed as smectite based on comparing diffraction patterns of air-dried and ethylene glycol treated samples where the 001 peak shifts to a very strong peak near 16.9Å. Illite is indicated by very weak 001, 002 and 003 reflections at 10.1, 5 & 3.34Å respectively. Kaolinite has reflections that coincide with odinite, but it was not confirmed here by other analyses. XRD patterns show no evidence for glauconitic minerals in the MGB. Peaks are not apparent at approximately 10.5 and 3.3, which would indicate glauconitic minerals.

Random oriented powders from MGB clay pellet separates yielded peaks at 7.249Å and 3.566Å (001 and 002) (Figure 11). The odinite-1M data base peaks are very similar at 7.150Å and 3.580Å [45]. Eleven peaks were indexed and compared with the odinite-1M data base in unit cell calculation (Table 1). This refinement process indicates that odinite is a reliable interpretation for the 7.2Å clay peaks that appear in all the x-ray diffractograms in this study.

X-ray diffraction analyses of randomly oriented powders, in a pellet-matrix mix, in the location of 060 were evaluated. In analyses of verdine facies minerals, Odin [16] found wide and asymmetric peaks between 1.56Å (59.23° 2θ) and 1.51Å (61.35° 2θ) in the vicinity of the 060 diffraction peak. This peak gradually decreases towards 1.50Å. Results for the three central MGB samples, comprised of pellets and clay matrix, show the 121- (*hkl*) peak of quartz at 1.54Å (59.970° 2θ) [46] which partially masks the reflections for verdine minerals. The broad peak, described by Odin [16] is subtle but present in MGB diffractograms which indicates the presence of verdine minerals (Figure S3). A 060 diffraction peak for kaolinite (62.31° 2θ) [33] is not present which

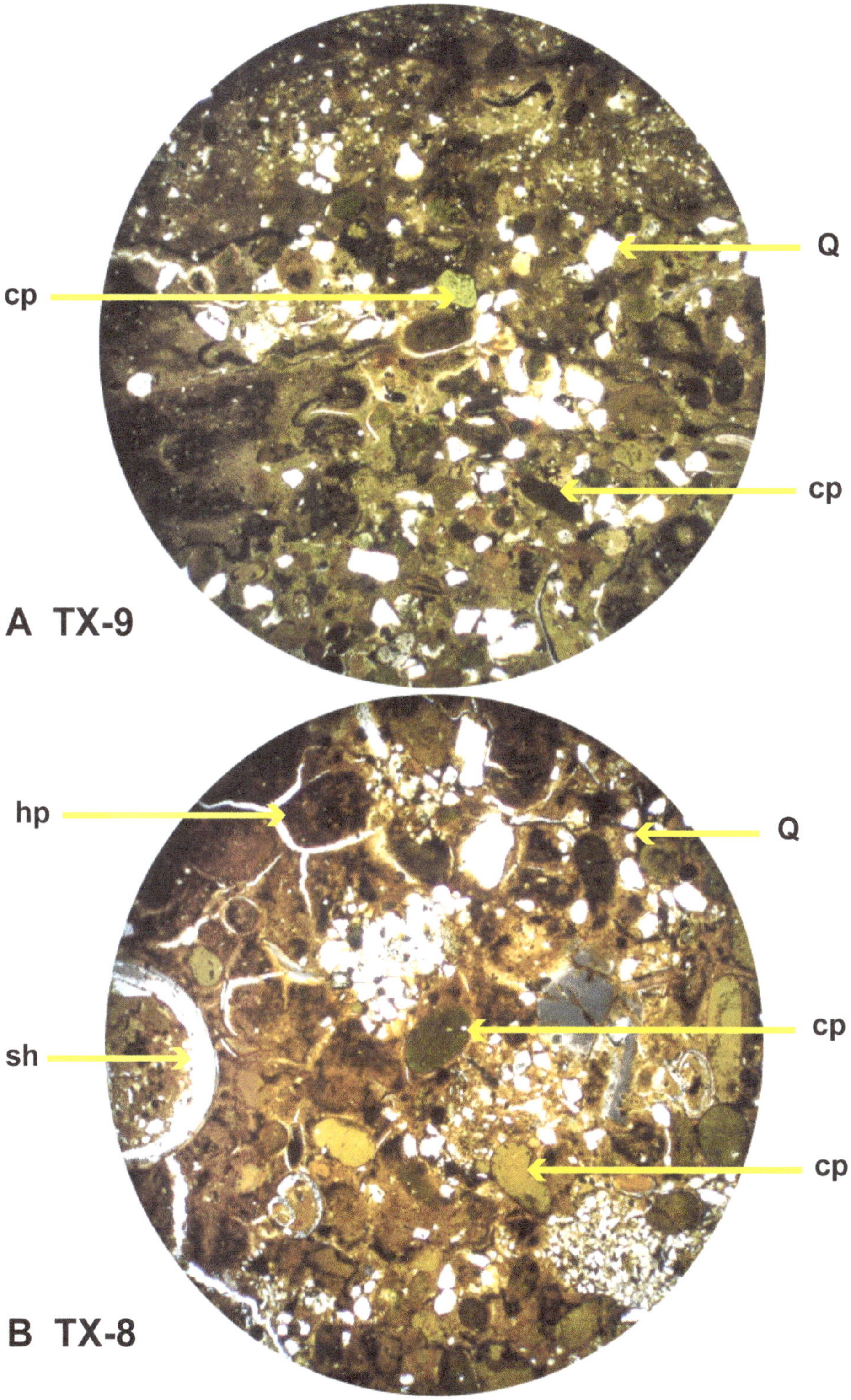

Figure 8. Photomicrographs from central MGB, A) TX-9, B) TX-8, cp = clay pellet, hp = heterogeneous pellet, sh = shell, Q = quartz.

is consistent with EMPA and QEMSCAN results that show a lack of kaolinite in MGB samples.

The electron microprobe results of 31 EMPA spot analyses of MGB clay pellets show consistent composition among the important cations, Representative EMPA analyses are given in Table 2. The maximum, minimum and mean of the 31 analyses are compared with published results (Table 3). The data were converted to apfu (atoms per formula unit), and they are illustrated adjacent to samples #699 and #508 (odinite in the type locality) [16,24] and presented in Figure 12. The comparison shows fairly

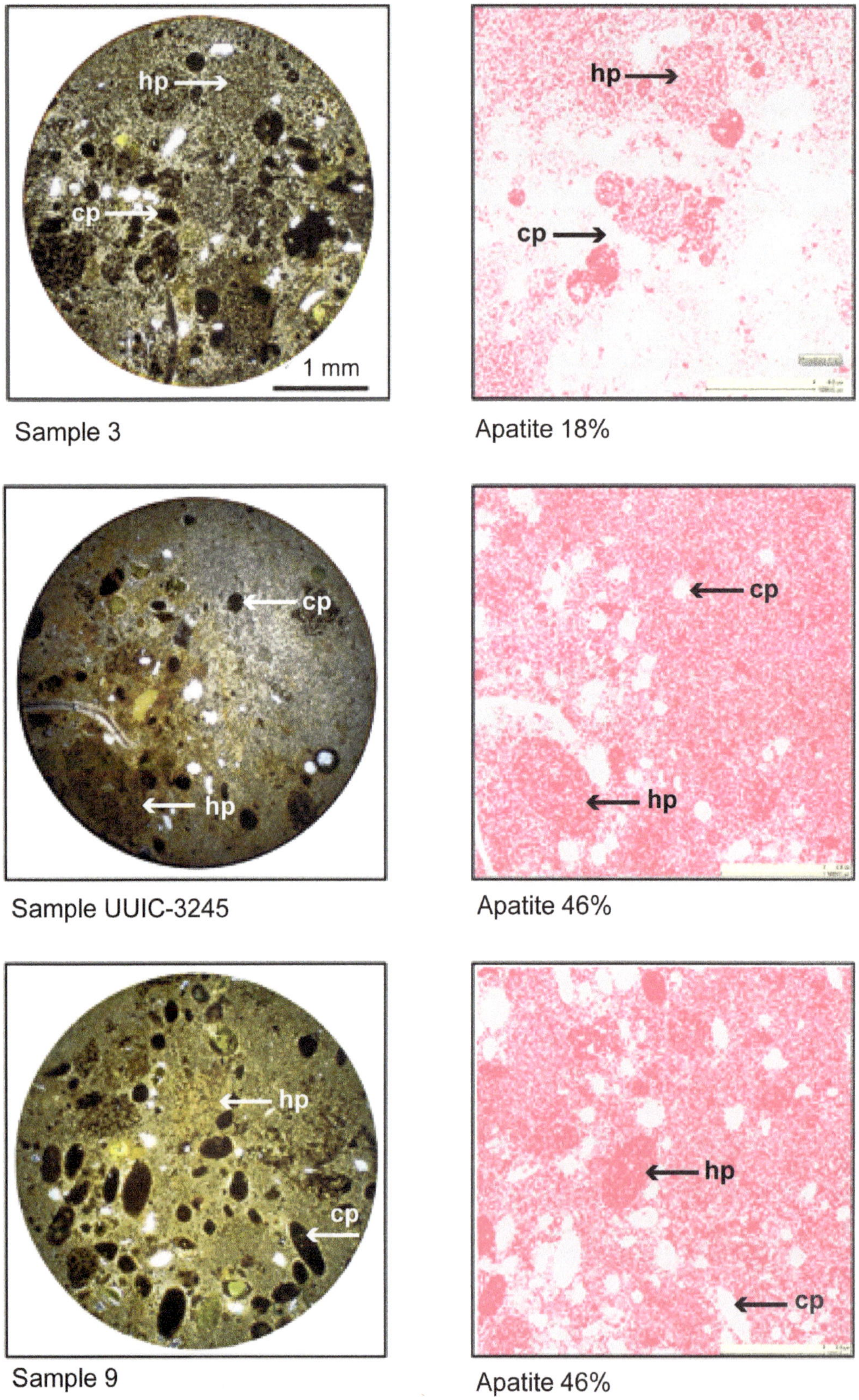

Figure 9. Top of MGB concretionary burrow fill, photomicrographs and QEMSCAN (false color) images, cp = clay pellets, hp = heterogeneous pellets, pink = apatite cement.

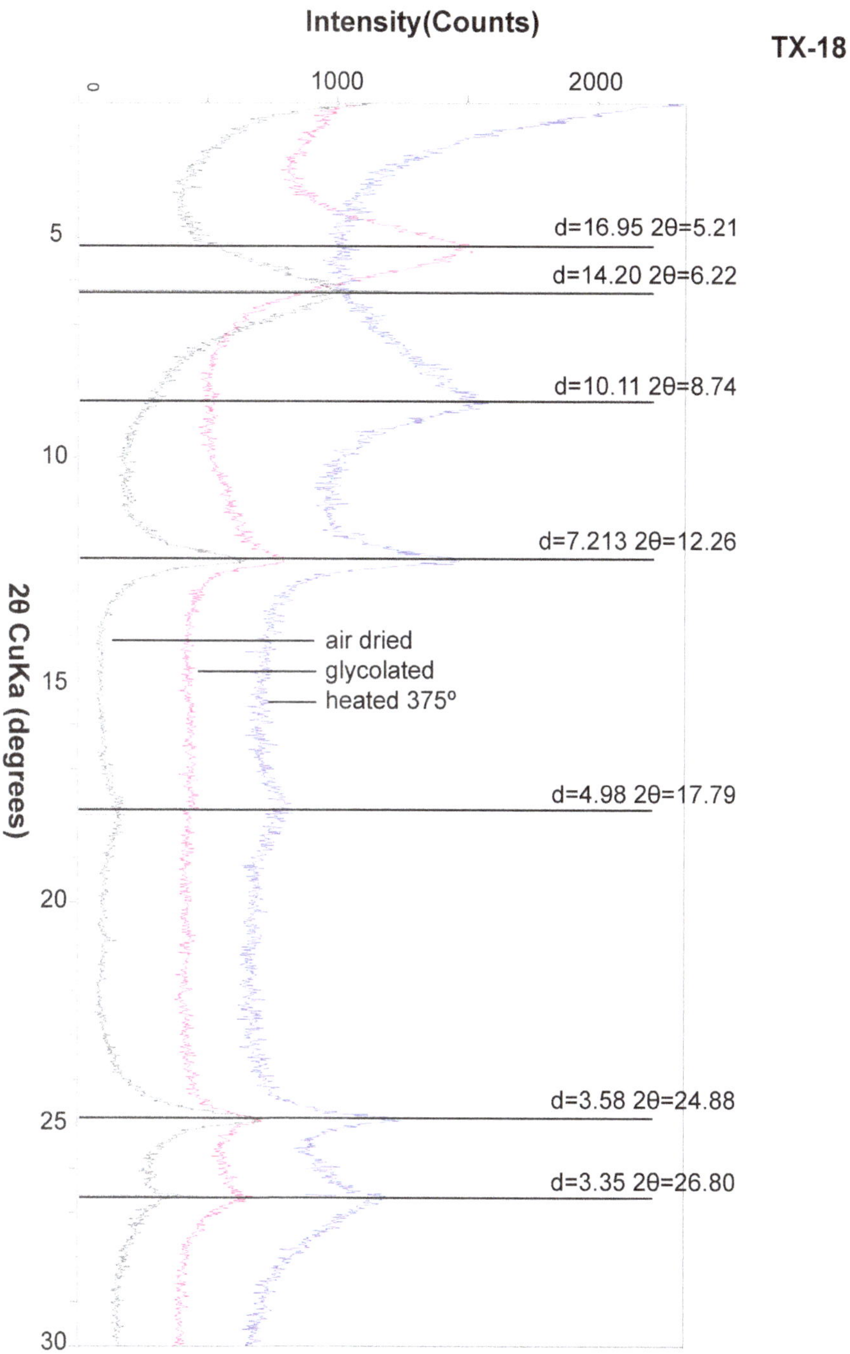

Figure 10. X-ray diffraction patterns of oriented, 2 μm clay fraction, from bulk sample TX-18, a mixture of 7 and 14Å clays (air dried, glycolated and heated to 375°).

consistent apfu results for the important cations. The formula calculated herein follows Bailey's [24] procedure but fixes the atomic Fe^{3+}/Fe^{2+} ratio at exactly 3.0. The following average structural formula calculated for MGB clay pellets is: $(Fe^{3+}_{0.89}$ $Mg_{0.45}$ $Al_{0.67}$ $Fe^{2+}_{0.30}$ $Ti_{0.01}$ $Mn_{0.01})$ $_{\Sigma=2.33}$ $(Si_{1.77}$ $Al_{0.23})$ $O_{5.00}$ $(OH)_{4.00}$. Odin's formula for #699 is slightly more Mg rich, and Fe, Al poor. This identifies odinite as the dominant clay mineral in MGB clay pellets. Raw chemical data for odinite analyses in the MGB clay pellets are presented in Table S1.

Glauconitic mineral grains, although small and very sparse, were the focus of some spot analyses in an attempt to locate a potassium X-ray signal. The glauconitic mineral analyses (>2 wt.% K_2O) were omitted from the above structural formula calculation for odinite. MGB clay pellets, for the most part, are low in K_2O, with an average of 1.5 wt. %. The few glauconitic grains encountered were small and fractured. Three analyses showed K_2O >6 wt. %, which qualifies as mature glauconite. Thus, the Main Glauconite Bed actually contains very little glauconite (3% by QEMSCAN). Raw data for glauconitic minerals and other clays are presented in Table S2.

Several clay types were characterized in a study of Claiborne Group iron-rich authigenic clays by Huggett et al. [11,21]. They

Table 1. Unit cell calculation from XRD data (odinite-1M (monoclinic)).

	Sample TX-MGB	SD	PDF# 00-048-1857	SD
a =	5.35	±0.01	5.38	±0.01
b =	9.38	±0.03	9.34	±0.02
c =	7.42	±0.03	7.39	±0.02
Beta =	104.79	±0.40	104.03	±0.25

Eleven peaks were indexed. MGB pellets compared with ICDD. All units are in angstroms (Å).

used the general term serpentine for 7Å iron-rich mixed layer clays, including odinite. The average chemistry of clay pellets from the MGB is very similar to the average of 53 analyses for pellets from the upper 8 m of the Crockett Formation, which includes the MGB, from the same locality as this study [11]. All MGB clay pellet analyses from this study, including glauconitic grains, are displayed as total Fe_2O_3 plotted against Al_2O_3 together with the published results (Figure13). The compositional variation of MGB clay pellets is evident, where they are seen to lie within the compositional field of authigenic iron-rich clays from other Claiborne Group clays [11]. Also shown are published analyses of other clays including nontronite, berthierine, Fe-smectite, vermiculite and kaolinite. These, along with the MGB clay pellets, tend to lie along a general trend between nontronite and kaolinite, and are distinctly lower in Al_2O_3 than berthierine (Figure 13) [11].

Mössbauer spectra were obtained on clay separates from the rock matrix and pellets (Figure S4), and Mössbauer parameters are provided in Table 4. Peak area ratios indicate a predominance of Fe^{3+} (63–81% of total Fe). Ferric/ferrous ratios in clays in fecal pellets are lower than in matrix clays from the same sample, indicating that the biogenic nature of the pellets provides micro-reducing environments on the seafloor favorable for the formation of verdine minerals. Figure 14 shows a plot of isomer shift vs. quadrupole splitting for the MGB clays together with representative published data for glauconite. For the MGB clays Fe^{3+} values of the isomer shift fall between 0.30 and 0.40. Of the eight samples, four had doublet pairs for Fe^{3+} resulting in a bimodal distribution in QS of 1.02–1.24 and 0.66–0.84. In the case of the other four samples in which there was a single doublet for Fe^{3+}, the range in values for QS falls within the lower range, 0.70–0.83.

Collectively, these data suggest the presence of two slightly different octahedral sites for Fe^{3+}, possibly suggesting some trioctahedral character in the clays. The difference in the average of the quadrupole splitting between the two sites, 0.41 mm/s, is similar to that reported for glauconite by Ali [47] but shifted to higher values by about 0.25 mm/s. Based on peak areas (Table 4), Fe^{3+} is more abundant in the site characterized by lower quadrupole splitting. The isomer shift for Fe^{2+} ranges from 1.13–1.25 mm/s, and the data fall into two clusters, especially defined by QS. There were no doublet pairs fit to the Fe^{2+} spectra. Nonetheless, the distinct bi-modal array of IS and QS values suggests again the presence of two octahedral sites for Fe^{2+}. Because of its substantial Fe content, there is considerable literature on Mössbauer spectra of glauconite, and a number of workers have allocated Fe to *cis*-M2 and *trans*-M1 sites [47–51]. However, Rancourt and co-workers [52–54] have argued that these sites may not be positively delineated, and that the spectral data may be recording local distortion environments. Dyar [55] suggested that hydrogen content may be responsible for variation around octahedral sites because of the difference in the location of hydroxyls around M1 and M2 sites. We do not have direct determination of H_2O contents of MGB clays, but they do have measureable F and Cl, and both vary in concentration. Substitution of F and Cl, and possibly O in OH sites could change the geometry of the adjacent *trans* and *cis* sites and be responsible for the bimodal behavior of MGB clays.

The Mössbauer Fe^{3+}/Fe^{2+} ratio is diagnostic for identifying verdine clay minerals. The average Fe^{3+}/Fe^{2+} ratio in MGB samples of matrix and pellets is 73:27%. A similar ratio of 73.6:26.4% was calculated for phyllite-V clay (later named odinite) in modern sea- floor clays [16]. The Fe^{3+}/Fe^{2+} ratio of clay matrix is slightly higher than that of clay pellets: 78:23%, versus 68.5:31.5%, respectively. The biogenic nature of the pellets provides micro-reducing environments on the seafloor favorable for the formation of verdine minerals.

QEMSCAN reports chemical data according to the SIP mineral definition. The O&G SIP tests for clays in the following order: glauconite, illite, kaolinite, nontronite, smectite and chlorite. Odinite is not defined in the O&G SIP algorithm, but its composition strongly overlaps with chlorite, and odinite is known to alter to chlorite [24]. Therefore, the O&G SIP was adjusted to accommodate the MGB clay samples by changing the mineral name from chlorite to odinite for this study. During QEMSCAN analysis of MGB clays, some 50% ended up in chlorite groups by default and are regarded as odinite.

Odinite and small amounts of smectite, illite and glauconitic minerals (<3 area %) are displayed in various shades of green in digital false color on QEMSCAN textural maps (Figures 15, 16, 17). Textural and compositional variation in the central MGB (Figure15) shows that clay matrix is predominantly odinite and illite, while clay pellets are predominantly odinite and smectite. Figure 16 demonstrates that odinite is dominant in both the matrix and clay pellets with minor amounts of illite, smectite and glauconitic minerals. Glauconitic minerals occur in fairly pure grains, although very few in number (about 3 area %). QEMSCAN analyses from the central MGB show that quartz grains average 18 area % and calcite averages about 3 area %, occurring as shell fragments.

Concretionary burrow fill at the top of the MGB is predominantly siderite and apatite cement, in various quantities. Sample UUIC 3245 had 46:43 area % apatite-siderite, while sample 9 had 46:39 apatite-siderite. Other concretionary burrows are dominated by siderite. Textural map, Figure 17, shows an 8:1 mixture siderite-apatite. False color pink represents the P_2O_5-rich mineral, apatite. Clay pellets remain unaltered even though they are incorporated in the concretionary burrow fill.

The identification of odinite, a key verdine mineral, is supported by all of the above analytical results. The unit cell calculation shows lattice parameters that are uniquely odinite. It is the dominant clay type in the MGB clay pellets. The structural formula calculated from EMPA data is consistent with that of odinite. X-ray diffraction results suggest an odinite, smectite, illite clay mixture. Verdine facies clays besides odinite include smectite and Fe^{3+} chlorite, sometimes referred to as swelling chlorite [16,24]. The MGB clay is moderately high in Fe, and Mössbauer spectroscopy found an approximate Fe^{3+}/Fe^{2+} ratio of 3:1, which is consistent with odinite. QEMSCAN identified the dominate clay type of the MGB as chlorite-like, which has close similarity to odinite.

Siderite and apatite. Concretions at the top of the MGB are composed of a fine-grained apatite-silicate mixture. Analyses of siderite are presented in Table 5. Despite attempts with the

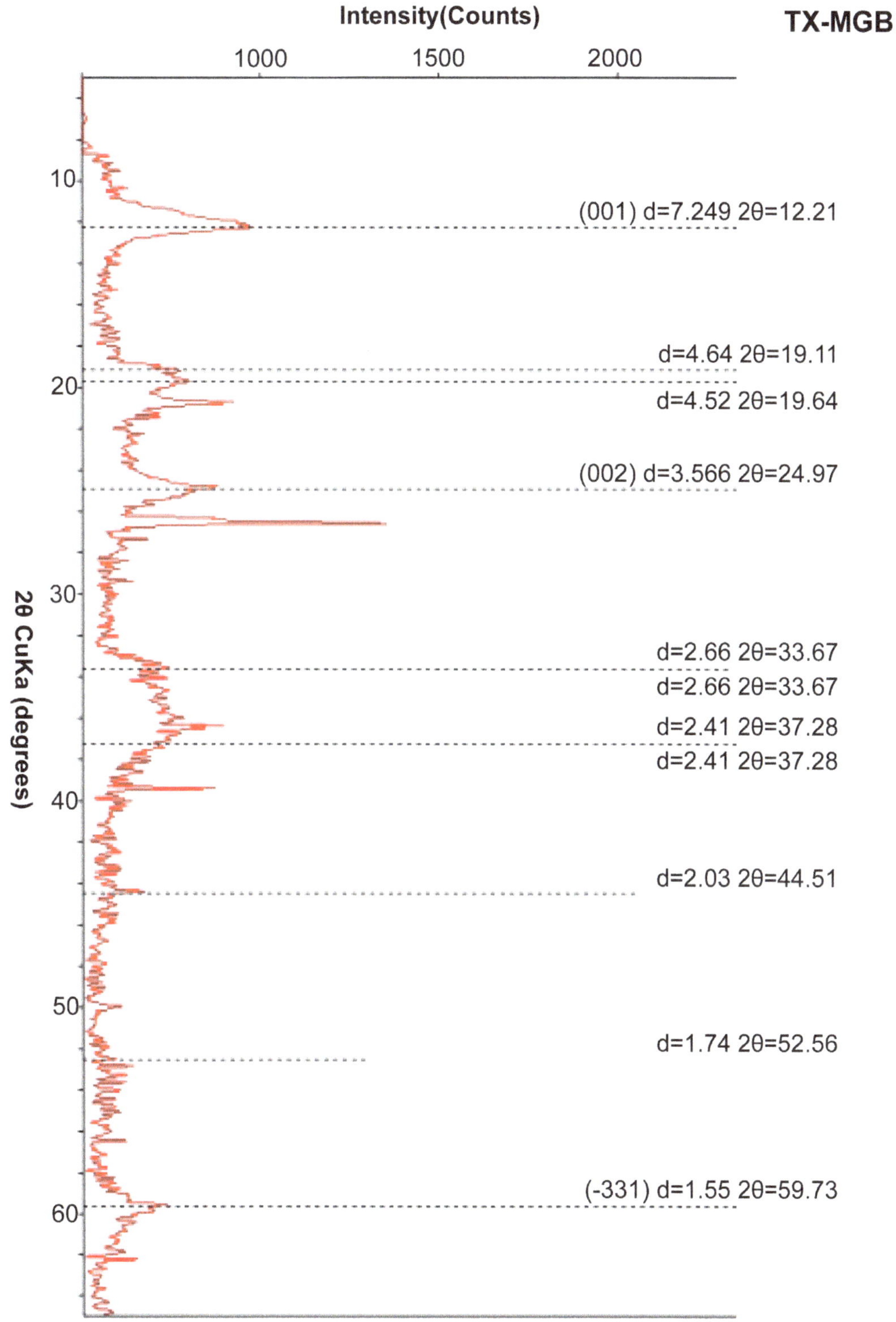

Figure 11. X-ray diffraction pattern of an unoriented powder mount of concentrated clay pellets, TX-MGB. Eleven labeled peaks were indexed in unit cell calculation for odinite-1M. Unlabeled peaks are quartz. Peak at 2θ = 59.73 is quartz overlapped by odinite.

electron microprobe to locate regions without a phosphorus X-ray signal, when measured, phosphorus was always detected, ranging in three samples from 0.8 to 4.0%, which corresponds to about 2–10% apatite. A small amount of Al_2O_3 was also recorded, and probably represents a minor amount of clay. Mole percent carbonate end-members are calculated with $CaCO_3$ adjusted for the Ca-equivalent of apatite as determined from P_2O_5 concentrations when measured (Table 5). Mole fraction $FeCO_3$ ranges from 70–75%, consistent with the observation of Mozley [56] that marine siderite has <95 atom% Fe and that Mg>Ca. Huggett

Table 2. Representative analyses from MGB clay pellets, 11 each from the central MGB and concretionary burrow fill at the top of the MGB.

	Central MGB -->				Top MGB -->				
sample #	TX-8	TX-18	TX-18	5c	9	9	W	TX-17	TX-17
n*	6	2	2	1	3	1	1	2	4
SiO2	33.9	33.2	35.5	33.9	33.7	34.4	34.8	35.7	32.8
Al2O3	15.8	12.0	14.8	14.5	13.2	12.7	13.1	13.4	15.4
Fe2O3T	31.8	33.6	26.1	22.8	37.4	33.2	38.1	31.3	25.5
Fe2O3	*23.8*	*25.2*	*19.6*	*17.1*	*28.0*	*24.9*	*28.5*	*23.5*	*19.1*
FeO	*7.15*	*7.55*	*5.88*	*5.13*	*8.41*	*7.48*	*8.56*	*7.05*	*5.74*
MgO	6.63	4.63	5.41	6.73	4.17	4.09	4.31	4.60	6.72
MnO	0.18	0.10	0.06	0.10	0.06	0.07	0.04	0.08	0.15
TiO2	0.30	0.24	0.24	0.22	0.16	0.20	0.07	0.23	0.21
CaO	1.03	1.12	1.37	0.42	0.84	0.85	0.36	0.87	0.55
Na2O	0.08	0.05	0.06	0.10	0.10	0.09	0.14	0.06	0.09
K2O	0.34	0.50	0.63	0.90	1.12	1.76	1.45	0.78	0.46
H2O	11.9	11.1	11.4	10.9	11.6	11.3	11.9	11.6	11.1
F	0.03	0.52	0.19	0.36	0.02	0.16	0.00	0.41	0.23
Cl	0.06	0.02	0.07	0.14	0.30	0.00	0.18	0.01	0.19
Sum with FeO+Fe2O3	**101.2**	**96.3**	**95.1**	**90.5**	**101.8**	**98.0**	**103.5**	**98.3**	**92.8**

*Number of spot analyses in one pellet.

et al. [11] report a slightly higher cation mole fraction of 0.79 for siderite from the MGB from a partial analysis.

Analyses of concretions consisting of a fine-grained mixture of apatite and one or more alumino-silicates are provided in Table 6. Assuming that all P is contained in apatite, and accounting for the equivalent CaO, the composition of the silicate component (neglecting halogens) can be estimated. The calculated silicate component is similar in composition to MGB clays with respect to total Fe_2O_3 and Al_2O_3, although with slightly lower SiO_2, suggesting that the silicate in the phosphate-bearing cement is the same or very similar to the clay in the pellets and the matrix of the MGB.

Table 3. Maximum, minimum, mean and standard deviation of chemical analyses from MGB pellets and published odinite.

	MGB pellets				Odinite (samples 699 & 508)			
Oxide Wt %	Min	Max	Mean	SD	Min	Max	Mean	SD
			n = 31				n = 6	
SiO2	28.0	36.1	33.6	1.62	33.5	37.5	35.6	1.71
Al2O3	11.5	16.5	14.4	1.41	11.0	12.2	11.8	0.67
Fe2O3T	22.8	38.2	30.1	4.57	25.1	26.4	26.2	0.51
Fe2O3	*17.1*	*28.6*	*22.6*	*3.43*	*17.9*	*19.5*	*20.4*	*3.01*
FeO	*5.13*	*8.59*	*6.76*	*1.03*	*6.10*	*6.50*	*6.25*	*0.15*
MgO	3.95	7.90	5.71	1.16	9.70	11.0	9.78	0.68
MnO	0.04	0.31	0.14	0.07	0.00	0.33	0.22	0.17
TiO2	0.07	0.62	0.22	0.11	0.40	0.50	0.35	0.18
CaO	0.36	1.37	0.88	0.27	0.13	0.50	0.22	0.15
Na2O	0.02	0.16	0.08	0.04	0.00	0.20	0.07	0.10
K2O	0.26	1.76	0.71	0.40	0.35	1.30	0.63	0.44
H2O	10.5	12.0	11.4	0.41	10.0	15.0	11.6	1.81
F	0.00	0.82	0.20	0.18				
Cl	0.00	0.60	0.09	0.12				
Sum*	**76.9**	**115.3**	**96.7**		**89.1**	**104.6**	**96.8**	

*Sum with FeO+Fe2O3.

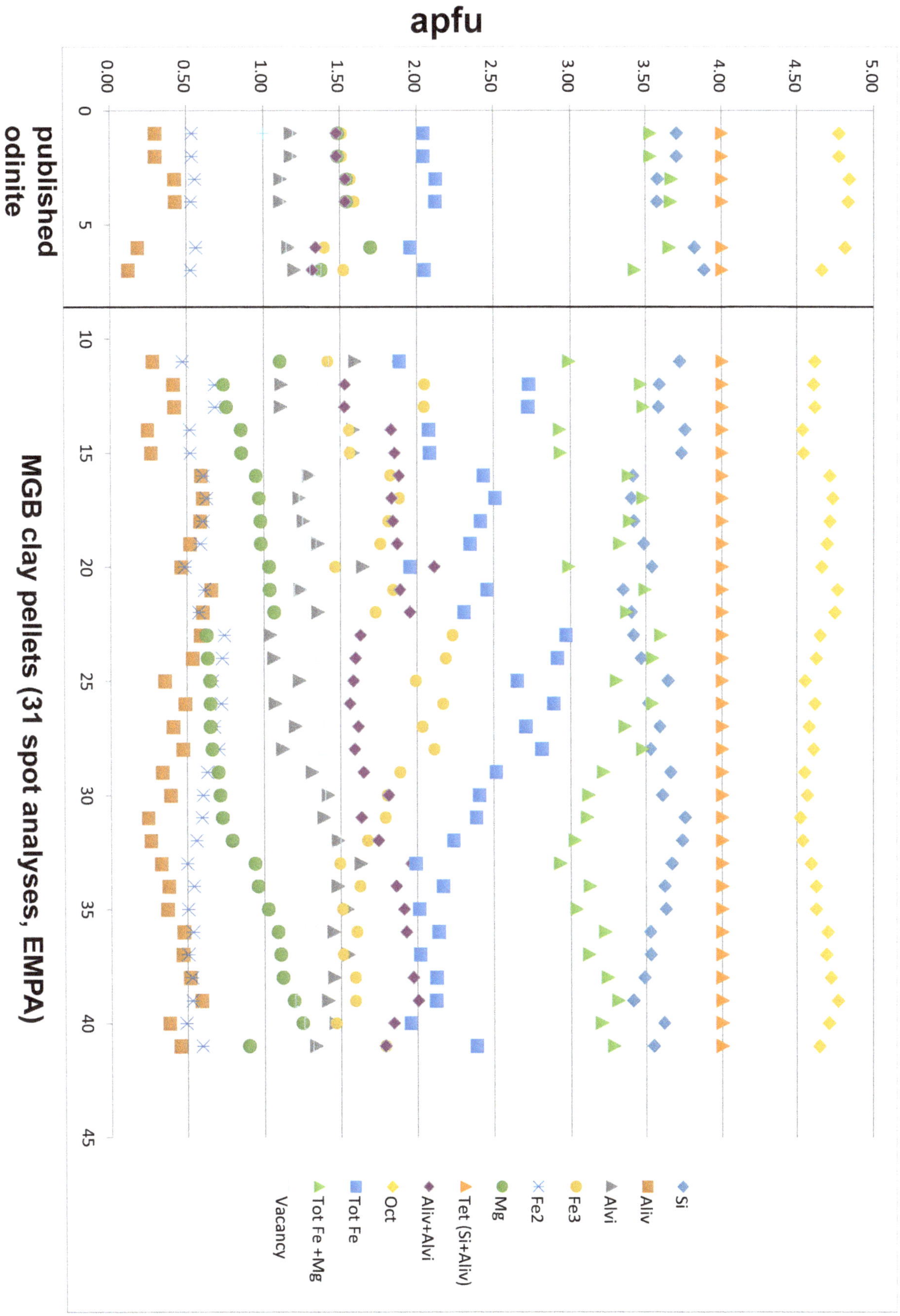

Figure 12. Chemical analyses (apfu), published [16,24] and MGB clay pellets, used to calculate mineral formula.

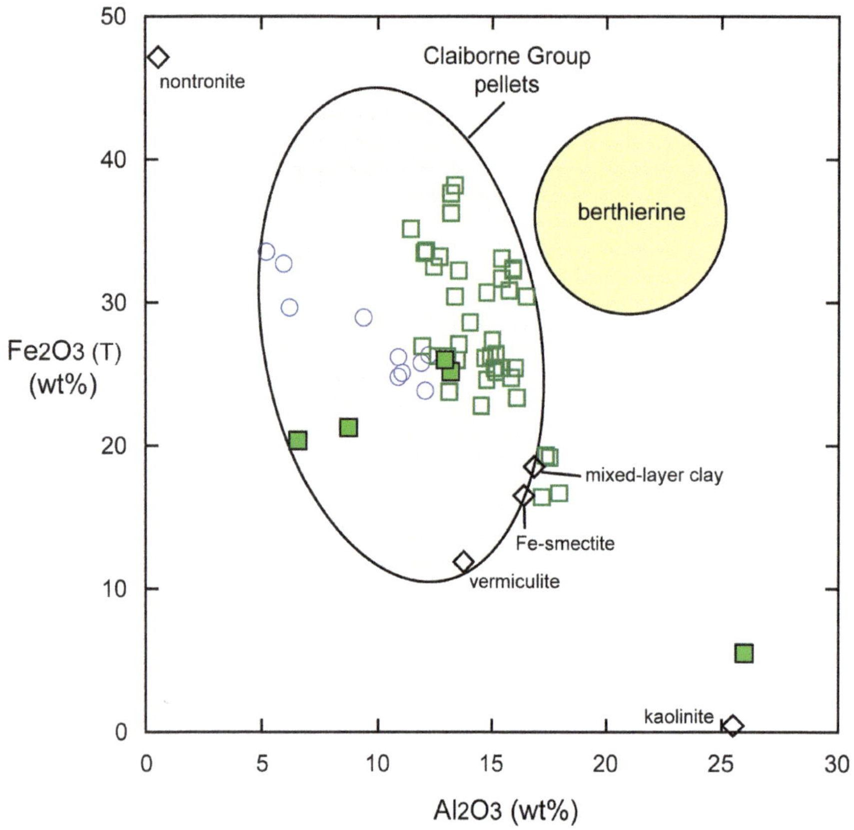

Figure 13. MGB clays plotted within the compositional field of iron-rich clays from the Claiborne Group. Open squares (green) are from the MGB and solid squares are glauconitic minerals from the MGB. Open circles are published analyses of odinite [16,24]. The range of compositions of clay in pellets from the Claiborne Group is indicated by the large oval [11,61]. The field of berthierine is from Brindley [61].

Table 4. Mössbauer parameters of 7Å clay mixture.

Sample		5c	5c	TX-9	TX-9	TX-18	W	W	W
		matrix	pellets	matrix	pellets	pellets	matrix	m & p	pellets
ferric	IS	0.31	0.36	0.30	0.36	0.33	0.32	0.29	0.37
	QS	0.70	0.83	0.82	0.79	0.83	0.66	0.84	0.83
	Width	0.49	0.62	0.34	0.33	0.54	0.42	0.38	0.59
	Area %	81.3	66.4	46.3	30.2	77.1	65.8	39.0	68.3
ferric	IS			0.36	0.40		0.36	0.37	
	QS			1.20	1.07		1.23	1.24	
	Width			0.43	0.59		0.38	0.45	
	Area %			26.2	33.0		14.4	28.7	
ferrous	IS	1.21	1.13	1.25	1.12	1.21	1.25	1.22	1.14
	QS	2.31	2.61	2.33	2.60	2.25	2.31	2.28	2.59
	Width	0.40	0.22	0.26	0.32	0.30*	0.40*	0.30*	0.30*
	Area %	18.7	33.6	27.5	36.8	22.9	19.8	32.3	31.7
	% Fe^{3+}	**81**	**66**	**72**	**63**	**77**	**80**	**68**	**68**
	% Fe^{2+}	**19**	**34**	**28**	**37**	**23**	**20**	**32**	**32**

IS: Isomer shift (mm/s).
QS: Quadrupole splitting (mm/s).
Width: FWHM (mm/s).
Spectrum temperature 295 K.
*parameter fixed.

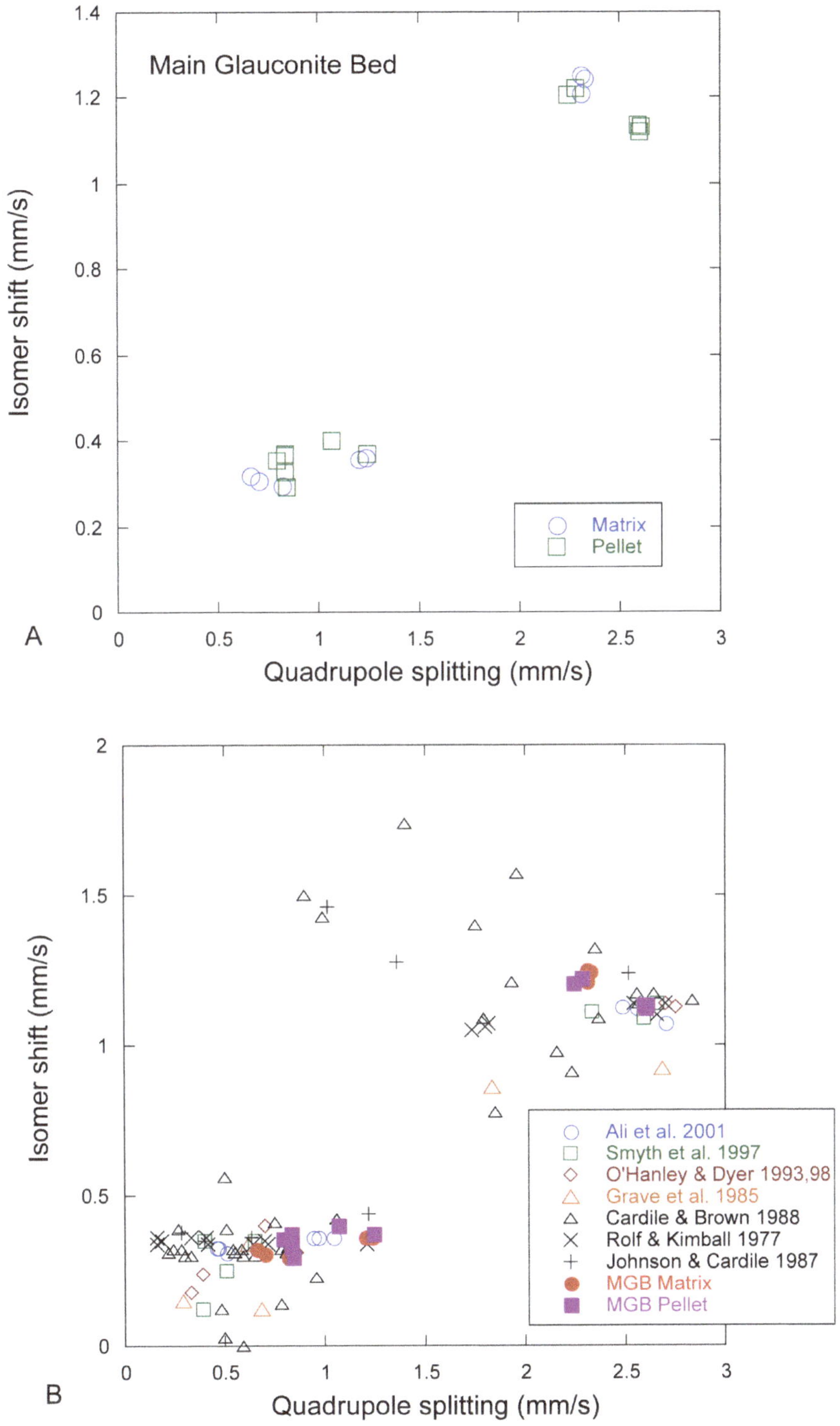

Figure 14. Plot of isomer shift vs. quadrupole splitting (in mm/s). A) Clays from the MGB (this study). B) MGB clays compared to published analyses of glauconite.

Discussion

The detailed mineralogy and geochemistry of the MGB determined in this study relate to paleoecology, sedimentology and stratigraphic findings from previous studies. When integrated with the ichnology of the MGB, a coordinated approach to understanding the depositional environment in a sequence stratigraphic context is possible. Verdine minerals, specifically odinite, were identified in the clay pellets. Thornton, in a 1994 master's thesis [10], also noted that the MGB clays had the characteristics of odinite. Whether the MGB is comprised of odinite or authigenic verdine minerals that were altered, the paleoenvironmental implications are the same. It is presumed that the present-day verdinization environment is key to understanding

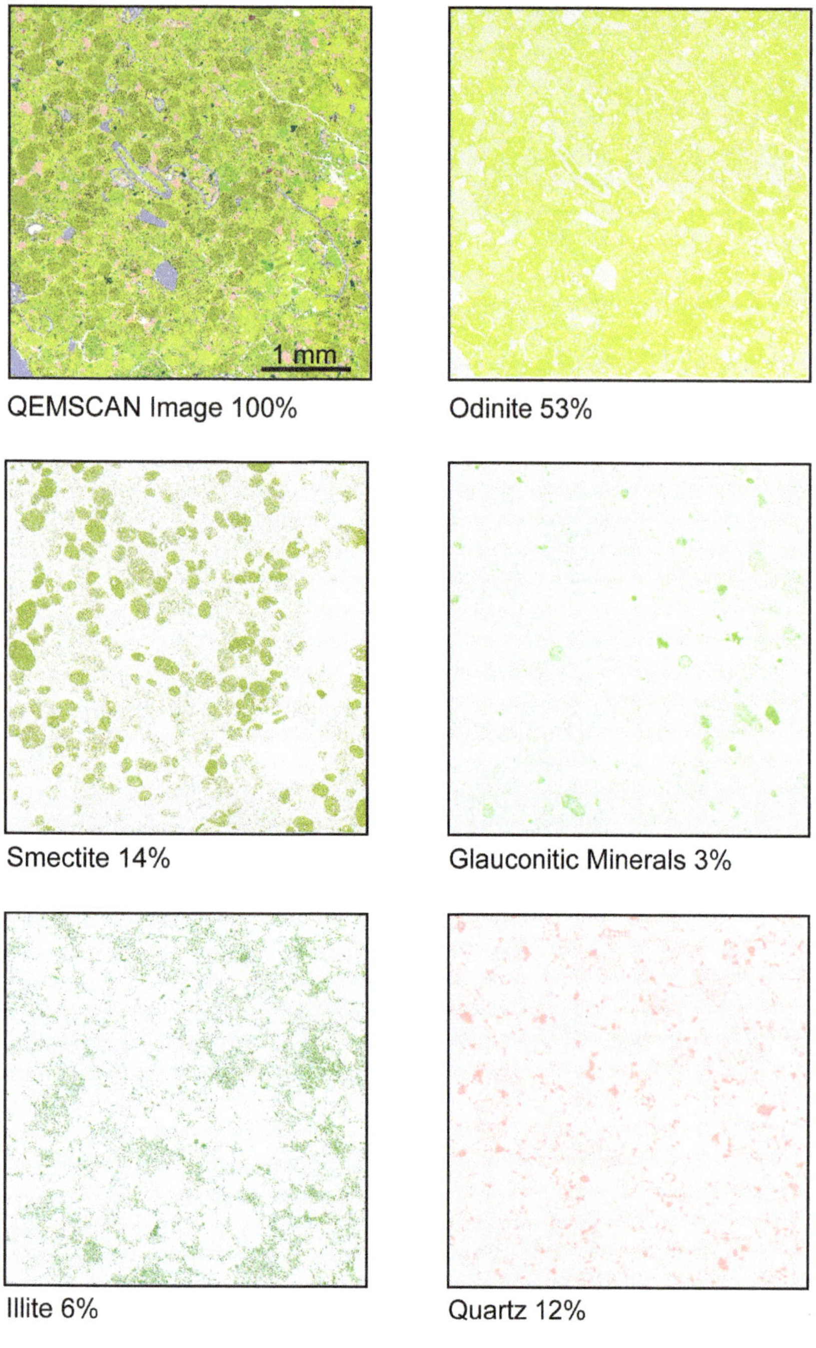

Figure 15. Central MGB QEMSCAN images with specific mineral area %, sample TX-18.

the Middle Eocene continental shelf green marine clay environments.

Pelleted Component

The mineralogy of the clay pellets and their occurrence in the central MGB and in concretionary burrow fill at the top of the MGB has implications for understanding depositional processes. Odinite-rich clay pellets are abundant in the central MGB. They are interpreted as fecal pellets based on their consistent shape, varied size, appearance and iron oxidation state. Their varied size is attributed to multiple producers, which reflects a diverse faunal community. Fecal pellets originally must have contained organic matter, resulting in locally more reducing conditions than the surrounding environment [24]. The pellets provided a favorable site for verdinization, because verdine minerals tend to form within sheltered, reducing, granular microenvironments. Clay pellets of the MGB have a lower Fe^{3+}/Fe^{2+} ratio than the clay matrix (Table 4), and consequently they reflect more reducing conditions, as would be expected for fecal material.

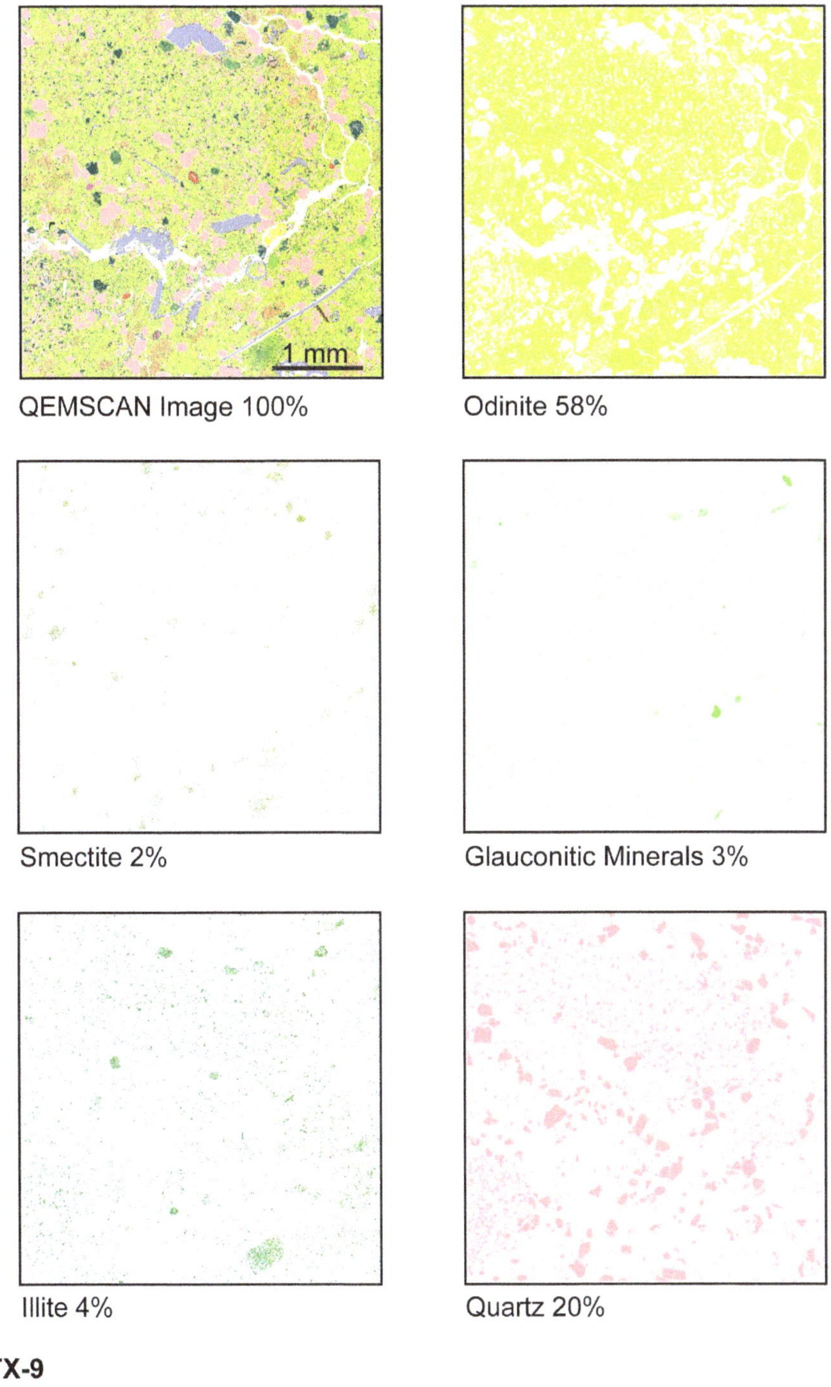

Figure 16. Central MGB QEMSCAN images with specific mineral area %, sample TX-9.

Clay pellets occur in both the central bed and in the concretionary burrow fill at top of the MGB. In the central bed, pellets are not evenly distributed throughout, suggesting that they may have been reworked by gentle winnowing and episodic storms. Clay pellets at the top of the MGB could have been introduced into the open burrows along with other detritus, possibly reworked from the central MGB, or they could have been left behind by occupants in the burrow fill. The clay pellets do not show characteristics of those produced by decapod crustacean burrowers.

The larger, heterogeneous pellets composed primarily of the apatite-cementing agent, can be seen in both the photomicrographs and the QEMSCAN mineral images of concretionary burrow fill. In comparison with the clay pellets, the apatite pellets are larger and more mineralogically heterogeneous (Figure 9). These pellets also occur in the central MGB as a loosely compacted unaltered mixture of clay and tiny clasts. At top of the MGB, these pellets were incorporated into the apatite and siderite-cementing agents of the concretionary burrow fill. Based on their affinity for phosphate minerals and their shape, they are

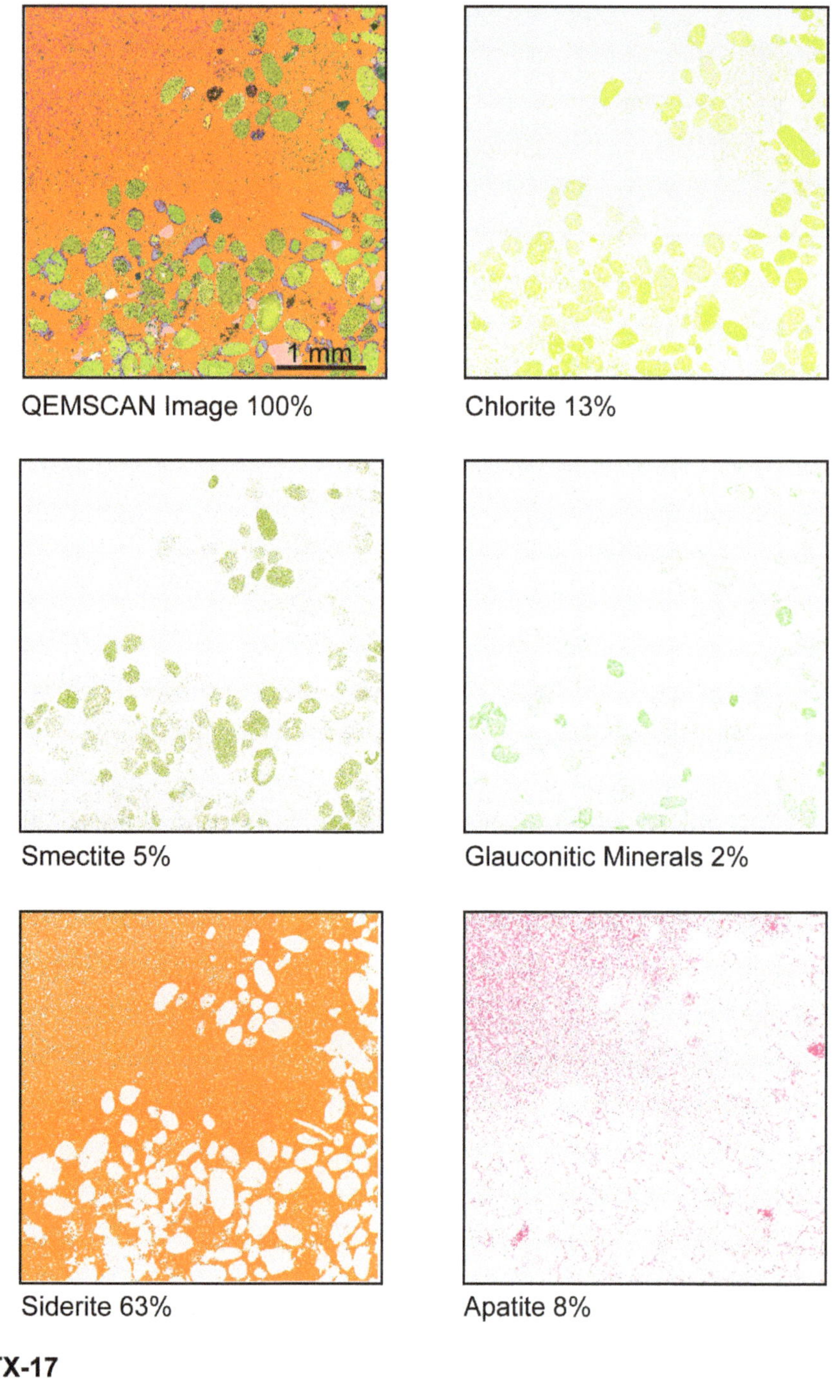

Figure 17. Top of MGB QEMSCAN images with specific mineral area %, sample TX-17.

interpreted as fecal in origin. Like the clay pellets, the apatite pellets could have been left behind by occupants in the burrow fill. They do not show the morphologic characteristics typical of pellets produced by decapod crustacean burrowers.

A very few fragmented glauconitic mineral grains (3 area %) occur in both the central and top of the MGB. Their existence is confirmed by electron microprobe on the basis of $K_2O > 2$ wt. % and by QEMSCAN analyses based on SIP definition. The fragmented nature of the grains lends support to the hypothesis that some winnowing and reworking of sediments may have occurred, and that glauconitic minerals were transported from deeper environments where they originated. Glauconitic minerals, once formed, are very stable in the marine environment, and they may well survive transport of appreciable distance. The glauconitic grains may be more brittle than the non-glauconitic clay pellets, and so they tend to fracture more readily. Glauconitic grains found in the concretions were likely added with clastics in the burrow fill.

Table 5. Concretionary burrow fill, MGB top, siderite and mole percent carbonate end-members calculated with $CaCO_3$.

Sample	3	9	3245	1 plug 4	1 plug 8	Avg	SD
n*	1	3	1	1	1	7	
SiO_2	2.30	2.50	1.90	1.16	2.42	2.05	0.55
TiO_2	0.00	0.05	0.03	0.06	0.04	0.03	0.02
Al_2O_3	1.60	0.90	1.00	0.50	1.14	1.01	0.40
FeO**	37.3	41.3	43.2	38.7	38.2	39.7	2.20
MnO	0.74	0.65	0.62	1.02	0.75	0.76	0.16
MgO	6.00	4.76	4.90	4.63	5.23	5.10	0.55
CaO	4.35	8.81	7.45	4.65	3.89	5.83	1.94
Na_2O	0.17	0.15	0.06	0.12	0.20	0.14	0.06
K_2O	0.21	0.08	0.06	0.05	0.19	0.12	0.08
P_2O_5	0.76	4.04	2.66				
Total	53.4	63.3	61.9	50.9	52.0	54.7	5.3
% apatite	1.8	9.6	6.3				
Mol %							
$CaCO_3$***	8.1	8.2	8.8	11.0	9.4	13.1	2.0
$MgCO_3$	20.2	15.4	15.2	15.3	17.5	15.9	2.0
$MnCO_3$	1.4	1.2	1.1	1.9	1.4	1.4	0.3
$FeCO_3$	70.3	75.2	75.0	71.7	71.7	69.6	2.3

*Spot analyses in one grain.
**Total Fe as FeO.
***$CaCO_3$ corrected for apatite.

Table 6. Concretionary burrow fill, MGB top, a fine-grained mixture of apatite and one or more alumino-silicates.

Sample	3	3	9	3245	Avg	SD
n*	1	2	2	2	7	
SiO_2	11.6	17.0	11.3	16.5	14.1	3.07
TiO_2	0.20	0.19	0.15	0.17	0.18	0.02
Al_2O_3	5.60	8.20	5.71	8.97	7.12	1.72
Fe_2O_3**	15.7	13.6	12.1	15.9	14.3	1.82
MnO	0.19	0.07	0.09	0.07	0.11	0.06
MgO	1.68	1.61	0.89	1.56	1.43	0.37
CaO	26.9	25.2	34.4	27.8	28.6	4.05
Na_2O	0.35	0.41	0.46	0.30	0.38	0.07
K_2O	0.34	0.45	0.31	0.47	0.39	0.08
P_2O_5	19.8	18.6	23.9	19.4	20.4	2.36
F	1.96	1.90	0.01	0.02	0.97	1.10
Cl	0.03	0.03	2.36	2.00	1.11	1.25
sum	84.2	87.3	91.7	93.2	89.1	4.12
Less O=F,Cl	0.83	0.81	0.54	0.46	0.66	0.19
Total	**83.4**	**86.5**	**91.2**	**92.7**	**88.5**	**4.29**
% apatite	46.8	44.0	56.5	45.9	48.2	4.9

*Spot analyses in one grain.
**Total Fe as Fe_2O_3.

Sequence Stratigraphy

In a sequence stratigraphic context, green minerals often are associated with condensed sections and transgressive-systems-tracts. A transgressive phase on a sediment-starved sea floor offers favorable conditions for the glauconitization or verdinization process. A regressive phase, on the other hand, may introduce a more energetic, oxidizing environment of higher sediment influx, which would inhibit clay authigenesis [25,29]. Green minerals of the verdine facies suggest the shallow inner shelf, more proximal zone of the transgressive surface and transgressive-systems-tract.

The MGB is part of a regionally mapped transgressive sequence during the Middle Eocene in the northwest Gulf of Mexico. The siliciclastic depositional center was associated with the Houston Embayment (Figure 1) [57]. Deposition occurred during an interval of fluctuating sea level in sediment-starved conditions, interfingering with episodes of higher sedimentation rate. The basal contact of the MGB is sharp, irregular and burrowed. The central MGB has verdine facies clays that indicate shallow marine conditions with transgressive characteristics in proximity to river mouth influx (Figure 7). The upper contact of the MGB marks a boundary between two contrasting facies. It is irregular due to burrowing [2,13]. The burrows are interpreted as firmground trace fossils that often occur in temporarily dewatered and compacted sediment. The top of the MGB may highlight a coplanar transgressive surface–sequence boundary in a parasequence. It is possible that the top of the MGB could have been exposed in an intertidal environment. Then the open burrows were filled with detritus during renewed submergence. Subsequent apatite and siderite cementation of the firmground would have occurred in a subaqueous setting.

The Crockett Formation was deposited in a complex marginal marine environment with dynamic sea level fluctuation. It represents a shallow-water transition zone in an overall upward deepening sequence. While odinite formed *in situ* associated with the fecal pellets of the central MGB, the few glauconitic mineral grains in the MGB may be explained by detrital reworking. Grains can be transported widely in transgressive systems, and glauconite is stable geochemically in the marine environment.

Paleoenvironment

The paucity of glauconitic minerals indicates that seafloor conditions during deposition of the MGB were unsuitable for their formation, or that the glauconitization process was incomplete or arrested. This might mean a number of things. For example, perhaps sea-floor conditions were not consistently marine, too close to river mouth influx, too close to shore, or too shallow water depth. Sediment influx can inhibit the glauconitization process by not allowing enough time for glauconitic minerals to mature. The few, small glauconitic mineral grains in the MGB were often fractured, which indicates they may be of allochthonous origin, so it appears that the glauconitization process did not occur during deposition of the MGB.

On the other hand, verdinization apparently did occur. The presence of odinite, especially in the clay pellets, indicates a set of sea-floor environmental conditions pertaining to geochemistry, temperature, oxidation state, energy and latitude. The central MGB reflects depositional processes in a low energy, somewhat sheltered, shallow, warm, mostly marine environment. Sedimentation rate was sufficiently slow in proximity of continental input of iron and gentle winnowing currents to provide Eh and pH conditions suitable for verdine mineral authigenesis [16,24,32].

The change in mineralogy from the central bed to the top of bed indicates a change in the sea-floor environment. The central MGB was deposited in a near shore, subaqueous setting, while it is possible that the top of the MGB, with its concretionary burrow fill comprised of pellets and detritus, may have been in an intertidal environment. This interpretation is supported by characteristic burrowing patterns at top of the MGB. It represents an interval of non-deposition, dewatering, compaction and burrowing. Ultimately, the open burrows excavated by infaunal crustaceans were filled with sediment and bioclasts, indicating return to a subaqueous environment in moderate energy where precipitation of siderite and apatite cementing agents occurred.

A coordinated approach to understanding the paleoenvironment of the MGB incorporates verdine facie characteristics, ichnological signature, and findings from previous MGB studies. During a period of regional transgression, relatively high sea level, and some lateral continuity of facies, several coastal siliciclastic environments may apply. These include protected shoreface settings, large scale normal marine lagoons, interdeltaic bays, sounds, or embayments. Tropical seagrass beds demonstrate high molluscan diversity, such as that observed in the MGB. Seagrasses were present in the Gulf and Caribbean during the warmer climatic conditions of Middle Eocene [58]. Seagrass environments are recognized only rarely in the geological record due to the low preservation potential of soft plants, but they probably were much more abundant than is normally realized. Benthic foraminiferal evidence would be instructive, however that was outside the scope of this study. In general, the setting of the MGB is characterized as shallow inner shelf, proximal to some terrestrial influence.

Conclusion

The MGB clays at Stone City Bluff of Middle Eocene, east-central Texas are characterized as verdine facies clay, largely comprised of the mineral odinite. Odinite is an iron and magnesium 1:1, 7Å clay. It is the main constituent of the clay fecal pellets and matrix of the central MGB. Few occurrences of odinite, other than modern odinite, are described in the literature. The MGB provides an informative example of odinite in the geologic record.

Odinite was established as the main constituent in clay pellets of the MGB through several analytical methods. X-ray diffraction found a 7.2Å clay mixture. Unit cell calculation indicated odinite-1M to be the most reliable interpretation. EMPA analysis of clay pellets, directed by QEMSCAN textural maps, identified the chemical composition. Mössbauer spectral analysis found the oxidation state at 3:1 Fe^{3+}/Fe^{2+} ratio diagnostic of odinite. Thus, essential chemical data was obtained to calculate the average structural formula for clay pellets of the MGB as follows: $Fe^{3+}_{0.89}$ $Mg_{0.45}$ $Al_{0.67}$ $Fe^{2+}_{0.30}$ $Ti_{0.01}$ $Mn_{0.01})$ $_{\Sigma=2.33}$ $(Si_{1.77}$ $Al_{0.23})$ $O_{5.00}$ $(OH)_{4.00}$. Odin and Bailey [16,24] first described odinite in 1988, and they calculated a similar mineral formula that was slightly more Mg rich and Fe, Al poor.

Documentation of odinite and verdine facies clays in the MGB is an important finding of this report. The environmental conditions associated with modern verdine facies clay occurrences include tropical latitudes, nearby runoff with iron influx, and water depth between 15 and 60 m (locally in 5 m depth). The characteristic sea-floor conditions are normal to nearly normal salinity, positive Eh, elevated sea-floor temperature and common association with fecal pellets. Circulating, winnowing currents are required to stir the sediment slightly and provide oxygenation. It is presumed that paleoenvironmental conditions were comparable during the Middle Eocene accumulation of authigenic verdine clays in the MGB.

The contrasting mineralogy between the central and the top of the MGB reflects a change in environmental conditions during its deposition. The central MGB has abundant pellets, comprised mostly of odinite, with lenses of shell concentrations and laminated sand. In contrast, the top of the MGB is a zone of concretions that are filled burrows comprised of pellets and detritus with siderite and apatite that formed as cementing agents. Glauconitic minerals are very sparse in both the central and top of the MGB and are probably allochthonous. A change in environmental conditions is reflected in a composite ichnofabric where deeper water biogenic activity in the central MGB is replaced by shallow water to intertidal burrowers in a firmground at the top of the MGB. The firmground was subsequently submerged and cemented.

The paleoenvironmental interpretation for the geologic section at Stone City Bluff has been the subject of numerous studies. This contribution of detailed mineralogy and geochemistry, although focused on the fecal pellets of the MGB, offers key information about processes during its deposition. With the identification of authigenic odinite or its alteration products, this study complements findings of previous studies. The MGB represents an ancient verdine facies paleoenvironment, which has implications for paleoecology, sedimentation rate, sequence stratigraphy and paleoclimate. It provides a coordinated approach to paleoenvironmental understanding of the dynamic depositional environment of the MGB at Stone City Bluff.

Supporting Information

Figure S1 X-ray diffraction pattern of oriented 2 μ clay from bulk sample TX-8, (air dried, glycolated, heated to 375° and heated to 500°C).

Figure S2 X-ray diffraction pattern of oriented 2 μ, clay from bulk sample TX-9, (air dried, glycolated, heated to 375° and heated to 500°C).

Figure S3 X-ray diffraction patterns of randomly oriented powders TX-8, TX-18, and TX-9. Minerals of the verdine facies at the 060 diffraction peak are between1.56Å (59.23° 2θ) and 1.51Å (61.35° 2θ). Quartz at the 121- (*hkl*) peak is 1.54Å (59.970° 2θ).

Figure S4 Mössbauer Spectrum TX-9 & W, matrix & pellets.

Table S1 EMPA analyses for all odinite raw data in oxides and apfu.

Table S2 EMPA analyses for all other clays, including glauconitic minerals in oxides.

Acknowledgments

The authors thank TE Yancey, JE Warme, CJ Flis, and JE Flis for access to critical samples and for comments about sedimentological and paleoecological implications. J Allen, NF Dahdah, MA Gorenc, PD Pahnke, WT Parry, WD Mace, and CG Jones provided valuable technical assistance. QEMSCAN Model EVO 50, SERIAL # E430 located at Energy and Geoscience Institute (EGI) was used for the QEMSCAN analyses, and the microprobe and XRD analyses were accomplished in the Department of Geology and Geophysics at the University of Utah.

Author Contributions

Conceived and designed the experiments: SCH AAE. Performed the experiments: SCH BPN EUP CDB MDD. Analyzed the data: SCH BPN EUP CDB. Contributed reagents/materials/analysis tools: BPN EUP MDD. Wrote the paper: SCH BPN EUP AAE CDB MDD.

References

1. Stenzel HB (1934) Decapod Crustaceans from the Middle Eocene of Texas. Journal of Paleontology 8: 38–56.
2. Stenzel HB, Krause EK, Twining JT (1957) Pelecypoda from the type locality of the Stone City beds (middle Eocene) of Texas. Austin: Bureau of Economic Geology, University of Texas. 237 p., 222 p. of plates p.
3. Stanton RJ, Jr., Warme JE (1971) Stop 1: Stone City Bluff: Trace fossils. A field guide to selected localities in Pennsylvanian, Permian, Cretaceous and Tertiary rocks of Texas and related papers. In: Perkins BF, editor. SEPM: School of Geoscience Louisiana State University Misc Pubs. 2–10.
4. Stanton RJ, Jr. (1979) The Stone City Formation: in. Claiborne sediments of the Brazos Valley, southeast Texas: Field Trip Guidebook of the Houston Geological Society, Houston: 73–77.
5. Yancey TE, Davidoff AJ (1994) Paleogene sequence stratigraphy of the Brazos River section, Texas. Gulf Coast Association of Geological Societies Field Trip Guide: 104.
6. Yancey TE (1995) Depositional trends in siliciclastic deposits of the Stone City transgressive systems tract, middle Eocene, Texas. Gulf Coast Association of Geological Societies Field Guide: 581–586.
7. Davidoff AJ, Yancey TE (1993) Eustatic cyclicity in the Paleocene and Eocene: data from the Brazos River Valley, Texas. Tectonophysics 222: 371–395.
8. Davidoff AJ, Yancey TE (1993) Relating sequence stratigraphy to lithostratigraphy in siliciclastic-dominated shelf settings, Paleogene, central-east Texas. Transactions of the Gulf Coast Association of Geological Societies 43: 97–108.
9. Zuschin M, Stanton RJJ (2002) Paleocommunity reconstruction from shell beds: a case study from the Main Glauconite Bed, Eocene, Texas. Palaios 17: 602–614.
10. Thornton CA (1994) Sediment diagenesis, fossil preservation, and depositional environment in the Stone City/Lower Cook Mountain transgression (Middle Eocene, Southeast Texas): A test of chemical taphofacies in the rock record [Master of Science]: Texas A&M University. 189 p.
11. Huggett JM, Gale AS, McCarty D (2010) Petrology and palaeoenvironmental significance of authigenic iron-rich clays, carbonates and apatite in the Claiborne Group, Middle Eocene, NE Texas. Sedimentary Geology 228: 119–139.
12. Davidoff AJ, Yancey TE (1993b) Eustatic cyclicity in the Paleocene and Eocene: data from the Brazos River Valley, Texas. tectonophysics 222: 371–395.
13. Stanton RJ, Jr., Nelson PC (1980) Reconstruction of the trophic web in paleontology; community structure in the Stone City Formation (middle Eocene, Texas). Journal of Paleontology 54: 118–135.
14. Berg RR (1979) Stratigraphy of the Claiborne Group. Lower Tertiary of the Brazos River Valley: Guidebook of the Houston Geological Society.
15. Dyar MD, Gunter ME, Tasa D (2008) Mineralogy and optical mineralogy: Mineralogical Society of America Chantilly, VA.
16. Odin GS (1988) Green Marine Clays: Oolitic Ironstone Facies, Verdine Facies, Glaucony Facies and Celadonite-Bearing Facies - A Comparative Study. Amsterdam: Elsevier. 445 p.
17. Burst JF (1958) Mineral heterogeneity in glauconite pellets. American mineralogist 43: 481–497.
18. Bhattacharyya DP (1983) Origin of Berthierine in Ironstones. Clays and clay minerals 31: 173–182.
19. Triplehorn DM (1966) Morphology, internal structure, and origin of glauconite pellets. Sedimentology 6: 247–266.
20. McCarty DK, Drits VA, Sakharov B, Zviagina BB, Ruffell A, et al. (2004) Heterogeneous mixed-layer clays from the Cretaceous Greensand, Isle of Wight, southern England. Clays and clay minerals 52: 552–575.
21. Huggett JM, McCarty DK, Calvert CC, Gale AS, Kirk C (2006) Serpentine-nontronite-vermiculite mixed-layer clay from the Weches formation, Claiborne group, middle Eocene, northeast Texas. Clays and clay minerals 54: 101–115.
22. Sánchez-Navas A, Martín-Algarra A, Eder V, Reddy BJ, Nieto F, et al. (2008) Color, Mineralogy and Composition of Upper Jurassic West Siberian Glauconite: Useful Indicators of Paleoenvironment. The Canadian Mineralogist 46: 1249–1268.
23. Needham SJ, Worden RH, McIlroy D (2004) Animal-sediment interactions: the effect of ingestion and excretion by worms on mineralogy. Biogeosciences 1: 113–121.
24. Bailey SW (1988) Odinite; a new dioctahedral-trioctahedral Fe (super 3+)-rich 1:1 clay mineral. Clay minerals 23: 237–247.
25. Amorosi A (1995) Glaucony and sequence stratigraphy; a conceptual framework of distribution in siliciclastic sequences. Journal of Sedimentary Research, Section B: Stratigraphy and Global Studies 65: 419–425.
26. Huggett JM, Gale AS, Clauer N (2001) The nature and origin of non-marine 10 A clay from the Late Eocene and Early Oligocene of the Isle of Wight (Hampshire Basin), UK. Clay minerals 36: 447–464.
27. Hesselbo SP, Huggett JM (2001) Glaucony in ocean-margin sequence stratigraphy (Oligocene-Pliocene, offshore New Jersey, USA; ODP Leg 174A). Journal of Sedimentary Research 71: 599.
28. Kronen J, Glenn CR (2000) Pristine to reworked verdine: keys to sequence stratigraphy in mixed carbonate-siliciclastic forereef sediments (Great Barrier Reef). Special Publication-SEPM 66: 387–404.
29. Stonecipher SA (1999) Genetic Characteristics of Glauconite and Siderite: Implications for the Origin of Ambiguos Isolated Marine Sandbodies. Special Publication-SEPM 64: 191–204.
30. Thurman HV, Trujillo AP (1999) Essentials of Oceanography. Upper Saddle RiverNew Jersey 07458: Prentice-Hall, Inc. 527 p.
31. Dyar MD (1984) Precision and interlaboratory reproducibility of measurements of the Mössbauer effect in minerals. American Mineralogist 69: 1127–1144.
32. Ku TCW, Walter LM (2003) Syndepositional formation of Fe-rich clays in tropical shelf sediments, San Blas Archipelago, Panama. Chemical Geology 197: 197–213.
33. Moore DM, Reynolds RC, Jr. (1997) X-Ray Diffraction and the Identification and Analysis of Clay Minerals. book: 1–378.
34. Pestitschek B, Gier S, Essa M, Kurzweil H (2012) Effects of Weathering on Glauconite: Evidence from the Abu Tartur Plateau, Egypt. Clays and clay minerals 60: 76–88.
35. Ryan PC, Hillier S (2002) Berthierine/chamosite, corrensite, and discrete chlorite from evolved verdine and evaporite-associated facies in the Jurassic Sundance Formation, Wyoming. American Mineralogist 87: 1607–1615.
36. Pouchou JL, Pichoir F (1991) Quantitative analysis of homogeneous or stratified microvolumes applying the model "PAP" Electron probe quantitation. New York: Plenum Press.
37. Wivel C, Mørup S (1981) Improved computational procedure for evaluation of overlapping hyperfine parameters distributions in Mössbauer spectra. Journal of Physics E14: 605–610.
38. Dyar MD, Schaefer MW, Sklute EC, Bishop JL (2008) Mössbauer spectroscopy of phyllosilicates: Effects of fitting models on recoil-free fractions and redox ratios. Clay Minerals 43: 3–33.
39. Bradbury CD (2012) Rapid Reconnaissance Methods of Investigation for Trace Components in Deep-Sea Sediments [Masters]: University of Utah.
40. Ayling B, Rose P, Petty S, Zemach E, Drakos P (2012) QEMSCAN (Quantitative evaluation of minerals by scanning electron microscopy): capability and application to fracture characterization in geothermal systems. Proc, Thirty-Seventh Workshop on Geotherm Reserv Eng. Stanford, California: Stanford University.
41. Gottlieb P, Wilkie G, Sutherland D, Ho-Tun E, Suthers S, et al. (2000) Using quantitative electron microscopy for process mineralogy applications. JOM Journal of the Minerals, Metals and Materials Society 52: 24–25.
42. Haberlah D, Williams MAJ, Halverson G, McTainsh GH, Hill SM, et al. (2010) Loess and floods: High-resolution multi-proxy data of Last Glacial Maximum (LGM) slackwater deposition in the Flinders Ranges, semi-arid South Australia. Quaternary Science Reviews 29: 2673–2693.
43. Geological Society of America (2009) Rock color chart with genuine Munsell color chips.
44. Amorosi A (1997) Detecting compositional, spatial, and temporal attributes of glaucony: a tool for provenance research. Sedimentary Geology 109: 135–153.
45. (JCPDS) JCoPDS (2005) The International Center for Diffraction Data. Newtown Square, PA. pp. over 500,000 entries.
46. Zachariasen WH, Plettinger HA (1965) Extinction in quartz. Acta Crystallographica 18: 710–714.
47. Ali AM, Hsia Y, Liu R, Zhang J, Duan W, et al. (2001) A Mössbauer study of evolution of glauconite from Chinese seas. Spectroscopy letters 34: 701–708.
48. Johnston JH, Cardile CM (1987) Iron substitution in montmorillonite, illite and glauconite by 57 Fe Mossbauer spectroscopy. Clays and Clay Mineralogy 35: 170.
49. Rolf RM, Kimball CW, Odom IE (1977) Mössbauer characteristics of Cambrian glauconite, central USA. Clays and Clay Minerals 25: 131–137.
50. Drits V, Srodon J, Eberl D (1997) XRD measurement of mean crystallite thickness of illite and illite/smectite: Reappraisal of the Kubler index and the Scherrer equation. Clays and clay minerals 45: 461–475.
51. Cardile CM, Brown IWM (1988) An 57 Fe Mössbauer Spectroscopic and x-ray diffraction study of New Zealand Glauconites. Clay Minerals 23: 13.
52. Rancourt D (1994) Mossbauer spectroscopy of minerals. II. Problem of resolving cis and trans octahedral Fe2+ sites. Physics and Chemistry of Minerals 21: 250–257.
53. Rancourt D, Ping J, Berman R (1994) Mössbauer spectroscopy of minerals. Physics and Chemistry of Minerals 21: 258–267.

54. Rancourt D, Ping J, Boukili B, Robert J (1996) Octahedral-site Fe2+ quadrupole splitting distributions from Mössbauer spectroscopy along the (OH, F)-annite join. Physics and Chemistry of Minerals 23: 63–71.
55. Dyar MD (2002) Optical and Mössbauer spectroscopy of iron in micas. Reviews in mineralogy and geochemistry 46: 313–349.
56. Mozley PS (1989) Relation between depositional environment and the elemental composition of early diagenetic siderite. Geology 17: 704–706.
57. Galloway WE, Ganey-Curry PE, Li X, Buffler RT (2000) Cenozoic depositional history of the Gulf of Mexico basin. AAPG bulletin 84: 1743.
58. Eva AN (1980) Pre-Miocene Seagrass communities in the Caribbean. Palaeontology 23: 231–236.
59. Baker ET, Jr. (1995) Stratigraphic nomenclature and geologic sections of the Gulf Coastal Plain of Texas Austin, Texas: U.S. Department of the Interior, U.S. Geological Survey. 94–461.
60. Barnes VEPD (1995) Geologic Atlas of Texas, Austin Sheet. Francis Luther Whitney Memorial Edition ed. The University of Texas at Austin, Bureau of Economic Geology.
61. Brindley GW (1982) Chemical compositions of berthierines; a review. Clays and clay minerals 30: 153–155.

6

Connectivity between Migrating and Landlocked Populations of a Diadromous Fish Species Investigated Using Otolith Microchemistry

Ingrid Tulp[1]*, Marieke Keller[1], Jacques Navez[2,3], Hendrik V. Winter[1], Martin de Graaf[1], Willy Baeyens[2,3]

1 Institute for Marine Resources and Ecosystem Studies (IMARES), Wageningen University, IJmuiden, The Netherlands, **2** Department of Analytical and Environmental Chemistry, Vrije Universiteit Brussel, Brussels, Belgium, **3** Section de Pétrographie-Minéralogie-Géochimie, Koninklijk Museum voor Midden-Afrika, Musée Royal de l'Afrique Centrale, Tervuren, Belgium

Abstract

Smelt *Osmerus eperlanus* has two different life history strategies in the Netherlands. The migrating population inhabits the Wadden Sea and spawns in freshwater areas. After the closure of the Afsluitdijk in 1932, part of the smelt population became landlocked. The fresh water smelt population has been in severe decline since 1990, and has strongly negatively impacted the numbers of piscivorous water birds relying on smelt as their main prey. The lakes that were formed after the dike closure, IJsselmeer and Markermeer have been assigned as Natura 2000 sites, based on their importance for (among others) piscivorous water birds. Because of the declining fresh water smelt population, the question arose whether this population is still supported by the diadromous population. Opportunities for exchange between fresh water and the sea are however limited to discharge sluices. The relationship between the diadromous and landlocked smelt population was analysed by means of otolith microchemistry. Our interpretation of otolith strontium (^{88}Sr) patterns from smelt specimens collected in the fresh water area of Lake IJsselmeer and Markermeer, compared to those collected in the nearby marine environment, is that there is currently no evidence for a substantial contribution from the diadromous population to the spawning stock of the landlocked population.

Editor: Sadie Jane Ryan, SUNY College of Environmental Science and Forestry, United States of America

Funding: This research is part of the strategic research program KBIV "Sustainable spatial development of ecosystems, landscapes, seas and regions" which is funded by the Dutch Ministry of Economic Affairs, Agriculture and Innovation, and carried out by Wageningen University Research centre. Collection of some of the material was made possible by the National Program Sea and Coastal Research from The Netherlands Organisation for Scientific Research under contract nr 839.08.240. The research is also supported by the Dutch research program ANT (Autonome Neerwaartse Trends) lead by Deltares (The Netherlands) and by the Flemish Science Foundation (FWO) grant FWO G.0.096.08.N.10 via the acquisition of the LA-ICPMS instrument. The funders had no role in study design, data collection and analysis, decision to publish, or preparation of the manuscript.

Competing Interests: The authors have declared that no competing interests exist.

* E-mail: ingrid.tulp@wur.nl

Introduction

Smelt *Osmerus eperlanus* occurs along the coast of Europe from North West Russia in the north, including the White Sea and Baltic Sea, to the Bay of Biscay in the south [1]. Anadromous smelt resides in estuaries originally, but several populations have become landlocked due to geological processes or habitat alterations by man [2]. Many fish species are known for their large phenotypic plasticity in life history traits in relation to environmental characteristics. Similarly smelt has shown the ability to adapt to new circumstances in many locations. Smelt can show diverse life history traits among populations, ranging from migratory (diadromous) to resident in freshwater [3].

Smelt was abundant in the former Zuiderzee estuary (The Netherlands) before its closure with a large dam (Afsluitdijk) in 1932 (Fig. 1). After the Afsluitdijk was built, the Zuiderzee estuary was closed off from the sea resulting in the freshwater Lake IJsselmeer, in which a landlocked population of smelt developed. Migrating opportunities became severely reduced to instants when dam sluices are opened and water current is low enough to allow passage. Diadromous smelt still inhabit coastal waters and the Wadden Sea and can grow up to 25 cm and eight years old, and become mature at 3–4 years [2]. The landlocked smelt morphs become mature after only one year and stay smaller (10–12 cm) than anadromous smelt morphs [4]. Smelt spawn in early spring in freshwater, preferably in running water in tributaries of lakes or along shallow shores of lakes and rivers, on firm surfaces such as sand, gravel and stones [3]. The majority of landlocked smelt spawn only once and individuals bigger than 12 cm (and presumably older than 2 years) make up around 1% of the entire landlocked population (Drost & de Witte unpubl.).

The smelt populations of IJsselmeer and Markermeer have been declining strongly since 1990 [5]. In winter, the Lake IJsselmeer area is home to the largest population of waterfowl in Western Europe [6]. Based on the number of (among other groups) fish eating birds both lakes are protected as wetlands of international importance for water birds in the Ramsar Convention and have been assigned as Natura 2000 areas. Smelt fisheries used to be an important activity locally and are carried out during the spawning season, from mid-February till mid-March, when large numbers aggregate near hard substrate areas, such as dykes [7]. Because of

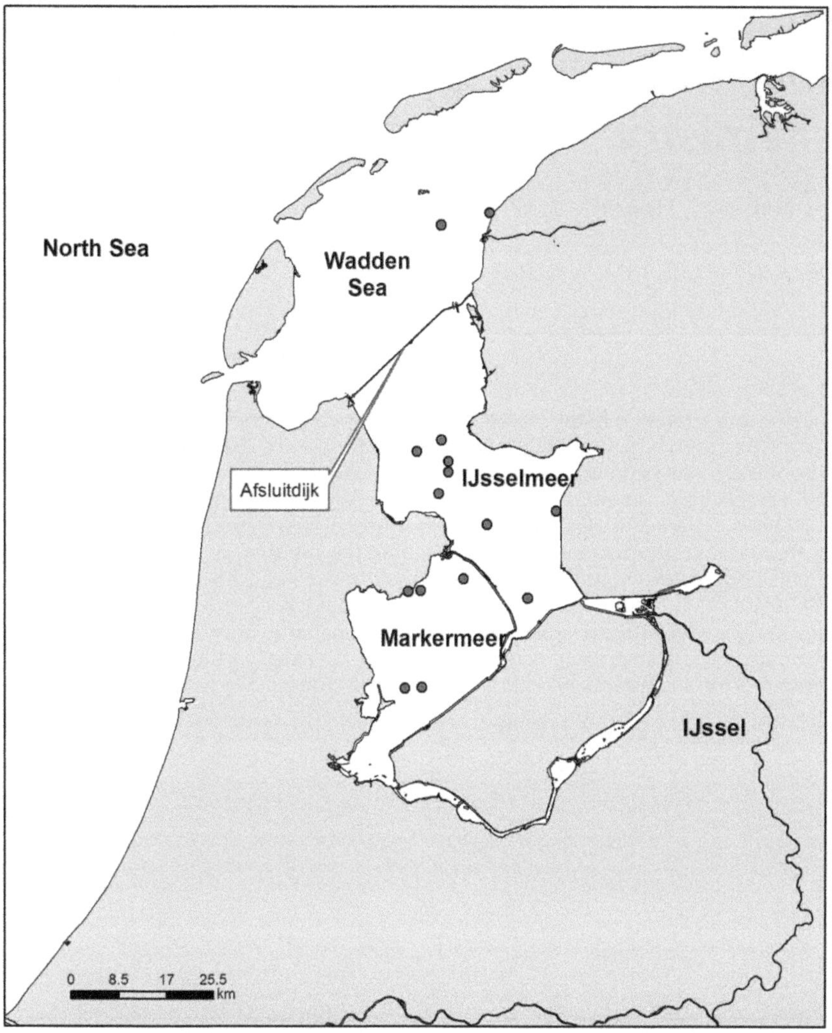

Figure 1. Map of the study area showing the locations where samples were taken.

low biomass, smelt fishery was closed during five years in the last decade (unpubl. data).

The life history of anadromous fish depends on both the quality of their habitat and the connectivity between habitats After the creation of both lakes two possible developments may have occurred: 1) smelt became separated in a landlocked and a diadromous population, with the landlocked population closing its life cycle within the lakes and the diadromous population using spawning sites elsewhere in areas adjoining the Wadden Sea; 2) the two populations are mixed and the diadromous population consists of the individuals washed out through the sluices; returning from the Wadden Sea to Lake IJsselmeer is limited to events when sluices in the dyke are opened.

Given the decline in the landlocked smelt and the potential for recovery of the population, the question we aim to answer is: do anadromous smelt still contribute to the total spawning biomass of smelt in Lake IJsselmeer and Markermeer? Could the landlocked smelt stock be resupplied with Wadden Sea smelt? Insight in the importance of the marine stock for the landlocked stock would contribute to the understanding of the population dynamics of smelt, and consequently of those bird species that rely on smelt as a food source.

A method commonly used to distinguish freshwater from marine origin in fish is comparing the Strontium (^{88}Sr) profiles along sagittal otoliths [8,9,10,11,12]. The utility of this method stems from the fact that otoliths are largely metabolically inert, deposit a thin layer calcium carbonate continuously with age and reflect ambient chemistry [13]. The origin of smelt collected at different locations in Lake IJsselmeer and Markermeer during the peak spawning period was investigated, by comparing the Strontium (^{88}Sr) signature of their sagittal otoliths with those of specimens collected in the Dutch Wadden Sea (Fig. 1). If the diadromous population still contributes to the landlocked population, we expected to find individuals with a marine signal in their ^{88}Sr otolith profiles in the spawning period. The ^{88}Sr concentrations in the edge of the otoliths (the part most recently deposited) of Wadden Sea individuals was used as a marine reference. These concentrations were compared to ^{88}Sr concentrations in specimens collected in both lakes at spawning time.

Methods

Study Area

The estuary Zuiderzee was closed off from the Dutch Wadden Sea sea in 1932 by a dam resulting in the formation of the freshwater lake IJsselmeer (Fig. 1). This was part of a larger plan with a threefold aim: protecting central Netherlands from the effects of the North Sea, increasing the Dutch food supply with new agricultural land and improving water management by creating a lake from the former uncontrolled salt water inlet. This lake is nowadays the largest freshwater body in the Netherlands. It is a shallow lake with a mean depth of 4 m and with up to 8 m deep channels. The lake was further divided by three land reclamation projects, reducing its surface with 40% to about 1900 km^2 [5]. The lake Markermeer with a surface area of 700 km^2, was created after the construction of a dam in 1975, with the objective of reclaiming Lake Markermeer, a plan that was later abandoned.

Lake IJsselmeer gets nutrient input from the River IJssel, a tributary of the River Rhine. Primary production and fish production in Lake Markermeer is lower than in Lake IJsselmeer [14,15]. Both lakes have multiple functions, fisheries, recreation, drinking water supply, transport and as a rest and forage area for birds [16]. Nutrient reductions in the early 1980s lead to a drop in phosphate levels. In Lake IJsselmeer retention time of water is around 6 months, whereas in Lake Markermeer it is around 12 months [5].

Migratory possibilities for smelt to pass the different dams are limited. In the Afsluitdijk, two sluice groups (one in the West and one in the East) control the outflow of fresh water and there are two shipping locks. These are the only direct connections between lake IJsselmeer with the IJssel and Rhine river basin. The westerly sluice complex has three groups of discharge sluices with five tubes in each group. The easterly sluice complex has two groups with five tubes each. Each tube is 12 m wide. Instead of the former opening of 35 km, nowadays only 25 sluices measuring 12 m form the connection between the Wadden Sea to IJsselmeer, a reduction of 99%. The discharge is governed by rain fall, so in spring discharge is generally high and chances for inland migration, especially for weak swimmers, are limited to periods at the end of discharge periods when the water level outside and inside the dike are at a similar level. Migration possibilities have not changed greatly since the closure in 1932. The nearest possibility for smelt to enter an unobstructed freshwater area and spawn is the Ems-Dollard estuary, which is ca 150 km East from the Afsluitdijk. Migration possibilities between Lake IJsselmeer to Lake Markermeer and reverse are limited to two complexes, at the North side there are two shipping locks and a discharge sluice (two tubes of 20 m wide) and at the South side there is a shipping lock and a discharge sluice (six tubes of 16 m wide).

Mean monthly background ^{88}Sr values vary between IJsselmeer and Markermeer, with lower levels during summer than winter and a generally lower level in Markermeer than in IJsselmeer (Fig. 2, data standard monitoring Rijkswaterstaat, concentration measured in surface water). Similar data for the Wadden Sea are not available.

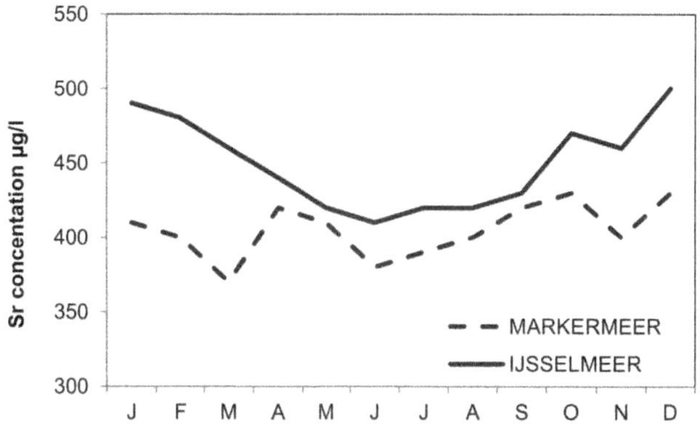

Figure 2. Seasonal variation of Strontium concentrations in Markermeer and IJsselmeer in 2009 (data Rijkswaterstaat).

Sample Collection

In March 2010 smelt were sampled with a 2,5 m beam trawl (rigged with fine-meshed cod-end; 0.5 cm) in Lake IJsselmeer/Markermeer. All known spawning aggregations were sampled (Fig. 1). Adult fish were collected from the Wadden Sea with a 3 m beam trawl (rigged with a fine-meshed cod-end; 20 mm) between April and October 2009 to serve as a marine reference. After capture fish were kept frozen and taken to the laboratory for later dissection. All fish were measured to the nearest mm. In total 110 smelt, 56 from IJsselmeer (collected in 4 hauls/locations) and 54 from Markermeer (4 hauls/locations, length 5–7 cm, two specimens of 11 and 16 cm), and 9 smelt (length 13–23 cm) from the Wadden Sea (2 hauls/locations) were collected. The permission to collect fish in the areas under study was issued by the Ministry of Economic Affairs in the Netherlands, department implementation fisheries regulation (nr 60898). In The Netherlands there is no need for special approval to catch or kill fish by an ethics committee (the experiments on animals act, BWBR0003081). Smelt is a fragile fish and most of them were already dead when taken on board. For the few fish still alive, suffering was ameliorated by immediate freezing.

The total length of all specimens were measured (± 1 mm) and otoliths were collected in the laboratory. One otolith from each specimen was cleaned and embedded in resin. The otoliths were placed on a first layer of nearly cured resin, after which the otoliths were fixed evenly with a second layer of resin. Embedded otoliths were sectioned through the nucleus. The thickness of the slides ranged between 0.5 and 0.6 mm. Three resin strips, containing 10 sectioned otoliths each, were glued to a glass slide with clear resin. No coverslip was used. The slides were stored in covered containers until analysed by laser ablation (LA) inductively coupled mass spectrometry (ICPMS).

Laser Ablation-ICP-MS Analyses

An ICP-MS (Thermo Scientific X-serie2) coupled to a laser (ESI New Wave UP-193 Fast Excimer ArF) was used to determine the concentration of ^{88}Sr along sagittal otoliths. The NIST612 reference material was used for the optimization of following instrumental parameters: the movement rate, the dwell time, the flow of Ar, the flow of He, the output and the repetition rate. The choices were made according to the intensities and stabilities measured. Logically, with a longer dwell time, the stability increases. The configuration set allowed a resolution of 3.7 µm, an ablation diameter of 25 µm (35 µm at the pre-ablation point) and a scan speed of 25 $\mu m s^{-1}$).

In order to better answer our questions, a quantitative analytical method was developed instead of working with $^{88}Sr/^{44}Ca$ ratios. To do this, calibration standards NIST 612 and 610 (lines of 600 µm) were chosen. Finally, to correct the drift and differences in efficiency, the ^{43}Ca was used as the internal standard. Given the size of smelt otoliths (ca 1 mm), the analysis is fast enough and the extent of standards was estimated every 10 samples. To calculate concentrations, normally the standards enclosing the samples were chosen. Given the stability of the measures ($r^2 = 0.99$), all the concentrations were calculated based on a line with all standards measured during one day. In preliminary tests, the resin itself was

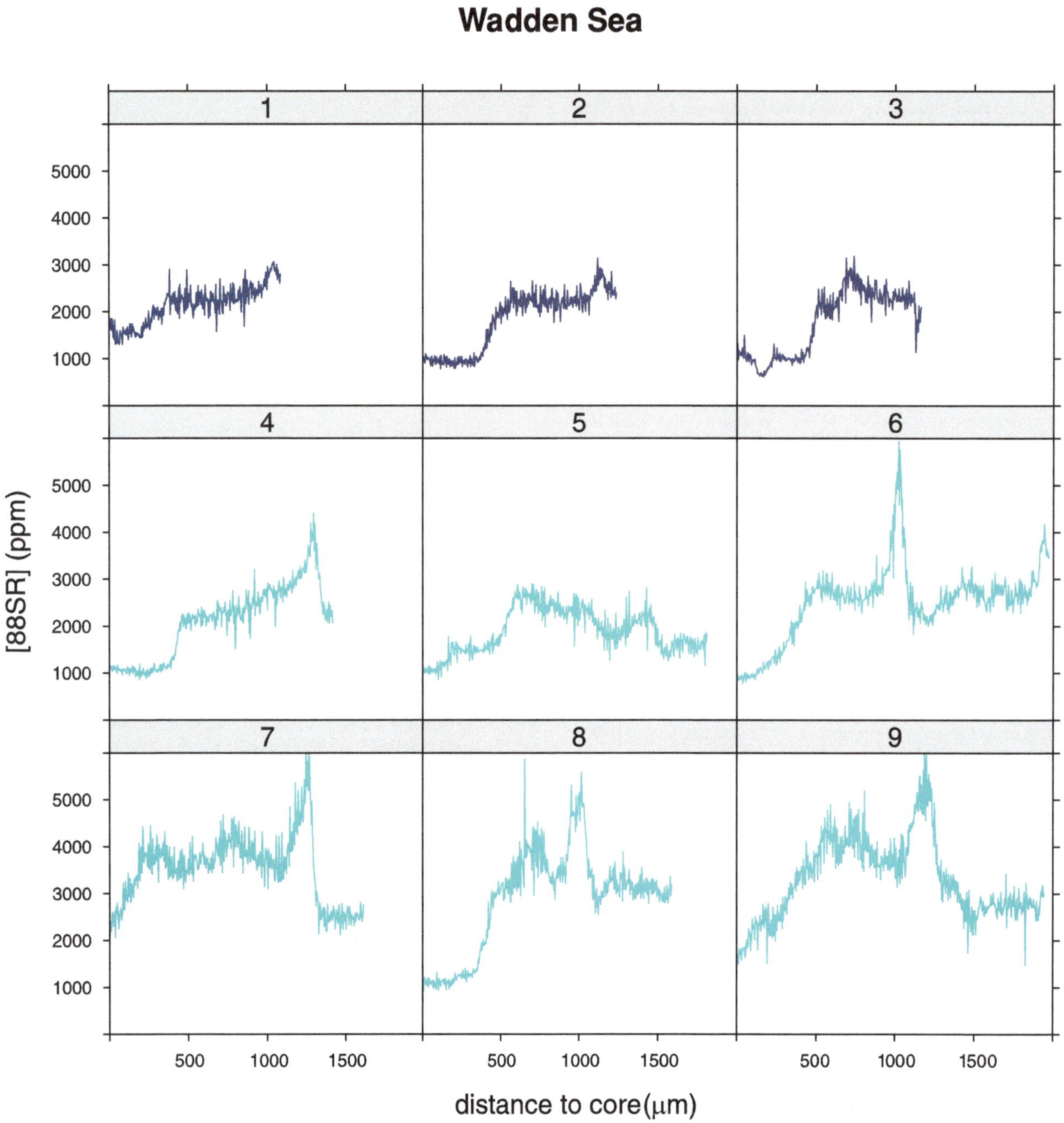

Figure 3. Strontium concentration profiles (in ppm) for smelt caught in the Wadden Sea, going from the nucleus (left) to the outer edge (right). The colours indicate the different hauls, all profiles in one colour originate from the same haul.

measured and it showed a very special signature, completely different from the otoliths. Such signatures were not observed in the otolith measurements. In addition, to avoid the risk of contamination by the resin or by dust, the line of measurement was situated in the middle of the sample and avoided the extreme edges. Moreover a pre-ablation was carried out. Tests indicated that a third ablation did not differ from the second one. Hence, all results refer to the second ablation measurement that was always preceded by a pre-ablation step. All the measures are in TRA mode (time resolved analyses). The laser was led through the core of the otolith.

Data Analyses

Of each profile cross-cutting the otolith from the rostrum, through the nucleus, to the post-rostrum, the part from the nucleus to the rostrum was used for analyses. ^{88}Sr at the cores show the conditions the fish encountered lived when they began to grow. Only the outer 50 measurements (roughly representing 185 μm) of Wadden Sea individuals were used to define the threshold for the marine signal, to ascertain that the ^{88}Sr concentration truly represented the marine environment and to avoid potential ontogenetic effects in ^{88}Sr deposition. Similarly the last 50 measurements (outer edge) of the otoliths of fish caught in the

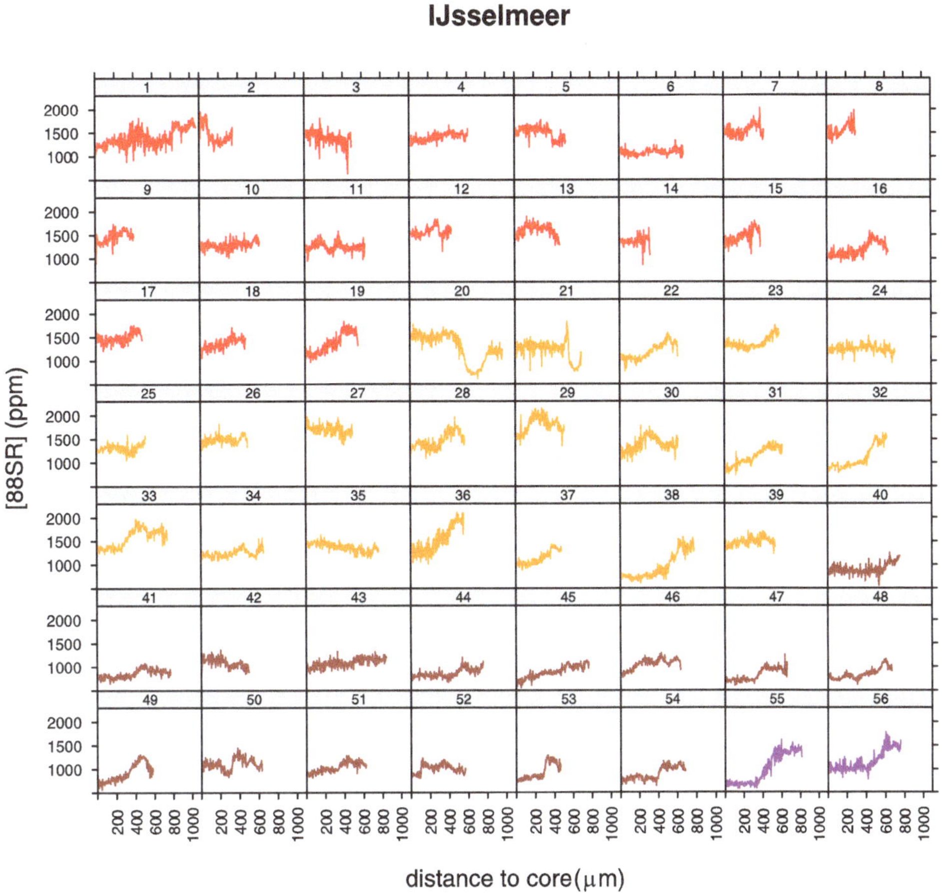

Figure 4. Strontium concentration profiles (in ppm) for smelt caught in Lake IJsselmeer, going from the nucleus (left) to the outer edge (right). The colours indicate the different hauls, all profiles in one colour originate from the same haul.

lakes were chosen to represent the conditions at time and shortly before time of capture.

Specimens originated from different hauls, but a varying number of individuals were collected from each haul. Therefore, the overall differences between the concentrations of ^{88}Sr concentration in the section near the edge of the otoliths from specimens collected in the Wadden Sea and both lakes were tested with a mixed effect model with individual nested within haul as random effect and location as fixed effect [17]. The random intercept allow for correlations between observations from the same site and individual. Because of heterogeneity in the data, ^{88}Sr concentrations were log-transformed prior to analyses.

All data analyses were carried out in R 2.15.1.

Results

^{88}Sr profiles of individuals caught in the Wadden Sea showed strong variations with ^{88}Sr levels at the outer edge generally varying between 2000 and 6000 ppm, but not in the core (Fig. 3). The core values were all below 2000 ppm. The level at the outer edge, the part of the otolith that was most recently deposited, was used as a reference for the marine environment. Eight out of the nine individuals showed mean values greater than 2000 in the final 50 measurements (mean of the final 50 measurements of the 9: 2260 ppm (range 1600–3100 ppm). Individual 5 from the Wadden Sea has a different profile from the other eight, showing values below 2000 ppm at the edge. The overall pattern of this individual is however different from IJsselmeer individuals, that never show values>2000 ppm. The individual profiles of all specimens collected in fresh water showed ^{88}Sr levels ranging between 500 and 2000 ppm. Profiles of specimens in Lake IJsselmeer were more variable than those from Lake Markermeer. In Lake IJsselmeer more than half of the individuals showed lowest values at the core and an increase in ^{88}Sr concentration going from the core to the edge, but never as high as the level identified as marine (2260 ppm) (Fig. 4). Individuals from the 3rd haul (brown, 40–54) had lower values at the core (500–1000) than most specimens from other hauls. Levels varied between 500–2000 ppm. The two largest animals (Fig. 4, individuals 20 and 21) had a distinct pattern: after constant flat profiles both showed a strong dip in ^{88}Sr concentrations to values<1000 ppm. Profiles from Lake Markermeer specimen's otoliths were generally flat, with no clear trend going from the core to the edge and no obvious differences between individuals originating from different hauls (Fig. 5). ^{88}Sr concentrations varied between 500–1200 ppm.

Mean ^{88}Sr concentrations at the otolith edge were significantly higher in the Wadden Sea than in both lakes, but also differed between the two lakes, with lowest values found in Lake Markermeer (Table 1, Fig. 6). At the core ^{88}Srconcentrations did not differ significantly between any of the areas (Table 1, Fig. 6). Although mean core values were generally lower at lake Markermeer the difference was not statistically significant (Fig. 6).

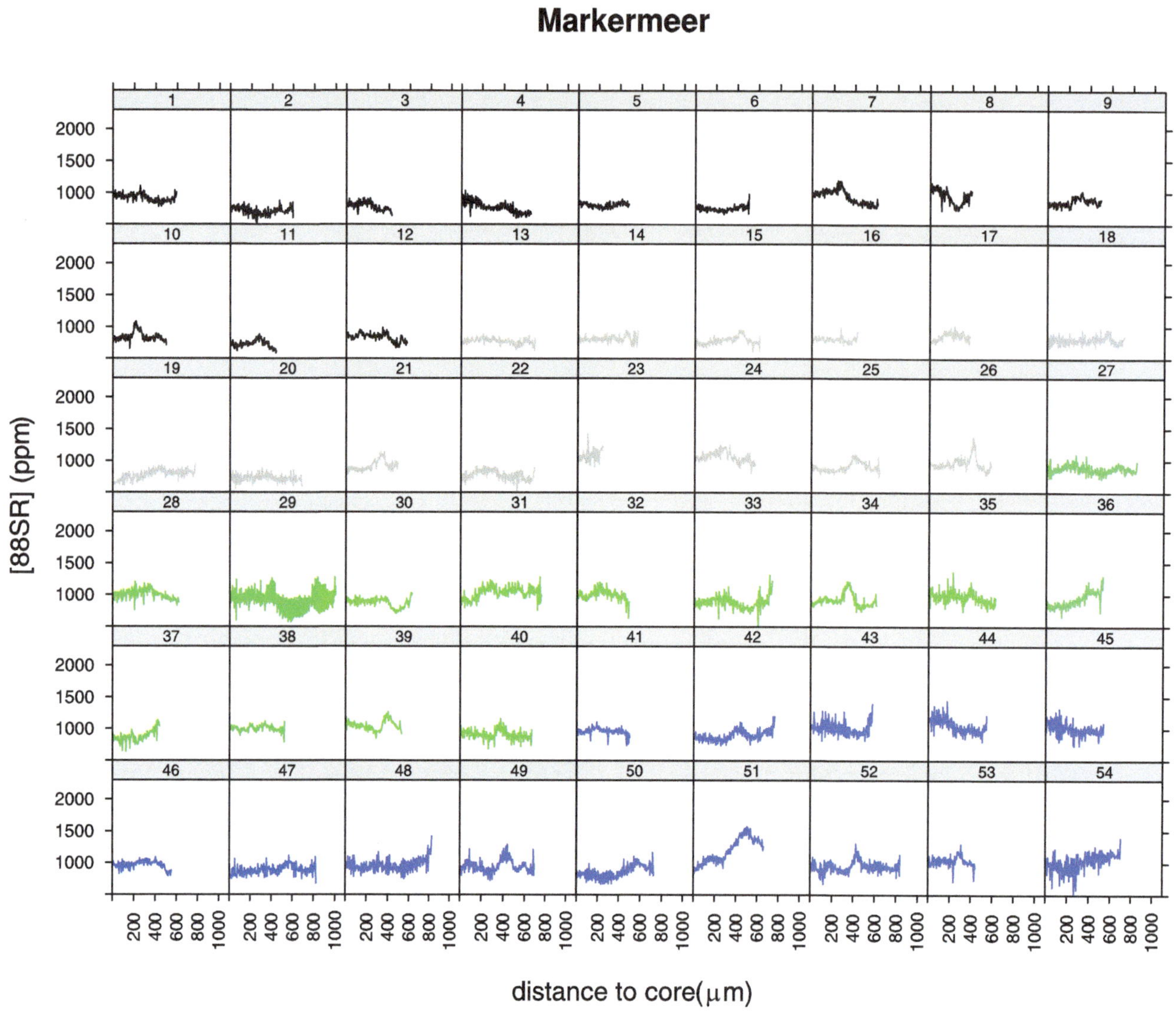

Figure 5. Strontium concentration profiles (in ppm) for smelt caught in Lake Markermeer, going from the nucleus (left) to the outer edge (right). The colours indicate the different hauls, all profiles in one colour originate from the same haul.

Discussion

Connectivity and Dispersal between Freshwater and Marine Environment

The ^{88}Sr concentrations in otoliths of individuals collected in the freshwater lakes were notas high as recorded in the otolith edges of specimens caught in the Wadden Sea. Therefore, based on our study we argue that there is no evidence of a substantial contribution from diadromous individuals to the landlocked smelt population. However, the possibility for a contribution can not be completely excluded. Observations near the discharge sluices,have shown that Wadden Sea smelt occasionally enters Lake IJsselmeer (Kruitwagen unpubl. data). Spawning generally takes place within a few weeks [18]. Due to ship time limitations the specimens were collected in a short period in March only, but covered all known spawning locations in Lake IJsselmeer/Markermeer. The sampling gear used is probably suboptimal for catching larger size classes of smelt if they had occurred. Large smelt were caught in fyke nets or gillnets in Lake IJsselmeer/Markermeer in the late 20th century. Currently, these size classes are never encountered on the market or auction anymore (unpubl.data). A scenario in which the larger diadromous smelt enter Lake IJsselmeer and possibly even Markermeer for a very short period and in low numbers is possibly. Oriadromous smelt may quickly pass Lake IJsselmeer and head to historically known spawning sites, such as River IJssel [19]. In such scenario the recruits still contribute to the population in Lake IJsselmeer. Given the fact that older and larger individuals in most fish species (and also shown in smelt) contribute disproportionally to the recruitment [20,21], a contribution from the diadromous to the landlocked morphs is possible, even if only a few larger sized individuals spawn in Lake IJsselmeer or Markermeer.Since landlocked smelt populations are generally less fecund than estuarine (migrant) smelt populations, a disproportional contribution from estuarine populations could even be expected [22].

The fact that smelt older than two years are rarely observed in Lake IJsselmeer can be explained if they pass the sluices and stay in the Wadden Sea without returning to Lake IJsselmeer. Annually (with a peak in October) large numbers 0 group smelt are passing the sluices to the Wadden Sea (195 tonnes, 65.000.000 individuals,

Table 1. Estimates, estimated degrees of freedom and significance for parameters testing area effects for log^{88}Sr concentrations at the core and the edge of the otoliths in mixed models with individual fish nested within sample as random effects.

response variable	parameter	estimate	standard error	df	t-value	P value
^{88}Sr at edge	intercept	3.1207	0.03942	6142	79.164	0.0000
	area: Markermeer	−0.1715	0.05524	9	−3.105	0.0126
	area: Wadden Sea	0.1866	0.05688	9	3.281	0.0095
^{88}Sr at core	intercept	3.0323	0.04071	6172	74.477	0.0000
	area: Markermeer	−0.0831	0.05672	9	−1.465	0.1767
	area: Wadden Sea	0.0171	0.05995	9	0.285	0.7815

which roughly equals one third of the total IJsselmeer population (Drost & de Witte unpubl.).

^{88}Sr levels for the Wadden Sea fish were highly variable. The Wadden Sea is a very dynamic habitat with strong variations in salinity (24–32 PSU [23]) caused by the tide and the outflow of freshwater through the sluices in the Afsluitdijk. Although ^{88}Sr values from the Wadden Sea are not available, they are likely to vary greatly and given the strong positive correlation between ^{88}Sr and salinity, to follow similar seasonal and daily fluctuations as salinity (http://wwwold.nioz.nl/nioz_nl/ccba2464ba7985d1eb1906b951b1c7f6.php). This high variability in salinity is likely causing the strong fluctuations in ^{88}Sr profiles from Wadden Sea individuals and a small overlap in ^{88}Sr values between Wadden Sea and Lake IJsselmeer measured in the outer 50 of the otoliths. Because Wadden Sea samples consisted of specimens that must have been 2–4 years old [18,22], we expected to find dips in the ^{88}Sr profile (other than the core area that was probably deposited in their early life in fresh water), potentially indicating a spawning event in freshwater. The absence of such dips could mean that either these individuals did not go to a freshwater area for spawning, did not spawn yet or did not grow (and feed) while being in their spawning areas. It is known that females leave the spawning areas earlier than males, but it is not known whether time spent in rivers differs between the sexes [24]. In all spawning does not take more than 1–4 weeks [21]. The very high concentrations of >3000 ppm (Fig. 3) may indicate the influence of North Sea water inflow of higher salinity.

Exchange between Freshwater Areas

The two individuals with a dip in their profile later in life (Fig. 4, 20 and 21) suggest that they moved towards water with a lower ^{88}Sr profile than Lake IJsselmeer. The ^{88}Sr level is similar to that found in smelt in Lake Markermeer. Another possibility is that they swam up the river IJssel, an important spawning location in former days when the lakes were not closed off from the sea yet [19]. There are records of smelt (ca 20 cm) swarming at the River IJssel mouth and migrating upstream for several tens of kilometres dating back to 1840s [19]. A dedicated search for smelt during the spawning period in the River IJssel mouth, and analyses of their otoliths ^{88}Sr profiles could clarify if this scenario currently still happens. These two individuals and the Lake IJsselmeer individuals 38 and 40–55 that show core values closer to the range of Markermeer individuals, could be an indication of exchange between Lake IJsselmeer and Lake Markermeer. The only group that has rather low ^{88}Sr values (around 1000 ppm) are the ones caught in the east of Lake IJsselmeer (3rd sample in Fig. 4); they cannot be distinguished from Markermeer individuals. The flat ^{88}Sr profiles found in Markermeer probably indicates these fish belong to a very local population. ^{88}Sr profiles were very similar within and between hauls. Similar profiles within individuals of one haul but different from another haul can be expected if fish stay together in one school for most of their life and encounter similar circumstances as they move around. Alternatively the similar patterns are caused by the ^{88}Sr signature of the specific region, which would indicate that smelt do not move around a lot, a finding which shows agreement with a closely related species the rainbow smelt *Osmerus mordax* [25,26,27]

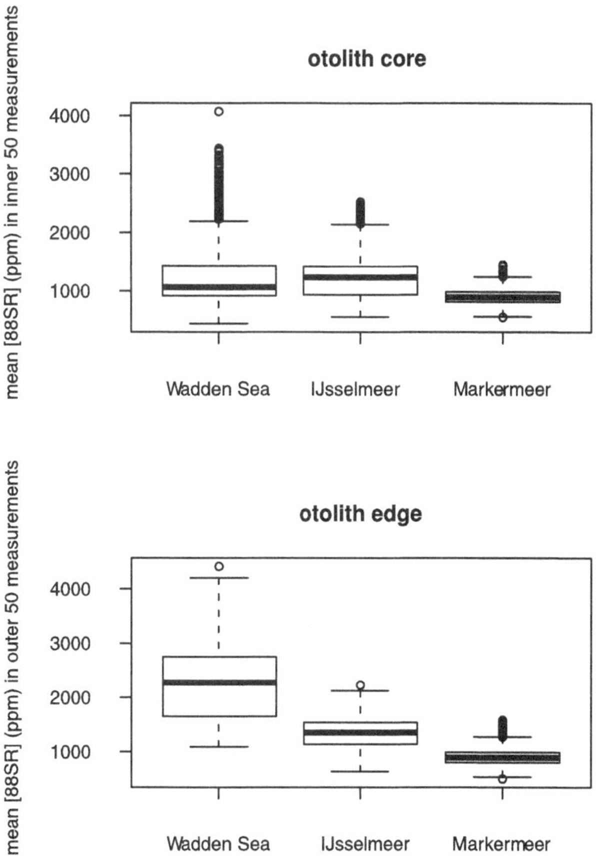

Figure 6. Mean Sr (in ppm) concentrations in otoliths of the inner 50 measurements (core) and the outer 50 measurements of smelt (edge) caught in the Wadden Sea, Lake IJsselmeer and Lake Markermeer.

Implications for Management of Landlocked Population

Our results show that under the current management there is no indication of a strong contribution of the Wadden Sea smelt to the populations of both lakes. Currently the possibilities for smelt to re-enter Lake IJsselmeer through the sluices at either Kornwerderzand or Den Oever are limited to periods when sluices are opened. Passage of the discharge sluices towards the Wadden Sea is only possible shortly after opening and towards Lake IJsselmeer shortly before closure. D iadromous smelt tends to aggregate in front of the sluices in March, where in 2009 ca 50 tonnes of adult fish were caught in the commercial smelt fisheries (Drost & de Witte unpubl.). Given the limited entrance possibilities, management to enhance the contribution of diadromous smelt to the Lake IJsselmeer population should concentrate on improving the opportunity to enter by adjusting the discharge regime or increasing the number of entrance points in the spawning period.

Acknowledgments

We want to thank the crew of the Tjeûkemar (property of the Netherlands Institute of Ecology, NIOO, Wageningen, The Netherlands) that helped in collecting the fish samples. H. van der Veer and H. Witte from the Royal Netherlands Institute for Sea Research kindly provided Wadden Sea specimens. N. van der Meeren (IMARES) prepared the otoliths.

Author Contributions

Conceived and designed the experiments: IT MK HW WB MDG. Performed the experiments: JN MK. Analyzed the data: IT JN. Contributed reagents/materials/analysis tools: MK JN WB. Wrote the paper: IT.

References

1. Lassalle G, Crouzet P, Rochard E (2009) Modelling the current distribution of European diadromous fishes: an approach integrating regional anthropogenic pressures. Freshwater Biology 54: 587–606.
2. Nellbring S (1989) The ecology of smelts (Genus Osmerus): a literature review. Nordic J Freshw Res 65: 116–145.
3. Kottelat M, Freyhof J (2007) Handbook of European freshwater fishes. Cornol, Switzerland: Publications Kottelat.
4. Willemsen J (1977) Population dynamics of percids in Lake IJssel and some smaller lakes in the Netherlands. J Fish Res Bd Can 34: 1710–1719.
5. De Leeuw JJ, Dekker W, Buijse AD (2008) Aiming at a moving target, a slow hand fails! 75 yeasr of fisheries management in lake IJsselmeer, the Netherlands Journal of Sea Research 60: 21–31.
6. van Roomen M, Hornman M, Flink S, Langendoen T, van Winden E, et al. (2012) Flyway-trends for waterbird species important in Lakes IJsselmeer and Markermeer. SOVON. 2012-22 2012-22.
7. Mous PJ (2004) Interactions between fisheries and birds in IJsselmeer, The Netherlands: Wageningen University, The Netherlands.
8. Borcherding J, Pickhardt C, Winter HV, Becker JS (2008) Migratory history of North Sea Houting *Coregonus oxyrinchus* (L.) caught in Lake IJsselmeer (The Netherlands) inferred from scale transects of 88Sr:44Ca ratios. Aquatic Sciences 69: 47–56.
9. Geffen AJ, Nash RDM, Dickey-Collas M (2011) Characterization of herring populations west of the British Isles: an investigation of mixing based on otolith microchemistry. ICES Journal of Marine Science: 1447–1458.
10. Morais P, Dias E, Babaluk J, Antunes C (2011) The migration patterns of the European flounder *Platichthys flesus* (Linnaeus, 1758) (Pleuronectidae, Pisces) at the southern limit of its distribution range: Ecological implications and fishery management. Journal of Sea Research 65: 235–246.
11. Arai T, Chino N (2012) Diverse migration strategy between freshwater and seawater habitats in the freshwater eel genus Anguilla. Journal of Fish Biology 81: 442–455.
12. Walther BD, Limburg KE (2012) The use of otolith chemistry to characterize diadromous migrations. Journal of Fish Biology 81: 796–825.
13. Secor DH, Hendersonarzapalo A, Piccoli PM (1995) Can otolith microchemistry chart patterns of migration and habitat utilization in anadromous fishes. Journal of Experimental Marine Biology and Ecology 192: 15–33.
14. Mous PJ, Van Densen WLT, Machiels MAM (2004) Vertical distribution patterns of zooplanktivorous fish in a shallow, eutrophic lake, mediated by water transparency. Ecology of Freshwater Fish 13: 61–69.
15. van Densen WLT, Cazemier WG, Dekker W, Oudelaar HGJ (1990) Management of the fish stock in lake IJssel, the Netherlands. In: Steinmetz B, Hughes RH, editors. Proceedings of a symposium organized by the European Inland Fisheries Advisory Commission. Goeteborg, Sweden: Pudoc. pp. 313–327.
16. Buijse AD, van Eerden MR, Dekker W, van Densen WLT (1993) Elements of a trophic model for IJsselmeer (the Netherlands), a shallow eutrophic lake. In: Christensen V, Pauly D, editors. Proceedings of the Trophic models of aquatic ecosystems, ICLARM Conference Proc. pp. 90–94.
17. Zuur AF, Ieno EN, Walker NJ, Saveliev AA, Smith GM (2009) Mixed effects models and extensions in ecology with R. New York: Springer.
18. Doherty D, McCharty TK (2004) The ecology and conservation of European Smelt (*Osmerus eperlanus*) from Waterford estuary, in southeastern Ireland. Biology & environment, Proceedings of the Royal Irish Academy 104B: 125–130.
19. van den Ende WP (1847) Verslagen van de Vereeniging tot Bevordering der Inlandsche Ichthyologie. Eerste deel. An. Nijhoff, Arnhem, the Netherlands.
20. Berkeley S, Hixon M, Larson R, Love M (2004) Fisheries sustainability via protection of age structure and spatial distribution of fish populations. Fisheries 29: 23–32.
21. Hutchinson P, Mills DH (1987) Characteristics of spawning-run smelt *Osmerus eperlanus* (L.), from a scottish river, with recommendations for their conservation and management. Aquaculture and fisheries management 18: 249–258.
22. Belyania TJ (1969) Synopsis of biological data on smelt, *Osmerus eperlanus* L. Rome: Food and Agriculture Organisation of the United Nation.
23. van Aken HM (2008) Variability of the salinity in the western Wadden Sea on tidal to centennial time scales. Journal of Sea Research 59: 121–132.
24. Quigley D, Igoe F, O'Connor W (2004) The European Smelt *Osmerus eperlanus* L. in Ireland: General biology, ecology, distribution and status with conservation recommendations Biology & Environment: Proceedings of the Royal Irish Academy 104: 57–66.
25. Bradbury IR, Campana SE, Bentzen P (2008) Otolith elemental composition and adult tagging reveal spawning site fidelity and estuarine dependency in rainbow smelt. Marine Ecology-Progress Series 368: 255–268.
26. Bradbury IR, Coulson MW, Campana SE, Baggs E, Bentzen P (2009) Postglacial recolonization and the loss of anadromy in rainbow smelt from coastal Newfoundland. In: Haro A, Smith KL, Rulifson RA, Moffitt CM, Klauda RJ et al., editors. Challenges for diadromous fishes in a dynamic global environment. pp. 79–96.
27. Bradbury IR, DiBacco C, Thorrold SR, Snelgrove PVR, Campana SE (2011) Resolving natal tags using otolith geochemistry in an estuarine fish, rainbow smelt *Osmerus mordax*. Marine Ecology-Progress Series 433: 195-U576.

7

Perceptions of Rule-Breaking Related to Marine Ecosystem Health

Matthew J. Slater[1*¤], Yunus D. Mgaya[2], Selina M. Stead[1]

1 School of Marine Science and Technology, Newcastle University, Newcastle upon Tyne, Tyne and Wear, United Kingdom, **2** Department of Aquatic Sciences and Fisheries, University of Dar es Salaam, Dar es Salaam, Tanzania

Abstract

Finding effective solutions to manage marine resources is high on political and conservation agendas worldwide. This is made more urgent by the rate of increase in the human population and concomitant resource pressures in coastal areas. This paper links empirical socio-economic data about perceptions of marine resource health to the breaking of marine management rules, using fisheries as a case study. The relationship between perceived rule-breaking (non-compliance with regulations controlling fishing) and perceived health of inshore marine environments was investigated through face-to-face interviews with 299 heads of households in three Tanzanian coastal communities in November and December 2011. Awareness of rules controlling fishing activity was high among all respondents. Fishers were able to describe more specific rules controlling fishing practices than non-fishers ($t = 3.5$, $df = 297$, $p < 0.01$). Perceived breaking of fishing regulations was reported by nearly half of all respondents, saying "some" (32% of responses) or "most" (15% of responses) people break fishing rules. Ordinal regression modelling revealed a significant linkage ($z = -3.44$, $p < 0.001$) in the relationship between respondents' perceptions of deteriorating marine health and their perception of increased rule-breaking. In this paper, inferences from an empirical study are used to identify and argue the potential for using perceptions of ecosystem health and level of rule-breaking as a means to guide management measures. When considering different management options (e.g. Marine Protected Areas), policy makers are advised to take account of and utilise likely egoistic or altruistic decision-making factors used by fishers to determine their marine activities.

Editor: Konstantinos I. Stergiou, Aristotle University of Thessaloniki, Greece

Funding: This work was funded by the Leverhulme Trust, UK limited. The funders had no role in study design, data collection and analysis, decision to publish, or preparation of the manuscript.

Competing Interests: The authors have declared that no competing interests exist.

* E-mail: m.slater@ncl.ac.uk

¤ Current address: Department of Marine Aquaculture, Institute for Marine Resources GmbH, Bremerhaven, Bremen, Germany

Introduction

Rule-breaking by resource users is a common explanation for fisheries management failing to achieve sustainability and conservation goals worldwide. An increasing amount of information shows significant and negative impacts of fishers' rule-breaking on fish stocks, national economies and global trade [1]. However, finding solutions to achieve compliance to support sustainable marine resources management remains a global unresolved challenge. Understanding the reasoning, perceptions, pressures and socio-economic drivers behind decisions made by fishers, at all scales, to break fishing rules remains central to achieving compliance and viable fish stock management [2,3]. Understanding and integrating the context-specific characteristics of rule-breaking allows management measures to be tailored to address rather than compound local issues.

Research to date has attempted to describe rule-breaking's justification through moral objections or through fisher's reasoned decisions to comply or not. In the first instance rule-breaking by fishers is favoured by their view that rules are unfair or illogical, that rules are contrary to their conscience, or that rule-breaking is morally acceptable due to higher needs, for example, one must catch fish by any means in order to feed one's family [4]. Equally, fishers may break rules set by authorities they do not respect or accept as ruling their fishing areas (lack of co-governance). Yet, despite the agreed importance of belief systems and moral aspects in rule-breaking, it is most commonly argued that fisheries rule-breaking stems from simple economic cost:benefit analysis [1,4]. Using similar assumptions to rational choice theory in criminology, Sumaila et al [1] argue that fishers decide as individuals in a calculated manner whether it fulfils their immediate economic needs and is worthwhile to risk breaking the rules controlling fishing [5]. When applying this model, fishers may decide whether the benefits of increased catch and/or reduced catch effort obtained through breaking rules controlling fishing will be greater than: 1) cost of being caught (fines/confiscation), including an inherent calculation of likelihood of being caught and cost of avoidance behaviour required; 2) perceived benefit of complying; 3) social implications of fisher's involvement in rule-breaking. Sumaila et al. [1], argue that in most cases this cost:benefit analysis favours the decision to break rules.

If these assumptions are correct, the above model predicts a global economic weighting towards rule-breaking which can be explained by the relatively low chances of being caught breaking rules due to the geographical size of marine areas required to be monitored, insufficient resources to enforce rules and lack of economic viability of enforcement [6]. While enforcement can be effective in smaller defined marine areas, such as no-take zones or small marine protected areas, most nations have insufficient resources and capacity to effectively enforce fishing laws over the majority of their territorial waters [1,7].

Increasingly draconian and complex laws along with heavy punishments also generally lead to a continuous cycle of increased rule-breaking among fishers due to confusion, resentment and loss of respect for rules [8]. Furthermore, there exists a lack of self-enforcement of fishing rules, i.e. fishermen are generally unlikely to inform on their rule-breaking colleagues. Commonly, with external actors unable to enforce rules, and a lack of enforcement by other fishers, individual fishers are solely responsible for deciding if a rule will be broken [8]. Thus studying socio-economic drivers that underpin individual decision-making behaviour can identify a range of factors which may be linked to likelihood of rule-breaking under different local conditions. This information can be applied to a wide range of marine management scenarios globally as part of using context-specific knowledge to promote and gain support for specific rules.

Where enforcement is ineffective or of secondary importance, individual fisher's compliance can be dependent on personal conviction on the one hand and on societal expectation on the other. Fishers may be more likely to comply if they feel social pressure or moral obligation to accept them [2,9]. If community and individual/household conditions are appropriate, then perceived and actual involvement of individual fishers in rule-making and management (i.e. co-management and co-enforcement) can be effective in making rules more logical and morally acceptable to fishers, and also increases awareness of rules [10]. Co-management has been credited with supporting positive effects in compliance within coastal resource management programmes worldwide, despite the measures taken often offering no economic benefits on the fishers who support them [11,12].

In the current research we argue, individual perceptions (what individuals perceive or believe they perceive as states, such as marine ecosystem health, or occurrences) while representative and subjective, can influence reasoning and ultimately behavioural choices. To illustrate, an awareness and acceptance of the problem related to a specific rule (in this case health of a threatened marine fish stock) has been shown to have a positive effect on willingness to comply with that rule, although the underlying stimuli can be complex [13,14]. Jager et al. [4] showed that concern about levels of fish depletion increased fishers' belief that rule-breaking was unacceptable in the case of fishing for food, although this did not extend to the fishers own target species when examined in view of expected fishing income from commercial stocks. Fishers' acceptance of a specific environmental problem or overall stock degradation and the clearly communicated association of that problem with fishing rules intended to counter it may increase individual willingness to comply. It is unclear whether this applies at a societal level, resulting in pressure or expectation to comply.

Effective fisheries management, including clear communication about enforcement efforts and regulatory measures can benefit from being formed on the basis of information about how fishers and societies they belong to perceive rule-breaking. The current study aims to contribute to understanding of the reasoning, perceptions and socio-economic drivers underlying fishers' decision-making about how they fish, specifically about breaking of fishing rules.

This paper reports the findings from interviews with 299 heads of households in three fishing communities in the United Republic of Tanzania. The perceptions of respondents, both fishers and non- fishers, within fishing communities, that is, the society fishers belong to, are analysed specifically to isolate the social variables statistically linked to perceived rule-breaking. Identifying relevant perceptions and associations to compliance allows a discussion of potential effects on behavioural choice. From a proactive management stance this knowledge can be used to map out scenarios to be presented to local communities to communicate possible consequences of their actions. This can allow informed choice, for example, about which conservation measures are more likely to have favourable outcomes and thus be supported. The overall aim is to understand relevant socio-economic variables that may find application in improving fisheries management, thus supporting recommendations for future use in management options and policy making.

Methods

2.1 Ethics statement

Ethics approval was sought and obtained for this study through submission of an ethics assessment to the Newcastle University Ethical Committee by which all methods including verbal consent were deemed acceptable and of no ethical concern. Researchers obtained verbal consent from participants before conducting household surveys. In this case consent was documented only in completion of the interview cover sheet (i.e. a survey ID was assigned and the interviewer recorded their own name and began interviewing). The verbal consent procedure ensured (as a minimum) participants were informed about the survey, its purpose, and how the data would be utilized and were informed they would receive reports on findings at community presentations. Written consent was not sought as it was possible that the low literacy level of a number of respondents would have made written consent unrepresentative of actual consent. Participant's names were not recorded nor were other personal details which could be used to identify individual respondents.

2.2 Survey design

Face-to-face interviews were conducted with 299 heads of households in three coastal villages - Ununio, Kunduchi and Buyuni -in the United Republic of Tanzania in November and December 2011 (Figure 1). Study sites varied according to economic status, proximity to the coast and dependence on small-scale fisheries. To summarise:

- Ununio is 20 kilometres north of Dar-es-Salaam (estimated population 1050; 168 households).
- Kunduchi is 15 km north of Dar-es-Salaam (estimated population 1370; 212 households).
- Buyuni is 50 km south of Dar-es-Salaam, (estimated population 1100; 204 households).

Ununio and Kunduchi are ('diffuse') peri-urban fishing villages on the outskirts of Dar es Salaam, with encroachment of urban expansion on land area for agriculture and construction [15]. Buyuni is rural, and requires 3–4 hours car travel to reach from Dar es Salaam. There is no known history of co-management or participation in rule-making within the three fishing villages, there is however a no-take Marine Protected Area on Mbudya Island in close proximity to Kunduchi which is partially operated by ex-fishers.

Villages were mapped using satellite images as a reference and ground-proofed by in-field observations to create full hand-drawn maps of all roads, pathways and dwellings. Maps were then used to systematically sample (numbered) households within each community. The head of household of every second house was interviewed.

Figure 1. Map of Tanzania with expanded area central coast of Tanzania indicating position of coastal villages from North to South: A. Ununio, B. Kunduchi, and C. Buyuni. (Source https://www.cia.gov/library/publications/cia-maps-publications/Tanzania.html).

'Household' was defined as a unit of people that share a house [16]. Heads of households were interviewed where possible as decision-makers considered to hold more detailed information about current livelihoods, income and socio-economic characteristics. If the head of household was not available at the second visit, interviewers requested to interview an adult from the household fully informed about the household's full range of income-related activities and livelihoods.

Data collected from the interviews described respondents' dependence on marine resources and perceptions of marine health. Respondents discussed their current economic status and employment along with awareness of rules to control fishing activity (open) and were asked questions focussed on whether people break these rules. Response options were not prescriptive or numerical but were designed to provide a simple ordinal-scale estimate of perceived levels of rule-breaking. Thus response options to the question "Do people break these rules?" were: Yes, most people break them; Yes, some people break them; Yes, but very few people break them; No, no-one breaks them.

Household income, possessions and utilities, such as electricity, mobile phone and other physical assets such as house construction types (e.g. cement) combined with land and house ownership were recorded and used as indicators of wealth.

2.3 Material Style of Life

A material style of life (MSL) unitary measure was calculated on the basis of rotated component weightings obtained using a factor analysis with Varimax rotation applied to all data for presence or absence of key assets [17]. Key assets examined were all household possessions and utilities examined in interviews. These were equivalent to the selection in previous analysis of fisher village survey data as detailed by Cinner et al. [17] with the additions made in the current study including land ownership, ownership of mobile phones, availability of water services and availability of electricity. Assets with weightings between 0.3 or −0.3 in first and second components after Varimax rotation were excluded. Remaining factor weightings for all individuals were summed to create a compact MSL score for each respondent household (Figure 2).

2.4 Statistical analysis and ordinal regression modelling

A cumulative link model [18] was fitted to the data with responses to the ordinal variable 'perception of rule-breaking' (for rules controlling fishing) as the dependent variable with four ordered response categories as follows:

4 - Yes, most people break the rules

3 - Yes, some people break them

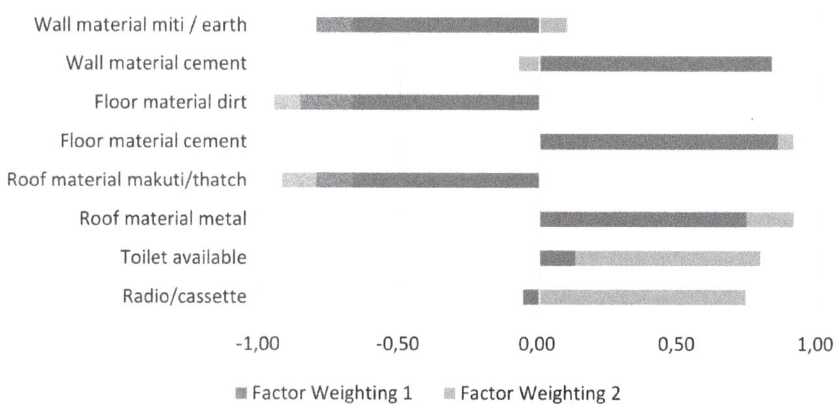

Figure 2. Factor weightings following Varimax Rotation, values indicate contribution of factors (if present) to overall Material Style of Life score, n = 299.

Table 1. Mean values of demographic data, response variables and key socio-economic indicators by village (n = 299, TZS = Tanzanian Shilling, HH = Household, Env. = environmental).

	Village			
	Buyuni	Kunduchi	Ununio	All villages
Factor/variable	Mean	Mean	Mean	Mean
Age	38 (13)	39 (13)	36 (13)	38 (13)
Years of education	5 (3)	5 (4)	6 (4)	5 (4)
Weekly income per person [23]	13,575 (12,298)	23,902 (22,128)	23,070 (22,515)	19,771 (19,686)
Total household inhabitants	6 (3)	5 (4)	6 (4)	5 (4)
Material Style of Life	−0.3 (1.9)	2.2 (0.8)	1.9 (1.2)	1.2 (1.8)
Social network strength	1(1)	1 (1)	1 (1)	1 (1)
	% responses	% responses	% responses	% responses
Respondent - Female	39.1	53.5	52.3	47.8%
HH main income fishery	35.2	68.1	46.5	49.5%
HH main income other	64.8	31.9	53.5	50.5%
Fishing reliable income? - yes	73.2	69.2	61.9	68.4
Immigrant – born elsewhere	55.5	78.2	63.6	65.6

2 - Yes, but very few people break them
1 – No, no-one breaks them

The following independent variables were selected following a critical appraisal of relevant literature and included in the cumulative link model, constructed using the ordinal package in R Statistical software (Table 1; Figure 3; Figure 4):

1. Is fishing a reliable source of income? (yes/no)
2. Estimate of environmental health of the sea (much better/better/same/worse/much worse)
3. Social network strength (Rating 1–3 depending on participation/decision-making in community)
4. Material style of life (MSL)
5. Weekly income per person in household (Tanzanian Shillings - TZS)
6. Occupation (fisher, farmer, businessperson, fishmonger, other)
7. Years of formal education (years)
8. Age (years)
9. Gender (male/female)
10. Immigrant (yes/no)

In order to compare the perceptions of fishers with that of the community as a whole, a separate cumulative link model was constructed incorporating only the responses of commercially active fishers (those who derive income from their catch) with the same variables excluding the variable Occupation.

Each ordinal regression model was refined by removal of the variable with the highest p-value (non-significant variables) stepwise and by refitting the model accordingly. The empirical identifiability (goodness of definition) was determined on the basis of the condition number of the Hessian; numbers below 10^4 were considered an acceptable level of definition.

Where applicable data were tested for normality and homogeneity of variances using Shapiro-Wilk and Levene's test respectively

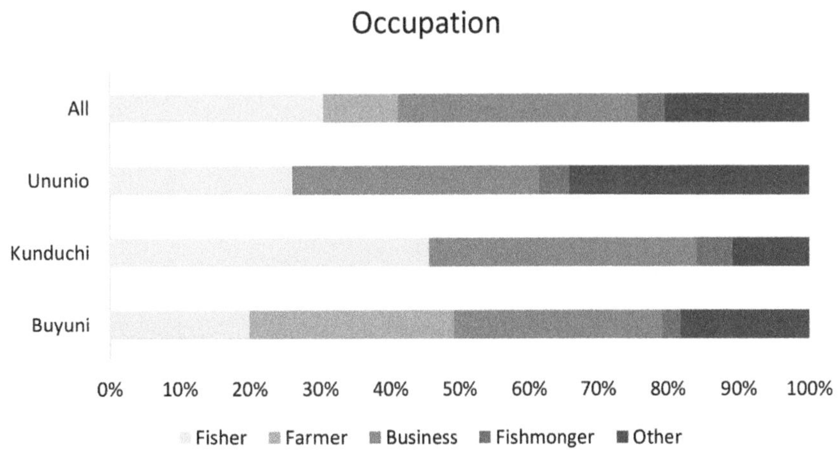

Figure 3. Distribution of occupations by village and all respondents, n = 292.

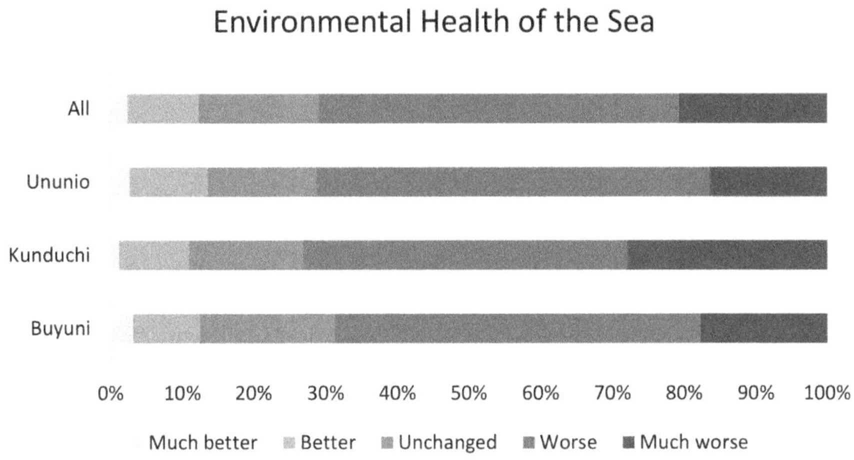

Figure 4. Distribution of individual response by village and all respondents for estimate of environmental health of the sea, n = 250.

before application of parametric statistical tests. An independent samples t-test was applied to compare means of rule awareness between fishers and non-fishers.

Results

The following results section begins by establishing the level of dependence on marine resources and socio-economic status of the interviewed villages. Comparative awareness of rules controlling fishing and respondents' perceptions on the level of rule-breaking amongst community members are then introduced. Finally an ordinal regression model reveals how explanatory variable(s) retained in the model explain levels of perceived rule-breaking.

3.1 Marine resource dependence and socioeconomic status of communities

All villages exhibited high dependence on marine and coastal resources, with between 35.2% (Buyuni) and 68.1% (Kunduchi) of households in each village ranking fishing or fish marketing as their main income source (Table 1; Figure 3). Respondents in peri-urban villages Kunduchi (TZS23,902±22,128 SD) and Ununio (TZS 23,070±22,515 SD), reported higher per capita income than the rural villagers of Buyuni (TZS 13,575±12,298SD) however values did not differ significantly. Peri-urban villages had more developed infrastructure – piped water and electricity - due to their proximity to urban networks (Table 1). Respondents from Ununio had the highest mean number of years of education (6.07±3.9 SD) and social network (measured as combined score for participation in groups and involvement in decisions at group and village level) strength score (1.1±1.0 SD). The majority of respondents (68.4%) felt that fishing was a reliable source of income. More than half of respondents (65.6%) were not born in the village they were currently living in and 70.9% of respondents felt that the environmental health of the sea was worse or much worse than in previous years (Figure 4).

3.2 Awareness of rules controlling fishing activity

The majority (86.0%) of respondents described at least one rule controlling fishing activity (Table 2). In open and multiple response questioning, fishers described significantly ($t = 3.5$, $df = 297$, $p < 0.01$) more specific rules on average (mean = 1.7±0.1 SE) than non-fishers (mean = 1.4±0.1 SE). The most common rule named by all respondents was the ban on use of explosives in fishing (136 mentions/29.7% of total mentions) followed by need for an appropriate fishing license (105 mentions/22.9% of total mentions) for artisanal and commercial fisheries and limits to net mesh size for net fishing (81 mentions/17.7% of total mentions). Bans on fishing with poisons such as cyanide (13 mentions/2.8% of total mentions) and the recent ban on beach seine fishing methods (13 mentions/2.8% of total mentions) were mentioned least frequently (Table 2).

3.3 Perceptions on breaking of rules controlling fishing activity

When asked whether people break the afore-mentioned rules, 41.0% of respondents answered that 'only very few people or individuals' break the rules, 31.5% of respondents answered that 'some people break the rules', 14.9% answered that 'most people break the rules' and the remaining 12.6% responded that 'no-one breaks fishery rules and regulations' (Table 2). A total of 77 interviewees or 26% of total interviewees were not willing or able to provide a response to this question.

3.4 Ordinal regression modelling

Stepwise removal of explanatory variables to refine the ordinal regression model resulted in a final model - for all respondents - which retained three significant explanatory variables:

1. Estimate of environmental health of the sea (much better/better/same/worse/much worse)
2. Immigrant (yes/no)
3. Gender (male/female)

Respondent's estimate of environmental health of the sea exhibited the greatest effect size and significance in the final ordinal regression model (Table 3). The final model predicts perception of rule-breaking to be significantly higher among respondents who; 1) think the sea is less healthy, 2) are immigrants, and 3) are male (Table 3).

Stepwise removal of explanatory variables to refine the second ordinal regression model, using responses of fishers only (100% male respondents), resulted in a final model - for all respondents - which retained two significant explanatory variables:

1. Estimate of environmental health of the sea (much better/better/same/worse/much worse)

Table 2. Rule awareness (n = 299), rules mentioned/described (n = 299) and respondent perceptions of rule-breaking (n = 222).

Variable	Response category	No. of responses (n)	% responses
Number of rules described	0	42	14.0
	1	85	28.4
	2	147	49.2
	3	21	7.0
	4	4	1.3
Rule description/mentions	No explosives	136	29.7
	Licensing	105	22.9
	Net mesh limits	81	17.7
	No spears/guns	19	4.1
	No beach seine	13	2.8
	Equipment bans	18	3.9
	Fish size limit	19	4.1
	No poison fishing	13	2.8
	Methods bans	17	3.7
	Others	37	8.1
Do people break fishing rules?	Yes, most of them	33	14.9
	Yes, some	70	31.5
	Yes, but very few	91	41.0
	No, no-one	28	12.6

2. Immigrant (yes/no)

The results of the final and refined model predicts perception of rule-breaking to be significantly higher among immigrant fishers and those fishers who think the sea is less healthy (Table 4).

Discussion

The current research identifies for the first time a statistically significant linkage between increased individual perceptions of fisheries rule-breaking and individual perceptions of worsening marine environmental health, based on empirical evidence. The findings are valuable in identifying this relationship, dependent variables and perceptions that influence resource users' environmental problem awareness and thus may influence their behavioural patterns. This information about natural resource health, perceptions and the way in which this knowledge may influence decision-making can be used to improve management and policy formulation. This can result in targeted individuals being more likely to support rules, especially if they are tailored and explained within the context of local priorities derived from empirical data that local people were involved in providing and can relate to.

With governments around the world increasing implementation of fishery and marine management measures as conservation tools, this study is timely for policy advisers in offering advice on what variables should be determined to understand the likely efficacy of rules and measures controlling stakeholders. We predict that this information, if considered and built upon, will help the formulation of rules that are more likely to gain support and achieve greater compliance among communities socially and economically dependent on vulnerable marine resources such as over-exploited fisheries. Given the limited resources and capacity available for enforcement, a targeted approach is more likely to be successful in

Table 3. Ordinal regression model (logistic) all respondents' responses to perception of rule-breaking for rules controlling fishing.

(a)	Estimate	Standard Error	z value	Pr(>|z|)
Gender	0.7514	0.2730	2.7521	0.0059218
Environmental health of sea	−0.4894	0.1423	−3.4392	0.0005834
Immigrant status	0.7320	0.2749	2.6633	0.0077369
(b)	**Estimate**	**Standard Error**	**z value**	
1|2	−1.5763	0.8295	−1.9004	
2|3	0.2417	0.8195	0.2949	
3|4	2.5094	0.8457	2.9671	

(a) Retained explanatory variables all respondents (n = 201). (b) Threshold coefficients: (log-likelihood: −241.1900, AIC: 494.3801, Condition number of Hessian: 2288.399).

Table 4. Ordinal regression model (logistic) fishers' responses to perception of rule-breaking for rules controlling fishing.

(a)	Estimate	Std. Error	z value	Pr(>|z|)
Immigrant status	1.4543	0.4624	3.1453	0.0016591
Environmental health of sea	−0.5250	0.2121	−2.4749	0.0133282
(b)	**Estimate**	**Standard Error**	**z value**	
1|2	−2.0255	1.0790	−1.8772	
2|3	0.3114	1.0545	0.2953	
3|4	2.6657	1.1117	2.3979	

(a) Retained explanatory variables fishers (n = 80). (b) Threshold coefficients: (log-likelihood: −89.85624, AIC: 189.7125, Condition number of Hessian: 1294.610).

reconciling conservation with sustainable livelihoods and, more generally, marine management goals.

Coastal villages in this study in the United Republic of Tanzania, as in many other parts of the developing world, are still heavily dependent on marine resources for income generation and food security [19]. [19] Most respondents believe that the health of this pivotal marine resource is worsening and while most are aware of rules controlling fishing activity, nearly half of those interviewed perceive high levels of rule-breaking. Ordinal regression modelling revealed immigrant status, gender and individual perceptions of marine resource health are significantly linked to high levels of perceived rule-breaking. Investigation of the level to which these relationships exist in a coastal community could be used to better contextualise stakeholder-led processes that inform on developing rules to support conservation objectives. Significant associations are discussed herein from an implementation perspective, building on prior research relating compliance to problem awareness. Implications for informing future management and enforcement are provided.

The distribution of responses regarding perceived 'environmental health of the sea' clearly demonstrates recognition of marine ecosystem degradation and constitutes problem awareness [14] at the individual level among more than seventy percent of respondents. In addition to apparent problem awareness, the ordinal regression analysis applied reveals a significant linkage between awareness of this environmental problem and respondents' perception of increased levels of rule-breaking. The current study does not test willingness to comply or acceptability of rule-breaking among respondents *per se*. However, applying existing propositions from the literature regarding problem awareness at an individual and societal level, the environmental problem awareness observed has been shown to be significantly linked to perceptions of rule-breaking, and is likely to have a strong positive effect on longer-term decisions on compliance [13]. Amongst individual respondents who are fishers, perception of marine environmental degradation and its linkage to rule-breaking constitutes a positive opportunity for increased compliance dependent on personal conviction of the validity of rules [4,9,13]. Equally, those fishers who see deteriorating marine health as a threat to their livelihoods have an individual incentive to decide to comply based on egoistic value threat (sensu Stern [14]). These significant variables can be applied by policy-makers to tailor and contextualise rules to appeal to these perceptions in a socially and economically context-specific relevant setting thus facilitating communication of the problems and possible solutions.

In the current study the belief that marine health is worsening and the linkage between negative marine health perceptions and perceived high levels of rule-breaking are shared by the majority of fishers and non-fisher members of their communities. Thus, while individual fishers recognise threats directly affecting them, the question arises as to whether they will also develop a moral sense of duty to comply in order to protect fishing/coastal community well-being (social-altruistic values) and the environment (biospheric values) [14]. A moral sense of duty to comply on the basis of social-altruistic values can be augmented by peer-group expectation and solidarity, however it is unclear to what degree fishers and non-fishers publicly voice their concerns regarding marine health or expectations regarding compliance. Sutinen and Kuperan [2] argue those fishers who are sensitive to societal or moral expectations may be motivated to reduce rule-breaking. Even fishers who do not perceive marine health degradation or do not associate it with rule-breaking at an individual level, may be influenced by this perception at a community level if held widely enough. Policy-makers and managers tailoring rules and management measures within a community context can thus effectively exploit community expectations and problem awareness to improve compliance. This does not, however, imply improvement on the basis of increased legitimacy of the governing body creating those rules [4,9,20].

Without internal or external drivers to commit fishers to compliance, enforcement is unlikely to be economically viable or effective [4,6,13]. While some targeted enforcement may be effective this can only be carried out if the underlying socio-economic factors determining fishers' decisions to comply are understood and addressed [2,3]. The awareness of problems identified by respondents in the current study highlights an opportunity at local and wider contexts to influence future compliance through what Sutinen et al 1990 termed "greater use of the non-coercive factors of compliance (social pressure, self interest, inducement, and obligation)" by framing problems using context-specific management measure options on how they could be tackled. These non-coercive factors are made more powerful, wide-reaching and effective through increased community and individual participation and communication of perceptions. In the specific context of the area studied this would involve increasing community group activities, particularly those with a marine awareness focus and/or with an opportunity for members to express their concerns about or awareness or marine health deterioration. Non-coercive factors benefit also from increased knowledge of the rules and their environmental objectives through public awareness activities, and from activities to cement perceptions of the causal relationship between rule-breaking and marine environmental degradation [3]. In the context of the current study area, public awareness campaigns highlighting key rules and indicating effects of the most common forbidden marine activities on marine health would be valuable.

There is little available literature to explore the significance of immigrant status and gender in explaining rule-breaking the final ordinal regression model presented herein. Immigrant status has previously been linked to increased negativity of perceptions of personal health and well-being but has been shown not to increase rule-breaking behaviour significantly amongst minority immigrants in a (developed) urban context [21,22]. While our definition of "immigrant", not born in this village, is broad in the current study, immigrants were significantly more likely to perceive rule-breaking compared with individuals born in the village. It is possible that immigrants experience less sense of social cohesion and thus less compulsion to not admit rule-breaking by other community members, if they sense a lower social cohesion within their fishing groups they may also not adhere to the usual lack of self-enforcement among fishers [8]. Conversely, however, if immigrants are poorly integrated into coastal communities, then there may be a decreased value of social pressure and obligation to adhere to rules as non-coercive factors in driving compliance [1 2006]. The results indicate a clear need for future research to determine the underlying differences in perceptions of rule-breaking among immigrants. Our inferences remain equivocal about why women were significantly less likely to perceive rule-breaking when compared to men, however, this may be best or partly explained by individual experience of fishing being extremely low among female respondents. Nonetheless, gender differences in perceptions towards rule-breaking warrants further investigation especially in terms of how knowledge on the way information is passed between different generations and how this might be used by women to influence compliance behaviour within their households and communities.

Conclusions

The current study complements existing knowledge identifying a chain of linkages between environmental awareness and environmental compliance. It does so by relating, for the first time, individual perceptions of rule-breaking with perceived degradation of the marine environment. Empirical data analysis points out these key linkages, providing insights into complex compliance processes. This study shows how problem awareness and orientation can provide a focal point for policy makers to inform development of more effective marine management measures. The authors recommend wider application of socio-economic indicators to aid development of rules that are context-specific for those targeted and most likely to be impacted. This may help to identify self-regulation and compliance measures that can be adopted following promotion of existing non-coercive factors. Increasing awareness of environmental health through education programs to target wider understanding of linkages between marine health and compliance, combined with promoting social environmental responsibility, are actions recommended as part of longer term management strategies aimed at supporting sustainable use of marine resources.

Acknowledgments

The authors wish to thank Prof Tim Gray, Prof Stephen Rushton and Dr. Aileen Mill for assistance in preparing this manuscript. The authors wish to thank the reviewers of this manuscript for their insightful inputs.

Author Contributions

Conceived and designed the experiments: MS SS. Performed the experiments: MS. Analyzed the data: MS. Contributed reagents/materials/analysis tools: YM SS. Wrote the paper: MS SS YM.

References

1. Sumaila UR, Alder J, Keith H (2006) Global scope and economics of illegal fishing. Marine Policy 30: 696–703.
2. Sutinen JG, Kuperan K (1999) A socio-economic theory of regulatory compliance. International Journal of Social Economics 26: 174–193.
3. Sutinen JG, Rieser A, Gauvin JR (1990) Measuring and explaining noncompliance in federally managed fisheries. Ocean Development & International Law 21: 335–372.
4. Jagers SC, Berlin D, Jentoft S (2012) Why comply? Attitudes towards harvest regulations among Swedish fishers. Marine Policy 36: 969–976.
5. Cornish DB, Clarke RV (1986) Situational Prevention, Displacement of Crime and Rational Choice Theory. In: Heal K, Laycock G, editors. Situational Crime Prevention: From Theory into Practice. London: Home Office Research and Planning Unit. pp. 1–16.
6. Williamson OE (1975) Markets and Hierarchies: Analysis and Antitrust Implications: A Study in the Economics of Internal Organization. New York: Free Press.
7. Warner TE, Pomeroy RS (2001) Creating compliance: A cross-sectional study of the factors associated with marine protected area outcomes. Marine Policy 36: 922–932.
8. Jentoft S, Mikalsen KH (2004) A vicious circle? The dynamics of rule-making in Norwegian fisheries. Marine Policy 28: 127–135.
9. Nielsen JR, Mathiesen C (2003) Important factors influencing rule compliance in fisheries lessons from Denmark. Marine Policy 27: 409–416.
10. Pomeroy RS, Pollnac R, Katon BM, Predo CD (1997) Evaluating factors contributing to the success of community-based coastal resource management: the Central Visayas Regional Project-1, Philippines. Ocean & Coastal Management 36: 97–120.
11. Maliao RJ, Pomeroy RS, Turingan RG (2009) Performance of Community-based coastal resource management programs in the Philippines: A meta-analysis. Marine Policy 33: 818–825.
12. Olsen SB (2002) Assessing progress toward the goals of coastal management. Coastal Management 30: 325–345.
13. Hansla A, Gamble A, Juliusson A, Garling T (2008) The relationships between awareness of consequences, environmental concern, and value orientations. Journal of Environmental Psychology 28: 1–9.
14. Stern PC (2000) Toward a coherent theory of environmentally significant behavior. Journal of Social Issues 56: 407–424.
15. Iaquinta DL, Drescher AW (2000) Defining Periurban: Understanding Rural-Urban Linkages and Their Connection to Institutional Contexts. Tenth World Congress. Rio de Janeiro: International Rural Sociology Association.
16. Sesabo JK, Tol RSJ (2005) Factors affecting Income Strategies among households in Tanzanian Coastal Villages: Implications for Development-conservation initiatives. Hamburg: Working Paper FNU-70.
17. Cinner JE, Daw T, McClanahan TR (2009) Socioeconomic factors that affect artisanal fishers' readiness to exit a declining fishery. Conserv Biol 23: 124–130.
18. Agresti A (2002) Categorical Data Analysis. Wiley Press.
19. FAO (2010) The State of World Fisheries and Aquaculture - 2010 SOFIA. In: Department FaA, Rome: FAO. pp. 218.
20. Hauck M (2008) Rethinking small-scale fisheries compliance. Marine Policy 32: 635–642.
21. Nielsen SS, Krasnik A (2010) Poorer self-perceived health among migrants and ethnic minorities versus the majority population in Europe: a systematic review. Int J Public Health 55: 357–371.
22. Reijneveld SA, van Nieuwenhuijzen M, Velderman MK, Paulussen T, Junger M (2012) Clustering of health and risk behaviour in immigrant and indigenous Dutch residents aged 19–40 years. Int J Public Health 57: 351–361.
23. Al-Hafedh YS, Alam A, Buschmann AH, Fitzsimmons KM (2012) Experiments on an integrated aquaculture system (seaweeds and marine fish) on the Red Sea coast of Saudi Arabia: efficiency comparison of two local seaweed species for nutrient biofiltration and production. Reviews in Aquaculture 4: 21–31.

8

Characterisation of Host Growth after Infection with a Broad-Range Freshwater Cyanopodophage

Siobhan C. Watkins[1], James R. Smith[2], Paul K. Hayes[1], Joy E. M. Watts[1]*

1 School of Biological Sciences, University of Portsmouth, Portsmouth, United Kingdom, **2** School of Pharmacy and Biomedical Sciences, University of Portsmouth, Portsmouth, United Kingdom

Abstract

Freshwater cyanophages are poorly characterised in comparison to their marine counterparts, however, the level of genetic diversity that exists in freshwater cyanophage communities is likely to exceed that found in marine environments, due to the habitat heterogeneity within freshwater systems. Many cyanophages are specialists, infecting a single host species or strain; however, some are less fastidious and infect a number of different host genotypes within the same species or even hosts from different genera. Few instances of host growth characterisation after infection by broad host-range phages have been described. Here we provide an initial characterisation of interactions between a cyanophage isolated from a freshwater fishing lake in the south of England and its hosts. Designated ΦMHI42, the phage is able to infect isolates from two genera of freshwater cyanobacteria, *Planktothrix* and *Microcystis*. Transmission Electron Microscopy and Atomic Force Microscopy indicate that ΦMHI42 is a member of the *Podoviridae*, albeit with a larger than expected capsid. The kinetics of host growth after infection with ΦMHI42 differed across host genera, species and strains in a way that was not related to the growth rate of the uninfected host. To our knowledge, this is the first characterisation of the growth of cyanobacteria in the presence of a broad host-range freshwater cyanophage.

Editor: Paul Hoskisson, University of Strathclyde, United Kingdom

Funding: Funding provided by EU FP7 KBBE award 265409: MicroAqua. The funders had no role in study design, data collection and analysis, decision to publish, or preparation of the manuscript.

Competing Interests: The authors have declared that no competing interests exist.

* E-mail: joy.watts@port.ac.uk

Introduction

Studies of viruses in the environment, particularly those infectious to prokaryotes, have led to important recent discoveries, including the existence of virophages [1] and viruses infectious to archaeal hosts [2], however, we still have a limited understanding of the relationships between viral lineages, or the interactions between viruses and their hosts within the environment [3].

Arguably, the most comprehensively studied group of non-medically or agriculturally important viruses, are the marine cyanophages [4]. Cyanophages infect cyanobacteria, which are globally important microorganisms found in many habitat types [5–7]. Cyanobacteria are major ecosystem engineers and are thought to have been responsible for the initial production of free oxygen in the Earth's atmosphere [8]. Present day cyanobacteria are particularly important in aquatic environments, where they can be the dominant primary producers, with some taxa able to reduce nitrogen from the atmosphere and incorporate it into organic matter [9]. Cyanophages are known to influence cyanobacterial community composition, population diversity and functional activity in the oceans by contributing to host-specific mortality, thereby perturbing both primary production and biogeochemical cycling. In addition to their role in host mortality, phages can act as transducing agents, transferring genetic material between bacterial lineages, and so influencing bacterial genotypic and thus phenotypic diversity [10]. It is becoming increasingly evident, therefore, that viruses in the environment have a significant impact on microbial ecology and the role of marine cyanophages as agents of influence in marine cyanobacterial systems is important.

Cyanophages in freshwater systems have received less attention than their marine counterparts. Some freshwater and marine cyanophages may share common ancestry [11–12], however, cyanophages that are distinct from marine equivalents have also been described from freshwater environments [13]. The members of a number of freshwater cyanobacterial taxa produce toxins and form surface blooms in eutrophic waters that can have disastrous implications for aquatic life and water quality [14]. The control of such harmful cyanobacterial blooms (HCBs) through the application of lytic cyanophages has been considered as a potential management strategy, but deploying cyanophages as a management tool poses a number of practical problems [15], including the intensity of water treatment necessary following a sudden release of previously cytosolic toxins into the water column, the practical aspects of generating and applying large quantities of virions to a pond, lake or river, and the public perception of this type of treatment strategy.

Other issues associated with the use of cyanophages to mediate HCBs relate to the intricacies of interactions between freshwater cyanobacteria and their phages in the environment. Bacteriophages have a range of strategies for interacting with their bacterial hosts, including the capacity to infect more than one strain or species, and the nature of that infection i.e. lytic or non-lytic. Previous work [15] has highlighted the prevalence of both narrow and broad host-range cyanophages obtained from freshwater samples: having either a narrow or a broad host-range can confer advantages depending on the environmental situation. For example, in low nutrient, sparsely populated waters, where phages may not come into contact with many potential hosts, a broad-host range is likely to be advantageous, whereas in areas where host abundance is high (such as in eutrophic freshwaters) it may be advantageous to infect a narrower range of hosts, but at high efficiency. The determination of a phage's host range is not possible without culture-based studies. This work presented here begins to extend our understanding of broad-host range cyanophages by examining the growth of diverse infected hosts under controlled conditions.

The determination of host range specificity is a well-established element of marine cyanophage characterisation [23–24], however, such studies tend to restrict characterisation to within, and between, *Prochlorococcus* and *Synechococcus* [25–26]. The ability of phages to infect a range of hosts is determined by a number of factors, including the presence of host restriction modification systems, phage genome uptake blocks and CRISPR/cas systems in the host, and attachment and adsorption mechanisms [27–30]. Cyanobacteria are an ancient lineage and their relationship with cyanophages is likely to have developed over the course of this long evolutionary history. This prolonged exposure of cyanobacteria to attack by cyanophages is likely to be recorded within cyanobacterial genomes [31], and will have contributed to the acquisition of robust defence and infection systems which may be collectively observed through direct examination of growth.

Host range is also important when considering a more fundamental problem surrounding the concept of phage-mediated control of HCBs, which is the genomic diversity of the bloom-forming cyanobacteria. The usual approach involves the isolation of clonal phages that can be applied directly to eutrophic waters to destroy HCBs. However, the genomic plasticity of target host cyanobacteria renders the possibility of identifying a single cyanophage, or a few cyanophages, that are lytic to all genotypes within a bloom very low. For example, *Microcystis* and *Planktothrix* communities have both been shown to encompass a variety of ecotypes that differ genetically and phenotypically, including in characteristics such as toxicity, presumably dependent on the ecological niche that they occupy [16–19]. Therefore, a single cyanophage is unlikely to be able to act as a lytic agent against a genetically diverse and constantly changing cyanobacterial population i.e. a cyanophage infectious or lytic against one ecotype of *Microcystis*, may not be infectious to another [20]. Furthermore, rates of co-antagonistic evolution observed between phages and bacteria will likely influence defence and attack strategies very quickly. The presence of temperate viruses, lysogens and pseudo-lysogeny in the environment [21–22], add a further level of complexity to the study of the ecology of freshwater viruses and their hosts. The standard methods for analysis of phage growth kinetics rely heavily on plaque assay based methods such as the one-step growth curve [32]. While plaque assays are possible for some planktonic cyanobacteria, such as members of the genera *Anabaena* and *Microcystis*, members of the ecologically important genus *Planktothrix* display gliding motility, which precludes plaque formation on solid media. Initial genetic characterisation of cyanophages has been performed previously using primers designed for the amplification of structural genes similar to those found in the myovirus coliphage T4. Myoviruses, however, are just one component of the virioplankton and, moreover, they are known to be widely divergent [33], and so current PCR-based molecular analyses impose significant constraints on the characterisation of newly isolated freshwater cyanophages. Therefore, the standard method for genetic characterisation of new phages is, increasingly, *de novo* whole genome sequencing (WGS) [34,35,36]. While WGS is a powerful technique, inaccessibility due to cost and the inherent difficulties associated with genome assembly and analysis, can be issues. These collective limitations, i.e. the inability to rely on plaque assays and lack of primers for rapid initial genetic characterisation, are, in part, responsible for the paucity of information about freshwater cyanophages.

Classification of phage morphology is typically assessed using transmission electron microscopy (TEM), however, preparation and visualisation of phages can be difficult [37]. Atomic force microscopy (AFM) is slowly emerging as an alternative to TEM for the determination of viral morphology, however to date it has only been used to image marine cyanophages [38]. AFM provides the potential to convert highly resolved digital images to quantitative data [39].

In this study we provide an initial characterisation of host growth after infection of cyanobacterial strains from multiple genera with ΦMHI42, a freshwater cyanophage originally isolated on *Microcystis aeruginosa* BC84/1. Using a combination of both TEM and AFM, we show that ΦMHI42 is a member of the Podoviridae. We demonstrate that this cyanophage has a broad host-range, in that it is able to infect both *Planktothrix* and *Microcystis*, and that the impact of infection on the growth of the host is variable between strains and across genera. To the best of our knowledge, this is the first description of the growth characteristics of cyanobacterial host cultures following infection with a broad host range freshwater cyanophage.

Materials and Methods

Host Strains and Growth Conditions

All cyanobacterial cultures used in this study were non-axenic. *Microcystis aeruginosa* BC84/1 (University of Bristol, UK), was grown in the open laboratory in BG11 medium [40] at room temperature (~20°C) and away from direct sunlight, at a constant incident irradiance of between 10–15 μmoles m^{-2} s^{-1}. Additional cyanobacterial strains used for host range assays were grown at constant temperature under cool white fluorescent light in incubators as follows: *Microcystis aeruginosa* 1450/8 (CCAP, Argyll, Scotland), was grown in BG11 medium at 20°C at an incident irradiance of 8–10 μmoles m^{-2} s^{-1}; *Planktothrix rubescens* 9316 and *Planktothrix agardhii* 137 (University of Bristol, UK), which were grown in Oscillatoria medium [41] at 20°C at incident irradiance of 2–3 μmoles m^{-2} s^{-1} for *P. rubescens* and 6–7 μmoles m^{-2} s^{-1} for *P.agardhii*; *Anabaena flos-aquae* (University of Bristol, UK) was grown in Oscillatoria medium at 20°C at an incident irradiance of 10–15 μmoles m^{-2} s^{-1}.

Initial Cyanophage Isolation

A surface water sample (5 L) was collected from a fishing lake in Hayling Island, Hampshire, UK (SZ 73658 98822) on 19th September 2011. No official permission was required to sample as the lake is a public water body. Aliquots of *M. aeruginosa* BC84/1, in early exponential growth (1.5 ml), were mixed with 200 μl of the water sample and left for 24 hours, under standard host growth conditions. The host/sample mixtures were then mixed with 1.5 ml sterile low-melting temperature agarose (0.8% w/v) at 37°C, and poured onto the surface of a BG11 agar (0.8% w/v) plates, incubated (as above) and checked for the formation of plaques. Plaques (in agar plugs) were removed from the plate and stored in cyanophage (CP) buffer (5 mM $MgCl_2$, 5 mM $CaCl_2$, 10 mM NaCl, 10 mM HEPES, pH 7), at room temperature in the dark, before being subjected to three rounds of purification (plating followed by plaque isolation, as described above).

Host Range Determination

Samples of phage-induced cyanobacterial lysate (200 μl) were combined with exponential phase cyanobacterial cultures (1.5 ml) in a multiwell tissue culture plates (well volume 3 ml). Additional growth medium was added (1 ml) and the plates were incubated as above and checked daily for the presence of cleared lysates. The lysate from cleared wells was removed and centrifuged at 20 000×g for two hours. Subsequently, the supernatants were collected and stored at room temperature in the dark. Where possible, cyanobacterial lawns (*M.aeruginosa*) were infected with ΦMHI42 and plaque assays which had been left to progress to the extent that no cyanobacterial host remained visible on the plate were kept for further examination. The resulting overgrowth of bacteria, associated with non-axenic growth, was examined for further plaque formation, and therefore the potential for infective capacity on heterotrophs.

Transmission Electron Microscopy (TEM)

Purified viral lysate was applied to pioloform coated copper grids and left to dry at room temperature. Samples were positively stained with 2% (w/v) uranyl acetate [42] and observed at 80 kV using a Jeol JEM 2100 Transmission Electron Microscope.

Atomic Force Microscopy

Poly-L-lysine (5 μL, 0.01% (w/v), mol wt 7000–150000) was placed on freshly cleaved muscovite mica (Agar Scientific, Stansted, Essex, UK), left for 2 minutes and the mica was then dried in a stream of N_2 at room temperature. Viral lysates were diluted (1:10) in ultrapure water (18 MΩ; Sigma-Aldrich) and placed (5 μL) on the mica surface, left for 2 minutes after which glutaraldehyde (5 μL; 4% (w/v) in 0.2 M sodium cacodylate) was added. After a further 2 minutes, the surface was extensively rinsed with ultrapure water and dried as above. The mica was then mounted on a nickel disc (1 cm diameter) with double-sided adhesive tape and placed on the AFM scanner. AFM studies were performed using a Multi-Mode/NanoScope IV Scanning Probe Microscope (Bruker, Santa Barbara, CA, USA) in air under ambient conditions (T = 24°C, RH = 45%) using the J-scanner (max. xy = 200 μm). Scanning was performed in Tapping Mode using a Si cantilever with integrated tip (t_{nom} = 3.6–5.6 μm, l_{nom} = 140–180 μm, w_{nom} = 48–52 μm, υ_{nom} = 200–400 kHz, υ_{meas} = 333.7 kHz, k_{nom} = 12–103 Nm^{-1}, R_{nom} < 10 nm; Model: OTESP, Bruker, France) and an RMS amplitude of 2.0 V. Images were subsequently processed (plane fitted and shaded using an artificial light source) using NanoScope analysis software (V 1.40, Bruker, CA, USA).

PCR Amplification

Viral-induced host cell lysates were used as a template for amplification reactions performed using published protocols and primer sets previously employed to characterise isolated and environmental samples of marine cyanophage (Table 1).

Growth of Hosts in the Presence of Cyanophage

The growth of cyanobacterial hosts was assessed using a turbidometric method; these data were used to compare growth characteristics of each culture after infection with ΦMHI42 [47]. For each host strain, nine replicate cultures were inoculated with ΦMHI42 at a multiplicity of infection (m.o.i.) of 5 or 0.5: phage titre was determined using *M. aeruginosa* BC84/1 as host and performed according to standard method based on viral quantification by plaque forming units [32]. Absorbance of the cultures was measured at 660 nm and 750 nm at 24 hour intervals throughout the infection cycle: cultures were infected with ΦMHI42 at the start of exponential growth. Uninfected cultures were used as controls.

Table 1. Primer sets used in this study. Primer sequence and PCR conditions are provided in the references.

Primer set designation	Target	Reference
CAP	g23 gene (Major Capsid Protein)	[43]
CPS	g20 gene (Coliphage Portal Vertex Protein)	[44]
CPS8.1/1.1	g20 gene (Coliphage Portal Vertex Protein)	[45]
DNAPol	Podovirus DNA Polymerase	[46]

Results

Characterisation of ΦMHI42

TEM analysis revealed a cyanophage displaying characteristics most similar to those of the family *Podoviridae* (Fig. 1A). ΦMHI42 appears to possess a short, non-contractile tail, however, the capsid, with a diameter of approximately 100–120 nm, is considerably larger than for most previously described podoviruses (usually 40–60 nm). Capsid size was confirmed using AFM (Fig. 1B): lighter areas in the AFM image represent areas raised above the substrate surface. The image shows repeating, discrete units on the surface of the capsid, which had an average diameter of 34 nm (340 Å), as determined with the Nanoscope analysis package.

It was not possible to generate PCR amplicons from ϕMHI42-induced cell lysates using marine cyanophage-specific primer sets (Table 1) which is consistent with previous studies [15]. The highly degenerate primer set CPS 8.1/1.1 (45) primed the amplification of a fragment of about the expected size, but this amplicons lacked any sequence similarity to any previously characterised bacteriophage derived nucleotide or protein sequences (data not shown).

Host Growth Characteristics after Infection with ΦMHI42

In host range specificity tests, ΦMHI42 was used to infect cultures of *Microcystis*, *Planktothrix* and *Anabaena*, and also tested for infective capacity against heterotrophic bacteria associated with the non-axenic cultures. Cultures of both *Microcystis* and *Planktothrix* were susceptible to lytic infection, but associated bacteria and cultures of *Anabaena* were not. Host growth profiles (Fig. 2) demonstrate the progression of infection in replicate cultures established from the same inoculum and using a m.o.i. of 5. Growth of the hosts in the uninfected controls progressed as expected, where the large error bars are associated with the variability in growth commonly seen in cultures where the cells (*Microcystis*) or filaments (*Planktothrix*) form clumps.

The growth profile of *M. aeruginosa* BC84/1 infected with ΦMHI42 (Fig. 2A) shows a clear onset of lysis after about 400 hours and subsequent drop in cell concentration (absorbance) across all replicates, yielding a profile similar to that observed for fast-growing lytic phage-infected bacteria, for which this technique has been previously used. This profile results from the step-wise release of phage from the host cell [48]. Although it is difficult to

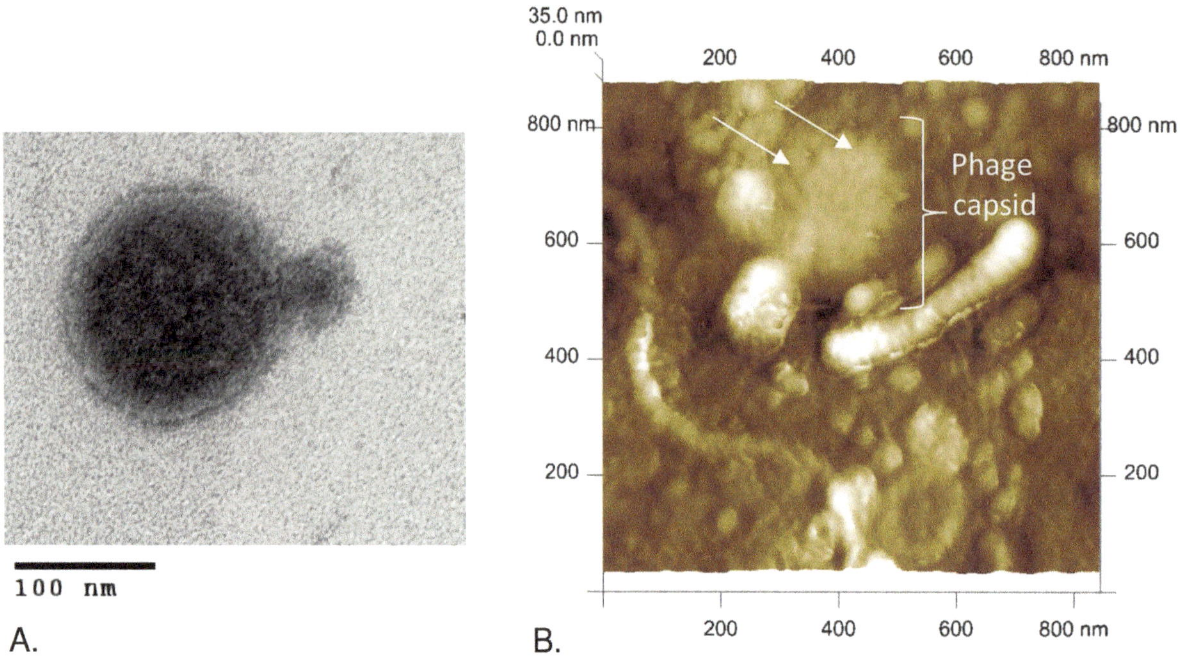

Figure 1. ϕMHI42 viewed at high magnification using: A. TEM and B. AFM. A shows a single phage, similar in morphology to the family *Podoviridae*. B shows a single phage among associated cell debris, the presence of a viral capsid confirms size and morphology of the structures observed in A. Repeating units on the capsid are highlighted with arrows.

determine a latent period from the growth curve generated, it is possible to compare growth characteristics between different host strains infected with ΦMHI42. Infection of BC84/1 with ΦMHI42 resulted in a halt in exponential growth of the host after approximately ten days. In contrast to BC84/1, infection of *M. aeruginosa* 1450/8 with ΦMHI42 resulted in retardation of host growth as opposed to catastrophic lysis: the infected *M. aeruginosa* 1450/8 continued to grow, but at a reduced rate and achieving a lower cell concentration in stationary phase in comparison to the control cultures (Fig. 2B). At this time, the cultures did not appear to have become scenescent, however, samples removed from the cultures also subsequently produced lysis in a culture of BC84/1. The growth profile on this host was suggestive of a chronic infection strategy. Assessment of exponential growth rates (r) also showed differences in infection profile between the two strains of *M. aeruginosa* (Fig. 3A & B). Analysis of r values of all replicates showed that BC84/1 cultures infected with ϕI42 had significantly higher exponential growth rates than the uninfected controls. This can also be seen on the lysis profile (Fig. 2A). *M. aeruginosa* 1450/8 replicate cultures also demonstrated a significant, but less pronounced, difference in exponential growth rate between infected and control cultures.

Laboratory cultures of *Planktothrix* species often demonstrate variable growth patterns, such as a tendency to clump, in ways that seem to be unrelated to changes in standard growth conditions. For this reason the growth curves of the uninfected cultures (Fig. 2C & D) were less reproducible than for *Microcystis*. Infection by ΦMHI42 resulted in almost immediate effect on the growth of the host strains (Fig. 2C & 2D). The onset of lysis was rapid and highly reproducible across all replicates. Cultures of *P. agardhii* 137 required approximately five days after infection to stop growing completely (Fig. 2C), considerably shorter than for *Microcystis*. Statistical analyses of exponential growth rates of replicate infected and triplicate control cultures were difficult to interpret due, again, to variability in host growth. Assessment of *P. agardhii* 137 showed a significant increase in growth rates of the infected group compared to the control group (Fig. 3C), with evidence for increased variability in growth rate between infected replicates versus the control. The growth profile generated for *P. rubescens* 9316 (Fig. 2D) is misleading: due to the rapid growth of heterotrophic bacteria in response to lysis of *P. rubescens*, absorbance begins to increase after hour 400, despite complete lysis of the red-pigmented cyanobacterium. No significant difference in exponential growth rates between infected and uninfected cultures was observed (Fig. 3D), possibly because of the overgrowth by other bacteria.

Effect of Lowering Multiplicity of Infection

To assess the effect of m.o.i on lysis profiles, growth of *Planktothrix* was also assessed at a nominal m.o.i. of 0.5. At this reduced m.o.i. there was no impact on the growth of the host (data not shown), which is consistent with the results from previous studies [49].

Discussion

The genetic richness and diversity in freshwater bacteriophage communities may equal or even surpass, that found in marine environments [13,50–51]. Bacteriophages are known to act as important agents of mortality, as genetic reservoirs and transfer agents within bacterial communities, the impact of which will depend on their ability to infect a range of species and host ecotype, and also on whether they adopt lytic or lysogenic modes of infection. A number of effective deep sequencing techniques have been developed for the characterisation of viral communities in the environment, however, it is only by isolation and laboratory propagation that host range can be determined and some lifestyle characteristics, such as the ability to lysogenise the host, assessed.

The infection characteristics of phages and their hosts can be determined using a number of standard techniques, the majority of

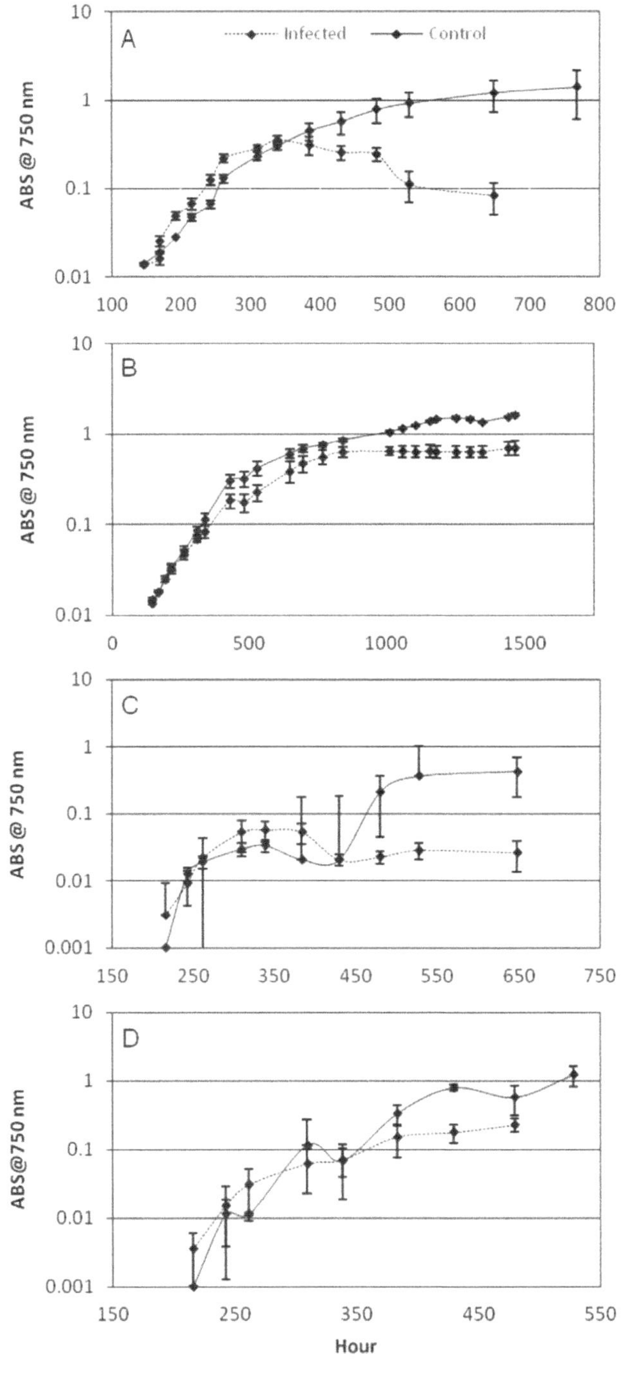

Figure 2. Lysis profiles generated by infecting five strains of cyanobacteria with ϕMHI42, as determined by measuring absorbance at 750 nm – *M. aeruginosa* BC84/1 (A), *M. aeruginosa* 1450/8 (B), *Planktothrix agardhii* 137 (C), *Planktothrix rubescens* 9316 (D). Profiles begin at point of infection and are plotted on a logarithmic scale. Dashed lines are infected culture series, error bars represent standard deviation over three replicate cultures. The solid line represents triplicate control cultures, which were inoculated with the host but not subsequently infected with phage.

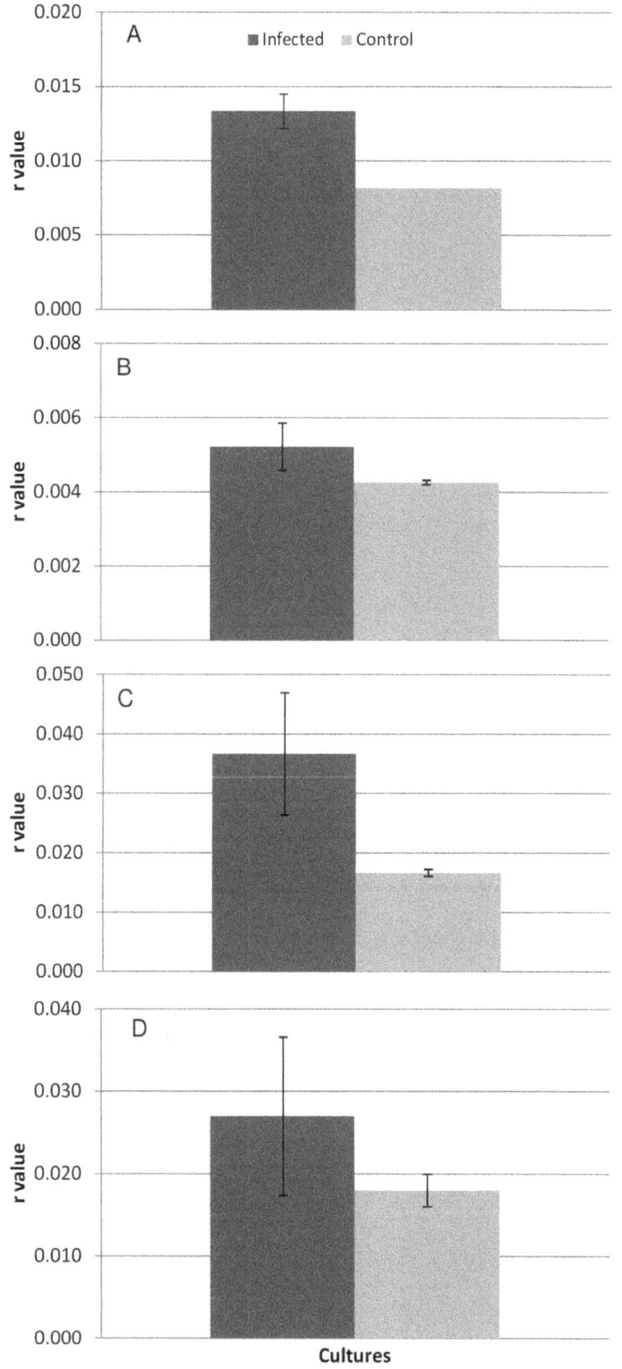

Figure 3. Statistical analysis based on mean r values for replicates cultures of each host strain infected by ΦMHI42 - *M. aeruginosa* BC84/1 (A), *M. aeruginosa* 1450/8 (B), *Planktothrix agardhii* 137 (C), *Planktothrix rubescens* 9316 (D). Error bars represent standard deviation. Growth rates were assessed for significant difference with the ANOVA test and differences separated with a Bonferroni corrected t-test.

which are based on the development of viral plaques on lawns of host cells [52]. While such techniques generally work well with bacteria that grow reliably on solid media and in axenic culture, the characterisation of phages of hosts that cannot be grown on plates or in the absence of other microorganisms presents challenges over and above those normally experienced when working with fastidious and/or slow growing bacteria. Although generally amenable to growth in laboratory culture, cyanobacteria have relatively long generation times and some, such as *Planktothrix*, clump in liquid culture and exhibit gliding motility.

The isolation of freshwater cyanophages on *Planktothrix* and other hosts does not, however, appear to be problematic, but, sustained propagation and maintenance of viability of freshwater cyanophages is challenging, particularly in relation to maintaining phage viability [53–54]. The issues around sustained viability are not uncommon, but they add a further level of complexity when attempting to characterise phage [55].

These challenges are reflected in the lysis profiles generated from infecting four strains of cyanobacteria with ΦMHI42. Results that are typical of lytic infection were obtained using the original host, *M. aeruginosa* BC84/1, a unicellular cyanobacterium that has been in culture since 1984. BC84/1 was the only strain in this study on which plaques form reliably, and so was used to generate the viral inocula used for all infections. The time taken for cultures to die, estimated from the growth curve, reflects the inherent slow growth rate of the host ($r = 0.0082 \pm 0.0\ hr^{-1}$). Of interest is the effect that exposure to ΦMHI42 appears to have on the growth rate of the host. For *M. aeruginosa* BC84/1, during the exponential period prior to lysis the average growth rate increased in comparison to the uninfected controls from $0.0082 \pm 0.0\ hr^{-1}$ to $0.013 \pm 0.0012\ hr^{-1}$. A similar result was observed for *M. aeruginosa* 1450/8 on a smaller, but nonetheless significant scale (from $0.0043 \pm 0.00006\ hr^{-1}$ to $0.013 \pm 0.0049\ hr^{-1}$), and which also demonstrated the slowest growth rates of all the host strains. It has been previously documented that phage infection can induce alterations in metabolic activity of their hosts [56], particularly in processes such as photosynthesis [57] which may ultimately be beneficial to the phage. That higher growth rates in these infected hosts relates directly to the presence of this phage requires confirmation. *M. aeruginosa* 1450/8 produced a very different growth profile in comparison to *M. aeruginosa* BC84/1 when incubated in the presence of ΦMHI42: there was no defined onset of lysis, and while there was a discernible colour change in the infected cultures (from blue-green to pale yellow), suggesting the establishment of lysis, a clear lysate was not produced (the lysate was, however, infectious to, and consequently produced lysis in, BC84/1). This lack of defined lysis continued in the infected cultures even through the period when the parallel controls went through stationary phase and senescence. In further comparison to BC84/1, 1450/8 grew particularly unreliably on solid medium, and was often outcompeted by heterotrophic bacteria associated with the non-axenic culture: plating appeared to strongly select for their growth as opposed to that of the cyanobacterium.

The growth of *M. aeruginosa* 1450/8 uninfected controls was more reproducible than for the other host strains, evident from the small size of the error bars, but, infection appeared to slightly increase the levels of variation between replicates. The instability observed during the infection cycle of *M. aeruginosa* 1450/8 may have resulted solely from the overgrowth by the normally background-level bacterial contaminants present in these non-axenic cultures, however, it may also relate to the unstable nature of the type of phage growth that allows the simultaneous survival of the host and the production of viable virions. Retardation of growth in *M. aeruginosa* 1450/8 in response to infection, as opposed to acute lysis, may suggest a state of chronic infection or pseudo-lysogeny [58]. The marked difference between the growth of the two strains of *Microcystis* in the presence of ΦMHI42 may relate to a number of factors associated with genetic diversity within the genus [59], however, differences in phage resistance between these two strains could also be a relic of their historical exposure to very different phage-related selection pressures imposed prior to isolation into culture [60–61]. ΦMHI42 was initially isolated on *M. aeruginosa* BC84/1, and was purified by continuous rounds of infection on this strain. It is possible that this purification process selected for variants of ΦMHI42 that were better adapted for growth on *M. aeruginosa* BC84/1 than the original isolate, but given that ΦMHI42 retained the ability to infect both *Planktothrix* and a further *Microcystis* this seems unlikely.

Despite having been isolated on *Microcystis* BC84/1, ΦMHI42 was more virulent against *Planktothrix*, a filamentous cyanobacterium. Previously, it has not been possible to examine a clonal phage in relation to *Planktothrix* [15], due to its capacity to glide and therefore inability to produce plaques on solid medium. In this case we were able to use the lysis profiling technique as ΦMHI42 had already been rendered clonal on its original host, *M. aeruginosa* BC84/1. While the rate of lysis is expected to be related to the normal growth rate of the host [32,62], the time taken for the two *Planktothrix* cultures to stop growing was significantly shorter than was observed for the *Microcystis* infections. It is possible that the observed differences are related to the filamentous nature of *Planktothrix*, with rapid transmission of infection between cells along the length of the filament. If this is the case, then it would also have implications for the spread of infection in the natural environment. While assessment of host growth rate was performed for both *Planktothrix* cultures, it was difficult to draw conclusions from the data collected for *P.rubescens 9316*. For infected cultures of *P. agardhii* 137, however, infection did produce a significant increase in rate of growth compared to the control, although with more variability between replicate infected groups than was observed for *Microcystis*.

That four different strains of cyanobacteria can exhibit different responses to infection by the same cyanophage, is likely to be related to the concepts of 'ecotypes' and niche separation in the environment [63]. The existence of ecotypes has been clearly demonstrated in cyanobacteria, where strains with similar or identical 16S rRNA gene sequences can demonstrate considerable differences in phenotype [64]. In part, the niche occupied by particular ecotypes can be defined by the composition of phage community, i.e. resistance or susceptibility of individual cyanobacterial lineages to phage communities will determine whether or not they are able to grow at that location. The interaction between the host and any infecting phage is dynamic and often described as an "arms race" [65] where the level of exposure to the host and to other phages infecting the same host may all play a role in phage infection dynamics, which in turn will influence the evolution of host defence against phage attack. How the host response to attack by phage is likely to have developed is less clear, and further investigation is warranted to assess the presence of host restriction modification enzymes [66–67] or CRISPR/cas systems [68–69] in resistant strains.

The presence and overgrowth of contaminating bacteria in response to lysis of the dominant member of a non-axenic culture highlights further difficulties associated with examining cyanobacterial infections and subsequent physiological aspects of phage characterisation. Cyanobacteria obtained from culture collections are rarely axenic, and growth conditions selecting against the growth of the cyanobacterium will often result in rapid growth of the contaminants [70–71]. Non-axenic overgrowth was observed in three of the four cyanobacterial cultures used for lysis profiles, *P.agardhii* being the only culture not demonstrating this trait.

In the environment, harmful cyanobacterial blooms are likely to be affected by viral-induced host mortality, and the capacity of phages to drive genetic diversification through horizontal gene transfer will influence the generation of new cyanobacterial ecotypes and the genotypic composition of toxic blooms populations. Genetic diversification may be mediated across genera by broad host-range phages. The nature of interactions between generalist phages, specialist phages and their hosts are not well characterised, and neither are the impacts that these may have on the phage communities that could influence the formation, structure and disruption of blooms. If a phage, such as ΦMHI42, is capable of removing hosts from multiple genera, this may have impact on the general level of phage diversity. Similarly, if a phage is capable of inducing a chronic growth phase in a host present in the environment, this suggests resistance to such a phage will be generated quickly. Therefore, it is likely that the composition of both the phage and host communities is extremely dynamic, but it seems certain that interactions between the two will be important in shaping the ecology of freshwaters.

Acknowledgments

We thank Annette Richer (University of Bristol) for supplying cyanobacterial cultures and her advice regarding their growth, and Simon Cragg (University of Portsmouth) for help with operation of the TEM.

Author Contributions

Conceived and designed the experiments: SCW JEM PKH. Performed the experiments: SCW JRS. Analyzed the data: SCW JRS JEM PKH. Contributed reagents/materials/analysis tools: SCW JRS JEM PKH. Wrote the paper: SCW JRS JEM PKH.

References

1. La Scola B, Desnues C, Pagnier I, Robert C, Barrassi L, et al. (2008) The virophage as a unique parasite of the giant mimivirus. Nature 455: 100–4.
2. Rice G, Tang L, Stedman K, Roberto F, Spuhler J, et al. (2004) The structure of a thermophilic archaeal virus shows a double-stranded DNA viral capsid type that spans all domains of life. Proc Natl Acad Sci USA 101: 7716–20.
3. Suttle CA. (2005) Viruses in the sea. Nature 437: 356–61.
4. Angly FE, Felts B, Breitbart M, Salamon P, Edwards RA, et al. (2006) The marine viromes of four oceanic regions. PLoS Biol 4: 2121–2131.
5. Acinas SG, Haverkamp THA, Huisman J, Stal LJ. (2009) Phenotypic and genetic diversification of *Pseudanabaena* spp. (cyanobacteria). ISME J 3: 31–46.
6. Zhong F, Gao Y, Yu T, Zhang Y, Xu D, et al. (2011) The management of undesirable cyanobacteria blooms in channel catfish ponds using a constructed wetland: Contribution to the control of off-flavor occurrences. Water Res 45: 6479–88.
7. Pepe-Ranney C, Berelson WM, Corsetti FA, Treants M, Spear JR. (2012) Cyanobacterial construction of hot spring siliceous stromatolites in Yellowstone National Park. Environ Microbiol 14: 1182–1197.
8. Berman-Frank I, Lundgren P, Falkowski P. (2003) Nitrogen fixation and photosynthetic oxygen evolution in cyanobacteria. Res Microbiol 154: 157–64.
9. Zehr JP. (2011) Nitrogen fixation by marine cyanobacteria. Trends Microbiol 19: 162–73.
10. Rohwer F, Thurber RV. (2009) Viruses manipulate the marine environment. Nature 459: 207–12.
11. Short CM, Suttle CA. (2005) Nearly identical bacteriophage structural gene sequences are widely distributed in both marine and freshwater environments. Appl Environ Microbiol 71: 480–486.
12. Dreher TW, Brown N, Bozarth CS, Schwartz AD, Riscoe E, et al. (2011) A freshwater cyanophage whose genome indicates close relationships to photosynthetic marine cyanomyophages. Environ Microbiol 13: 1858–74.
13. Chénard C, Suttle CA. (2008) Phylogenetic diversity of sequences of cyanophage photosynthetic gene psbA in marine and freshwaters. Appl Environ Microbiol 74: 5317–24.
14. Paerl HW, Fulton RS, Moisander PH, Dyble J. (2001) Harmful freshwater algal blooms, with an emphasis on cyanobacteria. ScientificWorldJournal 1: 76–113.
15. Deng L, Hayes PK. (2008) Evidence for cyanophages active against bloom-forming freshwater cyanobacteria. Freshw Biol 53: 1240–1252.
16. Sogge H, Rohrlack T, Rounge TB, Sønstebø JH, Tooming-Klunderud A, et al. (2013). Gene flow, recombination, and selection in cyanobacteria: Population structure of geographically related *Planktothrix* freshwater strains. Appl Environ Microbiol 79: 508–15.
17. Rohrlack T, Skulberg R, Halstvedt CB, Utkilen HC, Ptacnik R, et al. (2008) Oligopeptide chemotypes of the toxic freshwater cyanobacterium *Planktothrix* can form subpopulations with dissimilar ecological traits. Limnol Oceanogr 53: 1279–1293.
18. Briand E, Escoffier N, Straub C, Sabart M, Quiblier C, et al. (2009). Spatiotemporal changes in the genetic diversity of a bloom-forming *Microcystis aeruginosa* (cyanobacteria) population. ISME J 3: 419–29.
19. Wilson AE, Sarnelle O, Neilan BA, Salmon TP, Gehringer MM, et al. (2005) Genetic variation of the bloom-forming cyanobacterium *Microcystis aeruginosa* within and among lakes: Implications for harmful algal blooms. Appl Environ Microbiol 71: 6126–6133.
20. Jenkins CA, Hayes PK. (2006). Diversity of cyanophages infecting the heterocystous filamentous cyanobacterium *Nodularia* isolated from the brackish Baltic Sea. J Mar Biol Assoc UK 86: 529–536.
21. Huang S, Wang K, Jiao N, Chen F. (2012) Genome sequences of siphoviruses infecting marine *Synechococcus* unveil a diverse cyanophage group and extensive phage-host genetic exchanges. Environ Microbiol 14: 540–558.
22. McDaniel L, Breitbart M, Mobberley J, Long A, Haynes M, et al. (2008) Metagenomic analysis of lysogeny in Tampa Bay: Implications for prophage gene expression. PLoS One 3: 1–9.
23. Sullivan MB, Waterbury JB, Chisholm SW. (2003) Cyanophages infecting the oceanic cyanobacterium *Prochlorococcus*. Nature 424: 1047–51.
24. Marston MF, Sallee JL. (2003) Genetic diversity and temporal variation in the cyanophage community infecting marine *Synechococcus* species in Rhode Island's coastal waters. Appl Environ Microbiol 69: 4639–4647.
25. Wilson WH, Joint IR, Carr NG, Mann NH. (1993) Isolation and molecular characterization of five marine cyanophages propagated on *Synechococcus* sp. strain WH7803. Appl Environ Microbiol 59: 3736–43.
26. Lu J, Chen F, Hodson RE. (2001) Distribution, isolation, host specificity, and diversity of cyanophages infecting marine *Synechococcus* spp. in river estuaries. Appl Environ Microbiol 67: 3285–3290.
27. Lenski RE, Levin BR. (1985) Constraints on the coevolution of bacteria and virulent phage: a model, some experiments, and predictions for natural communities. Am Nat 125: 585–602.
28. Mann NH. (2003) Phages of the marine cyanobacterial picophytoplankton. FEMS Microbiol Rev 27: 17–34.
29. Hoskisson PA, Smith MCM. (2007) Hypervariation and phase variation in the bacteriophage "resistome". Curr Opinion Microbiol 10: 396–400.
30. Labrie SJ, Frois-Moniz K, Osburne MS, Kelly L, Roggensack SE, et al. (2012) Genomes of marine cyanopodoviruses reveal multiple origins of diversity. Environ Microbiol 15: 1356–76.
31. Kuno S, Yoshida T, Kaneko T, Sako Y. (2012) Intricate interactions between the bloom-forming cyanobacterium *Microcystis aeruginosa* and foreign genetic elements revealed by diversified CRISPR signatures. Appl Environ Microbiol 78: 5353–60.
32. Hadas H, Einav M, Fishov I, Zaritsky A. (1997) Bacteriophage T4 development depends on the physiology of its host *Escherichia coli*. Microbiology 143: 179–85.
33. Petrov VM, Nolan JM, Bertrand C, Levy D, Desplats C, et al. (2006). Plasticity of the gene functions for DNA replication in the T4-like phages. J Mol Biol 361: 46–68.
34. Yoshida T, Nagasaki K, Takashima Y, Shirai Y, Tomaru Y, et al. (2008). Ma-LMM01 infecting toxic Microcystis aeruginosa illuminates diverse cyanophage genome strategies. J Bacteriol 190: 1762–72.
35. Holmfeldt K, Solonenko N, Shah M, Corrier K, Riemann L, et al. (2013). Twelve previously unknown phage genera are ubiquitous in global oceans. Proc Natl Acad Sci 110: 12798–12803.
36. Ponsero AJ, Chen F, Lennon JT, Wilhelm W. (2013). Complete genome sequence of cyanobacterial Siphovirus KBS2A Genome Announc 1: 3–4.
37. Ackermann HW, Tiekotter KL. (2012). Murphy's law-if anything can go wrong, it will: Problems in phage electron microscopy. Bacteriophage 2: 122–129.
38. Kuznetsov YG, Chang S-C, Credaroli A, Martiny J, McPherson A. (2012) An atomic force microscopy investigation of cyanophage structure. Micron 43: 1336–42.
39. Sikora AE, Smith JR, Campbell SA, Firman K. (2012). AFM protein–protein interactions within the EcoR124I molecular motor. Soft Matter 8: 6358–6363.
40. Rippka R, Deruelles J, Waterbury JB, Herdman M, Stanier RY. (1979) Generic assignments, strain histories and properties of pure culture of cyanobacteria. J Gen Microbiol 111: 1–61.
41. Booker MJ, Walsby AE. (1979) The relative form resistance of straight and helical blue-green algal filaments. Eur J Phycol 14: 141–150.
42. Ackermann HW. (2009) Phage classification and characterisation. In: Clokie MRJ & Kropinski AM, editors. Bacteriophages: Methods and Protocols 1: Isolation, Characterization and Interactions. New York: Humana Press. 175–202.
43. Cook A. (2003) Studies on the genome of the marine cyanophage S-PM2. Ph.D Thesis. University of Warwick, UK.
44. Fuller NJ, Wilson WH, Joint IR, Mann NH. (1998) Occurrence of a sequence in marine cyanophages similar to that of T4 g20 and its application to PCR-based detection and quantification techniques. Appl Environ Microbiol 64: 2051–2060.
45. Sullivan MB, Coleman ML, Quinlivan V, Rosenkrantz JE, Defrancesco AS, et al. (2008) Portal protein diversity and phage ecology. Environ Microbiol 10: 2810–23.
46. Avrani S, Wurtzel O, Sharon I, Sorek R, Lindell D. (2011) Genomic island variability facilitates *Prochlorococcus*-virus coexistence. Nature, 474: 604–8.
47. Abedon ST, Hyman P, Thomas C. (2003) Experimental examination of bacteriophage latent-period evolution as a response to bacterial availability. Appl Environ Microbiol 69: 7499–7506.
48. Delbrück M. (1940) The growth of bacteriophage and lysis of the host. J Gen Physiol 23: 643–60.

49. Hyman P, Abedon ST. (2009) Practical methods for determining phage growth parameters. In: Clokie MRJ & Kropinski AM, editors. Bacteriophages: Methods and Protocols 1: Isolation, Characterization and Interactions. New York: Humana Press. 175–202.
50. Baker AC, Goddard VJ, Davy J, Schroeder DC, Adams DG, et al. (2006) Identification of a diagnostic marker to detect freshwater cyanophages of filamentous cyanobacteria. Appl Environ Microbiol 72: 5713–9.
51. Kimura S, Sako Y, Yoshida T. (2013) Rapid *Microcystis* cyanophage gene diversification revealed by long- and short-term genetic analyses of the tail sheath gene in a natural pond. Appl Environ Microbiol 79: 2789–95.
52. Van Twest R, Kropinski AM. (2009) Bacteriophage enrichment from water and soil. In: Clokie MRJ & Kropinski AM, editors. Bacteriophages: Methods and Protocols 1: Isolation, Characterization and Interactions. New York: Humana Press. 15–22.
53. Dunbar J, White S, Forney L. (1997) Genetic diversity through the looking glass: effect of enrichment bias. Appl Environ Microbiol 63: 1326–31.
54. Amann RI, Ludwig W, Schleifer KH. (1995) Phylogenetic identification and in situ detection of individual microbial cells without cultivation. Microbiol Rev 59: 143–69.
55. Hugenholtz P, Goebel BM, Pace NR. (1998) Impact of culture-independent studies on the emerging phylogenetic view of bacterial diversity. J Bacteriol 180: 4765–4774.
56. Ankrah NYD, May AL, Middleton JL, Jones DR, Hadden MK, et al. (2013). Phage infection of an environmentally relevant marine bacterium alters host metabolism and lysate composition. ISME J.
57. Shan J, Jia Y, Clokie MRJ, Mann NH. (2008). Infection by the 'photosynthetic' phage S-PM2 induces increased synthesis of phycoerythrin in Synechococcus sp. WH7803. FEMS Microbiol Lett 283: 154–161.
58. Miller RV, Day MJ. (2008) Contribution of lyosgeny, pseudolysogeny, and starvation to phage ecology. In: Abedon ST, editor. Bacteriophage ecology. Cambridge: Cambridge University Press. 114–143.
59. Frangeul L, Quillardet P, Castets A-M, Humbert J-F, Matthijs HCP, et al. (2008) Highly plastic genome of *Microcystis aeruginosa* PCC 7806, a ubiquitous toxic freshwater cyanobacterium. BMC Genomics 9: 274.
60. Tanabe Y, Kasai F, Watanabe M M. (2009) Fine-scale spatial and temporal genetic differentiation of water bloom-forming cyanobacterium *Microcystis aeruginosa*: revealed by Multilocus Sequence Typing. Environ Microbiol Rep 1: 575–82.
61. Humbert JF, Duris-Latour D, Le Berre B, Giraudet H, Salençon MJ. (2005) Genetic diversity in *Microcystis* populations of a French storage reservoir assessed by sequencing of the 16S-23S rRNA intergenic spacer. Microb Ecol 49: 308–14.
62. Middelboe M. (2000) Bacterial growth rate and marine virus–host dynamics. Microb Ecol 40: 114–124.
63. Cohan FM. (2002) What are bacterial species? Annu Rev Microbiol: 457–87.
64. Rocap G, Distel DL, Waterbury JB, Chisholm SW. (2002) Resolution of *Prochlorococcus* and *Synechococcus* ecotypes by using 16S-23S ribosomal DNA internal transcribed spacer sequences. Appl Environ Microbiol 68: 1180–1191.
65. Weitz JS, Hartman H, Levin SA (2005). Coevolutionary arms race between bacteria and bacteriophage. Proc Natl Acad Sci USA 102: 9535–9540.
66. Zhao F, Zhang X, Liang C, Wu J, Bao Q, et al. (2006). Genome-wide analysis of restriction-modification system in unicellular and filamentous cyanobacteria. Physiol Genomics 24: 181–90.
67. Snyder L. (1995) Phage-exclusion enzymes: a bonanza of biochemical and cell biology reagents? Mol Microbiol 15: 415–20.
68. Sorek R, Kunin V, Hugenholtz P. (2008) CRISPR - a widespread system that provides acquired resistance against phages in bacteria and archaea. Nat Rev Microbiol 6: 181–6.
69. Kuno S, Yoshida T, Kaneko T, Sako Y (2012). Intricate interactions between the bloom-forming cyanobacterium Microcystis aeruginosa and foreign genetic elements revealed by diversified CRISPR signatures. Appl Environ Microbiol 78: 5353–5360.
70. Garcia-Pichel F, Prufert-Bebout L, Muyzer G. (1996) Phenotypic and phylogenetic analyses show *Microcoleus chthonoplastes* to be a cosmopolitan cyanobacterium. Appl Environ Microbiol 62: 3284–91.
71. Nelissen B, De Baere R, Wilmotte A, De Wachter R. (1996) Phylogenetic relationships of nonaxenic filamentous cyanobacterial strains based on 16S rRNA sequence analysis. J Mol Evol 42: 194–200.

9

Functional Tradeoffs Underpin Salinity-Driven Divergence in Microbial Community Composition

Chris L. Dupont[1*§], John Larsson[2*§], Shibu Yooseph[3], Karolina Ininbergs[2], Johannes Goll[4], Johannes Asplund-Samuelsson[2], John P. McCrow[1], Narin Celepli[2], Lisa Zeigler Allen[1], Martin Ekman[2], Andrew J. Lucas[5], Åke Hagström[6], Mathangi Thiagarajan[4], Björn Brindefalk[2], Alexander R. Richter[3], Anders F. Andersson[7], Aaron Tenney[3], Daniel Lundin[7], Andrey Tovchigrechko[4], Johan A. A. Nylander[8], Daniel Brami[3], Jonathan H. Badger[3], Andrew E. Allen[1], Douglas B. Rusch[4], Jeff Hoffman[1], Erling Norrby[9], Robert Friedman[1], Jarone Pinhassi[10], J. Craig Venter[1], Birgitta Bergman[2]

1 Microbial and Environmental Genomics, J. Craig Venter Institute, San Diego, California, United States of America, **2** Department of Ecology, Environment and Plant Sciences, Stockholm University, Stockholm, Sweden, **3** Informatics Group, J. Craig Venter Institute, San Diego, California, United States of America, **4** Informatics Group, J. Craig Venter Institute, Rockville, Maryland, United States of America, **5** Marine Physical Laboratory, Scripps Institution of Oceanography, University of California San Diego, San Diego, California, United States of America, **6** Swedish Institute for the Marine Environment (SIME), University of Gothenburg, Gothenburg, Sweden, **7** KTH Royal Institute of Technology, Science for Life Laboratory, School of Biotechnology, Solna, Sweden, **8** Department of Biodiversity Informatics, Swedish Museum of Natural History, Stockholm, Sweden, **9** Center for History of Science, The Royal Swedish Academy of Sciences, Stockholm, Sweden, **10** Centre for Ecology and Evolution in Microbial Model Systems, Linnaeus University, Kalmar, Sweden

Abstract

Bacterial community composition and functional potential change subtly across gradients in the surface ocean. In contrast, while there are significant phylogenetic divergences between communities from freshwater and marine habitats, the underlying mechanisms to this phylogenetic structuring yet remain unknown. We hypothesized that the functional potential of natural bacterial communities is linked to this striking divide between microbiomes. To test this hypothesis, metagenomic sequencing of microbial communities along a 1,800 km transect in the Baltic Sea area, encompassing a continuous natural salinity gradient from limnic to fully marine conditions, was explored. Multivariate statistical analyses showed that salinity is the main determinant of dramatic changes in microbial community composition, but also of large scale changes in core metabolic functions of bacteria. Strikingly, genetically and metabolically different pathways for key metabolic processes, such as respiration, biosynthesis of quinones and isoprenoids, glycolysis and osmolyte transport, were differentially abundant at high and low salinities. These shifts in functional capacities were observed at multiple taxonomic levels and within dominant bacterial phyla, while bacteria, such as SAR11, were able to adapt to the entire salinity gradient. We propose that the large differences in central metabolism required at high and low salinities dictate the striking divide between freshwater and marine microbiomes, and that the ability to inhabit different salinity regimes evolved early during bacterial phylogenetic differentiation. These findings significantly advance our understanding of microbial distributions and stress the need to incorporate salinity in future climate change models that predict increased levels of precipitation and a reduction in salinity.

Editor: Jose Luis Balcazar, Catalan Institute for Water Research (ICRA), Spain

Funding: This work was supported by the Beyster Family Fund of the San Diego Foundation; Life Technologies Foundation; National Science Foundation (grant 0850256); BalticSea2020 Foundation; O. Engkvist Byggmästare Foundation; Science for Life Laboratory, Stockholm; Baltic Ecosystem Adaptive Management Program; EcoChange. The funders had no role in study design, data collection and analysis, decision to publish, or preparation of the manuscript.

Competing Interests: The authors have declared that no competing interests exist.

* E-mail: cdupont@jcvi.org (CLD); john.larsson@su.se (JL)

§ These authors contributed equally to this work.

Introduction

Metagenomic surveys of the surface ocean reveal a continuum in bacterial community composition with phylogenetic variations at the genus or family level [1–4] while phylogenetic marker surveys of salinity gradients have highlighted changes at higher taxonomic levels [5–10]. Despite large population sizes and high dispersal rates, the boundary between freshwater and marine environments appears to constitute an almost insurmountable transition barrier for bacteria [8], yet the underlying reasons have remained enigmatic. A few studies have compared genomic characteristics of microbial communities at a limited number of salinities, with subtle changes observed in functional gene repertoires [11–13]. While highly significant, these studies examined disconnected microbial communities and used different extraction procedures, which limit comparisons [14].. To surpass such shortcomings, we examined functional adaptations to salinity along a natural continuous gradient, spanning the freshwater to marine range, which allowed direct comparative metagenomic analyses, and used robust multivariate statistical methods, fully incorporating potentially co-varying environmental variables.

The Baltic Sea, one of the largest brackish ecosystems in the world (Fig. 1A), is intercontinental, non-tidal, and subject to constant freshwater input from major rivers in the north and saline water inflows from the North Sea resulting in a stable north-south salinity gradient of 2-10 practical salinity units (PSU) (Fig. 1B). The Baltic Sea is a 'young' water environment with the brackish state existing for about 4000 years. The low to moderate salinity levels are known to have imposed strong evolutionary constraints upon multicellular eukaryotes with limited species richness relative to adjacent fresh- and marine waters [15] while consequences for the microbial communities have remained largely unknown. The first large scale molecular (16S rRNA gene) survey of bacterial diversity in the Baltic Sea reported a community structured by salinity gradients with specific microbes at each salinity [16]. This contrasts with estuaries exhibiting faster turnover times and a mix of freshwater and marine bacteria at intermediate salinities [5]. Presumably, the long residence time in the Baltic Sea (3–30 years, [17]) allows for local adaptations along the salinity gradient. Thus the Baltic Sea provides an ideal natural model system for examining how salinity governs both microbial community composition and metabolic function. In an attempt to understand the evolutionary separation of freshwater and marine microbes we sequenced the genomes of microbial communities along the Baltic Sea salinity gradient and linked changes in functional potential to environmental parameters. Our results indicate that large-scale changes in core metabolic pathways determine which organisms succeed at different salinity regimes and that such profound changes explain the deep phylogenetic divide between freshwater and marine microbiomes.

Materials and Methods

Samples were collected during the European leg of The Sorcerer II Global Ocean Sampling expedition in July 2009 with parallel environmental characterization using an integrated CTD-oxygen, chlorophyll a, and pH sensor package. For each country, a Marine Science Research Permit was submitted under the U.N. Conference on Law of the Sea to the U.S. State Department which contacted the Ministry of Foreign Affairs in the respective country. Submissions were then distributed to the appropriate agency in each country. For Sweden, permission was granted by the Swedish Coast Guard. For Denmark, permission was granted by the Danish Ministry of Foreign Affairs. For Germany, permission was granted by the German Federal Maritime and Hydrographic Agency (BSH). In addition, the Access and Benefit Sharing Focal Point was contacted under the U.N. Convention on Biological Diversity to ascertain whether additional permissions were required to access genetic resources. Separate applications were not required in the visited countries.

A total of 21 microbial communities were sampled at 11 locations within the brackish Baltic Sea area (Fig. 1A, Table S1). Eight locations were sampled at two depths (including seven surface/chl a maximum pairs and one chl a maximum/suboxic zone pair), one location was sampled at three depths (surface/chl a maximum/72 m) and at two locations only surface samples were taken. Environmental metadata and coordinates for each sampling location are given in Table S1. Sample preservation, extraction, and sequencing was conducted as described previously [3]. Reads were annotated using the JCVI metagenomic annotation pipeline [18], along with APIS [3], and fragment recruitment [2]. A global assembly of all sequence data was performed using the Newbler Assembler [19], with a final assembly of 490 million basepairs, a N50 of 1.5 Kbp, and a largest contig of 106 Kbp. Approximately 15,000 16S rRNA subunit amplicons were also sequenced for each location and size fraction and phylogenetically annotated as described previously [16]. All sequence and annotation data is available at CAMERA [20] under accession number CAM_P_0001109 and Metarep [21] (metarep.jcvi.org). Detailed descriptions of community composition determinations and regularized canonical correlation analysis (RCCA) are available in the Supplementary information.

Results and Discussion

We collected aquatic samples from 21 freshwater, marine, and brackish sites within the Baltic Sea area, including the freshwater arctic lake Torne Träsk. The transect spans the Bothnian Bay, Bothnian Sea and the large Baltic Proper sub-basins, in turn semi-connected to the marine waters of the Swedish west coast, here referred to as Baltic West (Fig. 1A). Interpolated depth profiles of temperature, oxygen, salinity, pH, and chlorophyll a fluorescence (chl a) place the samples in context of a basin wide salinity gradient (Fig. 1B). The majority of the samples were from the relatively young (<30 years) surface layer of the Baltic Sea. Two samples were taken below the halocline and thus capture older waters: GS678 represents an oxygen minimum sample taken at 74 m within the Baltic Proper and GS689 at 72 m depth in the oxygenated waters in fjord Gullmarn on the Swedish west coast. Comparisons with the long-term Baltic Sea monitoring program verified that the conditions were typical for the sampled locations at that time of year (Fig. S1). Samples were fractionated into particle size classes of 0.1–0.8, 0.8–3.0 and 3.0–200 μm, each of which was shotgun-sequenced using pyrosequencing. The dataset comprises 7.6 billion nucleotides in 20.9 million reads (avg. length = 324 bp), with 22.9 million translated amino acid sequences (Table S2). In addition, approximately 15,000 16S rRNA subunit sequences were obtained for each size-fractionated sample.

Microbial community composition in the Baltic Sea varied dramatically relative to the considerably more geographically distant ocean sites in previous Global Ocean Sampling (GOS) datasets, both in terms of magnitude and taxonomic hierarchy [2,4,22]. Specifically, the most abundant bacterial phyla and classes changed in a continuous fashion along the transect (Fig. 2A) while classes of viruses and phytoplankton divisions were equally dynamic but exhibited a more stochastic pattern (Fig. S2). These trends were robust to annotation biases as 16S rRNA gene amplicon sequencing, the AMPHORA2 classification system [23] (31 core genes) and automated phylogenetic inference system (APIS [3], all genes) showed strong correlations (Fig. S3). Examining the individual size fractionated metagenomes revealed that the changes in bacterial community composition occur almost entirely in the 0.1–0.8 and 0.8–3.0 μm size fractions, which likely contain free-living bacteria, and not the 3–200 μm size fraction (Fig. S4), despite the salinity being equal. Bacteria found in these larger size fractions have substantially larger average genome sizes relative to those found in the smaller size fractions (see Table S3 and [3]), which allows for an expanded regulatory and metabolic capacity [24]. We interpret our results to suggest that expanded metabolic and regulatory potential relax the influence of environment on genomic contents.

In order to investigate the environmental effect on the microbial community composition we used regularized canonical correlation analysis (RCCA), a multivariate method that analyzes the correlation between sets of several independent (e.g. environmental data) and dependent (e.g. metagenomic signatures) variables

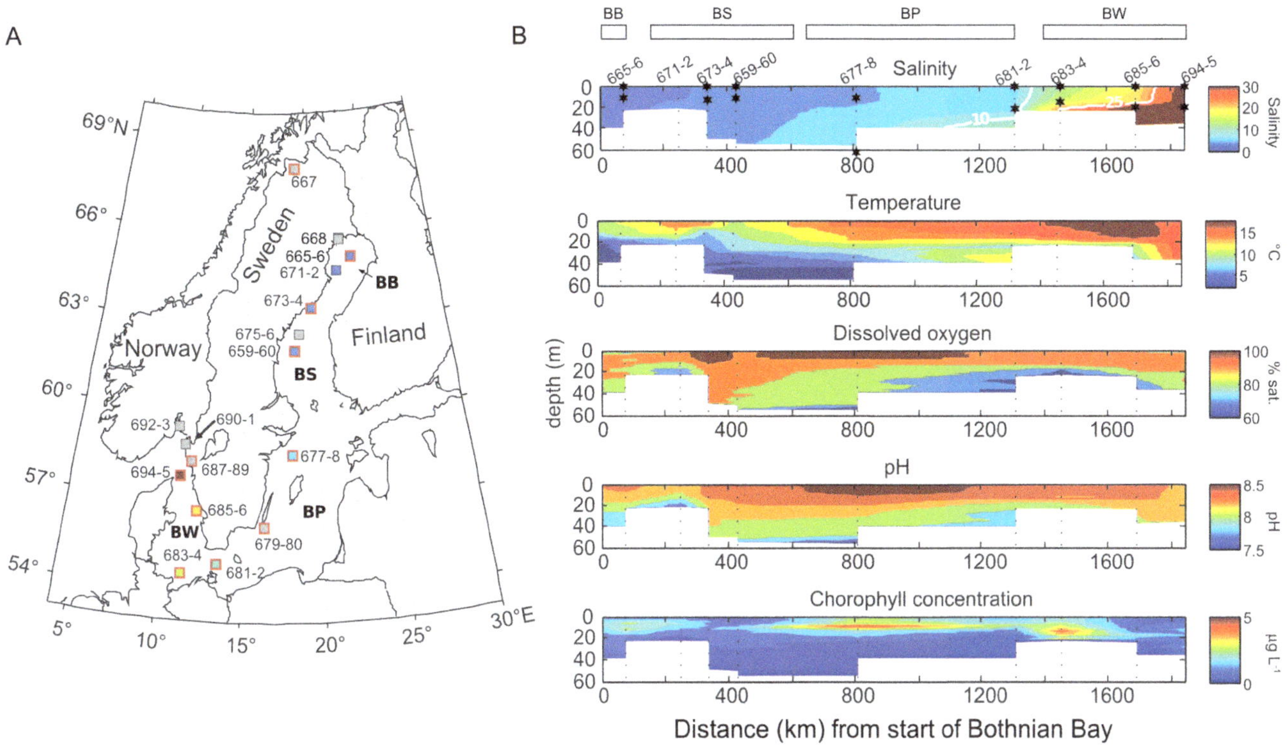

Figure 1. The Baltic Sea region and sampling locations. A) Squares with red outlines indicate sites with metagenomic sequencing. 16S rRNA gene tag sequencing was conducted at all sites. B) Temperature, salinity, pH, oxygen, and chlorophyll a measurements across the Baltic Sea transect. Only stations shown on the section were used for section interpolation. The transition from limnic to marine conditions along the 1800 km axis of the Baltic Sea leads to an extensive area of slowly varying brackish conditions. Suboxic, low-pH conditions exist throughout the central basin below 50–60 m. All environmental metadata is in table S1. Sub-basin abbreviations are marked on the map in bold font (BB – Bothnian Bay, BS – Bothnian Sea, BP – Baltic Proper, BW – Baltic West).

[25,26]. Briefly, it calculates linear combinations of variables (canonical variates) in each set and finds the maximum (canonical) correlation between the combinations. It simultaneously measures the relationship between and within the variable sets and weights the contribution of the variables in each set to the global relationship. This makes the method robust to dependencies among the variables. RCCA of our environmental and metagenomic data showed that the first linear combination of environmental variables (canonical variate) was primarily and negatively associated with salinity ($r = -0.976$) at the phylum level (Fig. 2B) for combined size fractions, with similar results at the genus level (Fig. S5). Thus, the changes in bacterial community composition were primarily driven by changes in salinity. Organismal phyla/classes preferentially found in low-salinity habitats were actinobacteria, chloroflexi, betaproteobacteria, and the eukaryote ciliate oligohymenophora. In contrast, gammaproteobacteria, alphaproteobacteria, bacteroidetes, and N4-like viruses (*Podoviridae*) were positively associated with higher salinity (Fig. 2B). The second highest canonical variate (canonical correlation 0.734) associated positively with the N:P nutrient ratio and negatively with chl a and oxygen. Organisms mostly associated with this environmental combination were aquificae, cyanobacteria, and phycoviridae. These results recapitulate the trends observed for single or multiple classes of bacteria in previous surveys [5–7,9,10,16], while adding new insight to viruses and eukaryotes. Our experimental design also allowed us to relate these distribution patterns to potential patterns in functional genes, which was not done in the previous studies. Due to the general lack of statistically significant dependence of eukaryotes and viruses on measured environmental variables we focused the functional analysis on bacteria. We also note that bacterial reads comprised 80.1% of the complete dataset and thus provide the coverage necessary for detailed within taxonomy analyses not possible with viruses or eukaryotes.

A RCCA analysis of environmental variables and KEGG functional modules of bacteria, including metabolic pathways, transporters, and two-component transcriptional regulation systems showed that salinity was the primary constituent of the first environmental canonical variate M1 ($r = -0.986$) while M2 correlated positively with N:P ratio ($r = 0.770$) and negatively with total phosphorous ($r = -0.716$). The linear combination of modules that maximized correlation with M1 (canonical correlation 0.997) separated a majority of individual KEGG modules (Fig. 3A). This analysis shows that, as was the case for community composition, salinity was the main determinant of nearly all large-scale differences in the metabolic potential of the Baltic Sea bacterial communities. Importantly, these results show correlations between combinations of variables (environmental and metagenomic), thereby minimizing confounding effects potentially introduced by co-variation between salinity and other environmental variables. Similar M1 or salinity-driven KEGG module correlations were observed for both 0.1 and 0.8 µm size fractions (Fig. S6A–B), while bacteria in the largest size fraction (3.0–200 µm), with larger genomes and perhaps attached to particles or aggregates, displayed slightly different profiles, including a better correlation with oxygen and chl a (Fig. S6C–D). This latter result is consistent with the general lack of a salinity-dependent biogeography for the largest size fraction noted above (Fig. S4). Thus there appears to be a scenario where increased genome size,

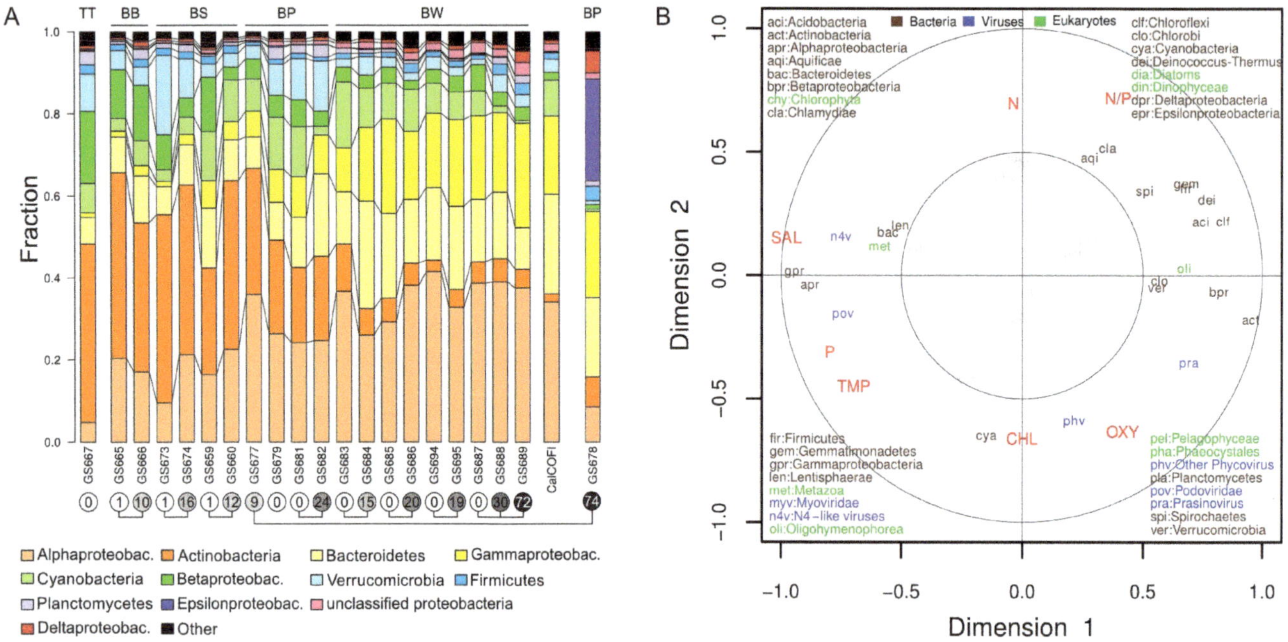

Figure 2. Salinity-driven shifts in bacterial community composition. A) Bacterial communities characterized by AMPHORA2. The 12 most abundant groups are shown with "Other" defined as the remaining taxonomic groups. Sequences have been combined for all non-viral size fractions (0.1, 0.8 and 3.0 μm) with normalization for genome equivalents. Sampling depth (m) is indicated within circles below sample names, with interconnecting circles indicating samples from the same location. The corresponding basin for samples is shown in the top margin: TT = Torne Träsk, BB = Bothnian Bay, BS = Bothnian Sea, BP = Baltic Proper, BW = Baltic West. A merged set of metagenomes from the coastal eastern Pacific (CalCOFI) was included for comparison. B) RCCA analysis of community composition. The biplot shows correlations between environmental (red) and phylogenetic (color coded by superkingdom) variables with the corresponding first and second canonical variates. The first two canonical variates explain 65% of the environmental variables and 49% of the organismal variables. Biplots for a RCCA at the genus level is shown in figure S5. The outer circle (1) and inner circles (0.5) display the amount of variance explained by the linear combinations of variables. The variance explained for points within the inner circle is less than 25%, thus these have been dimmed.

potentially facilitated by a higher trophic status, overrules the effects of salinity observed in pelagic bacteria. Overall, these results are highly different from previous RCCA analysis of surface marine metagenomes, all collected from sites with PSU values greater than 30, where nitrate and temperature were more important factors in governing community structure [22]. Notably, our use of KEGG modules allows for an explicit exploration of linked pathways and metabolic changes.

Strikingly, we found that salinity strongly influenced the usage of central analogous metabolic pathways, each of which shares products or substrates with another salinity-influenced pathway (Fig. 3, Fig. S7). Analogous metabolic pathways achieve approximately the same result, although using different intermediate metabolites, energetics, and genes. For instance, the H^+-translocating (NDH) and the Na^+-translocating (NQR) NADH:quinone dehydrogenases, key enzymes in bacterial respiration, were more abundant in low and high salinity environments, respectively (Fig. 3B, Fig. S7). Biosynthetic pathways for two major isoprenoid quinones, menaquinone (MK) and ubiquinone (UQ) were also found to correlate with low and high salinity environments, respectively. The NQR dehydrogenase effectively exports sodium from the cell and has been identified in a limited number of cultivable marine bacteria [27]. In addition, the energetics of the NQR dehydrogenase makes it dependent on the high-potential UQ electron acceptor [28], as reflected in our results (Fig. 3). NDH dehydrogenases generate a proton motive force across the cytoplasmic membrane (i.e. 1 1/3 more ATP per quinone reduction) and use the low-potential MK electron acceptor. Consequently, both the NQR and UQ pathways dominate at higher salinities, while NDH and MK pathways at low salinity. Similar trends were observed for the two non-overlapping biochemical modules for glycolysis, the Embden-Meyerhof (EMP) and Entner-Doudoroff (ED) pathways, and the non-mevalonate and mevalonate biosynthetic pathways used for isoprenoid biosynthesis (Fig. 3, Fig. S7). The pathways of glycolysis and isoprenoid biosynthesis generate the basic building blocks required for the two quinone biosynthesis pathways, though the explicit linkages noted here are novel (Fig. S7). Overall, these salinity-associated differences were found in metabolic pathways that are interlinked by substrates within each salinity regime. Consistent with changes in glycolytic pathways observed here, a recent comparison of marine and freshwater metagenomes showed significant overrepresentation of the EMP enzyme 6-phosphofructokinase in freshwater metagenomes [13].

While Figure 3B presents the abundance of pathways along the entire salinity gradient, some pathways are selected for in specific salinity ranges. For instance, changes in dehydrogenase utilization and use of glycolytic pathways were most evident at salinity levels below 15 PSU, while quinone and isoprenoid biosynthesis changed in a linear fashion across the entire salinity gradient. These results provide insight to the distribution of isoprenoid and glycolysis pathways in natural and predominantly uncultivated microbes across salinity gradients, which was recently reviewed for cultivated organisms without regard to environmental parameters [29,30]. The metabolic units we associated with high salinity (e.g. UQ quinones and ED glycolysis) and low salinity (e.g. MK quinones and EMP glycolysis) have been linked to aerobic and anaerobic strategies, respectively, in cultivated prokaryotes

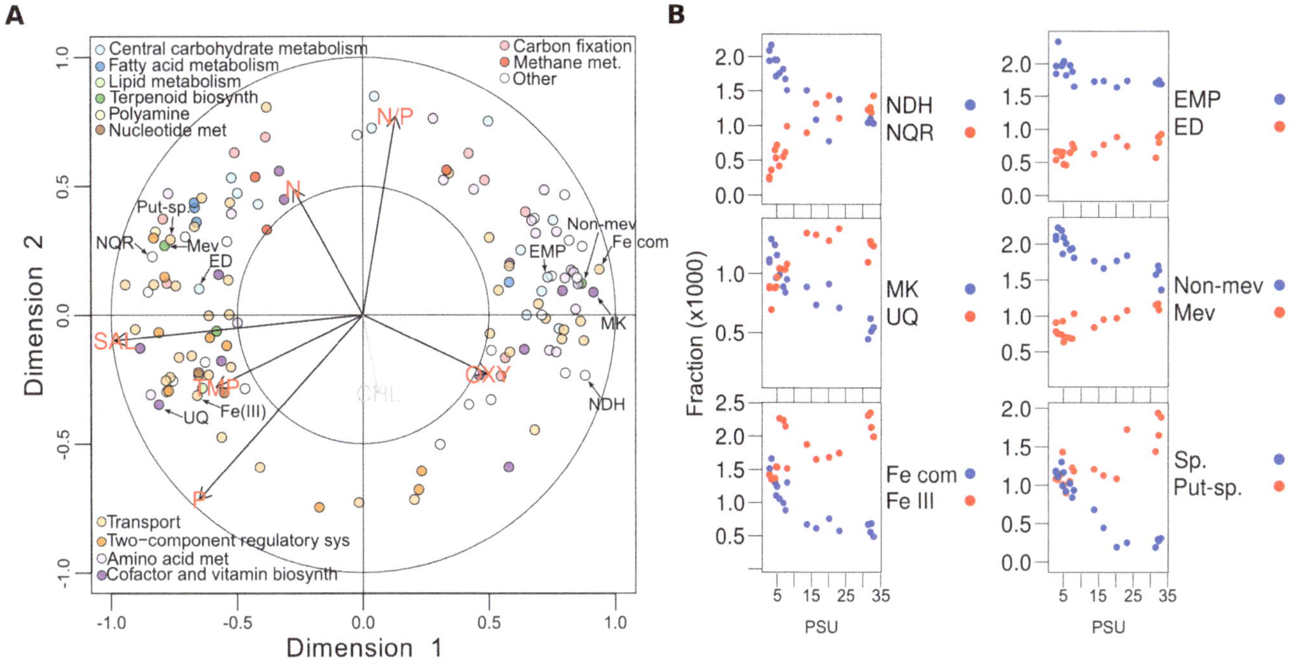

Figure 3. Salinity and functional potential. A) KEGG module and metadata correlations with first and second canonical variates for the bacterial communities in combined size fractions. Text labels and associated arrows indicate the position of analogous modules shown in panel B. The outer circle (1) and inner circles (0.5) display the amount of variance explained by the linear combinations of variables. The variance explained for points within the inner circle is less than 25%, thus these have been dimmed. Biplots for each size fraction and the combined communities are shown in fig. S6. B) Plots showing the distribution of analogous central metabolic pathways highlighted in panel A along the salinity gradient. Note that the spermidine transport system ('Sp.') is not shown in panel A due to its non-normal distribution.

[30,31]. It has been suggested that this differential usage of quinones is due to their respective midpoint potentials, but examples of MK usage despite unfavorable redox potential have been reported [31]. Furthermore, while there is strong evidence that cultivated obligate anaerobes almost exclusively rely on the EMP pathway there is great variability in the choice of glycolytic pathway in cultivated facultative and obligate aerobes [30] that will be dominant in our oxic sample set. Our data indicate a weak negative relationship between oxygen and salinity (Fig. 3A) and it is thus unlikely that the increased usage of the EMP pathway in low-salinity environments is due to a higher proportion of anaerobic prokaryotes. The preferential usage of the EMP pathway, which yields more ATP compared to the ED analog, and the NDH dehydrogenase, which is used in ATP-producing oxidative phosphorylation, in low-salinity environments (Fig. 3) instead indicate a reliance on these pathways for ATP-production in freshwater bacteria. Potentially, marine bacteria rely to a greater extent on other sources of ATP, such as proteorhodopsins. Taken together, our cultivation-independent analysis shows that salinity is the main driver of changes related to respiratory and glycolytic pathways in the naturally occurring bacterial populations.

Some modules lacking analogous pathways were found preferentially in marine or freshwater samples, including the *com* system of competence-related DNA uptake (freshwater) and glutathione biosynthesis (marine). The increased abundance of genes for DNA uptake in freshwater samples indicates a higher level of gene-sharing between bacteria in these environments, or an increased use of naked DNA as a source of nutrients. In addition, surface-attached microbial communities have been shown to be enriched in gene-exchange functions [32] and the higher particle concentration often associated with brackish and low-salinity estuarine waters may thus facilitate cell-to-cell interactions and DNA exchange opportunities. Moreover, several ABC transporter systems with similar substrates, including those for Fe-complex (low salinity) vs. Fe (III) (high salinity) and for the polyamines spermidine (low salinity) and putrescine (high salinity), also exhibited salinity dependent abundances (Fig. S8). Not unexpectedly, import systems for the osmoprotectants glycine betaine (GBT) and choline, and lipoprotein exporters were enriched in marine sites with the converse scenario for the FtsX/FtsE putative transporter previously implicated in salinity tolerance [33]. Further, sodium export and zinc transport systems showed a strong correlation with salinity, corroborating recent findings from comparisons of freshwater and marine metagenomes [13]. However, our results show that transporters for osmolytes exhibited the greatest increase in abundance at levels greater than 20 PSU, thereby offering clarity into adaptations across continuous changes in salinity.

To some extent, phylogeny and functional potential are interlinked and one possible explanation for some of the results invokes salinity selecting for different bacterial phyla or classes for reasons beyond the metabolic changes. Glutathione biosynthesis provides a clear example of this; actinobacteria have replaced glutathione biosynthesis with mycothiol synthesis, and the decreased abundance of glutathione biosynthesis pathways at low salinities mirrors phylogenetic choice. Collating the phylogenetic and KEGG annotations prior to RCCA analyses on phylogenetic subgroups, we found salinity to be a major determinant of functional variations within the nine most abundant organismal groups that have enough coverage for statistical analyses (Fig. S9). In four of these nine groups (actinobacteria, bacteroidetes, cyanobacteria and verrucomicrobia), salinity was the primary constituent of the first environmental variate, while for three additional groups (betaproteobacteria, firmicutes and gammaproteobacteria) salinity was the secondary constituent. For instance, the Fe (III) and Fe-complex transporters were correlated with salinity for the actinobacteria, bacteroidetes and firmicutes. Non-mevalonate isoprenoid biosynthesis and the EMP glycolytic pathway were preferred at low salinity in the firmicutes, verrucomicrobia, actinobacteria and bacteroidetes with shifts towards the ED pathway at high salinity for the latter two groups. We also found preferential usage of the MK synthesis pathway at low salinity for the gammaproteobacteria and finally an increased abundance of the H^+ translocating dehydrogenase at lower salinity in the actinobacteria, betaproteobacteria, gammaproteobacteria, verrucomicrobia and bacteroidetes with a strong shift towards the Na^+ translocating enzyme at higher salinity in the latter. Many of these sub-phyla metabolic changes represent deviations from the predominant metabolic repertoire of each individual phyla based on genomic surveys [29,30]. While not all changes in central analogous pathways were observed in all bacteria, we stress that salinity was a major determinant of variations in function, both at the community level (Fig. 3) and within the most abundant bacterial phyla and classes (Fig. S9).

To further examine environmental influence on function within a taxa, we examined the phylogenetic distribution and functional potential of the most abundant order in our dataset, the alphaproteobacteria of the SAR11 cluster (order *Pelagibacterales* [34]), found across the whole Baltic Sea transect (Fig. S10A). The most abundant bacterial order in open oceans [35], *Pelagibacterales* comprise several identified subgroups (ecotypes) [34,36]. Most SAR11 sequences from the low salinity (<10 PSU) Baltic Sea samples clustered with clade IIIa *Ca.* Pelagibacter IMCC9063 (isolated from brackish waters, Svalbard [37]) and to a lesser degree with HIMB114 (Fig. 4A). Lake Gatun SAR11 sequences formed a freshwater sister group, LD12 or IIIb, while sequences from the coastal Pacific and the marine Baltic West clustered within the marine SAR11 Ia sub-clade. Sequences from mid to high salinity environments in the Baltic Sea formed a sister group to Ia, potentially Ib, a sub-clade without a reference genome. Numerous analyses indicate the Baltic populations as unique species from currently cultivated strains. For example, reads from low salinity sites were aligned to the IMCC9063 genome at 70–80% nucleotide identity (Fig. S10B), consistent with it being a new species (generally 75% average nucleotide identity). Further, the phylogenetic placement (Fig. 4A) shows a high amino-acid diversity of individual core proteins. This whole-genome and multiple protein analysis of SAR11 corroborated and expanded upon previous biogeographical findings from 16S rRNA gene based analyses [7], and illustrated the existence of novel species within the freshwater/brackish SAR11 clades.

Organisms in the order *Pelagibacterales* have small genomes and share a large core genome [34], indicating that variations in the pan-genome of SAR11 would explain its salinity-associated distribution. We therefore analysed variations in functional potential within the dataset for the reads and assembled contigs classified as *Pelagibacterales*, an analysis facilitated by the wealth of *Pelagibacterales* genomes. The SAR11 assembly pool consisted of 20,309 contigs with an N50 length of 1,730 bp and a maximum length of 28,089 bp. Large differences in functional potential were evident for SAR11 populations along the Baltic Sea salinity gradient (Fig. 4B, Fig. S10C). The data showed different strategies for amino acid, nucleotide and cofactor utilization, as well as for osmolyte synthesis between the low- and high-salinity SAR11 populations, the former adapted to *de novo* synthesis and the latter to uptake of these compounds from the surrounding medium (e.g. synthesis vs. uptake of glycine; Fig. 4B). An overrepresentation of amino-acid import systems was recently reported in high- compared to low-salinity environments [13] and our data suggest that these differences may be driven by large SAR11 populations with contrasting metabolic strategies. The mevalonate and non-mevalonate pathways for isoprenoid biosynthesis were positively and negatively correlated with salinity, respectively, in the read-based SAR11 analysis, mirroring the community level pattern (Fig. 3, Fig. S7). In addition, SAR11 transport systems for polyamines, GBT and Fe (III) were positively correlated with salinity. An assembly-driven analysis corroborated the read-based results, and also highlighted differences in ascorbate biosynthesis pathways between the low and high-salinity populations (Fig. S10C). Glycolysis, a trait completely lost in some SAR11 strains [38], via both the EMP and ED pathways were negatively correlated with salinity in the Baltic Sea dataset ($R = -0.89$ and $R = -0.87$, $p < 2.2e^{-16}$). In contrast to other proteobacteria, *Pelagibacterales* did not change their preference for NAD dehydrogenase or quinones. The correlation between type I dehydrogenase and salinity was not statistically significant within the *Pelagibacterales* and the NQR type was missing altogether. This analysis of functional potential within the SAR11 lineage shows that numerous pathways are differentially abundant at the varying salinity regimes. Essentially, the low-salinity Baltic Sea populations of SAR11 have acquired multiple energy generating catabolic pathways not found in marine SAR11 populations (Fig. 4B, Fig. S10). Overall, the phylogenetic shifts in SAR11 populations (Fig. 4A) are underpinned by large-scale metabolic changes, and this genetic flexibility has likely helped SAR11 overcome salinity gradients and contributed to its ecological success.

Summary

This dataset differs from previous metagenomic surveys examining salinity in its resolution of a continuous gradient, consistent wet-lab and informatics techniques, which is important for comparative genomics [14], its focus on the most geographically prevalent salinity gradient, that is the range between freshwater and marine systems, and a multivariate statistical analysis that reveals systematic restructuring of the successful bacterial central metabolism across this gradient.

The metabolic pathways we identified as "salinity influenced" are primordial parts of bacterial metabolism and require a considerable set of different genes, which offers a plausible explanation for the deep phylogenetic separation between the organisms that dominate at the various salinity regimes. We propose that metabolic adaptations dictate which bacteria succeed at a given salinity. Moreover, the use of pathways with a higher ATP-yield (i.e. a proton versus sodium pumping complex I and the EMP versus the ED glycolytic pathway) at low salinity could impact bacterial growth efficiency, i.e. the amount of carbon assimilated into biomass relative to that released as CO_2. Growth efficiencies vary from 0.05–0.5 and help determine if an aquatic environment is either a sink or source of CO_2 to the atmosphere [39,40]. Previous surveys across salinity gradients have shown conflicting results in the relationship between salinity and bacterial respiration [40–43] and our findings stress the need for further work to determine to what extent salinity influences CO_2 assimilation. This will help ascertain whether salinity should be incorporated in future climate change models that currently predict an increase in precipitation and subsequent reduction in salinity levels [44,45].

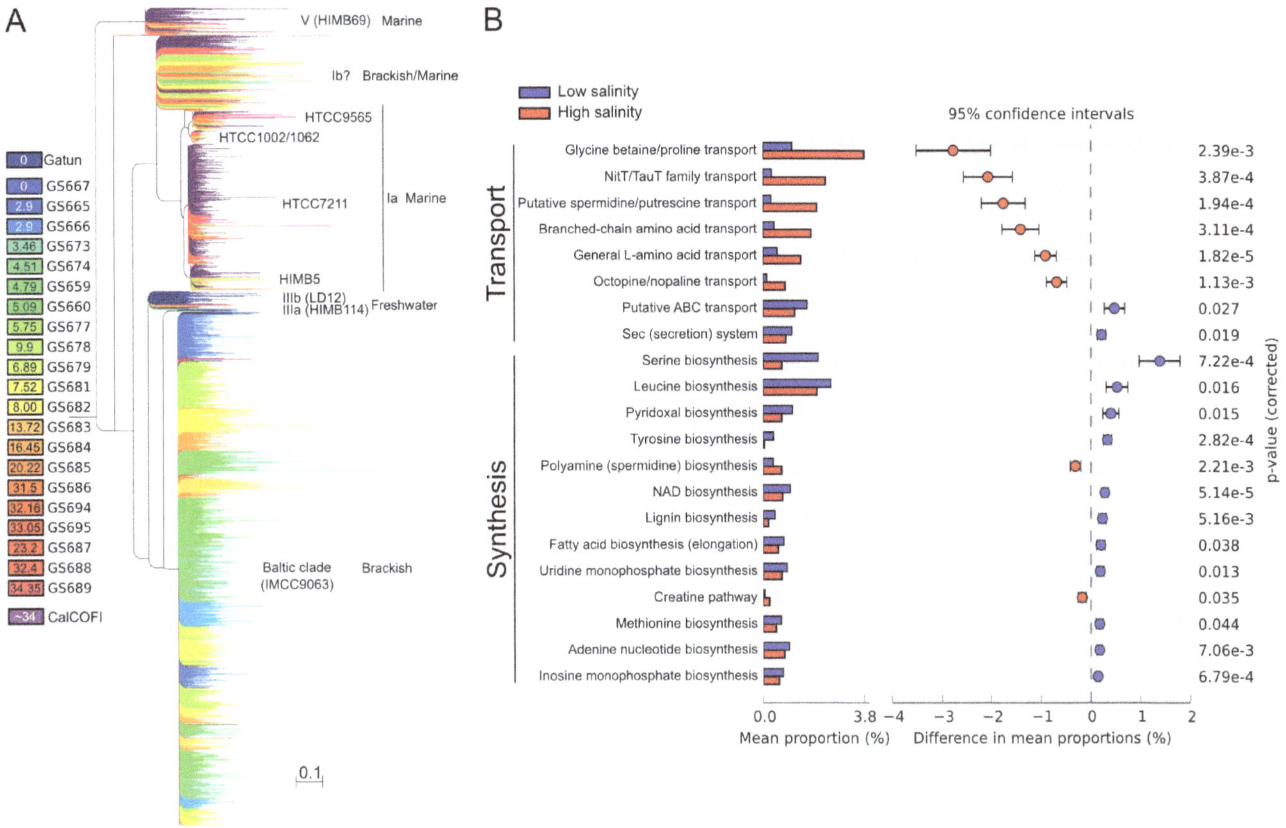

Figure 4. Baltic Sea SAR11 phylogeny and function. A) Metagenomic sequences classified as SAR11 placed onto a reference tree built from 31 concatenated core genes (see Methods). The Rickettsiales sister group (not shown for clarity) of *Candidatus* Pelagibacter was chosen as outgroup from AMPHORA2 reference trees. Terminal branches are colored according to the legend in the left margin. Sequences with likelihood weights<0.9 were pruned from the tree. Branches corresponding to sequences from the freshwater Lake Gatun and coastal California metagenomes are colored purple and dark blue, respectively. Numbers in the boxes indicate salinity (PSU). B) Functional differences between the low and high salinity SAR11 populations. Plot of differences in KEGG categories related to biosynthesis ("Synthesis") and transport ("Transport") of amino acids, nucleotides and cofactors. Bar plots show mean proportions for each category in the different populations and the confidence interval. Welch's t-test p-values (corrected using the Bonferroni method) are shown in the right margin of the plot.

Supporting Information

Figure S1 Seasonal variations in metadata variables at or near sites along the Baltic Sea transect. Seasonal measurements were acquired from the SMHI and HELCOM annual archives with values from December 1898 to January 2012. The closest set of monitoring stations were selected for each site in the transect (limited to a maximum distance of 5–30 km and maximum sampling depth difference of 10 m) and values were averaged for these stations. The x-axis corresponds to one year (Jan–Dec) and the blue line shows the average of each month. Standard deviation is indicated by the yellow area around each line. Number of total measurements for each month is shown in brackets on the x-axis. The measured data during sampling (June/July 2009) is shown as a red circle. Since no monitoring data was available for Torne Träsk (GS667) only the data measured during the 2009 sampling is shown (left). For the rest of the sampling sites, a plot shaded in red indicates that no adjacent monitoring data was present. The corresponding Baltic Sea sub-basin for each set of samples is shown above the figure: TT = Torne Träsk, BB = Bothnian Bay, BS = Bothnian Sea, BP = Baltic Proper, BW = Baltic West.

Figure S2 Taxonomic composition of eukaryotes. (A) and virus (B) along the Baltic Sea transect. All data based on APIS. Samples are arranged from north (low-salinity) to south (high-salinity) and the corresponding Baltic Sea sub-basin for each set of samples is shown below each panel: BB = Bothnian Bay, BS = Bothnian Sea, BP = Baltic Proper, BW = Baltic West.

Figure S3 Correlation plots of bacterial taxonomic composition as assayed using AMPHORA, APIS and 16S rRNA gene amplicon sequencing analysis.

Figure S4 Taxonomic composition of bacteria in the A) 0.1–0.8 μm, B) 0.8–3.0 μm, and C) 3.0–200 μm size fractions at each site. The same bacterial phyla/classes as in Figure 2A are shown. Sampling depth (m) is indicated within circles below sample names, with interconnecting circles indicating samples from the same location. The corresponding basin for samples is shown in the top margin: TT = Torne Träsk, BB = Bothnian Bay, BS = Bothnian Sea, BP = Baltic Proper, BW = Baltic West. CalCOFI represents samples taken from the coastal eastern Pacific.

Figure S5 Regularized canonical correlation analysis of the abundance of bacterial phylotypes at the genus level. For ease of visualization, each genus is color coded according to Phylum. The analysis recapitulates many of the trends from the Phyla level analysis, with actinobacteria, verrucamicrobia, and many bacteroidetes being found at low salinity, while gamma and alpha proteobacterial are found at high. This analysis does provide more detailed information, such as the alphaproteobacteria in the SAR11 lineage, HIMB114, being found at low salinity that is recapitulated by other analyses (Fig. 5A).

Figure S6 A) Correlation of functional modules against first variant M1 for both 0.1 and 0.8 um communities. B) A sample by sample comparison of the 1st dimension of 0.1 (x axis) and 0.8 (y axis) fractions. C) Module and metadata correlations with first and second canonical variates for the bacterial communities found in the 3.0–200 µm fraction. D) A sample by sample comparison of the 1st dimension of 0.8 and 3.0 fractions.

Figure S7 A schematic of community scale metabolic adaptations across salinity gradients. Note that while the metabolic pathways are not expressly linked, they share substrates and products and thus have the potential.

Figure S8 Distribution of prokaryotic-type ABC transporters in the Baltic dataset (see Methods for details). Heatmap color scaling is based on rows. *"Putative salt transporter" has been suggested to be involved in salt transport and/or cell division.

Figure S9 Regularized canonical correlation analysis of functional modules within the 9 most abundant bacterial phyla or classes in the total dataset. For each plot, the outer circle (1) and inner circles (0.5) display the amount of variance explained by the linear combinations of variables. The variance explained for points within the inner circle is less than 25%, thus these have been dimmed. Labelled arrows show the correlation coefficient of environmental variables. Each point represents a KEGG functional module and is colored according to its parent metabolic category (see KEGG Metabolic Category legend). Specific modules discussed in the text are indicated in each plot by numbers (1–13) and described in the KEGG Module legend with colored circles matching their parent metabolic category.

Figure S10 A) Annotations of SAR11 by APIS. SAR11 (Pelagibacterales) are found across the entire transect, but in a species and clade specific manner. Clade IIIa (IMCC9063) is primarily found in the low salinity environments while clade Ia (Can. Pelagibacter ubique and HTCC7211) dominates the high salinity environment. B) Fragment recruitment plots of different Pelagibacterales genomes. While IMCC9063 is the most similar to those found in the low salinity waters (cold colors), the population in this region is only 70–80% average nucleotide identity and should be considered a distinct species and possibly sub-clade. C) Frequency of functional modules related to transport and biosynthesis in the SAR11-like combined assembly. Contigs were divided into brackish and marine bins based on read proportion from samples in the different environments (brackish: GS665-GS682, marine: GS683-GS695).

Table S1 Sites and environmental data for the sites where metagenomic sequencing was conducted.

Table S2 Number of peptides generated from pyrosequencing reads for each filter at each site, along with the portion that could be annotated using the JCVI metagenomic annotation pipeline.

Table S3 Bacterial genome size estimates for samples.

Acknowledgments

The authors thank the captain and crew of the Sorcerer II for logistical support and Dr. Mark Adams for constructive comments and discussions on the manuscript.

Author Contributions

Conceived and designed the experiments: CLD JL SY KI ÅH AEA JH EN RF JP JCV B. Bergman. Performed the experiments: CLD JL SY KI JG JAS JPM NC LZA ME AJL MT B. Brindefalk AR AA A. Tenney DL A. Tovchigrechko JAAN DB JB AEA DBR. Analyzed the data: CLD JL SY KI JG JAS JPM NC LZA ME AJL MT B. Brindefalk AR AA A. Tenney DL A. Tovchigrechko JAAN DB JB AEA DBR JP B. Bergman. Wrote the paper: CLD JL SY KI JG NC LZA ME B. Brindefalk JP B. Bergman.

References

1. Venter JC, Remington K, Heidelberg JF, Halpern AL, Rusch D, et al. (2004) Environmental Genome Shotgun Sequencing of the Sargasso Sea. Science 304: 66–74. doi:10.1126/science.1093857.
2. Rusch DB, Halpern AL, Sutton G, Heidelberg KB, Williamson S, et al. (2007) The Sorcerer II Global Ocean Sampling Expedition: Northwest Atlantic through Eastern Tropical Pacific. PLoS Biol 5: e77. doi:10.1371/journal.pbio.0050077.
3. Zeigler Allen L, Allen EE, Badger JH, McCrow JP, Paulsen IT, et al. (2012) Influence of nutrients and currents on the genomic composition of microbes across an upwelling mosaic. ISME J 6: 1403–1414. doi:10.1038/ismej.2011.201.
4. Hewson I, Paerl RW, Tripp HJ, Zehr JP, Karl DM (2009) Metagenomic potential of microbial assemblages in the surface waters of the central Pacific Ocean tracks variability in oceanic habitat. Limnol Ocean 54: 1981–1994. doi:10.4319/lo.2009.54.6.1981.
5. Crump BC, Armbrust EV, Baross JA (1999) Phylogenetic analysis of particle-attached and free-living bacterial communities in the Columbia river, its estuary, and the adjacent coastal ocean. Appl Environ Microbiol 65: 3192–3204.
6. Zwart G, Crump BC, Agterveld MPK, Hagen F, Han S (2002) Typical freshwater bacteria: an analysis of available 16S rRNA gene sequences from plankton of lakes and rivers. Aquat Microb Ecol 28: 141–155. doi:10.3354/ame028141.
7. Campbell BJ, Kirchman DL (2013) Bacterial diversity, community structure and potential growth rates along an estuarine salinity gradient. ISME J 7: 210–220. doi:10.1038/ismej.2012.93.
8. Logares R, Bråte J, Bertilsson S, Clasen JL, Shalchian-Tabrizi K, et al. (2009) Infrequent marine–freshwater transitions in the microbial world. Trends Microbiol 17: 414–422. doi:10.1016/j.tim.2009.05.010.
9. Glöckner FO, Fuchs BM, Amann R (1999) Bacterioplankton Compositions of Lakes and Oceans: a First Comparison Based on Fluorescence In Situ Hybridization. Appl Environ Microbiol 65: 3721–3726.
10. Lozupone CA, Knight R (2007) Global patterns in bacterial diversity. Proc Natl Acad Sci 104: 11436–11440. doi:10.1073/pnas.0611525104.
11. Debroas D, Enault F, Jouan-Dufournel I, Bronner G, Humbert J-F (2011) Metagenomic Approach Studying the Taxonomic and Functional Diversity of the Bacterial Community in a Lacustrine Ecosystem. In: Bruijn FJ de, editor. Handbook of Molecular Microbial Ecology II.John Wiley & Sons, Inc. pp.287–293.
12. Dinsdale EA, Edwards RA, Hall D, Angly F, Breitbart M, et al. (2008) Functional metagenomic profiling of nine biomes. Nature 452: 629–632. doi:10.1038/nature06810.
13. Eiler A, Zaremba-Niedzwiedzka K, Garcia MM, McMahon KD, Stepanauskas R, et al. (2013) Productivity and salinity structuring of the microplankton

revealed by comparative freshwater metagenomics. Environ Microbiol. doi:10.1111/1462-2920.12301.
14. Morgan JL, Darling AE, Eisen JA (2010) Metagenomic Sequencing of an In Vitro-Simulated Microbial Community. PLoS ONE 5: e10209. doi:10.1371/journal.pone.0010209.
15. Remane A (1934) Die Brackwasserfauna. Verhandlungen Dtsch Zool Ges 1934.
16. Herlemann DP, Labrenz M, Jürgens K, Bertilsson S, Waniek JJ, et al. (2011) Transitions in bacterial communities along the 2000 km salinity gradient of the Baltic Sea. ISME J 5: 1571–1579. doi:10.1038/ismej.2011.41.
17. Reissmann JH, Burchard H, Feistel R, Hagen E, Lass HU, et al. (2009) Vertical mixing in the Baltic Sea and consequences for eutrophication – A review. Prog Oceanogr 82: 47–80. doi:10.1016/j.pocean.2007.10.004.
18. Tanenbaum DM, Goll J, Murphy S, Kumar P, Zafar N, et al. (2010) The JCVI standard operating procedure for annotating prokaryotic metagenomic shotgun sequencing data. Stand Genomic Sci 2: 229–237. doi:10.4056/sigs.651139.
19. Margulies M, Egholm M, Altman WE, Attiya S, Bader JS, et al. (2005) Genome sequencing in microfabricated high-density picolitre reactors. Nature 437: 376–380. doi:10.1038/nature03959.
20. Seshadri R, Kravitz SA, Smarr L, Gilna P, Frazier M (2007) CAMERA: A Community Resource for Metagenomics. PLoS Biol 5: e75. doi:10.1371/journal.pbio.0050075.
21. Goll J, Rusch DB, Tanenbaum DM, Thiagarajan M, Li K, et al. (2010) METAREP: JCVI metagenomics reports–an open source tool for high-performance comparative metagenomics. Bioinforma Oxf Engl 26: 2631–2632. doi:10.1093/bioinformatics/btq455.
22. Raes J, Letunic I, Yamada T, Jensen LJ, Bork P (2011) Toward molecular trait-based ecology through integration of biogeochemical, geographical and metagenomic data. Mol Syst Biol 7.
23. Wu M, Scott AJ (2012) Phylogenomic analysis of bacterial and archaeal sequences with AMPHORA2. Bioinforma Oxf Engl 28: 1033–1034. doi:10.1093/bioinformatics/bts079.
24. Molina N, van Nimwegen E (2009) Scaling laws in functional genome content across prokaryotic clades and lifestyles. Trends Genet 25: 243–247. doi:10.1016/j.tig.2009.04.004.
25. Patel PV, Gianoulis TA, Bjornson RD, Yip KY, Engelman DM, et al. (2010) Analysis of membrane proteins in metagenomics: Networks of correlated environmental features and protein families. Genome Res 20: 960–971. doi:10.1101/gr.102814.109.
26. Gianoulis TA, Raes J, Patel PV, Bjornson R, Korbel JO, et al. (2009) Quantifying environmental adaptation of metabolic pathways in metagenomics. Proc Natl Acad Sci U S A 106: 1374–1379. doi:10.1073/pnas.0808022106.
27. Kogure K (1998) Bioenergetics of marine bacteria. Curr Opin Biotechnol 9: 278–282.
28. Neehaul Y, Juárez O, Barquera B, Hellwig P (2012) Thermodynamic Contribution to the Regulation of Electron Transfer in the Na+-Pumping NADH:Quinone Oxidoreductase from Vibrio cholerae. Biochemistry (Mosc) 51: 4072–4077. doi:10.1021/bi300343u.
29. Odom AR (2011) Five Questions about Non-Mevalonate Isoprenoid Biosynthesis. PLoS Pathog 7. doi:10.1371/journal.ppat.1002323.
30. Flamholz A, Noor E, Bar-Even A, Liebermeister W, Milo R (2013) Glycolytic strategy as a tradeoff between energy yield and protein cost. Proc Natl Acad Sci: 201215283. doi:10.1073/pnas.1215283110.
31. Søballe B, Poole RK (1999) Microbial ubiquinones: multiple roles in respiration, gene regulation and oxidative stress management. Microbiology 145: 1817–1830.
32. Ganesh S, Parris DJ, DeLong EF, Stewart FJ (2014) Metagenomic analysis of size-fractionated picoplankton in a marine oxygen minimum zone. ISME J 8: 187–211. doi:10.1038/ismej.2013.144.
33. De Leeuw E, Graham B, Phillips GJ, ten Hagen-Jongman CM, Oudega B, et al. (1999) Molecular characterization of Escherichia coli FtsE and FtsX. Mol Microbiol 31: 983–993.
34. Grote J, Thrash JC, Huggett MJ, Landry ZC, Carini P, et al. (2012) Streamlining and Core Genome Conservation among Highly Divergent Members of the SAR11 Clade. mBio 3.
35. Morris RM, Rappé MS, Connon SA, Vergin KL, Siebold WA, et al. (2002) SAR11 clade dominates ocean surface bacterioplankton communities. Nature 420: 806–810. doi:10.1038/nature01240.
36. Vergin KL, Beszteri B, Monier A, Thrash JC, Temperton B, et al. (2013) High-resolution SAR11 ecotype dynamics at the Bermuda Atlantic Time-series Study site by phylogenetic placement of pyrosequences. ISME J 7: 1322–1332. doi:10.1038/ismej.2013.32.
37. Oh H-M, Kang I, Lee K, Jang Y, Lim S-I, et al. (2011) Complete genome sequence of strain IMCC9063, belonging to SAR11 subgroup 3, isolated from the Arctic Ocean. J Bacteriol 193: 3379–3380. doi:10.1128/JB.05033-11.
38. Schwalbach MS, Tripp HJ, Steindler L, Smith DP, Giovannoni SJ (2010) The presence of the glycolysis operon in SAR11 genomes is positively correlated with ocean productivity. Environ Microbiol 12: 490–500. doi:10.1111/j.1462-2920.2009.02092.x.
39. Rivkin RB, Legendre L (2001) Biogenic carbon cycling in the upper ocean: effects of microbial respiration. Science 291: 2398–2400. doi:10.1126/science.291.5512.2398.
40. Del Giorgio PA, Cole JJ (1998) Bacterial Growth Efficiency in Natural Aquatic Systems. Annu Rev Ecol Syst 29: 503–541. doi:10.1146/annurev.ecolsys.29.1.503.
41. Chin-Leo G, Benner R (1992) Enhanced bacterioplankton production and respiration at intermediate salinities in the Mississippi River plume. Mar Ecol Prog Ser 87: 87.
42. Ram AP, Nair S, Chandramohan D (2003) Bacterial growth efficiency in the tropical estuarine and coastal waters of Goa, southwest coast of India. Microb Ecol 45: 88–96.
43. Langenheder S, Kisand V, Wikner J, Tranvik LJ (2003) Salinity as a structuring factor for the composition and performance of bacterioplankton degrading riverine DOC. FEMS Microbiol Ecol 45: 189–202.
44. Meier HM (2006) Baltic Sea climate in the late twenty-first century: a dynamical downscaling approach using two global models and two emission scenarios. Clim Dyn 27: 39–68.
45. Meier HEM, Hordoir R, Andersson HC, Dieterich C, Eilola K, et al. (2012) Modeling the combined impact of changing climate and changing nutrient loads on the Baltic Sea environment in an ensemble of transient simulations for 1961–2099. Clim Dyn 39: 2421–2441. doi:10.1007/s00382-012-1339-7.

Impacts of Climate-Change-Driven Sea Level Rise on Intertidal Rocky Reef Habitats Will Be Variable and Site Specific

Jaqueline Thorner[1,2]*, Lalit Kumar[1], Stephen D. A. Smith[2]

1 University of New England (UNE), School of Rural and Environmental Sciences, Armidale, New South Wales, Australia, **2** National Marine Science Centre, Southern Cross University (SCU), Coffs Harbour, New South Wales, Australia

Abstract

Intertidal rocky reefs are complex and rich ecosystems that are vulnerable to even the smallest fluctuations in sea level. We modelled habitat loss associated with sea level rise for intertidal rocky reefs using GIS, high-resolution digital imagery, and LIDAR technology at fine-scale resolution (0.1 m per pixel). We used projected sea levels of +0.3 m, +0.5 m and +1.0 m above current Mean Low Tide Level (0.4 m). Habitat loss and changes were analysed for each scenario for five headlands in the Solitary Islands Marine Park (SIMP), Australia. The results indicate that changes to habitat extent will be variable across different shores and will not necessarily result in net loss of area for some habitats. In addition, habitat modification will not follow a regular pattern over the projected sea levels. Two of the headlands included in the study currently have the maximum level of protection within the SIMP. However, these headlands are likely to lose much of the habitat known to support biodiverse assemblages and may not continue to be suitable sanctuaries into the future. The fine-scale approach taken in this study thus provides a protocol not only for modelling habitat modification but also for future proofing conservation measures under a scenario of changing sea levels.

Editor: Inés Álvarez, University of Vigo, Spain

Funding: University of New England funded the research (www.une.edu.au). The funders had no role in study design, data collection and analysis, decision to publish, or preparation of the manuscript.

Competing Interests: The authors have declared that no competing interests exist.

* E-mail: jduraubu@myune.edu.au

Introduction

There is strong consensus that sea levels will rise by as much as 1 m by 2100 in response to a warming climate [1]. While sea levels have varied by >120 m during the glacial/interglacial cycles, there was little change over the past several thousand years [2]. However, this condition of relative stasis changed during the 19th and early 20th centuries when the first indications of a rising sea level became evident [2,3]. Recent sea level rise (SLR) is a result of both ocean thermal expansion and the melting of glaciers and ice caps, and accelerated rates of SLR are likely to be largely attributable to the liberation of water comprising polar ice sheets [2,4–6]. According to Church *et al.* [7] an upper limit of 0.88 m in mean sea level is expected over the next hundred years. However, recent data acquired from satellite-altimeter and tide-gauge sources suggest sea levels are rising at over 3 mm/yr [8] with an acceleration rate of 0.013 mm/yr [9]. Current rates of SLR are at the upper limit of the projections of the Third Assessment Report of the Intergovernmental Panel on Climate Change (IPCC) [7], and there is strong concern that the contribution of shrinking ice sheets will push these even higher, especially if greenhouse gas emissions continue to increase [2]. Most of the climate models indicate that the increase in temperature over Greenland is one to three times higher than the global average. Ice sheet models project that a local warming of more than 3°C, if sustained over a longer term, would lead to the complete melting of the Greenland ice sheet, adding around 7 m to the sea level [10]. Also, the West Antarctic Ice Sheet could potentially contribute about 6 m to the sea level. The entire Antarctic ice sheet holds enough water to raise global sea levels by 62 m [5]. As current predictive models have a high level of uncertainty, there is concern that the estimates of mean SLR by 2100 could be considerably underestimated [7,11]. Future SLR is not expected to be globally uniform due to ocean circulation, wind pressure patterns and geological uplift [7,11–14]. Effects operating over local or regional scales, such as storm surges and spring tides, will add further variability [11].

While the magnitude of SLR is clearly difficult to predict with certainty, there is little doubt that intertidal habitats are likely to be the first to experience sea level rise-related effects [15]. Rocky shores occupy 30% of the world's coastline [16] and are complex and rich ecosystems supporting high biodiversity [17]. As one of the most accessible, diverse marine habitats, rocky shores are important features for education, recreation and harvesting [17–19]. When compared to other intertidal habitats such as beaches, mangroves and estuaries, and due to their geological nature, erosion processes are relatively slow on intertidal rocky reefs obviating the creation of new habitat over short periods of time: this exposes the biota to the very real threat of habitat loss with consequent reduction in biodiversity due to "coastal squeeze" [19]. Indeed, habitat loss and change is potentially one of the greatest threats to biodiversity conservation in a changing world [20,21], and is a key challenge for the management of biodiversity through static systems such as Marine Protected Areas (MPA) [22,23].

Predicting the magnitude of changes to intertidal rocky reefs, however, relies not only on accurate predictions of SLR, but also on the development of technologies to map and quantify habitats and environmental processes at appropriate spatial scales, in an efficient and cost-effective way [24]. Tools such as remote sensing and modelling, are becoming increasingly available [8,25–27] and are already being applied to the assessment of risk associated with SLR [28]. To date, LIDAR (Light Detection and Ranging) has the highest horizontal resolution and vertical accuracy [24,29] and is the only technology suitable for analyzing sea-level changes in the range predicted for the next 100 years [30,31].

Following the development and assessment of a combined approach using LIDAR, high-resolution digital imagery and GIS [31], this study evaluated the fine-scale consequences of different projections of sea level on the range of intertidal habitats for five headlands in the Solitary Islands Marine Park (SIMP), northern NSW, Australia. Specifically, we evaluated seven categories of rocky shore habitat (lower shallow pools, upper shallow pools, deep pools, upper boulder fields, lower boulder fields, upper platform and lower platform) and the current area of each habitat type was compared to future modelled scenarios to determine the likely changes to these intertidal rocky ecosystems and the possible consequences for biodiversity conservation.

Study Area

The SIMP is located on the mid-north-coast of NSW, Australia. The five headlands chosen for this research were: Arrawarra Headland (153°12′07′′E-30°3′33′′S); Oceanview Headland (153°12′14′′E-30°04′01′S); Mullaway Headland (153°12′15′′E-30°4′33′′S); Woolgoolga Headland (153°12′17′′E-30°06′30′′S); and Flat Top Point (153°12′26′′E-30°07′48′′S) (Figure 1). The region has a 2 m, semi-diurnal tidal cycle, a Mean Sea Level of 0.9 m, a Mean Low Tide Level of 0.4 m, and a Mean High Tide Level of 1.4 m [32]. Each of the five headlands is composed of metamorphic greywacke deposits from the late Carboniferous, approximately 280–350 million years old [33]. The headland landscape comprises cliffs, bedrock and scree (boulder fields) with variable wave-driven erosional patterns and areas of sand accumulation. The region is renowned for the overlap of tropical and temperate currents [34], facilitating a high biodiversity due to the presence of tropical, temperate and endemic species [35–38]. There have been few published studies on the biodiversity and ecology of rocky shore habitats within the region, but those that have been conducted, indicate a high diversity, and distribution patterns that are largely driven by habitat type and exposure to wave energy [36]. The total intertidal rocky shore mapped area for each headland was: Arrawarra Headland (19.129 m^2); Oceanview Headland (10.490 m^2); Mullaway Headland (16.134 m^2); Woolgoolga Headland (41.173 m^2); and Flat Top Point (13.458 m^2).

Methods

Habitat Mapping

Three-dimensional digital habitat maps were compiled for each of the five intertidal rocky reefs. The maps were obtained by coupling topographic LIDAR data (wavelength = 1550 μ; Leica ASL60 system) and high-resolution imagery in ArcGIS software [31,39]. The data resolution per pixel was 0.1 m (z), 0.3 m (x,y) for the LIDAR, and 0.1 m for the digital imagery. Intertidal rocky reef habitats were classified into seven types of habitats (Table 1) which recognized broad habitat type [36] and tidal height. Mean Sea Level (0.9 m) was used to differentiate between upper and lower shore levels. Ground-truthing of the remote-sensing data was performed along the headlands at very low tides (<0.4 m) on days with low swell (<1 m).

Sea Level Rise Modelling

Flooding maps were created as a layer over the five original habitat maps using ArcMap [39]. SLR projections were based on the predictions from the 2007 IPCC report [30]. The basic tide level for the analysis was Mean Low Tide Level (MLTL) (0.4 m) with flooding maps created for SLR of 0.3 m, 0.5 m and 1.0 m above current levels [32]. Changes in habitat area under each SLR scenario were calculated using ArcMap [39] for each headland, and data were summarized both as habitat maps and as relative percentage change for each habitat type at each headland. It should be noted that we modelled only eustatic change: storm surges or extreme events were not taken into consideration. Predictions were made under the assumption that tidal pattern will not change over the next 100 years (1 m SLR). The ratio of upper and lower shore area was calculated in order to evaluate the change in relative distribution of habitat under the different sea levels for the five headlands. This was done because lower shores generally have higher species richness and changes in the availability of habitat on upper and lower shores are therefore likely to influence patterns of biodiversity [40]. The quantity, mean size (±SE) of pools in the upper and lower shallow pool habitats were calculated in order to evaluate the change in habitat quality over the prediction period.

Results

Change in the intertidal rocky reef landscape

The results from the SLR projections show a dramatic change in the rocky shore landscape under a 1-m SLR, with an overall decline in habitat area for all habitat types with rare exceptions (Figure 2). Deep pools will be by far the most affected habitat type, disappearing from all headlands except Arrawarra under 1-m SLR. Upper shallow pools will be the second most affected habitat followed by lower boulder fields. Lower platform and lower shallow pool habitats will have an irregular pattern of change, with alternating increases and decreases in habitat area over the projection period. However, lower shallow pools will suffer a large loss under 1-m SLR, whereas lower platform habitat will suffer the least overall loss. Upper boulder field will be the second least affected habitat type followed by upper platform habitat. The change in the relative distribution of upper and lower habitats will also be irregular over the projected sea levels (Table 2). Arrawarra Headland and Ocean View Headland will show an initial increase in upper shore area with a subsequent decline at 0.3 m and 0.5 m SLR, nevertheless the final model predicts a larger upper shore area under 1 m SLR than at present. All the remaining headlands show a steady increase in the area of lower shore relative to the upper shore as sea level rises.

Upper Shallow Pool Habitat

Our modelling indicated that there will be an overall decline in the total area of upper shallow pool habitat at each of the five headlands under 1-m SLR. A reduction by >50% of habitat area will occur under 0.3-m SLR at all headlands except Ocean View Headland. At Mullaway Headland and Flat Top Point, the upper shallow pool habitat will completely disappear under 1-m SLR, whilst at Arrawarra Headland and Woolgoolga Headland, only 2% of the habitat will remain. Habitat loss will be lowest at Ocean View Headland which will maintain nearly half of its upper shallow pool habitat area under 1-m SLR. However, the habitat

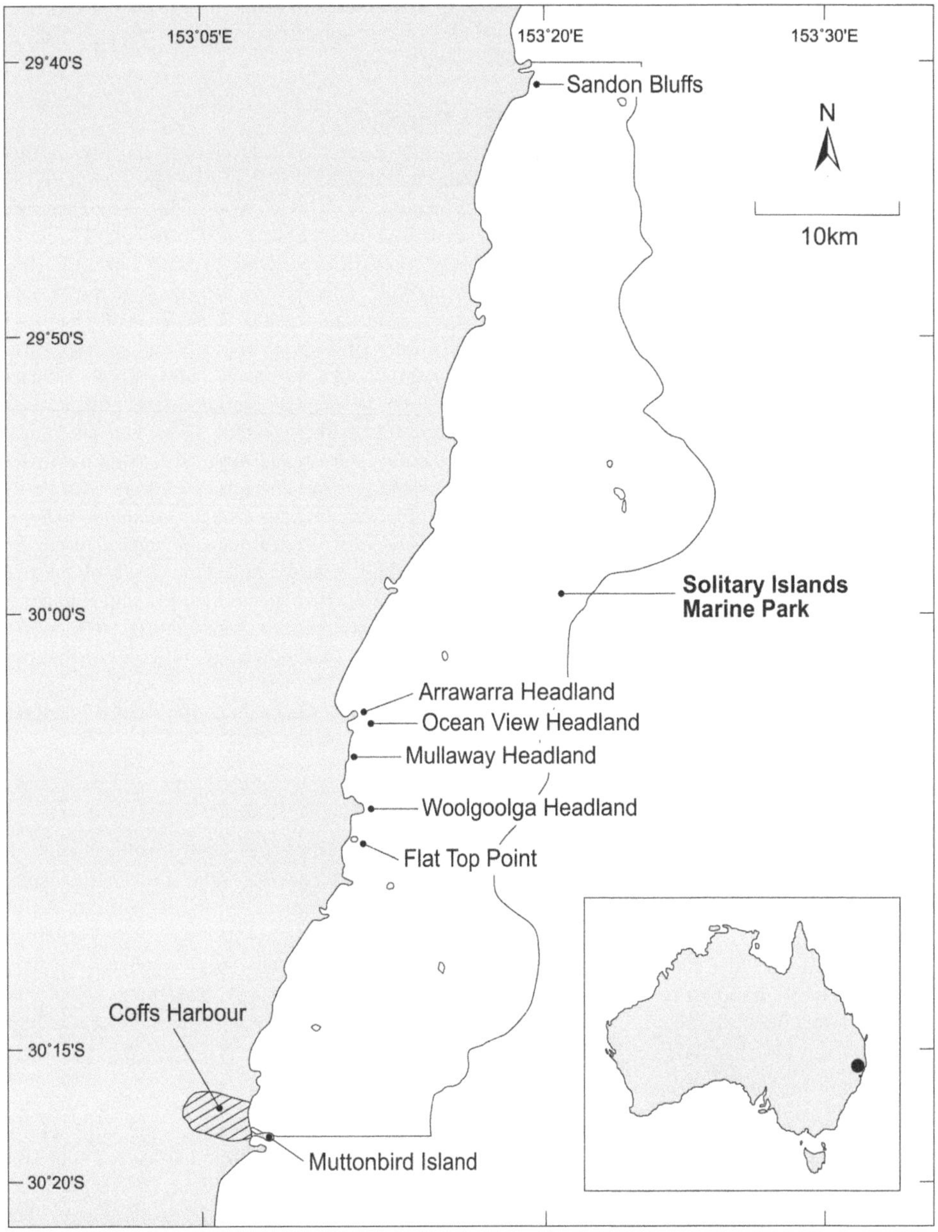

Figure 1. Solitary Island Marine Park map showing the five headlands analyzed in this study. Arrawarra Headland; Ocean View Headland; Mullaway Headland; Woolgoolga Headland and Flat Top Point.

quality tends to change with SLR, with a decrease in the number of pools for all headlands. The mean size of pools (m^2) will be reduced at Arrawarra Headland, Mullaway Headland and Flat Top Point whereas Ocean View Headland and Woolgoolga Headland will show a variable change in pool size as sea level rises (Table 3).

Lower Shallow Pool Habitat

Lower shallow pool habitat shows an irregular pattern of modification under 1-m SLR. At all headlands, with the exception of Woolgoolga Headland, habitat extent will be reduced by >80% under 1-m SLR. At Mullaway and Woolgoolga headlands, SLR will initially lead to an increase in lower shallow pool habitat (under 0.5-m SLR), with a subsequent decline. Ocean View Headland shows a dramatic reduction, to nearly 29% of the current extent, under 0.3-m SLR, followed by an increase to 44% under 0.5-m SLR, and a further decline to 5% under 1-m SLR. Flat Top Point shows the greatest reduction of habitat (50%) between 0.3-m and 0.5-m SLR. The number of pools will be reduced at all headlands; however, the mean pool size will tend to a steady decrease only in Arrawarra Headland, Ocean View Headland and Flat Top Point. Mullaway Headland and Woolgoolga Headland will initially show an increase in mean pool size (m^2) as the sea level rises 0.3 m and 0.5 m, and a subsequent decrease under 1-m SLR.

Deep Pool Habitat

This habitat type is the most at risk at all headlands, with 100% loss under 1-m SLR, at all headlands except Arrawarra Headland. The rate of habitat loss will differ between headlands, with Ocean

Table 1. Habitat classification based on the tidal exposure and substrate type (Source: [31]).

Habitat Classification	Habitat Description (Mean Sea Level (MSL) =0.9-m)
Upper Boulder Field	≥2 boulders (>0.25-m) per m^2 and located above MSL
Lower Boulder Field	≥2 boulders (>0.25-m) per m^2 and located below MSL
Upper Shallow Pool	Rock pools no deeper than 0.5-m, with bottom composed of sand, cobble, boulder or a mix of these, and located above MSL
Lower Shallow Pool	Rock pools no deeper than 0.5-m, with bottom composed of sand, cobble, boulder or a mix of these, usually with high algal cover and located below MSL
Deep Pool	Rock pools deeper than 0.5-m, with bottom composed of sand, cobble, boulder or a mix of these (this habitat type only occurred at the seaward edge of the lower shore at the five study locations).
Upper Platform	Emerged bedrock located above MSL
Lower Platform	Emerged bedrock located below MSL

View, Mullaway and Woolgoolga losing this habitat at 0.5-m SLR and Flat Top Point only at 1-m SLR.

Upper Boulder Field Habitat

With the exception of Woolgoolga Headland, all headlands show a steady decline in the extent of upper boulder fields under 1-m SLR. At Woolgoolga Headland, habitat extent remains the same under 0.5-m SLR and then suffers a decrease of ~50% under 1-m SLR. Flat Top Point will lose its entire upper boulder field habitat under 1-m SLR. While there is an overall decline at all headlands, this habitat type will suffer the second lowest overall loss (after lower platform habitat).

Lower Boulder Field Habitat

Lower boulder field habitat shows an overall decline under 1-m SLR. At all headlands, nearly 50% of the habitat area is predicted to be lost under 0.3-m SLR. Flat Top Point shows the slowest decline in habitat area with around 30% remaining under 1-m SLR, whereas the other headlands will lose >85% of this habitat. Woolgoolga Headland will lose this entire habitat under 0.5-m SLR but will then recover 2% under 1-m SLR.

Upper Platform Habitat

There will be a decrease in the area of upper platform habitat at all headlands over the next century (1-m SLR). Mullaway Headland will experience by far the greatest loss with 54% remaining under 0.3-m SLR, and 6% remaining under 1-m SLR. All other headlands show greater declines under 0.5-m SLR, with the exception of Ocean View Headland, which will retain 63% of upper platform habitat at 1-m SLR.

Table 2. Relative upper/lower shore area modification for the four projected sea levels at the five headlands.

Headland	0m	+0.3m	+0.5m	+1m
Arrawarra Headland	0.5	0.7	0.9	0.6
Ocean View Headland	2.5	3.7	3.5	3.4
Mullaway Headland	1.3	0.8	0.6	0.4
Woolgoolga Headland	2.7	2	1.4	0.7
Flat Top Point	1.4	1.1	1.3	0.4

Lower Platform Habitat

Lower platform habitat shows an irregular pattern of change over the next century. Arrawarra Headland will suffer a steady decline with <30% of habitat left under 1-m SLR. Flat Top Point will have a reduction of 50% under 0.5-m SLR, but very little subsequent change up to 1-m SLR. Mullaway Headland and Woolgoolga Headland will have an initial increase in habitat area at 0.5-m SLR, however, Mullaway Headland subsequently lose 18% (relative to the current area), whereas Woolgoolga Headland show a net 15% increase, under 1-m SLR. Ocean View will have the slowest rate of change with 76% of habitat area remaining under 1-m SLR.

Discussion

It is clear from the results of this study that we can make few generalizations about the rate and extent of intertidal habitat loss resulting from SLR. While the extent of most habitats is ultimately reduced in the worst-case model (1-m SLR), the rate-of-change and final outcome are dependent on the habitat type and specific locality. In other words, the effects of SLR on intertidal rocky reefs will be variable at small spatial scales.

To date, most of the research conducted to evaluate the impact of SLR on intertidal ecosystems refers to mangroves [41], salt marshes [42–44] and sandy beaches [45–47]. Just a few studies have focused on the rocky shore ecosystem but have targeted exclusively large spatial scales, such as kilometers of coastal area [19]. By conducting studies at this level of spatial resolution (cm to m of habitat area), it has been possible to assess the relative vulnerability of different locations to habitat loss which may have important implications for conservation planning. Most headlands will show an increase in the ratio of lower shore to the upper shore, despite the overall reduction in intertidal area. Since lower shore habitats are known to be generally richer than the upper shore

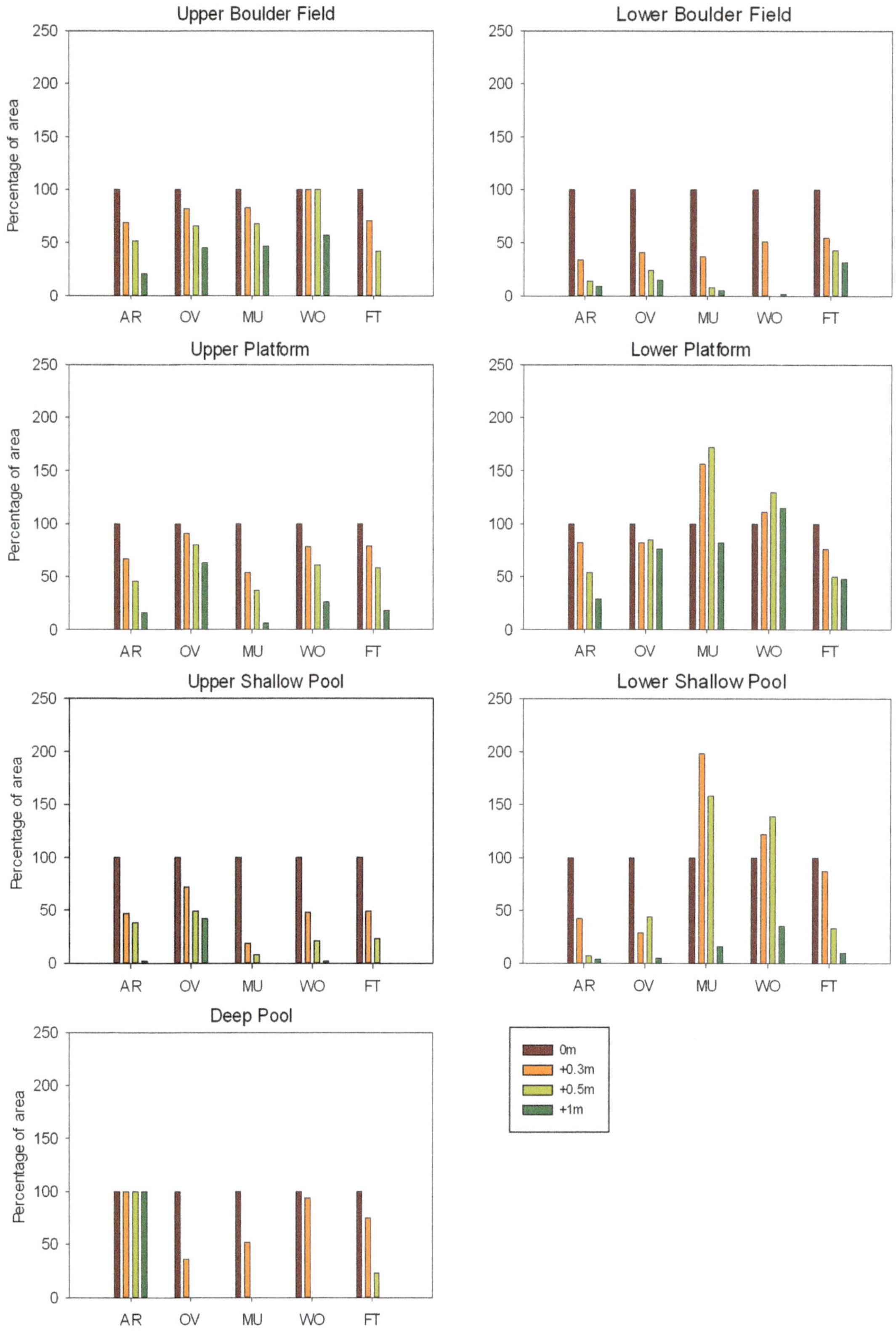

Figure 2. Habitats area change represented as percentage relative to current sea level at the five headlands. The four projected sea levels were 0 m,+0.3 m,+0.5 m and 1 m and the five headlands AR-Arrawarra Headland; OV-Ocean View Headland; MU-Mullaway Headland; WO-Woolgoolga Headland; FT-Flat Top Point.

habitats [40] this may help to ameliorate the potential loss of biodiversity associated with the overall decline in intertidal area.

Mean size and number of shallow pools will also show a variable, and headland-specific, change which would not have been

Table 3. Changes in the number of pools and their mean size (± SE) for the four projected sea levels at the five headlands (m^2).

Headland	Sea Level	Upper Shallow Pool		Lower Shallow Pool	
		n	Mean ± SE	n	Mean ± SE
Arrawarra Headland	0m	32	5.57±0.79	60	25.82±5.24
	+0.3m	21	3.96±0.66	32	20.14±5.73
	+0.5m	18	3.82±0.7	14	7.81±1.36
	+1m	1	4.34±4.34	17	3.79±0.74
Ocean View Headland	0m	17	10.21±3.34	19	10.73±3.45
	+0.3m	13	9.59±3.88	6	9.98±4.5
	+0.5m	9	9.31±5.36	8	11.20±3.71
	+1m	6	12.28±7.75	3	3.38±0.8
Mullaway Headland	0m	48	16.21±5.11	44	8.86±1.09
	+0.3m	13	11.63±5.73	46	16.77±5.12
	+0.5m	11	5.85±1.29	36	17.15±6.36
	+1m	–	–	11	5.85±1.29
Woolgoolga Headland	0m	106	12.42±2.21	61	11.75±1.79
	+0.3m	57	11.18±2.37	69	12.68±2.88
	+0.5m	29	9.66±2.54	72	13.81±3.07
	+1m	2	11.9±5.41	27	9.5±2.69
Flat Top Point	0m	24	9.85±2.09	20	27.73±5.38
	+0.3m	19	6.12±1.03	19	25.32±5.12
	+0.5m	10	5.44±1.13	15	12.13±3.14
	+1m	–	–	10	5.44±1.13

detected by modelling at a larger scale (km of coastline) [19]. The conservation of fine-scale spatial heterogeneity is a critical factor in marine reserve planning [48] and, consequently, predictive modelling of SLR at broader scales will fail to effectively incorporate this feature into conservation targets.

An important extension of this is that, depending on the relative loss of critical habitat, headlands currently gazetted as sanctuary zones (no take) may not be effective locations to meet conservation targets in the future. If the objective is to protect representative habitats that are likely to persist, there is little point targeting headlands that will show high rates of habitat loss and ultimately support little habitat. It is instructive to apply this concept to the current conservation status of the five headlands. Both Flat Top Point and Arrawarra Headland are currently afforded the highest level of protection under the zoning plan for the Solitary Islands Marine Park (Special Purpose and Sanctuary, respectively) [49]. However, they will show some of the greatest reductions in the extent of shallow pools (upper and lower) and upper boulder field habitat, which are associated with high biodiversity [36]. The sea level projections from the present research have shown that these are two of the most vulnerable types of habitat and could nearly disappear under 1-m SLR, potentially exposing the resident fauna to considerable impact.

Eighty percent of the world's oceanic coastlines comprise of rock platforms backed by steep cliffs [19]; this description typifies headland formation for much of the Australian coastline [50] suggesting profound changes in the landscape with direct consequences for biodiversity and recreational use. The greatest challenge for coastal managers will be to evaluate how to deal with the lack of predictability for climate change factors [20,21], since changes occur slowly and the effects also interact with other impacts already imposed on the environment such as pollution and anthropogenic disturbance [37,51].

Present conservation planning is based on current sea levels. This leads to two questions: i) Will present sanctuary zones be effective for ongoing conservation in a changing environment?; ii) Does loss of habitat area equate to a commensurate loss of biodiversity for intertidal rocky reef ecosystems? Clearly, there is a need to determine if targets that foster prioritization of sites where habitat loss will be lower will actually translate into similar representation for the biota. This also needs to be considered in terms of other factors determining ecosystem functioning (e.g. various processes that can vary over a range of spatial scales) [52]. Environmental management in a changing world must, therefore, not only take into account new scenarios expected to arise over the next decades, but also be flexible enough to handle sudden change in predictions, which can hamper long-term biodiversity conservation [53].

Acknowledgments

Coffs Harbour City Council provided the remote sensing data and we had the support of the National Marine Science Centre, Southern Cross University.

Author Contributions

Conceived and designed the experiments: JT LK SS. Performed the experiments: JT. Analyzed the data: JT. Contributed reagents/materials/analysis tools: JT LK. Wrote the paper: JT LK SS.

References

1. Willis JK, Church JA (2012) Regional sea-level projection. Science 336: 550–551.
2. Church JA, White NJ, Aarup T, Wilson WS, Woodworth PL, et al. (2008) Understanding global sea levels: past, present and future. Sustainability Science 3: 9–22.
3. Russell GL, Gornitz V, Miller JR (2000) Regional sea level changes projected by the NASA/GISS atmosphere-ocean model. Climate Dynamics 16: 789–797.
4. Cazenave A, Dominh K, Guinehut S, Berthier E, Llovel W, et al. (2009) Sea level budget over 2003–2008: A reevaluation from GRACE space gravimetry, satellite altimetry and Argo. Global and Planetary Change 65: 83–88.
5. Shepherd A, Wingham D (2007) Recent sea-level contributions of the Antarctic and Greenland ice sheets. Science 315: 1529–1532.
6. Horton R, Herweijer C, Rosenzweig C, Liu J, Gornitz V, et al. (2008) Sea level rise projections for current generation CGCMs based on the semi-empirical method. Geophysical Research Letters 35: L02715.
7. Church J, Gregory J, Huybrechts P, Kuhn M, Lambeck K, et al. (2001) Changes in sea level. In: Houghton JT, Ding Y, Griggs DJ, Noguer M, Van der Linden PJ, et al., editors. Climate Change 2001: The Scientific Basis: Contribution of Working Group I to the Third Assessment Report of the Intergovernmental Panel: 639–694.
8. Chust G, Galparsoro I, Borja Á, Franco J, Uriarte A (2008) Coastal and estuarine habitat mapping, using LIDAR height and intensity and multi-spectral imagery. Estuarine, Coastal and Shelf Science 78: 633–643.
9. Church JA, White NJ (2006) A 20th century acceleration in global sea-level rise. Geophysical Research Letters 33: L01602.
10. Gregory J, Huybrechts P (2006) Ice-sheet contributions to future sea-level change. Philosophical Transactions of the Royal Society A: Mathematical, Physical and Engineering Sciences 364: 1709–1732.
11. Senior C, Jones R, Lowe J, Durman C, Hudson D, et al. (2002) Predictions of extreme precipitation and sea-level rise under climate change. Philosophical Transactions of the Royal Society of London Series A: Mathematical, Physical and Engineering Sciences 360: 1301–1311.
12. Levitus S, Antonov J, Boyer T (2005) Warming of the world ocean, 1955–2003. Geophysical Research Letters 32: L02604.
13. Nicholls RJ, Wong PP, Burkett V, Codignotto J, Hay J, et al. (2007) Coastal systems and low-lying areas. In: Parry ML, Canziani OF, Palutikof JP, van der Linden PJ, Hanson CE, editors. Climate change 2007: impacts, adaptation and vulnerability. Contribution of Working Group II to the fourth assessment report of the Intergovernmental Panel on Climate Change. Cambridge, UK: Cambridge University Press. 315–356.
14. Meehl GA, Stocker TF, Collins WD, Friedlingstein A, Gaye AT, et al. (2007) Global climate projections. Climate change: 747–845.
15. Pethick J (2001) Coastal management and sea-level rise. Catena 42: 307–322.
16. Johnson ME (1988) Why are ancient rocky shores so uncommon? The Journal of Geology: 469–480.
17. Murray SN, Ambrose R, Dethier MN (2006) Monitoring rocky shores. Berkeley, CA: University of California Press.
18. Garcia A, Smith JR (2012) Factors influencing human visitation of southern California rocky intertidal ecosystems. Ocean & Coastal Management.
19. Jackson AC, McIlvenny J (2011) Coastal squeeze on rocky shores in northern Scotland and some possible ecological impacts. Journal of Experimental Marine Biology and Ecology 400: 314–321.
20. Hannah L (2008) Protected areas and climate change. Annals of the New York Academy of Sciences 1134: 201–212.
21. McLeod E, Salm R, Green A, Almany J (2008) Designing marine protected area networks to address the impacts of climate change. Frontiers in Ecology and the Environment 7: 362–370.
22. Mora C, Sale PF (2011) Ongoing global biodiversity loss and the need to move beyond protected areas: a review of the technical and practical shortcomings of protected areas on land and sea. Marine Ecology Progress Series 434: 251–266.
23. Hanski I (2011) Habitat loss, the dynamics of biodiversity, and a perspective on conservation. AMBIO: A Journal of the Human Environment 40: 248–255.
24. Runting RK, Wilson KA, Rhodes JR (2013) Does more mean less? The value of information for conservation planning under sea level rise. Global Change Biology 19: 352–363.
25. Adams J, Chandler J (2002) Evaluation of Lidar and Medium Scale Photogrammetry for Detecting Soft-Cliff Coastal Change. The Photogrammetric Record 17: 405–418.
26. Brock JC, Wright CW, Sallenger AH, Krabill WB, Swift RN (2002) Basis and methods of NASA airborne topographic mapper lidar surveys for coastal studies. Journal of Coastal Research: 1–13.
27. Populus J, Laurentin A, Rollet C, Vasquez M, Guillaumont B, et al. (2004) Surveying coastal zone topography with airborne remote sensing for benthos mapping. EARSeL eProceedings 3: 105–117.
28. Cooper HM, Chen Q, Fletcher CH, Barbee MM (2012) Assessing vulnerability due to sea-level rise in Maui, Hawai 'i using LiDAR remote sensing and GIS. Climatic Change: 1–17.
29. Hall RK, Watkins RL, Heggem DT, Jones KB, Kaufmann PR, et al. (2009) Quantifying structural physical habitat attributes using LIDAR and hyperspectral imagery. Environmental monitoring and assessment 159: 63–83.
30. Solomon S, Qin D, Manning M, Chen Z, Marquis M, et al. (2005) Climate change 2007: The physical science basis. Oxford: Taylor & Francis.
31. Thorner J, Kumar L, Smith SDA (2013) Fine scale 3D habitat mapping as a biodiversity conservation tool for intertidal rocky reefs. Journal of Coastal Conservation doi: 10.2112/JCOASTRES-D-12-00142.1.
32. Foremann MGG (1977) Manual for tidal heights analysis and prediction. Pacific Marine Science Report 77–10: 58 p.
33. Korsch R, Harrington H (1981) Stratigraphic and structural synthesis of the New England Orogen. Journal of the Geological Society of Australia 28: 205–226.
34. Malcolm HA, Davies PL, Jordan A, Smith SDA (2011) Variation in sea temperature and the East Australian Current in the Solitary Islands region between 2001–2008. Deep Sea Research Part II: Topical Studies in Oceanography 58: 616–627.
35. Malcolm HA, Jordan A, Smith SDA (2010) Biogeographical and cross-shelf patterns of reef fish assemblages in a transition zone. Marine Biodiversity 40: 181–193.
36. Smith SDA (2005) Rapid assessment of invertebrate biodiversity on rocky shores: where there'sa whelk there'sa way. Biodiversity and Conservation 14: 3565–3576.
37. Smith SDA, Rule MJ, Harrison M, Dalton SJ (2008) Monitoring the sea change: Preliminary assessment of the conservation value of nearshore reefs, and existing impacts, in a high-growth, coastal region of subtropical eastern Australia. Marine Pollution Bulletin 56: 525–534.
38. Harrison MA, Smith SDA (2012) Cross-shelf variation in the structure of molluscan assemblages on shallow, rocky reefs in subtropical, eastern Australia. Marine Biodiversity: 1–14.
39. ESRI (2010) ArcGIS Desktop. Redlands, CA: Environmental Systems Research Institute.
40. Menge BA, Branch GM (2001) Rocky intertidal communities. Marine community ecology: 221–251.
41. Gilman E, Ellison J, Coleman R (2007) Assessment of mangrove response to projected relative sea-level rise and recent historical reconstruction of shoreline position. Environmental monitoring and assessment 124: 105–130.
42. Morris JT, Sundareshwar P, Nietch CT, Kjerfve B, Cahoon D (2002) Responses of coastal wetlands to rising sea level. Ecology 83: 2869–2877.
43. Kirwan ML, Guntenspergen GR, D'Alpaos A, Morris JT, Mudd SM, et al. (2010) Limits on the adaptability of coastal marshes to rising sea level. Geophysical Research Letters 37: L23401.
44. Akumu CE, Pathirana S, Baban S, Bucher D (2011) Examining the potential impacts of sea level rise on coastal wetlands in north-eastern NSW, Australia. Journal of Coastal Conservation 15: 15–22.
45. Fish M, Cote I, Horrocks J, Mulligan B, Watkinson A, et al. (2008) Construction setback regulations and sea-level rise: mitigating sea turtle nesting beach loss. Ocean & Coastal Management 51: 330–341.
46. Brooks SM, Spencer T (2011) Shoreline retreat and sediment release in response to accelerating sea level rise: Measuring and modelling cliffline dynamics on the Suffolk Coast, UK. Global and Planetary Change 80: 165–179.
47. Bray MJ, Hooke JM (1997) Prediction of Soft-Cliff Retreat with Accelerating Sea-Level Rise. Journal of Coastal Research 13: 453–467.
48. Ackerly D, Loarie S, Cornwell W, Weiss S, Hamilton H, et al. (2010) The geography of climate change: implications for conservation biogeography. Diversity and Distributions 16: 476–487.
49. Marine Parks (2002) Solitary Islands Marine Park Zoning Plan. Marine Parks Authority New South Wales, NSW, Australia.
50. Hails J (1965) A critical review of sea-level changes in Eastern Australia since the Last Glacial. Australian Geographical Studies 3: 63–78.
51. Kappelle M, Van Vuuren MM, Baas P (1999) Effects of climate change on biodiversity: a review and identification of key research issues. Biodiversity & Conservation 8: 1383–1397.
52. Hawkins S, Sugden H, Mieszkowska N, Moore P, Poloczanska E, et al. (2009) Consequences of climate-driven biodiversity changes for ecosystem functioning of North European rocky shores. Marine Ecology Progress Series 396: 245–259.
53. Agardy T, Bridgewater P, Crosby MP, Day J, Dayton PK, et al. (2003) Dangerous targets? Unresolved issues and ideological clashes around marine protected areas. Aquatic Conservation: Marine and Freshwater Ecosystems 13: 353–367.

Spatial Access Priority Mapping (SAPM) with Fishers: A Quantitative GIS Method for Participatory Planning

Katherine L. Yates[1]*, David S. Schoeman[2]

1 Environmental Science Research Institute, University of Ulster, Coleraine, United Kingdom, 2 Faculty of Science, Health, Education and Engineering, University of the Sunshine Coast, Maroochydore DC, Queensland, Australia

Abstract

Spatial management tools, such as marine spatial planning and marine protected areas, are playing an increasingly important role in attempts to improve marine management and accommodate conflicting needs. Robust data are needed to inform decisions among different planning options, and early inclusion of stakeholder involvement is widely regarded as vital for success. One of the biggest stakeholder groups, and the most likely to be adversely impacted by spatial restrictions, is the fishing community. In order to take their priorities into account, planners need to understand spatial variation in their perceived value of the sea. Here a readily accessible, novel method for quantitatively mapping fishers' spatial access priorities is presented. Spatial access priority mapping, or SAPM, uses only basic functions of standard spreadsheet and GIS software. Unlike the use of remote-sensing data, SAPM actively engages fishers in participatory mapping, documenting rather than inferring their priorities. By so doing, SAPM also facilitates the gathering of other useful data, such as local ecological knowledge. The method was tested and validated in Northern Ireland, where over 100 fishers participated in a semi-structured questionnaire and mapping exercise. The response rate was excellent, 97%, demonstrating fishers' willingness to be involved. The resultant maps are easily accessible and instantly informative, providing a very clear visual indication of which areas are most important for the fishers. The maps also provide quantitative data, which can be used to analyse the relative impact of different management options on the fishing industry and can be incorporated into planning software, such as MARXAN, to ensure that conservation goals can be met at minimum negative impact to the industry. This research shows how spatial access priority mapping can facilitate the early engagement of fishers and the ready incorporation of their priorities into the decision-making process in a transparent, quantitative way.

Editor: Jan Geert Hiddink, Bangor University, United Kingdom

Funding: Research was carried out as part of a PhD, funded by the Department of Education and Learning, Northern Ireland. Additional field work resources and equipment were provided by the Agri-Foods and Biosciences Institute, Northern Ireland, as part of a service agreement with the University. The funders had no role in study design, data collection and analysis, decision to publish, or preparation of the manuscript.

Competing Interests: The authors have declared that no competing interests exist.

* E-mail: yates-k1@email.ulster.ac.uk

Introduction

As the human population grows, pressure on the marine environment escalates and impacts of our unsustainable use can be seen in biodiversity loss, reduced water quality and declining fish stocks [1–6]. Efforts to improve management of the marine environment, and the resources within it, are becoming increasingly spatial, specifically through the development of marine spatial planning (MSP) and the designation of marine protected areas (MPAs). Whilst these management tools offer many potential benefits [7–15], too often these fail to be delivered [16,17], or the planning process itself fails [18,19].

One reason for the failure of MSP and MPAs is lack of adequate stakeholder involvement, even though this is often stated as vital for a successful planning process [20]. Incorporating stakeholders is important because it allows the planner to better appreciate the context of the plan and the potential impacts of different planning options. This greater appreciation provides the opportunity to develop more inclusive plans, which take into account the needs of all parties. By so doing, a planner should be able to both reduce conflict and help to develop a sense of stakeholder ownership over the project. Compliance with a chosen plan can be enhanced by this perception of ownership and inclusiveness [21]. Engaging with stakeholders provides the opportunity to explain the need for management changes, increasing understanding amongst stakeholders, which can again improve compliance [21]. Involving stakeholders can also facilitate generation of information that may not have otherwise been available, such as data on the distribution of species. This information can be incorporated into the decision-making process, which enhances planners' abilities to find the most efficient solutions, thereby maximising benefits whilst minimising conflict and negative impacts. Stakeholder involvement, therefore, is vital not only to increase a project's chances of success, but also to enhance the quality of decisions and reduce implementation cost [20,22–25].

In the marine environment, one of the largest commercial stakeholder groups is the fishing community. They are also arguably the group that are most likely to suffer direct adverse economic and social impacts of spatial restrictions. For planners to be able to assess the impact of different options on the fishing community, they need spatial information on the relative importance of different parts of the sea. However reliable data for even the distribution of target species is all too often unavailable [26], with arguably even less known about the distribution of fishing effort, and practically nothing about priorities of the fishers themselves. To fill some of the gaps, researchers and managers have begun to involve fishers in participatory mapping studies, and the importance of such initiatives for effective management is increasingly accepted [27].

Participatory mapping projects have documented a diverse range of features, including: the locations of fishing grounds; the distribution of habitats and species; key management areas such as spawning grounds; and benthic and oceanographic features [26,28–36]. These projects provide invaluable information for marine management. However, the maps are of limited use to a planner trying to take into account stakeholder priorities. First, these maps are frequently qualitative, making transparent quantitative analysis of the different options difficult. Second, even when quantitative data on stocks are available, they do not explicitly provide information on fishers' access priorities: the distribution of resources does not necessarily relate directly or linearly to the importance of different areas to fishers or to the impact of restricting access to them. Nor should a planner assume to understand the ways in which other variables interact with the spatial distribution of the resource to determine the importance of an area to a fisher.

The development of vessel monitoring systems (VMS) attached to fishing boats has begun to provide very accurate, high-resolution quantitative data on the distribution of fishing effort [37]. VMS is very useful, but alone it still only provides information on where most fishing occurs, which is not necessarily the same as which parts are the most valuable to fishers. A more complete picture of the gross economic value can be obtained by combining VMS data with daily logbook entries, then mapping the catch value of different parts of the sea to the fishery [38].

In many situations, however, VMS data are still not available: at present, VMS are generally deployed only on larger vessels and in developed countries. Detailed landings data are not always collected, nor are daily logbook systems always in place. Indeed, even in developed countries such as the UK, the extent and spatial distribution of inshore fishing effort can be poorly known (pers. obs). When data are collected, they might still be inaccessible [39]. Thus, VMS-based methods are only possible sometimes, for some fisheries. Even then, resultant data infer the value of different areas to fishers, instead of documenting them. Alone, VMS data are not able to account for all the factors that might affect how important an area is for a fisher. Grounds closer to home, for example, could be more important than their gross value implies because of reduced costs required to access them in both economic (e.g. less fuel) and/or social (e.g. coming home each night) terms. Other areas might provide relatively low revenue, but still be very important because they are the only areas available at a given time of year, due to weather conditions; in such cases, maintaining access is necessary for maintaining a regular flow of income. Furthermore, VMS documents current spatial fishing patterns, which in many cases are already restricted by area-based quotas, by-laws, and existing protected areas. Therefore VMS may not actually represent fishers preferred fishing grounds. Finally, VMS methods do not facilitate the active involvement of the fishers; they thus neither meet the recognised and increasingly legislated need to involve stakeholders, nor provide the opportunity for stakeholders to feel part of the decision-making process.

Here a novel method, spatial access priority mapping (SAPM), is described and demonstrated for the Northern Irish fleet. SAPM requires no VMS, logbook or landings data, is suitable for single or multiple fisheries of any type, and can be performed using only basic features of standard spreadsheet and GIS software. It actively engages fishers, documenting rather than inferring their priorities, and produces quantitative maps that can be combined easily with other data if available and desired.

Study Area – the Northern Ireland Context

Marine management in Northern Ireland faces many challenges. Recent European legislation, such as the Marine Strategy Framework Directive, is setting new standards for the environmental status of coastal waters. A Marine Bill is currently pending in the Northern Ireland Assembly that will provide the Department of the Environment (DOE) with new powers to develop marine spatial plans and to designate marine conservation zones (MCZs), a new type of marine protected area. The European Common Fisheries Policy, which has significant impact on the Northern Irish Fleet, is under review. The Northern Ireland Inshore Fisheries Strategy is also under review by the Fisheries Division of the Department for Agriculture and Rural Development (DARD). In addition, there is significant focus on the development of marine renewable energy in Northern Irish waters [40–42].

Whilst these developments bring many potential benefits for the Northern Ireland community, they will likely also bring spatial restrictions for fishers. In order to properly take fishers into account, and to make the most efficient plans possible, planners need to understand which parts of the sea are most important to fishers. However data are patchy at best. VMS is currently only in place for vessels 15 m and longer. Associated data have been processed to produce effort maps for some of their range (Irish Sea, but not the Clyde or other areas) [43]. Information on the inshore fishing fleet, which relies on smaller vessels, is limited, with no current monitoring of effort either in terms of magnitude or distribution. The distribution and stock levels of *Nephrops* are regularly monitored [44], and there are annual surveys for scallops [45] and ground fish [46]. There is, however, currently no information on the status of lobster or crab stocks, despite these resources supporting almost a third of the active vessels in the fleet. Access to the data that do exist is also restricted, sometimes even between Government Departments (pers. obs.).

Issues between the Northern Ireland Government Departments, particularly poor communication and co-operation between DARD and DOE, have caused problems for marine management in Northern Ireland in the past [47]. With DOE responsible for developing the spatial management plans, and DARD responsible for managing the fisheries that these plans are likely to impact, there is potential for further conflict. Thus, there is a real need for universally available data and a transparent way of incorporating the priorities of the fishers into the planning process.

The majority of the Northern Irish fleet is based at three main ports: Kilkeel is the largest, followed by Portavogie and Ardglass. In addition, there are over twenty minor ports at which small numbers of pot boats are based (Figure 1). A few Northern Ireland registered boats also fish out of ports in neighbouring countries, mainly Greencastle in the Republic of Ireland. There are 367 vessels registered in the Northern Ireland fleet, 224 of which are officially recorded as active (DARD, pers.comm). The majority of those 224 vessels are under 15 m (127, 57%) and would thus not be covered by current VMS data.

The most important fishery in terms of revenue and employment is *Nephrops*, with 99 of the 224 vessels focusing virtually exclusively on *Nephrops*, and a further 11 mixed-fishery vessels having some *Nephrops* interest (DARD, pers.comm). Pot-fishing accounts for the second highest number of vessels, with 69 of the 224 registered as exclusively potting, and 8 mixed-fishery vessels having potting interest (DARD, pers.comm). There are two main Fish Producer Organisations (FPOs): the Anglo-Northern Irish

Figure 1. Map of the home ports from which commercially active, Northern Ireland registered vessels fish. Map of the home ports from which commercially active, Northern Ireland registered vessels fish Data obtained from the Department of Agriculture and Rural Development, Northern Ireland, 2012.

Fish Producer Organisation (ANIFPO); and the Northern Ireland Fish Producer organisation (NIFPO). There are two pot fishers associations, the North Coast Lobster Fishermen's Association and the North-East Lobster Fishermen's Co-Operative, and a Scallop Fishermen's Association. All of these representative bodies reported attempting to engage with the Government Departments to improve management.

Methods

Interviews

Research was approved by the School of Environmental Sciences Ethics Committee, University of Ulster, Coleraine. All interviews were conducted in accordance with ethics procedures, and research plans, including the consent procedure, were submitted and approved prior to the initiation of the project, in accordance with School and University policy. Prior to each interview, information sheets detailing the research were provided to fishers and discussed, and then verbal consent was obtained from participants. Written consent was not sought, partly to maintain the informal setting that assisted with the response rate, but mainly because the respondents were not required to give their names during the interview, thus requiring a signature (a form of identification) was considered inappropriate. Verbal consent was recorded in the response form. All interviews were conducted by the lead author, ensuring that the procedure was uniform for all respondents, and all individual responses were recorded in a confidential database. The pre-interview information sheet and questionnaire are available in Files S1 & S2.

Interviews consisted of two parts: a semi-structured questionnaire; and a mapping exercise. Each interview lasted between 30 minutes and two hours. The vast majority of fishers interviewed (>90%) were approached directly at ports. A number of other fishers responded to flyers that had been distributed by mail. All respondents were skippers or vessel owners. With the exception of three interviews conducted in fisher's homes (2.9%), all interviews were conducted in ports (97.1%): on vessels, at the quayside, in a cafe, or in the ANIFPO or NIFPO offices.

During the mapping exercise, fishers were provided with both paper admiralty charts and digitised admiralty charts within a GIS. Fishers indicated their priority areas (as polygons) directly onto the digitised charts. They were encouraged to identify as many or as few areas as they liked, and to cover as much area as they wanted. However, each respondent was advised that the greater the total area, the lower the priority that could be assigned to each unit of area; thus they were encouraged to be specific. Those fishers active in multiple fisheries were asked to indicate the target species of each area. They were given the option of assigning areas different levels of importance and/or indicating how much of the year they spent in each area, which was then converted into a percentage. This then became the importance value, such that an area that was fished for a fifth of the year received an importance value of 20. On completion of the mapping exercise, respondents were encouraged to review their results and to make changes as desired.

Generating the Maps

The priority (SAP km^{-2}) of each of the areas indicated by a particular respondent was calculated by multiplying the number of full-time crew on the associated boat by the importance value, and then dividing by the number of square kilometres the area covered (Figures 2 & 3 show illustrative examples). Part-time or seasonal crew were assigned a *pro rata* proportion of a full-time member's weighting, so that a boat that reported four full-time staff throughout the year and one extra full-time staff for six months of the year was recorded as having a total of 4.5 crew. Each individual (full time) fisher has a total SAP of 100, such that a boat with a crew of 4.5 has a total SAP of 450 associated with it, and an area that has an SAP of 10 km^{-2} represents a tenth of a fisher per km^2.

In order to create maps that represented the whole fleet, the results obtained from the individual respondents need to be scaled up (Figure 4 shows an illustrative example). Based on initial examination of data and discussions with the fishers, vessels were divided into six size categories. All vessels under 9 m were pot boats. With the exception of two specialist offshore crab boats, no vessels over 10 m fished pots. In the ≥9 m <10 m category there was a mix of pot boats and small dredge or trawl boats. Dredge and trawl boats of similar size generally have very similar numbers of crew, but pot boats of the same size tend to have fewer crew and are often single handed. A two-sided t-test demonstrated that there were significantly fewer crew on pot boats between 9 and 10 m than other gear types ($t = 3.05$, $df = 20$, $p = 0.007$). Consequently, this size category was further divided by gear: pots or not pots. Data was obtained for only one pelagic vessel, of which there are three in Northern Ireland; however, an additional category was assigned for pelagic vessels as they have much higher crew numbers (8–10) than any other type. Thus vessels were divided into a total of eight categories (Figure 5) and the average number of crew per vessel for each of these categories was calculated using the sample data.

A list of vessels at each port was provided by the fisheries division of DARD, along with the length and registered gear type of each vessel. The list showed 224 active registered vessels, but during the research a number of inaccuracies were found with this list and it was updated to include additional boats: 14 that were originally listed as 'no recorded fish effort this year', and 8 that were not listed. The final vessel list contained 246 vessels and was broken down by port, gear type and length. Using this and the calculated average crew per vessel, an estimate was made of the total number of fishers on all active vessels by each port, gear and length combination.

In each length/gear/port combination, the estimated number of crew was divided by the actual number of crew sampled in order to create a final weighting scheme, which was used to

Port A: Illustrative example of Basic SAPM, single areas, no weighting and no scaling

Port A has three small pot-fishing boats, all single-handed (1 fisher) and all in the same size category. All pot fishers at the port were interviewed so no scaling up is required. They all chose only one area.

Fisher 1 (yellow) chose a total area of $3km^2$. Fisher 2 (Blue) chose an area of $4km^2$ and Fisher 3 (red) chose an area of $10km^2$.

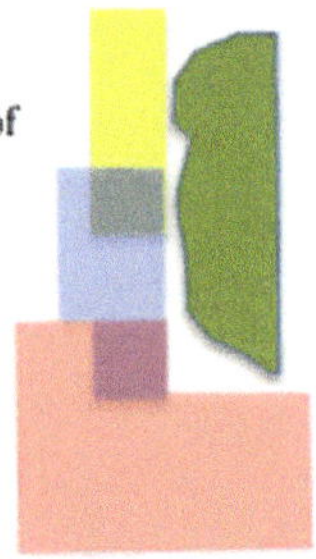

$$SAP = \frac{Crew \times Importance}{Area\ (km^2)}$$

As each fisher chose just one area, that area is inherently 100% of their importance, so their SAPs are as follows:

Fisher 1:
$(1\ crew \times 100\ importance) \div 3km^2 = 33.3\ SAP\ km^{-2}$
Fisher 2:
$(1\ crew \times 100\ importance) \div 4km^2 = 25\ SAP km^{-2}$
Fisher 3:
$(1\ crew \times 100\ importance) \div 10km^2 = 10\ SAP km^{-2}$

Each fisher has the same total priority, but these priorities are spread over different areas: the larger the area the lower the resultant SAP km^{-2}. This is justified because the value of a given area to a fisher depends on the total area they fish: Fisher 1 losing access to $1km^2$ of their fishing area would suffer a significantly larger impact than Fisher 2 losing access to $1km^2$ of their area.

To calculate the total SAP for pot fishers at Port A the individual SAPs are summed. Thus the $1km^2$ overlap of priority areas identified by Fishers 1 and 2 will have a SAP of 58.3 km^{-2} ($33.3km^{-2}$ + $25km^{-2}$) and the overlap in priority areas identified by Fishers 2 and 3 will have a total of SAP 35 km^{-2} (25 km^{-2} + 10 km^{-2}).

Figure 2. Illustrative example of basic SAPM. Illustrative example of basic SAPM, no weighting or scaling involved.

Port B: Illustrative example of SAPM with multiple areas and varying importance

Port B has two small pot-fishing boats, both are single-handed and both skippers have been interviewed. In this example, however, they each chose multiple areas and stated that those different areas have different levels of importance.

Fisher 1 chose two areas: $2km^2$ and $3km^2$. They stated that they spend 3 months (25% of the year) in the smaller area and 9 months (75% of the year) in the larger.
Smaller area:
$(1\ crew \times 25\ importance) \div 2km^2 = 12.5\ SAP\ km^{-2}$
Larger area:
$(1\ crew \times 75\ importance) \div 3km^2 = 25\ SAP\ km^{-2}$

Fisher 2 chose three areas: $4km^2$, $2km^2$ and $1km^2$ and stated they get 80% of their profit from the largest area and 10% from each of the other two.
Largest area:
$(1\ crew \times 80\ importance) \div 4km^2 = 20\ SAP\ km^{-2}$
Middle size area:
$(1\ crew \times 10\ importance) \div 2km^2 = 5\ SAP\ km^{-2}$
Smallest area:
$(1\ crew \times 10\ importance) \div 1km^2 = 10\ SAP\ km^{-2}$

Note that the total SAP for each fisher still adds up to 100.

Figure 3. Illustrative example of SAPM with multiple areas and varying importance. Illustrative example of SAPM with multiple areas and varying importance.

upscale the maps generated by respondents to represent the whole active fleet (Figure 4). Where no skipper of a vessel with a particular gear-length-port combination had been interviewed, responses from skippers at that port, using that gear, whose vessels most closely matched the required length, were upweighted appropriately (Figure 6 shows an illustrative example). On the very few occasions (3) where no skipper of vessels of that gear type and port had been interviewed, appropriate response(s) from the nearest port were used.

Priorities for each individual area were then multiplied by their corresponding weights to obtain a final weighted priority score. Resulting shape files were rasterised, at a resolution of 1 km^2, and all raster layers were combined using the raster calculator to construct the overall access priority map for the Northern Ireland fleet. Access priority maps for different sectors of the fleet were generated by selecting only the layers that related to that sector. To demonstrate the simplicity of the approach, all calculations were performed using tables and basic cell formulae in an Excel spreadsheet. Other freely available software such as R could be used as an alternative, making the analysis much quicker without changing the method. All spatial analysis was conducted in ArcMap 9.3.

Validation

In order to assess whether the sample of respondents was representative of the whole fleet, the three main characteristics of vessels were compared. A two-sided t-test was conducted to ascertain whether there was any significant difference in the vessel length distribution between the sample and the population. Chi-squared tests were used to test for any significant difference between the sample and the population in terms of the frequencies of gear types and registered ports. The smallest ports were grouped geographically to make this possible.

SAP maps from individual fishers within a given length-gear-port combination were overlaid to inspect for consistency of responses and verify the validity of up-scaling to represent all vessels. Variation is to be expected between different combinations, and even some minimal variation within combinations. However if the responses within a combination were very different up-scaling would not have been appropriate.

VMS-based effort maps (2009 and 2010) for boats 15 m and longer were provided by AFBI and were used to visually validate the areas indicated by those boats. It was not assumed that VMS is always an accurate representation of spatial access priorities. Indeed, as previously mentioned, an area may be fished for only a short time but may actually have a high value because of the quality of the catch or because it is the only site available at that time of year. Nevertheless, because it would be expected that all areas fished would have some priority associated with them, a visual inspection of the two sets of data provided a validity check

Port C: Illustrative example of SAPM using weighting to scale up sample result to represent all fishers

Port C has 11 *Nephrops* trawl boats in two different size classes. Interviews were obtained for six of the 11 boats. Each fisher chose one area.

NB. Average (mean) crew numbers are calculated based on results for the whole fleet (not just this port).

Size class (m)	Mean no. crew	No. registered active boats	No. with interview
10 - 14.9	3.0	6	3
15 - 19.9	3.5	5	3

Fisher 1:
14.5m boat, 3.5 crew, one area of 175 km^2

$$(3.5\ crew \times 100\ importance) \div 175km^2 = 2\ SAP\ km^{-2}$$

Fisher 2:
13.8 m boat, 3 crew, one area of 80km^2

$$(3\ crew \times 100\ importance) \div 80km^2 = 3.75\ SAP\ km^{-2}$$

Fisher 3:
14.9m boat, 3 crew, one area of 120km^2

$$(3\ crew \times 100\ importance) \div 120km^2 = 2.5\ SAP\ km^{-2}$$

In previous examples, all fishers were interviewed. Here three boats in the 10 - 14.9m class were included in interviews, but there are six registered at the port, so those three need to be scaled up to represent all six. This is done using a weighting, which is calculated by diving estimated number of crew by sampled number of crew.

$$Weighting\ (W) = \frac{estimated\ crew}{sample\ crew}$$

Estimated crew are those expected to be there: the mean number of crew for that gear-size combination multiplied by the no of registered, active vessels of that gear-size combination at the port. *Sample crew* is the total number of crew on all the vessels of that gear/size/port combination for which interview data has been collected.

$$W = \frac{3\ mean\ crew \times 6\ registered\ boats}{3.5 + 3 + 3\ sample\ crew} = \frac{18}{9.5} = 1.89$$

This weighting is then used to scale up the SAP's to account for the other vessels.

$$Weighted\ SAP = orginal\ SAP\ x\ W$$

Area 1: $2\ SAP\ km^{-2} \times 1.89 = 3.78\ SAP\ km^{-2}$
Area 2: $3.75\ SAP\ km^{-2} \times 1.89 = 7.09\ SAP\ km^{-2}$
Area 3: $2.5\ SAP\ km^{-2} \times 1.89 = 4.73\ SAP\ km^{-2}$

Weighting for 15 - 19.9m size class:
In this class interviews have been obtained for three boats, which have the following crew numbers: 3,4,3
$Estimated\ crew = 3.5\ mean\ crew \times 5\ boats = 17.5$
$Sample\ crew = 3 + 4 + 3 = 10$

$$W = \frac{17.5\ estimated\ crew}{10\ sample\ crew} = 1.75$$

Figure 4. Illustrative example of SAPM using weighting to scale up the sample to represent the population. Illustrative example of SAPM using weighting to scale up the sample (interviewed fishers) to represent the population (whole fleet).

for overall coverage. It would also be expected that a similar overall pattern would be evident.

Copies of the amalgamated maps for the whole fleet and for the four main fisheries, with the associated report [48], were made available to all members of the Northern Ireland fishing fleet to review. Digital copies were provided to the POs, fishers' associations and co-operatives, for distribution to their members. Hard copies were provided to the harbour offices of the POs and to harbour cafes, where they were made available for fishers to review. In addition, individuals were able to download digital copies of maps and reports from the University of Ulster website, the availability of which was advertised in the national fishing press. Feedback was provided in a number of ways: POs and associations solicited and passed on responses from members, some fishers responded using forms and prepaid envelopes that were provided along with the hard copies sent to POs and cafes, a number of individuals submitted responses by email, and others made direct contact with the lead author to discuss the results.

Results

Interviews were conducted in 20 different ports. The response rate (97%) was excellent. Fishers were very willing, often keen, to participate once the project was explained. Indeed, many expressed concern that those making the decisions understood neither the industry nor the fishers' needs, and were glad to have an avenue to express their opinions. In total, 103 individual skippers and owners of active vessels were interviewed. Only three skippers declined, two saying they were too busy and the other refusing to engage with the process. The respondents skippered and/or owned 118 vessels, representing 48% of the active fleet (updated list), and those vessels fish out of 25 different ports (Figure 7).

Overall, the sample is considered to be an accurate representation of the fleet (Tables 1& 2, Figure 7). The two-sided t-test on vessel length showed no significant difference between the sample and the population ($t = 0.06$, $df = 334$, $p = 0.995$). Chi-squared tests also showed no significant dependence between sample and fleet coverage either in terms of the registered port ($X^2 = 18.232$, $df = 19$, $p = 0.507$), or the type of gear used ($X^2 = 3.095$, $df = 9$, $p = 0.96$).

The map for the whole fleet produced by SAPM (Figure 8) shows the spatial heterogeneity in access priority and clearly highlights areas whose closure would likely cause greatest contention with the fleet. Unsurprisingly, results show that spatial priorities vary greatly between different fisheries (Figure 9) and, as closures are frequently only for certain types of gear, this can provide valuable input for planners and managers deciding what activities to restrict and where.

Comparisons among results from individual respondents showed a high degree of spatial overlap within a given length-gear-port combination: Figure 10 shows an example from a combination within the *Nephrops* fishery. Overlap was high across all of the three main ports for the *Nephrops* fishery, within each size category. Between size categories, overlap reduced, with smaller vessels having a corresponding reduced range from port. Overlap was also high across the three main ports for the scallop fishery (dredge boats). However, overlap across ports for pot-fishing was very low. This is unsurprising, as the small vessels generally have a

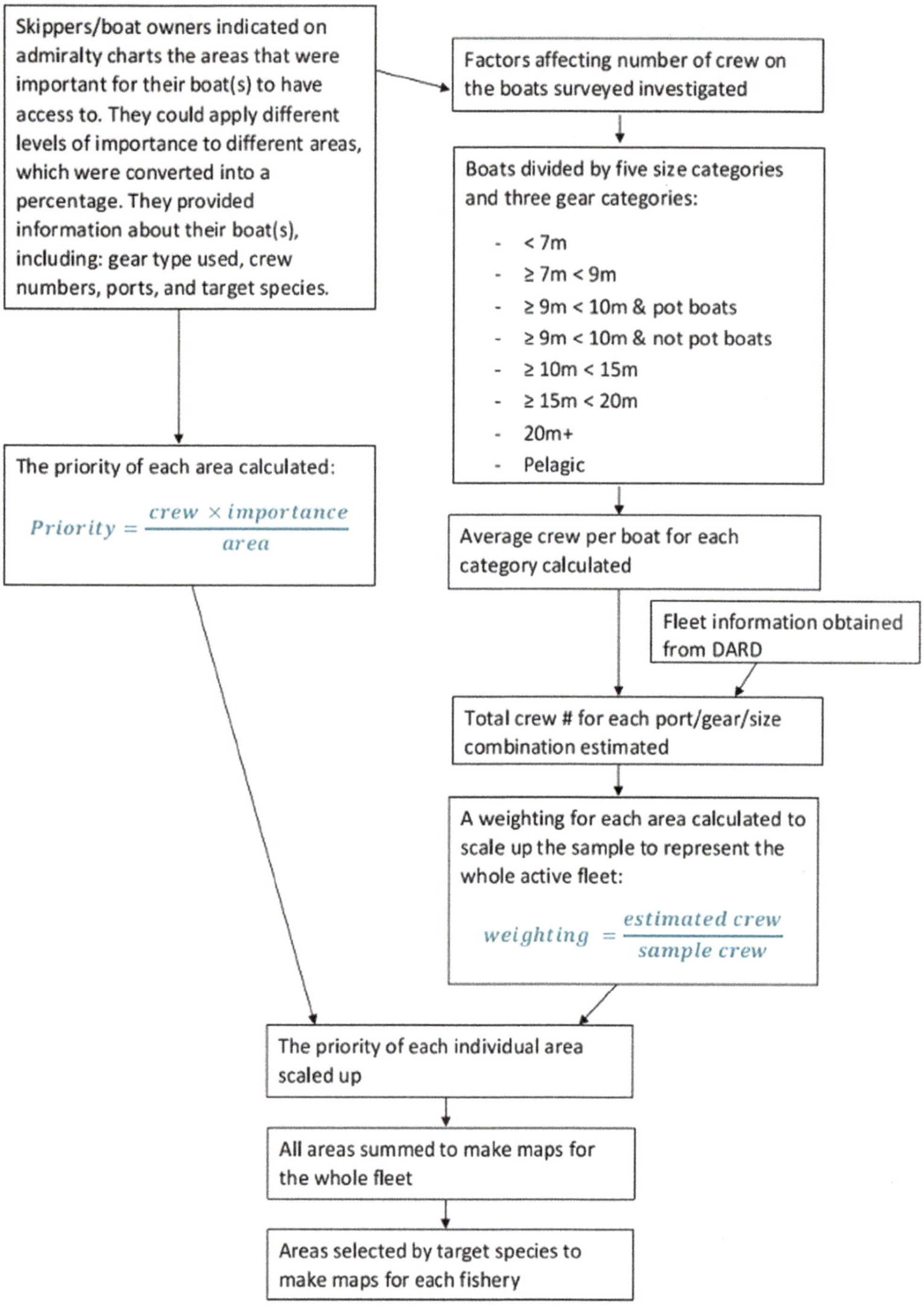

Figure 5. Flow diagram of the Spatial Access Priority Mapping Method. Flow diagram of the Spatial Access Priority Mapping Method use to map the access priorities of the Northern Ireland fleet in 2012. DARD stands for the Department of Agriculture and Rural Development, Northern Ireland.

very restricted range and fishers can be territorial between and sometimes even within ports. This stresses the importance of obtaining good spatial representation of the pot-fishing community, which can be difficult when small ports have just one or two fishers operating from them. In this study, the involvement of the Lobster Associations was invaluable in ensuring good coverage.

Overall, visual congruency between the access priority map for the over 15 m fleet and the VMS effort map is high, with all areas fished by the fleet being covered on the access priority map (Figure 11). The access priority map covers a larger total area than the VMS, which is unsurprising, and is most likely a combination of the inherently higher accuracy of the VMS and the wider scope of values that the access priority map captures.

Feedback on the initial report and maps was provided by 21 active skippers/owners, a 9% response rate, as well as by four retired fishers. Active skippers/owners represented all of the main

Port D: Illustrative example of SAPM when not all port/size/gear combinations have been sampled

Port D has 25 *Nephrops* trawl boats in five size categories.

NB. Average (mean) crew numbers are calculated based on results for the whole fleet (not just this port).

Size class (m)	Mean no. crew	No. registered active boats	No. with interview
10 - 14.9	3.5	6	3
15 - 19.9	4	2	0
20 - 24.9	4.2	12	5
25 - 29.9	4.5	5	3
30 - 34.9	4.6	1	0

As it has not been possible to sample all combinations, vessel(s) within the closest length class will have their weighting increased to represent the missing class. This is achieved by increasing their estimated crew.

Estimated crew is average number of crew for that gear-size combination multiplied by the number of vessels of that gear-size combination registered active at that port.

For the 15 - 19.9 m size category there are two 'closest' lengths, so both are used. Estimated crew for the class: mean crew × registered vessels = 4 × 2 = 8. Those estimated crew are divided evenly between the two closet classes, i.e. four extra crew are added to each.

Estimated crew for 10 - 14.9m class is therefore:

$$(mean\ crew \times registered\ boats) + additional\ crew = (3.5 \times 6) + 4 = 34\ Crew$$

Estimated crew for 20 - 24.9m class:

$$(mean\ crew \times registered\ boats) + additional\ crew = (4.2 \times 5) + 4 = 54.4\ Crew$$

For the 30 - 34.9 m class there is only one closest size class, 25 - 29.9 m, so the estimated crew, 4.6, is added entirely to that:

$$(mean\ crew \times registered\ boats) + additional\ crew = (4.5 \times 5) + 4.6 = 27.1\ Crew$$

This is in effect reallocating the estimated crew to classes for which sample data have been obtained. As estimated crew is used to calculate the weighting (see Box 3, example Port C), by proportionally increasing the estimated crew the SAP values are increased to take into account the missing vessels.

Interviewing every fisher would be the ideal, but rarely will that be possible. Collecting data for multiple vessels for each combination is the next best thing, but even that will not always be possible. There be just one or two vessels in a given combination and it might not be possible to find the skippers and/or they might not wish to partake.

Reallocating the estimated crew from unsampled combinations to the closest length class(es) of the same fishery (gear) and port allows the best possible estimation of their SAPs. Omission of those combinations altogether would lead to underrepresentation of those fishers and their fishery.

There will be occasions when reallocation to the closet length class may not be appropriate because of size based spatial restrictions. If there was, for example, a restriction in place that meant only vessels under 15m could fish within three miles of the shore, then it would be inappropriate to divide the estimated crew of the 15 - 19.9m class between both the 10 - 14.9m and 20 - 24.9m classes, as they are far less likely to prioritize areas they can not currently access. Thus, in this case, all the estimated crew should be allocated to the 20 - 24.9m class.

Figure 6. Illustrative example of SAPM when not all gear-port-size combinations have been sampled. Illustrative example of SAPM when not all gear-port-size combinations have been sampled.

fisheries and comprised: 8 trawl fishers; 9 pot fishers; 1 dredge/scallop fisher; and 3 fishers who targeted multiple fisheries. The response was very positive, with the result perceived by all respondents to be a good reflection of the fleet's priorities. However, many fishers expressed concern that Government Departments would not use fishers' inputs in forthcoming plans.

Data Sharing

An important aspect of this highly applied research was to provide data to feed into impending management changes. Once the maps, and associated report, had been reviewed by the fishing community and their comments had been added, it was formally presented to a group of representatives from the relevant Northern Ireland Government Departments: DARD, AFBI, DOE, the Northern Ireland Environment Agency and the Department of Enterprise, Trade and Investment (Marine Renewable Energy Division). In addition to the digital and hard copies of the report, they were also provided with the raster data layers for the five maps (the whole fleet and the four main component fisheries) for use in their GISs. All agencies stated intentions to use these data in forthcoming marine spatial planning and MCZ-designation activities.

The report and data layers were also provided to the UK Marine Management Organisation, Natural England, the UK Joint Nature and Conservation Committee, and the Council for Conservation and the Countryside (Northern Ireland). The report is freely available to download [47], and interested parties can request the data layers from the lead author. Data layers will be deposited in the UK data archive in 2014. Raw data and all information relating to individual fishers are confidential and will not be made available.

Discussion

More than 100 skippers and owners of Northern Ireland registered vessels were interviewed about their spatial access priorities: their boats represent almost half the active Northern Irish fleet. Data from these interviews was used to generate spatial access priority maps for both the whole fleet and for each of the main fisheries. These maps provide quantitative data on which parts of the sea are most important to the fishers, and they can be used to maximise the efficiency of protected area site selection and to minimise potential conflicts between conservation needs and those of the fishing industry. The maps can also be used to inform marine spatial plans, highlight areas of tension between fisheries and contribute to fisheries management plans.

This research demonstrates the willingness of Northern Ireland fishers to be involved in both research and management planning. The response rate was excellent, despite early concerns by many involved in local marine management that it would be difficult to

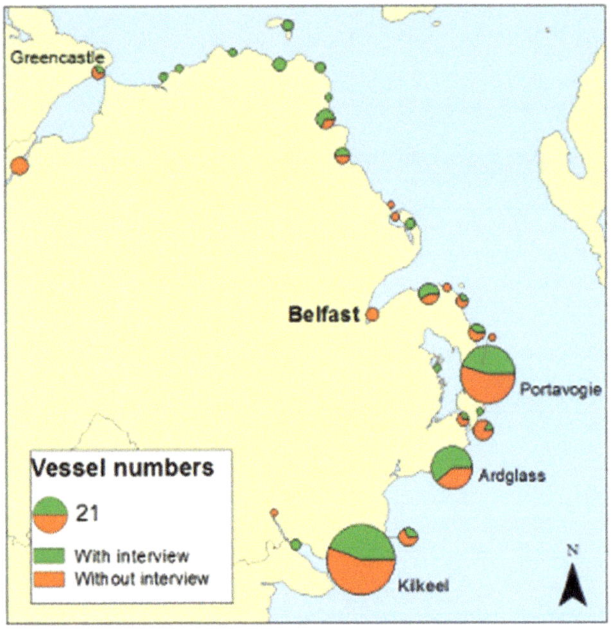

Figure 7. The proportion of vessels for which interviews were obtained or not obtained. Map of the Northern Irish ports, plus Greencastle, showing the proportion of vessels for which interviews were obtain for against those for which they were not. The size of the symbol for each port represents the number of commercially active, Northern Ireland fishing vessels registered there.

get the fishers to participate. In fact, fishers were extremely forthcoming with a range of information and many were pleased to have the opportunity to "have their say". The nature of the project may have been a factor in this. First, it was independent research, and thus it bypassed any historical animosity between the Government Departments and the fishers. Second, the research sought explicitly to map fishers' access priorities so that these could be taken into account in future planning processes; it was therefore in fishers' interests to take part. Third, the method was easy to explain to respondents: it was simple, intuitive and representative. This latter characteristic not only increases the chance of participation but also, increases the reliability of fishers' responses because they are far more likely to genuinely understand what is expected of them. Probably the main factor, however, was that the research went to them: interviews took place on fishers' boats, at the quayside, or in the cafes they frequent. This allowed the fishers to participate at a time and in a space that suited them, and it greatly reduced both the actual and opportunity costs of their participation. Indeed, many fishers complained that in order to participate they would normally have had to travel to meetings held in Belfast, at their own expense, and lose a day's fishing. That situation is not unique to Northern Ireland [31] and can lead to poor and/or skewed representation, with only certain fishers able or willing to contribute.

The response rate for the review of the results was much lower than for the face-to-face interviews. This is not surprising: response rates to distance surveys are normally much lower than face to face interviews [49], because in face-to-face interviews the researcher has the opportunity to explain the research and the value of the respondents' involvement. A survey discussing fishing revenue and expense data, distributed to 284 vessel owner/operators in Seattle, for example, had just 34 surveys returned, a 12% response rate [50]. It is also important to bear in mind that this response rate is artificially low because it is calculated by dividing the 21 respondents by all active skippers (246), when in fact it is unknown how many of those skippers were actually aware of the research, the review period or their opportunity to respond. The responses received also suggest that the lower response rate may in part reflect a general agreement with the results, combined with cynicism over the incorporation of the results into policy, making individuals less likely to take the time to respond.

The use of a both paper and digitised maps during the mapping process proved very effective, demonstrating the utility of this dual approach. Other mapping studies have annotated paper maps during interviews and then transformed them into digital form [36,51], but direct annotation into the GIS is more resource efficient and removes the possibility of introducing error or bias during a later digitising process. Use of digitised maps also provides much greater flexibility, allowing the interviewer to carry multiple maps and work at different scales, depending on the type of fisher being interviewed. A pot fisher, for example, may work an area smaller than 5 km^2 whereas a *Nephrops* fisher can be utilising the entire Irish Sea. Carrying around physical copies of sufficient maps, each with different resolutions to be appropriate for every fisher in each area is impractical, especially in the context of conducting interviews in small wheelhouses on boats. Digitised maps do not have this issue and can also be enhanced by adding or

Table 1. Comparision of the number of registered active Northern Ireland fishing vessels and the number with interviews, by size class.

Size, in meters	Number active	Number with interviews	% with interview
under 7	48	21	43.8
7 to <9	17	10	58.8
9 to <10	49	22	44.9
10 to <15	32	19	59.4
15 to <20	67	35	52.2
20 to <30	29	9	31.0
30+	4	2	50.0
Total	**246**	**118**	**47.9**

The total number of registered active Northern Ireland fishing vessels and the number with interviews, by size class. Fleet information supplied by the Department of Agriculture and Rural Development, Northern Ireland, and update during the research.

Table 2. Comparision of the number of registered active Northern Ireland fishing vessels and the number with interviews, by gear type.

Gear	Number active	Number with interview	% with interview
Dredge	5	4	80.0
Dredge & Nephrops trawl	23	13	56.5
Gill Net	13	3	23.1
Hand Fishing	1	0	0.0
Nephrops & Whitefish trawl	5	2	40.0
Nephrops trawl	104	49	47.1
Pelagic trawl	3	1	33.3
Nephrops trawl & Gill net	1	0	0.0
Pots	80	42	52.5
Pots & Nephrops trawl	4	1	25.0
Pots, Dredge & Nephrops trawl	2	2	100.0
Scallop Dredge	2	1	50.0
Seine netting	1	0	0.0
Pots & Dredge	2	0	0.0
Total	**246**	**118**	**47.9**

The total number of registered active Northern Ireland fishing vessels and the number with interviews, by each gear type. Fleet information supplied by the Department of Agriculture and Rural Development, Northern Ireland, and update during the research. Please note that many scallop dredgers are registered as dredge or dredge and nephrops.

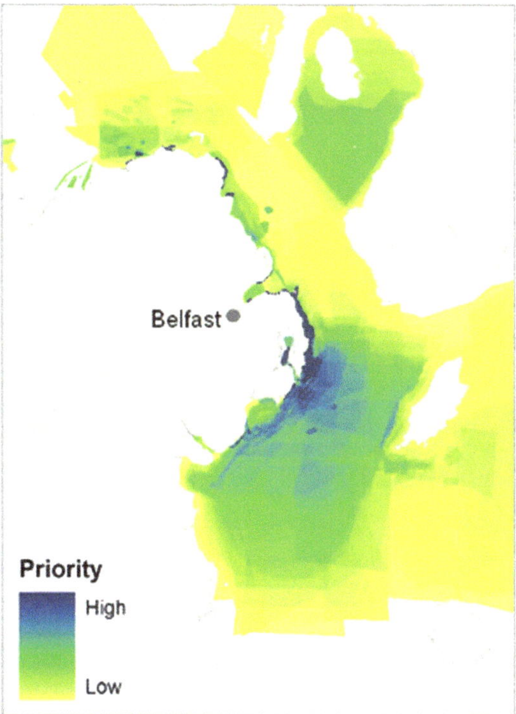

Figure 8. Map of the spatial access priorities of the Northern Irish fishing fleet. Map of the spatial access priorities of the Northern Irish fishing fleet generated from data obtained from 103 interviews with skippers and boat owners in 2012. It shows the main areas used by the fleet. There are a number vessels that have a much greater range than this. All areas were included in mapping process; however, for the purpose of maintaining an appropriate scale for visual display, they have not been shown.

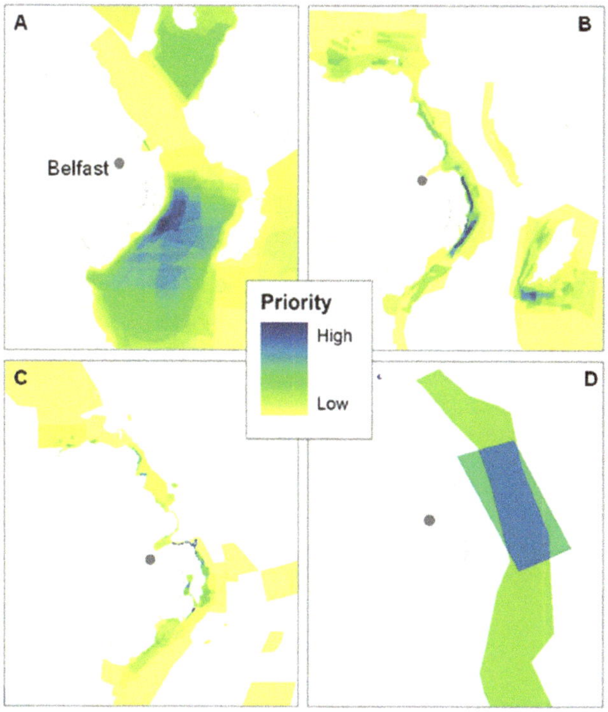

Figure 9. Maps of the spatial access priorities of the four main fisheries within the Northern Irish fishing fleet. Maps of the spatial access priorities of the four main fisheries within the Northern Irish fishing fleet: A) Nephrops, B) Pot fishing, C) Scallops and D) White fish. Maps were generated using data obtained from interviews with a total of 103 Skippers and boat owners in 2012. They show the main areas used by the fleet. There are a small minority of Nephrops vessels that have a greater range than this. All areas were included in mapping process; however, for the purpose of maintaining an appropriate scale for visual display, they have not been shown.

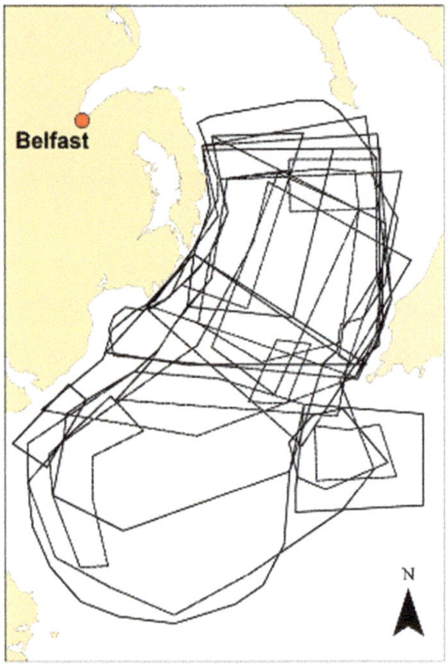

Figure 10. Spatial overlap of fishers' access priority areas within a given gear-length-port combination. Map of the access priority areas chosen by nine *Nephrops* fishers, from Northern Ireland. The nine fishers' vessels were within a given port-length combination and the map shows the spatial overlap in their responses.

highlighting reference points, such as lighthouses. Furthermore, when using digital maps, built-in tools such as "zoom" and "pan" have been shown to facilitate the mapping process by revealing more scale-dependent detail [26]. Fishers also appreciated the ability to quickly and accurately measure distances on the digital map when they were drawing on their areas. Having some broad-scale paper maps available did prove useful, mainly as a comfortable starting point for those less technologically inclined, and the use of a combination seems to provide the best of both worlds.

SAPM facilitated the documentation of fishers' actual access priorities, as opposed to interpreting their priorities from VMS-based mapping methods, and thus provides a fuller picture of which parts of the sea are most important to them. VMS alone maps only effort. Even when it is possible to incorporate log-book data with VMS data to estimate gross catch value of the different parts of the sea, this value is certainly not the only feature that makes an area important to fishers. Moreover, there are many boats that simply do not have VMS because they are too small. In Northern Ireland, over half the fleet would have their access priorities unaccounted for if planning were to use VMS-based maps alone.

Notwithstanding the limitations of VMS data, they remain vital for understanding the distribution of effort and are very important both for management and for informing the planning process. For example, they can be used to help predict the displacement impact of spatial restrictions. SAPM can be an important compliment to VMS based methods, when they are available, helping to compensate for their limitations. When they are not, it can be readily used as a standalone method.

This research also demonstrated how the SAPM method can be used to combine priorities across different types of fishery and to generate maps for the overall fleet, as well as for the separate fisheries. That SAPM can be used in any situation makes it suitable for enabling the direct comparison, and even combination, of the access priorities of multiple fleets, which otherwise might not have been possible due to differences in the associated data. Examples include: fleets that have VMS and fleets that do not, or fleets from different jurisdictions that access the same area but whose regulators have different fisheries data-capture and analysis methodologies.

Possibly the greatest advantage of SAPM over VMS is that it directly engages fishers in the process and allows them to contribute quantitative pre-planning data. The importance of stakeholder involvement is well recognised [20,23–25,33,52] and is increasingly becoming a legislative requirement, incorporated into both national and international policy [23,53]. Recent European Directives, such as the EU Public Participation Directive (2003/35/EC), have set new legal standards for stakeholder involvement, and almost every sustainability conference closes with a commitment to enhancing public participation [25].

Stakeholder involvement is sought because it has been shown to increase the quality and durability of environmental decisions, with the resultant plans better adapted to local socio-cultural and environmental conditions [23–25]. It can also reduce conflict and increase compliance [21,33]. Indeed stakeholder participation and involvement have been widely identified as essential components of effective ecosystem management [54,55], with differences in the extent and quality of participation being one of the key factors for variation in the quality of management approaches [20].

Stakeholder participation can also lead to the generation of additional data. Whilst undertaking SAPM, it would be relatively simple to gather other information, such as fishers' opinions of management options, or local ecological knowledge, which can enhance understanding of the ecosystems being managed. Both can provide highly valuable input to the planning processes. Inclusion of stakeholder knowledge such as this can also help to integrate stakeholders into the planning process and to promote a sense of ownership [56]. Certainly here in Northern Ireland, fishers were more than happy to contribute, and discussions indicated that the inclusion of their knowledge and opinions into the planning process would enhance buy-in.

Stakeholder involvement is not without its opponents, with counterarguments suggesting that the participatory process may not in reality lead to better-quality plans, and that the high levels of participation may actually increase conflict by having disputing parties at the negotiating table [20]. There are concerns that it will frustrate the planner by slowing down the process and that the strength of the final agreement will be diluted to balance competing interests [20]. Protected area designs that accommodate social needs may also result in outcomes that do not adequately conserve ecosystems [57]. Yet evidence suggests that this need not be the case. The incorporation of stakeholders during the planning process can increase the likelihood of plan implementation and the presence of certain stakeholders, particularly industry, has been shown to significantly increase the quality of local ecosystem plans [20]. Likewise, research shows that the use of fisheries data in MPA selection can reduce the negative impacts on the industry whilst still meeting conservation goals [58,59], and that incorporation of socio-economic considerations into systematic conservation planning can improve overall cost effectiveness [60]. It has also been frequently demonstrated that when fishers are involved in the research and decision-making process, management guidelines are more likely to be effective [26,61,62].

Another concern might be the credibility of stakeholder involvement, questioned on the basis that many may not have sufficient expertise to meaningfully engage in what are often

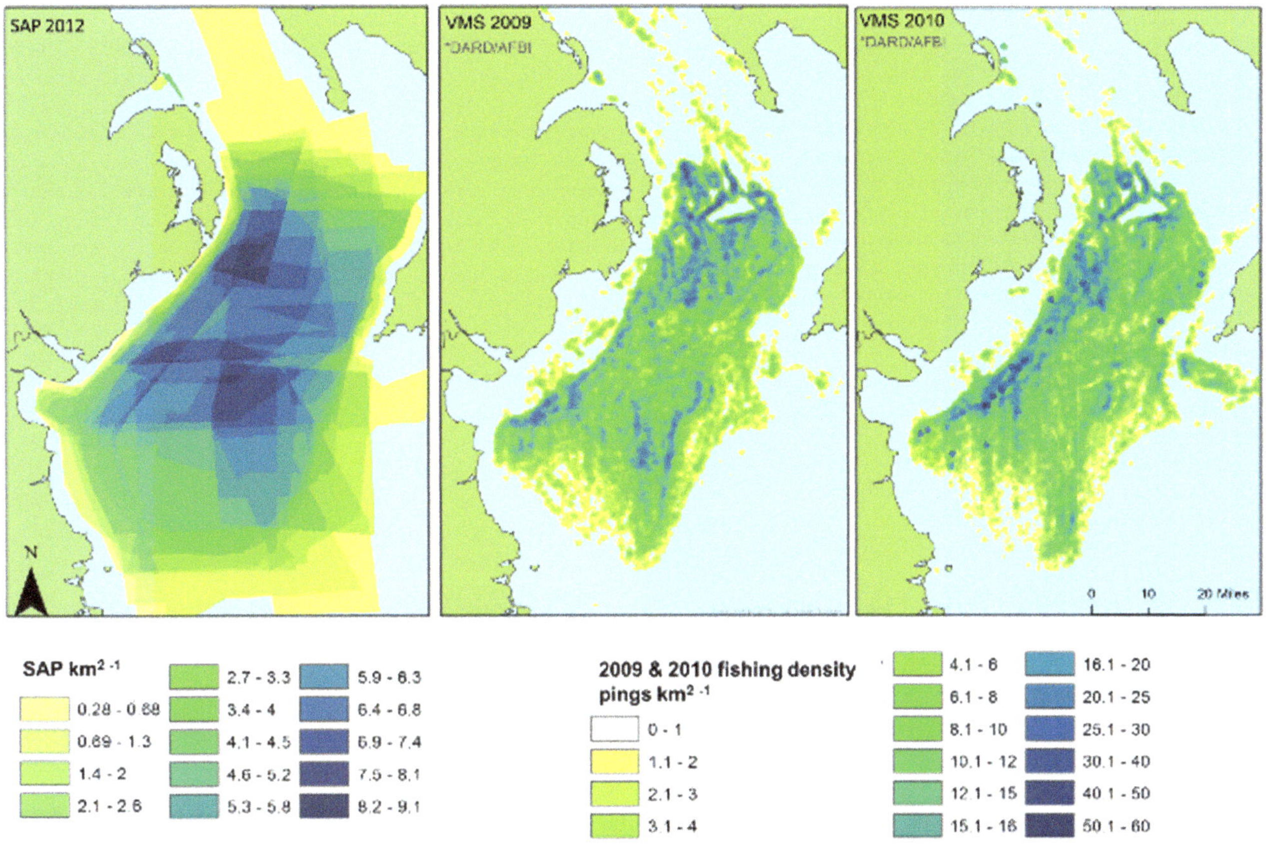

Figure 11. Spatial access priority map and VMS effort maps for the over 15 m *Nephrops* fleet. The spatial access priority map (2012) and the VMS effort maps (2009 and 2010) for over 15 m *Nephrops* fleet. VMS maps supplied by Laurence Rooney at AFBI. Processed VMS data were not available for all areas delineated by the fishers so this figure compares only main fishing areas where it was.

technical debates [63]. However, in this case it is actually the fishers that are the experts, often with decades of fishing experience. This acumen of fishers is increasingly recognised, with more and more studies documenting their extensive ecological knowledge [52,64,65] and surely who better to indicate which parts of the sea are most important to fishers than the fishers themselves?

With stakeholder consultation becoming more and more integrated into governance, the issue of consultation fatigue can also arise: when stakeholders are asked to take part in multiple or multi-stage participatory processes in which they perceive their involvement gains them little reward or capacity to influence the decisions being made [66]. Use of SAPM can help prevent that. It collects data in a systematic way, one of the recommendations for overcoming barriers to stakeholder participation [67]. Participation in SAPM is also of low cost to fishers and the method is readily explained. Resultant data are baseline preplanning data, so they can be used for multiple different planning projects, helping to avoid excessive or repeated demands on fishers' time. The maps generated can easily be used to demonstrate the results of the research and how that can feed into the planning process. They can also be readily incorporated into conservation planning software, such as MARXAN, where their influence on selecting potential MPA site can be illustrated.

As mentioned earlier, the cost of participation at centrally based stakeholder events can be a barrier to participation to fishers. Participation tends to be higher amongst those with more extreme positions and amongst those who are more dominant, influential and financially powerful within a community [68], so those fishers who do attend central events are unlikely to be truly representative of the fleet as a whole. By actually going to the fishers, SAPM can help to ensure a more diverse range of fishers is involved in the planning process. It also allows much larger numbers to partake: whilst it would be impractical for DOE to have 100 fishers around a planning table, SAPM allowed those fishers to contribute in a way that made them feel part of that process.

Other studies have looked at incorporating fishers' priorities into marine planning. In particular, a series of studies incorporating Californian fishers' knowledge and priorities into MPA selection has been conducted by Klein [58,69] and Scholz [31,32,70]. Fishers were asked to map their fishing grounds and to indicate relative importance [70], in a similar way to SAPM. The study area was then divided into 1381 planning units and the fishers data was used to calculate the relative importance of those planning units, across all fisheries and within a fishery [58]. Surrogate layers for relative fishing effort for each fishery were produced, and by summing the layers for the different fisheries they were able to develop an index of relative effort across all fisheries. In this method, each fishery receives equal weighting, such that a fishery that supported 10 fishers would have the same input into the final index as one that supported 1000. No weighting is applied to the individual fishers' areas either, so that results from a skipper whose vessel supported 10 crew would have the same weight as result from a skippers who vessel supported two.

This approach does increase the likelihood that all fisheries would be affected equally, which in this instance was a specific goal of the stakeholders (Klein, pers.comm). However, it is also often very desirable to minimise total negative impact on the fishing community and protect the greatest number of livelihoods. In later studies, Klein *et al* [69] and Scholz *et al* [32] incorporated a weighting system. They weighted the area(s) each fisher indicated by their vessel's revenue, which means that those fishers with vessels that caught more and made more profit had proportionally more input. This is similar to SAPM, where responses are weighted by the number of crew, so that those vessels with more crew have more total priority. Weighting by vessel revenue enabled Klein *et al* [69] and Scholz *et al* [32] to produce layers that showed the monetary value, in US dollars, of a given planning unit to an individual fisher. They summed those individual layers to create cost layers, which they used in Marxan to develop MPA planning options that minimised the negative impacts of spatial restrictions on the fishing industry.

The main difference between previous approaches and SAPM is that SAPM can allow an assessment of both relative and total access priority, and the production of surrogate cost layers, without the need for additional fisheries data. This is because SAPM uses information derived from interview (crew numbers), rather than revenue or landings, to weight the responses. Like the cost layer(s) created in both Klein *et al* [69] and Scholz *et al* [32], the SAPM generated surrogate cost layer(s) can be used in Marxan to minimise total negative impact of MPA designation on the fishing community. If the goal was to increase the likelihood that all fisheries are affected equally, like Klein *et al* [58], SAPM can also be used to map and assess relative priority, by dividing the SAP km^{-2} by the total SAP of a fishery and/or a fleet. Moreover, SAPM can also be used to assess the equality of spatial restrictions by analysing the SAPs of different port-gear-length combinations and comparing how much a given restriction or set of restrictions affects each group. When it is possible to access other data, such as revenue data, SAPM can be easily modified to incorporate it, see below for some possible adaptations to the method.

Another major difference is that SAPM specifically takes account of the different characteristics of the vessels (length/port/gear) when weighting and scaling responses. This is important because size of a vessel and the port from which it operates are both likely to have an impact on its preferred grounds, even when fishing for the same target species. Furthermore, the previous studies did not ascertain whether their samples (the fishers they interviewed) were representative of the fleet. Indeed, Scholz *et al* [32] state that to reach their sampling goal as efficiently and cheaply as possible they targeted fishers with the largest catches [32]. Whilst understandable, this suggests that, within a given fishery, the fishers with the smallest vessels had a lower chance of being interviewed. Smaller vessels tend to be more restricted in terms of fishing grounds and may have totally different spatial access priorities than larger vessels. By requiring a representative sample of the fleet SAPM helps to ensure all fishers are included, essential when aspiring to equitable stakeholder involvement.

Use of SAPM does not remove the need for subsequent engagement with the fishing community, nor will it remove all the potential conflict associated with MSP and MPA designation. It is a beginning, not an end: providing data to feed into a planning process and allowing a planner to assess the impact of different potential plans on fishers. Its use should improve the transparency of the planning process, enhance buy-in and increase trust.

Possible Adaptations to the Method

The weighting system used in SAPM is highly flexible, allowing for the incorporation of livelihoods or economic values beyond the vessels' crews. In Northern Ireland, for example, fish processors employ an estimated 500 full-time staff [43]. To incorporate these into the analysis, those 500 would be first separated into their associated fisheries. Then, based on the percentage of fishers within each length-gear-port combination for that fishery, they would be divided between each combination and added to the estimated number of crew. Doing this would proportionally increase the priority of fisheries with associated local processing, such as *Nephrops*, and cause a relative decrease in those without, reflecting the higher numbers of livelihoods that depend on those sectors. Similarly, if other data, such as landings, are available, the weighting system could also be modified accordingly. The total economic value of landings for a given fishery could be divided by the number of fishers in that fishery to obtain a value per fisher, which would then be multiplied by the weighting obtained from the standard method for each length/gear/port combination.

Future Research

This research did not aim to investigate the fishers' motivations for selecting different areas or for assigning priorities. Nor were they restricted to assigning priorities for a specific time span. It would be informative to expand future SAPM projects to include a debriefing session, where researchers explore with fishers their rationale and motivation for selecting areas and assigning priorities. An improved understanding of what makes areas important to fishers may help with predicting how they respond to spatial and temporal restrictions. Future SAPM projects could also incorporate the gathering of local ecological and historical knowledge, including: mapping where the fishers historically accessed resources and how their SAPs have changed over time, discussing when and why they changed, and comparing answers to legislative, ecological and climatic trajectories.

Conclusions

Spatial management can have many benefits for conservation and fisheries alike, but in order to realise them stakeholders need to be engaged in a strategic manner from an early stage. SAPM offers a tool that not only gathers quantitative data on fishers' access priorities, but also facilitates stakeholder engagement and increases the opportunity for dialogue between fishers and managers. The resultant maps provide a transparent, easy to demonstrate way of incorporating fishers' priorities, especially when combined with spatial planning software such as MARXAN. The use of SAPM should facilitate strategic conservation planning, maximising benefits whilst both minimising negative impacts on stakeholders and helping to ensure those impacts are spread equitably. We hope that the inclusive planning that SAPM encourages will lead to more successful solutions to the inevitable conflicts of interests that conservation planning encounters.

Supporting Information

File S1 Pre-interview Information sheet. Pre-interview information sheet provided to fishers prior to participation in a survey targeting skippers and owners of commercial fishing vessels registered in Northern Ireland. After the potential participant had read the information sheet, the content was discussed to ensure understanding and obtain verbal consent, before beginning the interview. Research was completed during 2012.

File S2 Questionnaire. Questionnaire used to guide semi-structured interviews of skippers and owners of Northern Ireland registered commercial fishing vessels. Interviews were preceded with pre-interview information sheets. 103 interviews were conducted in total, during 2012. The responses from some of these questions were used to develop Spatial Access Priority Maps, this study, whilst others were used to answer research questions explored in subsequent studies.

Acknowledgments

First and foremost the authors would like to thank all the fishers that generously gave up their time to make this research possible. Many thanks to everyone in ANIFPO, NIFPO and the North Coast Lobster Association; in particular we thank Mr Alan McCulla and Mr David Galbraith. Thanks also to members of DOE, NIEA, DARD and AFBI for their assistance, with special thanks to Laurence Rooney, at AFBI, for the provision of the VMS images. We would like to thank our colleagues Dr Thomas Smyth, Dr Sara Benettii and Dr Ruth Thurston for reviewing an earlier draft of the manuscript. We would also like to thank Dr N. Clare Eno and two other anonymous referees for their comments and suggestions. Input from all the reviewers greatly added to the final manuscript.

Author Contributions

Conceived and designed the experiments: KLY. Performed the experiments: KLY. Analyzed the data: KLY. Contributed reagents/materials/analysis tools: KLY. Wrote the paper: KLY. Introduced the first author to initial key contacts: DS. Provided assistance with the funding proposal that made this research possible: DS. Provided significant critical feedback on multiple drafts, improving the argument and the robustness of the discussion: DS. Corrected grammatical errors and made suggested improvements to prose: DS. Suggested the development of a flow chart to make the methods simpler to follow: DS.

References

1. Airoldi L, Beck MW (2007) Loss, status and trends for coastal marine habitats of Europe. Oceanography and Marine Biology: An Annual Review 45: 345–405.
2. Casey RA, Myers JM (1998) Near Extinction of a Large, Widely Distributed Fish. Science 281: 690–692. Available: http://www.sciencemag.org/cgi/doi/10.1126/science.281.5377.690.
3. Pauly D, Watson R, Alder J (2005) Global trends in world fisheries: impacts on marine ecosystems and food security. Philosophical transactions of the Royal Society of London Series B, Biological sciences 360: 5–12. Available: http://www.pubmedcentral.nih.gov/articlerender.fcgi?artid=1636108&tool=pmcentrez&rendertype=abstract.
4. Jackson JB, Kirby MX, Berger WH, Bjorndal KA, Botsford LW, et al. (2001) Historical overfishing and the recent collapse of coastal ecosystems. Science (New York, NY) 293: 629–637. Available: http://www.ncbi.nlm.nih.gov/pubmed/11474098.
5. Halpern BS, Walbridge S, Selkoe KA, Kappel CV, Micheli F, et al. (2008) A global map of human impact on marine ecosystems. Science (New York, NY) 319: 948–952. Available: http://www.ncbi.nlm.nih.gov/pubmed/18276889.
6. Worm B, Sandow M, Oschlies A, Lotze HK, Myers RA (2005) Global patterns of predator diversity in the open oceans. Science (New York, NY) 309: 1365–1369. Available: http://www.ncbi.nlm.nih.gov/pubmed/16051749.
7. Beukers-Stewart B, Vause B, Mosley M, Rossetti H, Brand A (2005) Benefits of closed area protection for a population of scallops. Marine Ecology Progress Series 298: 189–204. Available: http://www.int-res.com/abstracts/meps/v298/p189-204/.
8. Halpern BS, Warner RR (2002) Marine reserves have rapid and lasting effects. Ecology Letters 5: 361–366. Available: http://doi.wiley.com/10.1046/j.1461-0248.2002.00326.x.
9. Roberts CM, Hawkins JP (2000) Fully-protected marine reserves: a guide. WWF Endangered Seas Campaign, Washington, USA and Environment Department, University of York, UK. Avalible: http://billhutten.s3.amazonaws.com/fw/docs/345.pdf Accessed 2012 Jan 20.
10. Roberts CM, Bohnsack JA, Gell F, Hawkins JP, Goodridge R (2001) Effects of marine reserves on adjacent fisheries. Science (New York, NY) 294: 1920–1923. Available: http://www.ncbi.nlm.nih.gov/pubmed/11729316.
11. Worm B, Hilborn R, Baum JK, Branch TA, Collie JS, et al. (2009) Rebuilding global fisheries. Science (New York, NY) 325: 578–585. Available: http://www.ncbi.nlm.nih.gov/pubmed/19644114.
12. Hyrenbach KD, Forney KA, Dayton PK (2000) Marine Protected Areas and Ocean Basin Management. Aquatic Conservation: Marine and Freshwater Ecosystems 10: 437.
13. Heupel MR, Simpfendorfer CA (2005) Using Acoustic Monitoring to Evaluate MPAs for Shark Nursery Areas: The Importance of Long-Term Data. Marine Technology Society Journal 39: 10.
14. Rudd MA, Tupper MH (2002) The Impact of Nassau Grouper Size and Abundance on Scuba Diver Site Selection and MPA Economics. Coastal Management 30: 133.
15. Kelleher G, editor (1999) Guidelines for Marine Protected Areas. Gland, Switzerland: World Commision on Protected Areas. IUCN - The World Conservation Union. IUCN website. Avaliable: http://data.iucn.org/dbtw-wpd/edocs/PAG-003.pdf. Accessed 2012 Oct 10.
16. Jameson SC, Tupper MH, Ridley JM (2002) The three screen doors: Can marine "protected" areas be effective? Marine pollution bulletin 44: 1177–1183.
17. Hargreaves-Allen V, Mourato S, Milner-Gulland EJ (2011) A global evaluation of coral reef management performance: are MPAs producing conservation and socio-economic improvements? Environmental management 47: 684–700. Available: http://www.ncbi.nlm.nih.gov/pubmed/21365274.
18. Helvey M (2004) Seeking Consensus on Designing Marine Protected Areas: Keeping the Fishing Community Engaged. Coastal Management 32: 173–190. Available: http://www.tandfonline.com/doi/abs/10.1080/08920750490276236.
19. Weible CM (2008) Caught in a Maelstrom: Implementing California Marine Protected Areas. Coastal Management 36: 350–373. Available: http://www.tandfonline.com/doi/abs/10.1080/08920750802266387.
20. Brody SD (2003) Measuring the Effects of Stakeholder Participation on the Quality of Local Plans Based on the Principles of Collaborative Ecosystem Management. Journal of Planning Education and Research 22: 407–419. doi:10.1177/0739456X03253022.
21. Innes JE (1996) Planning through consensus building: A new view of the comprehensive planning ideal. Journal of the American Planning Association 62: 460–472.
22. Reed MS, Dougill AJ, Baker TR (2008) Participatory indicator development: what can ecologists and local communities learn from each other? Ecological applications: a publication of the Ecological Society of America 18: 1253–1269. Available: http://www.ncbi.nlm.nih.gov/pubmed/18686585.
23. Beierle TC (2002) The quality of stakeholder-based decisions. Risk analysis: an official publication of the Society for Risk Analysis 22: 739–749. Available: http://www.ncbi.nlm.nih.gov/pubmed/12224747.
24. Koontz TM, Thomas CW (2006) What Do We Know and Need to Know about the Environmental Outcomes of Collaborative Management? Public Administration Review 66: 111–121. Available: http://doi.wiley.com/10.1111/j.1540-6210.2006.00671.x.
25. Newig J (2007) Does public participation in environmental decisions lead to improved environmental qulaity? Towards an analytical framework. Research and Practice for a Sustainable Future 1: 51–71.
26. Moreno-báez M, Orr BJ, Cudney-bueno R, Shaw WW (2010) Using fishers' local knowledge to aid managment at regional scales: Spatial Distribution of small-scale Fisheries in the Northern Gulf of California Mexico. Bulletin of Marine Science 86: 1–16.
27. Maurstad A (2002) Fishing in murky waters: ethics and politics of research on fisher knowledge. Marine Policy 26: 159–166.
28. Aswani S, Hamilton RJ (2004) Integrating indigenous ecological knowledge and customary sea tenure with marine and social science for conservation of bumphead parrotfish (Bolbometopon muricatum) in the Roviana Lagoon, Solomon Islands. Environmental Conservation 31: 69–83. Available: http://www.journals.cambridge.org/abstract_S037689290400116X.
29. Aswani S, Lauer M (2006) Incorporating Fishermen's Local Knowledge and Behavior into Geographical Information Systems GIS for Designing Marine Protected Areas in Oceania. Human Organization 65: 81–102.
30. Hall GB, Close CH (2007) Local knowledge assessment for a small-scale fishery using geographic information systems. Fisheries Research 83: 11–22. Available: http://linkinghub.elsevier.com/retrieve/pii/S016578360600289X.
31. Scholz A, Bonzon K, Fujita R, Benjamin N, Woodling N, et al. (2004) Participatory socioeconomic analysis: drawing on fishermen's knowledge for marine protected area planning in California. Marine Policy 28: 335–349. Available: http://linkinghub.elsevier.com/retrieve/pii/S0308597X03001118.
32. Scholz AJ, Steinback C, Kruse S A, Mertens M, Silverman H (2011) Incorporation of spatial and economic analyses of human-use data in the design of marine protected areas. Conservation biology: the journal of the Society for Conservation Biology 25: 485–492. Available: http://www.ncbi.nlm.nih.gov/pubmed/21175829.
33. Reed MS (2008) Stakeholder participation for environmental management: A literature review. Biological Conservation 141: 2417–2431. Available: http://linkinghub.elsevier.com/retrieve/pii/S0006320708002693.
34. Gerhardinger LC, Hostim-Silva M, Medeiros RP, Matarezi J, Bertoncini ÁA, et al. (2009) Fishers' resource mapping and goliath grouper Epinephelus itajara (Serranidae) conservation in Brazil. Neotropical Ichthyology 7: 93–102.

Available: http://www.scielo.br/scielo.php?script=sci_arttext&pid=S1679-62252009000100012&lng=en&nrm=iso&tlng=en.
35. Des Clers S, Lewin S, Edwards D, Searle S, Lieberknecht L & Murphy D (2008) FisherMap Mapping the Grounds: recording fishermen's use of the seas. A report published for the Finding Sanctuary project. Finding Sanctuary website. Available: http://findingsanctuary.marinemapping.com/06_all%20project%20reports/Fishe rmap%20report%20November%202008.pdf. Accessed 2012 Aug 11.
36. Wheeler M, Chambers FMJ, Sims-Castley R, Cowling RM, Schoeman D (2008) From beans to breams: how participatory workshops can contribute to marine conservation planning. African Journal of Marine Science 30: 475–487.
37. Lee J, South AB, Jennings S (2010) Developing reliable, repeatable, and accessible methods to provide high-resolution estimates of fishing-effort distributions from vessel monitoring system (VMS) data. ICES Journal of Marine Science 67: 1260–1271. Available: http://icesjms.oxfordjournals.org/cgi/doi/10.1093/icesjms/fsq010.
38. Gerritsen H, Lordan C (2010) Integrating vessel monitoring systems (VMS) data with daily catch data from logbooks to explore the spatial distribution of catch and effort at high resolution. ICES Journal of Marine Science 68: 245–252. Available: http://icesjms.oxfordjournals.org/cgi/doi/10.1093/icesjms/fsq137.
39. Hinz H, Murray LG, Lambert GI, Hiddink JG, Kaiser MJ (2013) Confidentiality over fishing effort data threatens science and management progress. Fish and Fisheries 14: 110–117. doi:10.1111/j.1467-2979.2012.00475.x.
40. DETI (2011) Regional Locational Guidance (RLG) for Offshore Renewable Energy Developments in NI Waters. Department of Enterprise, Trade and Investment. Belfast. Belfast, Northern Ireland. Department of Enterprise, Trade and Investment website. Available:http://www.detini.gov.uk/regional_locational_guidance__rlg__for_offshore_renewable_energy_developments_in_ni_waters. Accessed 2012 Sep 15.
41. DETI (2009) Off shore wind and marine renewable energy in Northern Ireland strategic environmental assessment (SEA) - Non Technical Summary. Joint Nature Conservation Committee (JNCC) website. Available: http://jncc.defra.gov.uk/PDF/0955_Explanatory_Summary.pdf Accessed 2012 Nov 10.
42. The Crown Estate (2012) Successful bidders in Northern Ireland offshore energy leasing rounds. Crown Estate website. Avaliable: http://www.thecrownestate.co.uk/news-media/news/2012/northern-ireland-offshore-energy-successful-bidders/. Accessed 2012 Oct 5.
43. Cappell R, Nimmo F, Rooney L (2012) The value of Irish Sea Conservation Zones to the Northern Irish Fishing Industry. Report to: Seafish Northern Ireland Advisory Committee (SNIAC). POSEIDON Aquatic Resource Management Ltd. Seafish website. Available: http://www.seafish.org/media/Publications/Poseidon_NI_MCZ_valuation_final_report_August_2012__2_.pdf Accessed 2012 Oct 5.
44. ICES (2009) Report of the Benchmark Workshop on Nephrops (WKNEPH). Aberdeen, UK.
45. AFBI (n.d.) Inshore Fisheries - Scallops. Available: http://www.afbini.gov.uk/index/services/services-specialist-advice/coastal-science/inshore-fisheries/services-inshore-fisheries-bivalve-molluscs.htm. Agri-food and Biosciences Instiute website. Accessed 2012 Nov 10.
46. ICES (2012) ICES IBTSWG Report 2012 Report of the International Bottom Trawl Survey Working Group (IBTSWG): 27–30. March 2012, Lorient, France. ICES CM 2012/SSGESST: 03. 323 p.
47. Yates KL, Payo Payo A, Schoeman DS (2013) International, regional and national commitments meet local implementation: A case study of marine conservation in Northern Ireland. Marine Policy 38: 140–150. Available: http://linkinghub.elsevier.com/retrieve/pii/S0308597X12001285.
48. Yates K (2012) Mapping the spatial access priorities of the Northern Ireland Fishing Fleet. The Diverse Seas Project, University of Ulster, Northern Ireland. University of Ulster, Environmental Science Research Institute website. Available: http://www.science.ulster.ac.uk/esri/Katherine-Yates.html. Accessed 2013 Jun 12.
49. Kelley K, Clark B, Brown V, Sitzia J (2003) Good practice in the conduct and reporting of survey research. International journal for quality in health care: journal of the International Society for Quality in Health Care/ISQua 15: 261–266. Available: http://www.ncbi.nlm.nih.gov/pubmed/12803354.
50. Pacific States Marine Fisheries Commission (2002) West Coast Catcher Boat Survey Summary 1997–1998. Seattle. Pacific States Marine Fisheries Commission website. Avaliable : http://www.psmfc.org/efin/abstracts-data.html. Accessed 2013 Jan 20.
51. Close CH, Hall GB (2006) A GIS-based protocol for the collection and use of local knowledge in fisheries management planning. Journal of environmental management 78: 341–352. Available: http://www.ncbi.nlm.nih.gov/pubmed/16115723.
52. Thornton TF, Scheer AM (2012) Collaborative Engagement of Local and Traditional Knowledge and Science in Marine Environments: A Review. Ecology and Society 17. Available: http://www.ecologyandsociety.org/vol17/iss3/art8/.
53. Stringer LC, Reed MS, Dougill AJ, Seely MK, Rokitzki M (2007) Implementing the UNCCD: Participatory challenges. Natural Resources Forum 31: 198–211. Available: http://doi.wiley.com/10.1111/j.1477-8947.2007.00154.x.
54. Yaffee SL (1996) Ecosystem management in practice: the importance of human institutions. Ecological Applications 6: 724–727.
55. Duane T (1997) Community Participation in Ecosystem Management. Ecology Law Quarterly 24: 771–797.
56. Smith AH, Berkes F (1991) Solutions to the "Tragedy of the Commons": Sea-urchin Management in St Lucia, West Indies. Environmental Conservation 18: 131. Available: http://www.journals.cambridge.org/abstract_S0376892900021706.
57. Roberts CM, Andelman S, Branch G, Bustamante RH, Castilla JC, et al. (2003) Ecological criteria for evaluating candidate sites for marine reserves. Ecological ApplicationsVol13, suppl1 13: S199–S214.
58. Klein CJ, Chan A, Kircher L, Cundiff AJ, Gardner N, et al. (2008) Striking a balance between biodiversity conservation and socioeconomic viability in the design of marine protected areas. Conservation biology: the journal of the Society for Conservation Biology 22: 691–700. Available: http://www.ncbi.nlm.nih.gov/pubmed/18325043.
59. Richardson EA, Kaiser MJ, Edwards-Jones G, Possingham HP (2006) Sensitivity of Marine-Reserve Design to the Spatial Resolution of Socioeconomic Data. Conservation Biology 20: 1191–1202.
60. Moore J, Balmford A, Allnutt T, Burgess N (2004) Integrating costs into conservation planning across Africa. Biological Conservation 117: 343–350. Available: http://linkinghub.elsevier.com/retrieve/pii/S0006320704000023.
61. Cudney-Bueno R, Basurto X (2009) Lack of cross-scale linkages reduces robustness of community-based fisheries management. PloS one 4: e6253. Available: http://www.pubmedcentral.nih.gov/articlerender.fcgi?artid=2705799&tool=pmcentrez&rendertype=abstract.
62. Johannes RE, Freeman MMR, Hamilton RJ (2000) Ignore fishers ' knowledge and miss the boat. Fish and Fisheries 1: 257–271.
63. Fischer A, Young JC (2007) Understanding mental constructs of biodiversity: Implications for biodiversity management and conservation. Biological Conservation 136: 271–282. Available: http://linkinghub.elsevier.com/retrieve/pii/S0006320706005313.
64. Silvano RAM, Begossi A (2009) What can be learned from fishers? An integrated survey of fishers' local ecological knowledge and bluefish (Pomatomus saltatrix) biology on the Brazilian coast. Hydrobiologia 637: 3–18. Available: http://www.springerlink.com/index/10.1007/s10750-009-9979-2.
65. Silvano RAM, Valbo-Jørgensen J (2008) Beyond fishermen's tales: contributions of fishers' local ecological knowledge to fish ecology and fisheries management. Environment, Development and Sustainability 10: 657–675. Available: http://www.springerlink.com/index/10.1007/s10668-008-9149-0.
66. Burton P, Goodlad R, Croft J, Abbott J, Hastings A, et al. (2004) What works in community involvement in area-based initiatives? A systematic review of the literature. Available: http://webarchive.nationalarchives.gov.uk/20110218135832/http://rds.homeoffice.gov.uk/rds/pdfs04/rdsolr5304.pdf.
67. National Research Council (2008) Public participation in environmental assessment and decision making. National Acedemy Press. Washington, D.C. The National Academies Press Website. Avaliable: http://www.nap.edu/catalog.php?record_id=12434. Accessed 2012 Sep 12.
68. Turner M, Weninger Q (2005) Meetings with Costly Participation: An Empirical Analysis. Review of Economic Studies 553: 247–268.
69. Klein CJ, Steinback C, Watts M, Scholz AJ, Possingham HP (2010) Spatial marine zoning for fisheries and conservation. Frontiers in Ecology and the Environment 8: 349–353. Available: http://www.esajournals.org/doi/abs/10.1890/090047.
70. Scholz A, Birney K, Bonzon K, Chan C, Hamrin S, et al. (2006) Commercial fishing grounds and their relative importance off the Central Coast of California. Report to the California Marine Life Protection Act Initiative. Ecotrust. California Department of Fish and Wildlife website. Available: http://www.dfg.ca.gov/marine/pdfs/agenda_workshop041807_report.pdf. Accessed 2012 Feb 15.

12

Assessing the Exoproteome of Marine Bacteria, Lesson from a RTX-Toxin Abundantly Secreted by *Phaeobacter Strain DSM 17395*

Emie Durighello, Joseph Alexander Christie-Oleza, Jean Armengaud*

CEA, DSV, IBEB, Lab Biochim System Perturb, Bagnols-sur-Cèze, France

Abstract

Bacteria from the *Roseobacter* clade are abundant in surface marine ecosystems as over 10% of bacterial cells in the open ocean and 20% in coastal waters belong to this group. In order to document how these marine bacteria interact with their environment, we analyzed the exoproteome of *Phaeobacter strain DSM 17395.* We grew the strain in marine medium, collected the exoproteome and catalogued its content with high-throughput nanoLC-MS/MS shotgun proteomics. The major component represented 60% of the total protein content but was refractory to either classical proteomic identification or proteogenomics. We *de novo* sequenced this abundant protein with high-resolution tandem mass spectra which turned out being the 53 kDa RTX-toxin ZP_02147451. It comprised a peptidase M10 serralysin domain. We explained its recalcitrance to trypsin proteolysis and proteomic identification by its unusual low number of basic residues. We found this is a conserved trait in RTX-toxins from *Roseobacter* strains which probably explains their persistence in the harsh conditions around bacteria. Comprehensive analysis of exoproteomes from environmental bacteria should take into account this proteolytic recalcitrance.

Editor: Daniel Gillet, CEA (Atomic and Alternative Energies Commission), France

Funding: JA Christie-Oleza was supported by a fellowship from the Fundación Ramón Areces. The authors thank the Commissariat à l'Energie Atomique et aux Energies Alternatives for financial support. The funders had no role in study design, data collection and analysis, decision to publish, or preparation of the manuscript.

Competing Interests: The authors have declared that no competing interests exist.

* E-mail: jean.armengaud@cea.fr

Introduction

The extremely high diversity of marine microorganisms was shown ever since the first large metagenomic shotgun analysis carried out on total DNA extracted from filtered seawater samples collected from the Sargasso Sea near Bermuda [1]. Heterotrophic bacterioplankton communities in near-surface marine pelagic environments mostly comprise *Alphaproteobacteria* and *Gammaproteobacteria*, as well as *Sphingobacteria* and *Flavobacteria* (both belonging to the *Bacteroidetes* superphylum). As shown recently with environmental metagenomic data, prominent components belong to the SAR11, *Roseobacter* and SAR86 clades together with the OM60/NOR5 cluster [2]. The *Roseobacter* lineage is a phylogenetically coherent group of *Alphaproteobacteria* comprising up to 20% of marine microbial communities, especially in coastal and polar waters [3]. However, the *Roseobacter* clade is a physiologically heterogeneous group showing a high variety of lithoheterotrophic strategies. Despite a free-living lifestyle, *Roseobacter* strains are also found as planktonic or larger pluricellular symbionts [4].

Today, in-depth analysis of a microbial proteome relies on the use of high-resolution hybrid tandem mass spectrometers coupled to high-pressure liquid chromatography systems able to resolve complex peptide mixtures [5,6]. A comprehensive list of proteins and their quantitation in several physiological conditions by such shotgun approaches are a rather straightforward method [7,8]. Exoproteomes comprise the fraction of proteins secreted by cellular systems and proteins arising from cell lysis [9]. These are generally subjected to intense proteolytic activities and most components have a short half-life. Nevertheless, key components involved in nutrient imports, cell motility, signaling and interactions with other organisms should be refractory to proteolysis in order to be efficient. Analysis of bacterial secretomes is an important issue for the discovery of novel bioactive compounds and to determine how organisms interact within their environment. We have shown that microbial exoproteomes can be comprehensively analyzed by high throughput shotgun proteomics [10,11]. The use of the most recent generation of hybrid tandem mass spectrometers revealed a high diversity of protein virulence factors in the pathogenic *Bacillus cereus* bacteria [11]. To date, only a few exoproteome analyses have been reported on environmental avirulent isolates [10,12]. In this case, the data obtained helped elucidating the ecological distinctness among strains of the *Roseobacter* clade. Microbial exoproteomes are also insightful for understanding the molecular mechanisms of bacterial colonization of biotic or abiotic surfaces, i.e. bacterial adhesion and biofilm formation [13]. Their exploration gains ground with the current development of proteomic tools for environmental microbiology [14].

Phaeobacter gallaeciensis BS107 (previously *Roseobacter gallaeciensis* BS107) is the type species of the *Phaeobacter* genus [15]. This strain was obtained from larval cultures and collectors of the scallop *Pecten maximus* [16]. Other species of the genus have since been described: *Phaeobacter inhibens*, isolated from the German Wadden

Sea [15], *Phaeobacter daeponensis*, isolated from a tidal flat of the Yellow Sea at Daepo Beach in Korea [17], and *Phaeobacter arcticus*, a psychrophilic species isolated from marine sediments of the Arctic Ocean [18]. *Phaeobacter inhibens* was in-depth characterized with proteomics and metabolomics approaches highlighting its versatile metabolism [19,20,21,22]. The type species *Phaeobacter strain DSM 17395* was shown to be able to produce a new tropolone derivative, tropodithietic acid, which exhibits strong antibiotic properties against marine bacteria of various taxa and marine algae [23,24]. Additional studies led to the identification of nine new troponoid compounds belonging to the bacterial roseobacticide family [25]. More recently, it has been shown that *Phaeobacter strain DSM 17395* switches its secreted molecule metabolism to the production of potent and selective algaecides in response to p-coumaric acid, an algal lignin breakdown product that is symptomatic of aging algae. This switch converts *Phaeobacter strain DSM 17395* into an opportunistic pathogen of its algal host [26] and highlights the interest of the study of its secretome. The type strain *Phaeobacter gallaeciensis BS107T* has been deposited at public culture collections worldwide. Recently, it was shown that the Phaeobacter strain DSM 17395, previously annotated as *Phaeobacter gallaeciensis BS107T,* exhibits a much closer affiliation to *Phaeobacter inhibens* DSM 16374T and should thus be allocated to this species [27].

With the intention to further characterize the ecological distinctness of *Roseobacters* members and how they interact with their environment, we analyzed the secreted proteins of *Phaeobacter strain DSM 17395.* For this, we grew the strain in marine medium, collected the exoproteome and catalogued its protein content with high throughput nanoLC-MS/MS shotgun proteomics. Interestingly, the major protein in this exoproteome was recalcitrant to classical proteomic identification. Indeed it appears very abundant on SDS-PAGE migration but no protein was matching this band after proteomic analysis. A *de novo* sequencing strategy identified peptides of a RTX-like toxin which led to re-analyze proteomics data with no-enzyme specificity. This RTX-like toxin was then clearly identified. We found that this protein and its homologues exhibit unusually low content in basic residues. From these results, we proposed a general strategy to obtain more comprehensive exoproteome analyses of environmental bacteria.

Materials and Methods

P. gallaenciensis Growth and Exoproteome Preparation

Four flasks containing 40 mL of marine broth (Difco) were inoculated with *Phaeobacter strain DSM 17395* cells previously grown on marine agar plates. The cultures were incubated at 30°C under 180 rpm agitation until mid-exponential phase growth (OD_{600} = 0.6). Cultures were centrifuged at 3,000 *g* for 10 min at 20°C and supernatants were carefully filtered through two low protein-binding filters of 0.45 μm (Millex-Hv) and then 0.22 μm (Millex-GV) diameter pore (Millipore) in order to eliminate bacterial cells. Supernatants were frozen prior protein precipitation. Proteins from the supernatant were precipitated with tricholoroacetic acid as previously described [10]. The resulting protein pellets were dissolved into 90 μl of lithium dodecyl sulfate-β-mercaptoethanol protein gel sample buffer (Invitrogen), incubated at 99°C for 5 min and briefly centrifuged prior SDS-PAGE.

SDS-PAGE and Trypsin-*in gel* Proteolysis

A volume of 20 μl of the concentrated exoproteome of *Phaeobacter strain DSM 17395* was deposited per well onto 10% Bis-Tris NuPAGE gels (Invitrogen). SDS-PAGE was carried out using 1X 3-(N-morpholino) propanesulfonic acid solution (Invitrogen) in a XCell SureLock Mini-cell (Invitrogen) under a constant voltage of 200 V. At first, a long migration was carried out in order to fully resolve the exoproteome according to its molecular weight. For the proteomic shotgun analysis, the entire exoproteome was allowed to enter the polyacrylamide gel in a short migration (3 mm). Both gels were stained with SimplyBlue SafeStain, a ready-to-use Coomassie G-250 stain (Invitrogen). SeeBlue Plus2 (Invitrogen) was used as a molecular weight marker. Densitometry analysis of the long-migrated SDS-PAGE gel was carried out with the molecular imager GS-800 calibrated densitometer (BioRad) and the Quantity One software (version 4.6.9, BioRad) in order to establish the relative abundance of the main protein band. Densitometric relative ratio was obtained by comparing the density signal of each band to the total signal of the whole gel lane. These measurements were done in triplicate. Polyacrylamide gel bands (equivalent in volume to 50 μl) were cut and processed for *in*-gel proteolysis with Trypsin Sequencing Grade (Roche) followed by the ProteaseMax protocol (Promega) as previously described [11].

NanoLC-MS/MS Analysis

Peptide digests were resolved on an UltiMate 3000 LC system (Dionex-LC Packings) as previously described [28] prior to MS/MS measurements done with a LTQ-Orbitrap XL (Thermo-Fisher). For the FT/IT procedure, parameters used for the tandem mass spectrometry measurements were those previously described [29]. During the FT/FT procedure, the LC gradient used was a 90 min gradient with aqueous solvent A (0.1% HCOOH) and solvent B (0.1% HCOOH/80% CH_3CN) developed as follows: 5–60% B in 90 min, 60–90% B in 1 min, 90% B during 10 min, 90–5% B in 1 min and 5% B during 19 min. The "full scan" mass spectrum was measured from m/z 300 to 1800 th and the charge states considered were +2 and +3. A scan cycle was started with a high resolution full scan (resolution = 30,000) with the Orbitrap analyzer, followed by a high resolution MS/MS scan of the secondary ions obtained after fragmentation, again with the Orbitrap analyzer but with a slightly lower resolution (resolution = 15,000). A 30 sec dynamic exclusion was set. The activation type used was CID with a standard normalized collision energy set at 35.

MS/MS Database MASCOT Search and Protein Quantification by Spectral Count

Peak lists were generated with the MASCOT DAEMON software (v2.3.2, Matrix Science) from the LC-MS/MS raw data using default parameters. The MASCOT 2.3.02 search engine (Matrix Science) was used to search all MS/MS spectra against a database corresponding to a complete list of annotated Coding Domain Sequences (CDS) from *Phaeobacter strain DSM 17395* (NCBI RefSeq: NC_018290.1, NC_018291.1, NC_018288.1, and NC_018287.1). Bovine trypsin and the 22 most common keratin contaminants were included as a follow-up. Searches for peptides were first performed with the following parameters: mass tolerance of 5 ppm on the parent ion and 0.5 Da on the MS/MS (FT/IT procedure), static modifications of carboxamidomethylated Cys (+ 57.0215), and dynamic modification of oxidized Met (+15.9949). High-resolution MS/MS (FT/FT procedure) searches were performed with mass tolerance of 0.1 Da on the MS/MS. To process the data resulting from trypsin proteolysis, the maximum number of missed cleavage was set at 2. All peptide matches with a score above its peptidic identity threshold (set at $p<0.01$) with the ORF database and rank 1 were filtered with the IRMa 1.28.0 software [30]. A protein was only validated when at least two peptides had been assigned. In a second stage, we performed additional searches for both "semi-trypsin" and "no-enzyme" protease specificities. Protein abundance was evaluated in the shotgun analysis by MS/MS spectral counts as established by Liu and co-workers [31] and previously described [32]. Normalized Spectral count Abundance Factors (NSAF) was calculated as defined by Paoletti et al. [33].

De novo Assignment of MS/MS Spectra

Peak lists were submitted to Peaks Studio 5.3 software (Bioinformatics Solutions Inc). At first, a data refinement step was performed with a quality threshold set at 0.65. *De novo* sequencing was then carried out with the following parameters: a parent mass error tolerance of 5.0 ppm, a fragment mass error tolerance of 0.1 Da, cysteine carbamidomethylation (+57.02) as fixed modification, and methionine oxidation (+15.99) as variable modification. More stringent criteria were applied with a parent mass error tolerance of 2 ppm with high resolution mode for both MS and MS/MS. Successively, trypsin, semi-trypsin and no-enzyme were chosen as enzyme specificities. The list of assigned spectra obtained with MASCOT using *Phaeobacter strain DSM 17395* CDS database was compared with Peaks Studio results for each samples. Precursor mass spectra annotated with the Peaks Studio software but not interpreted with the Mascot engine gave candidates for manual annotation. Results were then sorted by their Average of Local Confidence (ALC) in order to choose the best spectra to annotate.

Results and Discussion

The most Abundant Protein within the Exoproteome of *Phaeobacter strain DSM 17395* is Recalcitrant to Standard Proteomic Identification

The resulting exoproteome of the marine bacterium *Phaeobacter strain DSM 17395* when grown in Marine Broth was collected and resolved by SDS-PAGE. Remarkably, an abundant protein band migrating at a molecular weight of about 55 kDa was observed in the protein profile. This 55 kDa band accounts for 60 (±2) % of the total exoproteome as estimated by densitometry (**Figure 1**). With the objective of identifying most of the components of this exoproteome, we analyzed the sample with the previously developed shotgun procedure [10]. The whole exoproteome was excised from a short SDS-PAGE migration (3 mm) as a single small polyacrylamide band. After trypsinization, the peptide mixture was injected into a C18 reversed phase chromatography column coupled to a high resolution LTQ-Orbitrap XL hybrid mass spectrometer. The acquired tandem mass spectra were assigned with the Mascot software using as database the complete list of annotated CDS from *Phaeobacter strain DSM 17395*. The dataset comprised 2321 MS/MS spectra, from which 1112 could be assigned to specific peptides from *Phaeobacter strain DSM 17395* (**Table S1 in File S1**). These spectra corresponded to 478 different peptides which validated the presence of 75 proteins with at least two different peptides (**Table S2 in File S1**). In terms of spectral counts, the first 15 proteins accounted for half of the total number of assigned spectra. The most detected proteins were an outer membrane porin (ZP_02147263), a flagellin-like protein (COG1344, ZP_02145485), a hemolysin-type calcium-binding protein (ZP_02144235), and the periplasmic phosphate-binding subunit of a phosphate ABC transporter (ZP_02143869) with 86, 55, 52, and 41 MS/MS spectra, respectively (**Table S2 in File S1**). Despite all these proteins were expected to be secreted, none was abundant enough (8% of total MS/MS spectra) as to be considered the highly secreted protein seen in **Figure 1**. Furthermore, their molecular weights (32, 28, 116, and 37 kDa, respectively) differ from that expected for this protein (≈55 kDa). Therefore, the shotgun results were in disagreement with the SDS-PAGE information and the abundant 55 kDa protein could not be identified. To solve this issue, we fully resolved the exoproteome onto a SDS-PAGE gel as shown in **Figure 1** and analyzed by nanoLC-MS/MS the protein content of the 55 kDa polyacrylamide band. Although only one major polypeptide was expected to be present in this polyacrylamide band, we evidenced 48 proteins identified with at least 2 peptides. **Table 1** reports the ten first proteins as ranked by their Mascot score (assigned MS/MS spectra and corresponding peptide characteristics are reported in **Table S3 in File S1**). Surprisingly we observed the protein GroEL Chaperonin (ZP_02145205) as the most detected protein with 61 MS/MS spectra although this protein is not expected to be secreted by bacteria. Another unexpected result was the presence in this polyacrylamide band of the 116 kDa protein ZP_02144235. Its tryptic sequence coverage showed that both the N- and C-termini were confirmed with proteomic-detected peptides, but the internal sequence was poorly covered (**Table S3 in File S1**). We presumed that this protein may be subjected to specific maturation such as a proteolytic cleavage. Again, none of the proteins shown in **Table 1** were in sufficient abundance as to be considered the abundant protein seen in **Figure 1**. Whether the expected abundant 55 kDa protein was encoded on the genome but missed during the annotation stage was also investigated by a new Mascot search using a six-frame translation of the whole nucleic acid sequence, as previously described [34]. This proteogenomic approach remained unsuccessful with no considerable findings (**Data not shown**). Therefore, the abundantly secreted protein of *Phaeobacter strain DSM 17395* remained unidentified with standard proteomic identification.

De novo Sequencing with High Resolution Data Revealed the Identity of the Abundantly Secreted Protein

We carried out a search to manually assign spectra that had not been automatically assigned to a tryptic peptide comprised in the

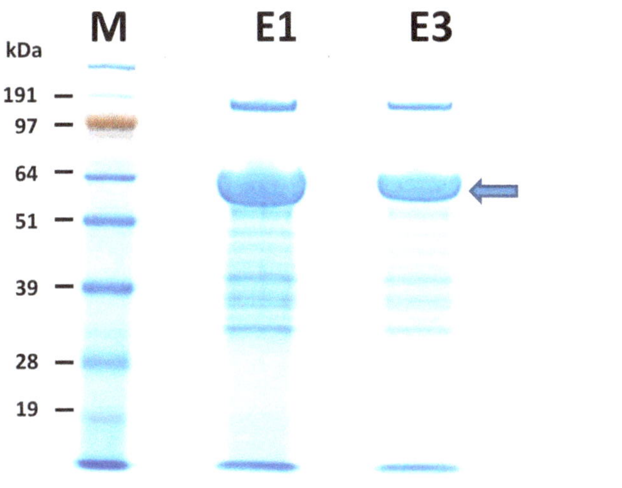

Figure 1. SDS-PAGE of the exoproteome of *Phaeobacter strain DSM 17395*. Exoproteins were resolved by a long migration on a 10% SDS-PAGE and stained with SimplyBlue SafeStain (Invitrogen). Lane **M**: SeeBlue Plus2 molecular weight range marker (Invitrogen). Lane **E1**: *Phaeobacter strain DSM 17395* exoproteome grown in Marine Broth (20 µg). Lane **E2**: *Phaeobacter strain DSM 17395* exoproteome grown in Marine Broth (8 µg). The 55 kDa major component is indicated with an arrow.

Table 1. List of the first ten proteins identified from the 55 kDa exoproteome band of *P. gallaeciensis* DSM[1].

Identification rank	Accession	Functional annotation	Number of residues	Protein Mascot Score	Molecular weight (Da)	Sequence Coverage (%)	Number of unique peptide	Spectral Count
1	ZP_02145205	Chaperonin GroEL	550	1632	57749	62	28	61
2	ZP_02146523	Oligopeptide/dipeptide ABC transporter*	518	929	57342	32	15	37
3	ZP_02144235	Hemolysin-type calcium-binding protein	1112	915	115931	15	13	35
4	ZP_02145689	Extracellular solute-binding protein, family 1	577	884	64572	36	15	33
5	ZP_02143510	Ser/Thr protein phosphatase/nucleotidase, putative	525	594	56013	36	12	25
6	ZP_02146184	Extracellular solute-binding protein, family 5	566	559	61051	31	11	21
7	ZP_02145456	Flagellar hook-associated protein	482	553	50979	28	9	16
8	ZP_02147263	Outer membrane porin	313	508	31831	28	9	21
9	ZP_02144479	Extracellular solute-binding protein, family 5	528	413	57617	17	7	20
10	ZP_02144175	Aconitate hydratase	895	431	96861	13	10	19

[1]Detected with at least three different peptides.
*Periplasmic component.

Phaeobacter strain DSM 17395 CDS database. For this *de novo* sequencing strategy, we reanalyzed the peptides from the 55 kDa protein band by nanoLC-MS/MS with a FT/FT procedure which enables getting higher resolution MS/MS spectra and better mass accuracy (15,000 in the orbitrap analyzer instead of the 3,000 previously done with the LTQ analyzer). From this new dataset acquisition, 246 MS/MS spectra were assigned directly with the Mascot software using the complete list of annotated CDSs from *Phaeobacter strain DSM 17395* as a database. The same dataset was also searched with the *de novo* sequencing software Peaks Studio 5.3 identifying 258 new MS/MS spectra that had not been assigned by Mascot. We manually annotated these spectra checking all peptide sequence possibilities. **Figure 2** shows the annotation of a representative MS/MS spectrum with a m/z ratio measured at 614.33826 amu corresponding to a di-charged peptide with a semi-tryptic sequence, DLVGDAGVNVLR. Peak assignment was confident with the detection of 10 monoprotonated b ions and 9 monoprotonated y ions, except that the second and eleventh residues could be either leucine or isoleucine. We could also assign the first residue with the detection of the y_{11} diprotonated fragment ion. A total of five sequences could be confidently assigned because of MS/MS mass accuracy. The Blast analysis of each of these five tentative peptides revealed strict identities to the ZP_02147451 protein from *Phaeobacter strain DSM 17395*. Curiously, all five *de novo* sequenced peptides corresponded to semi-tryptic peptides. With these data in hands, we finally supposed that the abundant exoprotein was ZP_02147451, annotated as a hemolysin-type calcium binding protein (RTX-toxin). Its theoretical molecular weight, 52624 Da, matched the expected molecular weight estimated from the SDS-PAGE gel.

High Resolution Data Allowed a High Coverage of ZP_02147451 with Non-tryptic Peptides

To confirm the identity of the protein and because *de novo* sequencing of some MS/MS spectra had shown the presence of abundant non-tryptic peptides, we re-analyzed the FT/FT experimental data with the Mascot software using "no-enzyme" as the enzyme parameter. **Table S4 in File S1** lists the resulting 228 MS/MS spectra assigned to 128 distinct peptides belonging to proteins of the *Phaeobacter strain DSM 17395* bacterium. As expected, less MS/MS spectra are assigned with the no-enzyme search mode than with the trypsin search mode, due to the loss of discrimination introduced by the larger search space. This time the hemolysin-type calcium binding protein ZP_02147451 was clearly the most abundant protein identified accounting 67 MS/MS spectra assigned (30% of the spectral counts). A total of 36 distinct peptides were found for this protein and allowed 51% coverage of the whole sequence (**Figure 3**). According to this "no-enzyme" Mascot search, 24 peptides were semi-tryptic, 10 items do not arise from a specific cleavage, and only 1 is a tryptic peptide despite the trypsinization protocol applied. Peptides were evenly distributed along the protein sequence except the peptidase unit region spanning residues 240 to 279 (**Figure 3**). **Table 2** shows the comparative list of proteins assigned in the two Mascot search modes ("trypsin" or "no-enzyme"). In order to compare results obtained in the two search modes we defined the Enzyme Unspecificity Factor (EUF) for each protein as the number of unique peptides in the no-enzyme MASCOT search divided by the number of unique peptides in the trypsin MASCOT search. As seen in the table, ZP_02147451 is the only protein having a higher number of unique peptides (18-fold more) when the "no-enzyme" condition is used. MASCOT search with the "no-enzyme" parameter does not change significantly the results of spectral counting when cellular proteome are analyzed as based on our

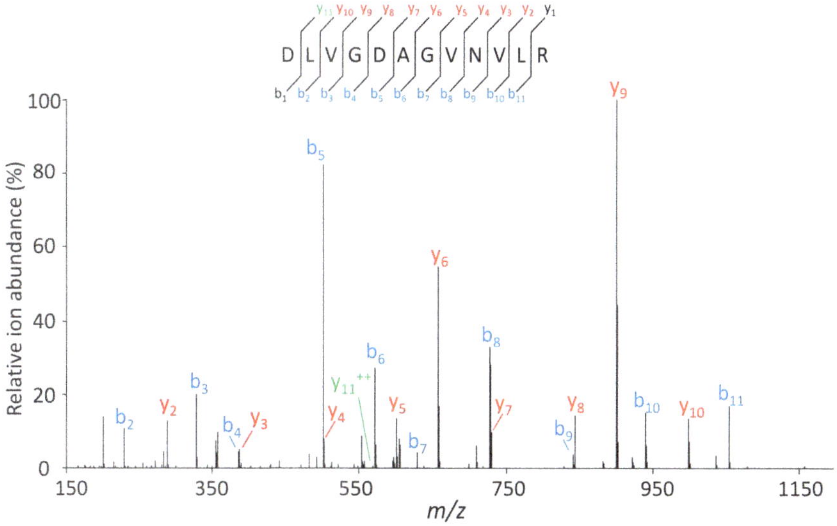

Figure 2. MS/MS spectrum of the semi-tryptic peptide [288–299] belonging to ZP_02147451. The MS/MS spectrum was acquired with a FT/FT procedure with an LTQ-Orbitrap XL hybrid mass spectrometer. The peptide sequence is shown on the top with the collision-induced fragmentation pattern. The *b* and *y* ions are shown in blue and red, respectively. The y_{11} di-charged ion is labeled in green.

previous data (Data not shown) from another *Roseobacter*, namely *Roseobacter denitrificans* [35].

The Recalcitrance of ZP_02147451 is Caused by an Extreme Lack of Basic Residues in its Sequence

The ZP_02147451 protein is a RTX-toxin with 533 amino acids that contains only a few basic residues (three lysines and six arginines). These residues are trypsin targets for proteolysis and, therefore, this RTX-like protein will derive in extremely large peptides after digestion. Considering the length of the protein and the position of these basic residues, only three tryptic peptides could be considered as detectable by the LC-MS/MS procedure with the LTQ-Orbitrap XL hybrid mass spectrometer (*m/z* range from 300 to 1800 th). When reconsidering all the peptides detected of ZP_02147451, less than half of the cleavages occurred at the C-terminus of arginines (30%) and lysines (12%) in accordance to the trypsin activity. Nevertheless, a large number of cleavages (20%) also arose at the C-terminus of asparagine (N). Other residues like glutamic acid (8%) and aspartic acid (7%) were also frequently found at the C-terminus of the detected peptides (**Figure S1**). Thus, acidic residues frequently appeared at this

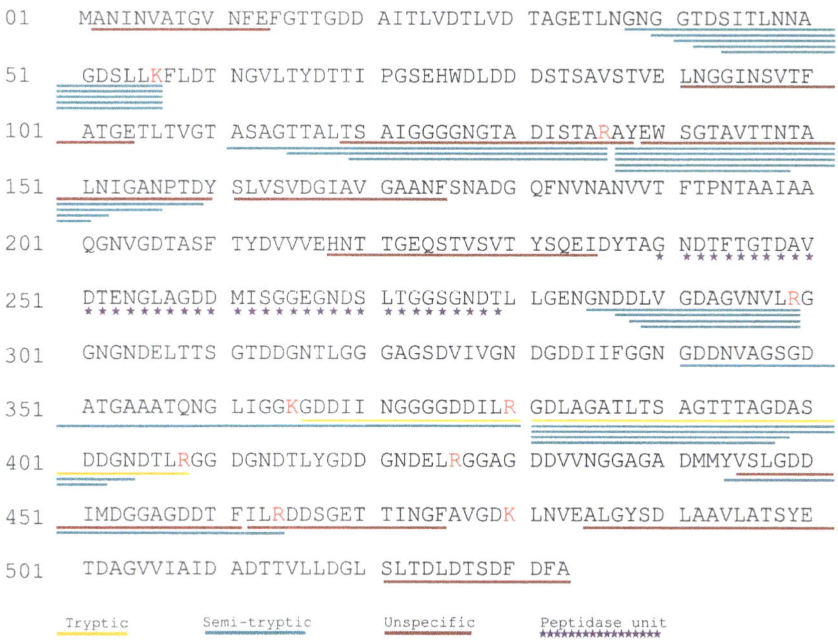

Figure 3. ZP_02147451 sequence coverage with non-tryptic, semi-tryptic, and tryptic peptides. The ZP_02147451 sequence is represented with its peptidase motif (residues 240 to 279) pointed out with purple stars. Peptides identified with the "no-enzyme" Mascot search are symbolized with a line underlining the sequence. Tryptic, semi-tryptic, and non-tryptic peptides are indicated in yellow, blue and red, respectively.

Table 2. List of proteins identified in the 55 kDa band using the high resolution FT/FT proteomic procedure[1].

			"Trypsin" as MASCOT search criterion					"No-enzyme" as MASCOT search criterion			
Accession	**Functional annotation**	**Molecular weight (Da)**	**Protein Mascot Score**	**Sequence Coverage (%)**	**Number of unique peptides**	**Spectral Count**	**EUF$**	**Protein Mascot Score**	**Sequence Coverage (%)**	**Number of unique peptides**	**Spectral Count**
ZP_02145205	chaperonin GroEL	57749	1118	50	22	38	**68**	905	37	15	20
ZP_02144235	Hemolysin-type calcium-binding protein	115931	685	13	12	48	**92**	711	15	11	40
ZP_02146523	oligopeptide/dipeptide ABC transporter*	57342	533	27	10	24	**70**	440	22	7	12
ZP_02145689	extracellular solute-binding protein, family 1	64572	354	19	7	12	**71**	293	13	5	7
ZP_02146184	extracellular solute-binding protein, family 5	61051	339	17	6	12	**83**	308	14	5	11
ZP_02144479	extracellular solute-binding protein, family 5	57617	338	18	6	10	**83**	302	15	5	6
ZP_02143510	Ser/Thr protein phosphatase/nucleotidase	56013	263	16	6	10	**83**	235	14	5	9
ZP_02143869	phosphate ABC transporter*	37145	221	9	2	3	**100**	221	9	2	3
ZP_02144175	aconitate hydratase	96861	232	9	6	9	**33**	112	3	2	2
ZP_02147263	outer membrane porin	31831	213	20	5	10	**40**	115	7	2	5
ZP_02145456	Flagellar hook-associated protein	50979	196	15	4	4	**75**	163	10	3	3
ZP_02145455	S-adenosyl-L-homocysteine hydrolase	45860	195	13	4	4	**100**	256	17	4	4
ZP_02143988	Hemolysin-type calcium-binding protein	30261	190	24	3	6	**100**	188	24	3	5
ZP_02144038	hypothetical protein RGBS107_15846	51273	152	12	4	8	**75**	127	7	3	5
ZP_02147451	Hemolysin-type calcium-binding protein	52624	163	8	2	4	**1800**	2463	51	36	67
ZP_02146962	extracellular solute-binding protein, family 1	47415	154	12	3	4	**67**	125	8	2	2

[1]Detected with at least three different peptides. The values are from a representative experiment.
*Periplasmic component.
$EUF stands for Enzyme Unspecificity Factor, defined as the ratio number of unique peptides (no-enzyme MASCOT search) per number of unique peptides (trypsin MASCOT search).

position although trypsin could not explain such cleavages. Taken together, the low number of detectable tryptic peptides of the RTX-toxin ZP_02147451 is the reason why we observed other proteolytic derived peptides. These alternative proteolytic activities are usually rare and masked by the highly abundant tryptic peptides. Furthermore, in standard analysis these peptides would not be considered during the search.

We subjected the whole exoproteome of *Phaeobacter strain DSM 17395* to a novel shotgun analysis with the FT/FT procedure in order to apply a *de novo* sequencing approach and evaluate alternative proteolysis patterns. The data were analyzed with MASCOT using a "no-enzyme" or "trypsin" parameter. **Table 3** shows the comparative list of proteins assigned with at least five different peptides in the "no-enzyme" Mascot analysis mode. Interestingly, with "trypsin" as criterion for the Mascot search, ZP_02147451 was detected with only one unique peptide (2 spectral counts). The same protein was detected with 12 peptides when the "no-enzyme" parameter was used. ZP_02147451 was the only protein that exhibited such a large increase of detected peptides when the non-tryptic search mode was used (12-fold, **Table 3**). Noteworthy, we did not find any specific proteolytic pattern in the other proteins because of the small number of extra non-tryptic peptides detected during the analysis.

RTX-toxins Exhibit Unusually Low Content in Basic Residues

After identifying the RTX-like protein ZP_02147451 by *de novo* sequencing we further investigated the reason of its recalcitrance to our standard proteomic approach. Within the 533 residue protein we only found nine basic amino acids. Thus, it results refractory to standard trypsin-based proteomic approaches because of the large tryptic peptides generated. These cannot be analyzed properly by mass spectrometers operated in classical conditions. We checked whether this was a specific feature for this protein compared to the whole *Phaeobacter strain DSM 17395* theoretical proteome. **Figure 4** shows the comparison for all the proteins with length above 100 residues. Indeed, ZP_02147451 is by far unusual in terms of basic residue content, *i.e.* 1 basic residue every 59 residues while the mean value is every 11 residues within the whole theoretical proteome of *Phaeobacter strain DSM 17395*. Noteworthy, we detected that other RTX-toxins encoded on the *Phaeobacter strain DSM 17395* genome are also unusually poor in basic residues. ZP_02146514, ZP_02146960, ZP_02146988, ZP_02146235, and ZP_02146693 are large RTX-toxin proteins (1725, 969, 296, 1112, and 937 amino acids respectively) which contain 1 basic residue every 30 residues in average (**Figure 4**). We wondered whether this characteristic is conserved among the other secreted proteins. **Figure 4** shows that the other proteins detected in the exoproteome of *Phaeobacter strain DSM 17395* do not exhibit such a specific trend, although most have a slightly lower RK content compared to the average. This trait is a common issue within the RTX-like proteins coded in *Roseobacter* members. The analysis of the 107 RTX-like proteins coded in 12 other sequenced *Roseobacter* strains revealed one basic residue every 27 amino acids (Data not shown). As an example, *Roseobacter* sp. MED193 exhibits two RTX-like proteins, ZP_01058551 and ZP_01057949, with an average tryptic peptide size of 80 and 73 residues, respectively. As a result, the secreted RTX-like proteins are generally refractory to trypsin digestion. We evidenced as illustrated in Figure 4 the presence in the exoproteome of a large protein with a low ratio KR/polypeptide length through the identification of three distinct peptides. This protein, ZP_02146938.1, is large (122 kDa) and annotated as the hypothetical protein RGBS107_08210. Its low content in basic residues and large molecular weight may suggest that it could also be a "RTX-toxin" far-related protein. Indeed, a Blast analysis confirmed such relationship.

Concluding Remarks

Secreted proteins play an essential role in successful adaptation of bacteria to their environment. These exoproteins are crucial for nutrient import, motility, and biofilm formation. In some cases, they are used by pathogens to adhere to and degrade cell walls from their targets and to suppress defense responses from their hosts [36]. With the development of novel high-throughput proteomic strategies and the use of novel generation of hybrid tandem mass spectrometers, novel insights were recently obtained regarding the secretome from various marine *Roseobacter* strains in several physiological conditions [10,12]. Here, we reported the analysis of the exoproteome of another *Roseobacter* member, namely *Phaeobacter strain DSM 17395*, which is known to produce potent and selective algaecides [25]. In this case, we found that the major secreted protein, representing 60% of the total exoproteome, was refractory to classical proteomic approach identification. This challenging issue made us changes our strategy and resort to *de novo* sequencing in order to identify the polypeptide. Interestingly, we found that the abundantly secreted protein was a RTX-like toxin, namely ZP_02147451, annotated as a hemolysin-type calcium-binding protein. This 533 amino acids polypeptide comprised a domain spanning from amino acids 239 to 279 and shared high similarities to the C-terminal peptidase M10 serralysin. Noteworthy, we did not find any specific proteolytic pattern for the other proteins because of the small number of extra non-tryptic peptides detected during the analysis. This indicates that no massive proteolysis occurred during exoproteome extraction as previously experienced [7,12,37]. As previously reported, the presence of highly secreted RTX-toxins seems to be a common issue in some *Roseobacter* strains [12]. In this sense, the main component of the exoproteome from *Ruegeria pomeroyi* DSS-3 was the PaxA RTX-toxin, observed whatever the culture condition experimentally tested. While some RTX-like proteins have been well characterized and considered as the main virulence factors in several well-known uropathogenic strains or different pathogenic *Vibrio* species [38] most of these toxin-like proteins remain uncharacterized in nonpathogenic environmental strains. This diverse group of RTX-like secreted proteins encoded in *Roseobacter* members is thought to have a potential effect on other members of the marine community in order to capture organic matter [39]. Interestingly, the secreted RTX-like proteins were depleted from the exoproteome of *R. pomeroyi* DSS-3 only in presence of a natural marine port community indicating a potential effect on its environment [7].

We concluded that the low content in basic residues detected in RTX-like proteins from *Roseobacter* strains could be linked to the function of these proteins rather than to the fact that they are secreted. Because these proteins encompass peptidase domains, their content in basic residues could be constrained to a low level for a higher resistance to the proteolysis of trypsin-like proteases or autoproteolysis as RTX-toxins may autoprocess [40,41]. Because they are secreted to the marine environment in high amounts, at least ZP_02147451 from *Phaeobacter strain DSM 17395* and PaxA from *R. pomeroyi* DSS-3, their function should be vital for bacteria from the *Roseobacter* clade. The present study shows that usual proteomic parameters for the interpretation of MS/MS data are misleading when exoproteomes are being analyzed. Here, trypsin-specific searches for MS/MS assignment did not identify correctly the ZP_02147451 RTX-toxin and led to false quantitative results regarding the main exoproteome components. Therefore, we propose for further exoproteome studies to analyze by tandem

Table 3. List of proteins identified in whole exoproteome of *Phaeobacter strain DSM 17395* using the FT/FT procedure[1].

			"Trypsin" as MASCOT search criteria						"No-enzyme" as MASCOT search criteria				
Accession	**Functional annotation**	**Molecular weight (Da)**	**Protein Mascot Score**	**Sequence Coverage (%)**	**Number of unique peptides**	**Spectral Count**	**NSAF (%)**	**EUF[$] (%)**	**Protein Mascot Score**	**Sequence Coverage (%)**	**Number of unique peptides**	**Spectral Count**	**NSAF (%)**
ZP_02144235	Hemolysin-type calcium-binding protein	115932	1109	20	17	70	2.4	88	1043	18	15	59	2.8
ZP_02145485	flagellin-like protein	28065	1057	66	16	63	9.0	106	1157	63	17	47	9.4
ZP_02146962	extracellular solute-binding protein, family 1	47415	888	46	13	34	2.9	92	853	44	12	32	3.8
ZP_02143869	phosphate ABC transporter*	37146	862	49	15	33	3.6	87	816	50	13	23	3.5
ZP_02147451	**Hemolysin-type calcium-binding protein**	**52624**	**82**	**3**	**1**	**2**	**0.2**	**1200**	**825**	**21**	**12**	**17**	**1.8**
ZP_02145205	chaperonin GroEL	57750	837	45	15	23	1.6	87	839	41	13	19	1.8
ZP_02147263	outer membrane porin	31831	798	34	12	156	19.7	108	848	36	13	100	17.6
ZP_02144302	glutathione synthetase	39420	622	46	10	20	2.0	80	544	34	8	17	2.4
ZP_02146401	phosphonate ABC transporter*	32128	539	49	10	19	2.4	80	513	35	8	14	2.4
ZP_02147202	TRAP dicarboxylate transporter	36347	548	38	9	23	2.5	89	522	33	8	22	3.4
ZP_02143908	TRAP transporter solute receptor	34084	541	35	11	26	3.1	82	498	35	9	16	2.6
ZP_02146891	Amino acid ABC transporter*	35345	512	41	9	16	1.8	89	471	41	8	14	2.2
ZP_02146522	Hemolysin-type calcium-binding region	31794	509	33	7	21	2.7	100	501	33	7	14	2.5
ZP_02146523	oligopeptide/dipeptide ABC transporter*	57342	499	26	9	18	1.3	78	441	21	7	14	1.4
ZP_02144316	putative iron ABC transporter*	35198	486	42	9	16	1.8	78	428	42	7	14	2.2
ZP_02146543	hypothetical protein RGBS107_00865	47171	479	31	8	17	1.5	75	430	23	6	12	1.4
ZP_02144531	OmpA/MotB	22083	434	82	9	11	2.0	56	325	48	5	6	1.5
ZP_02143879	Extracellular ligand-binding receptor	39919	408	47	8	14	1.4	75	337	36	6	9	1.3
ZP_02146376	putative sugar ABC transporter*	34311	399	41	8	19	2.2	75	345	41	6	12	2.0
ZP_02146184	extracellular solute-binding protein, family 5	61051	377	20	6	10	0.7	100	373	20	6	9	0.8
ZP_02144107	hypothetical protein RGBS107_16191	40145	368	38	7	11	1.1	86	341	28	6	10	1.4
ZP_02146027	EF hand domain protein	17617	363	53	6	10	2.3	67	294	36	4	7	2.2
ZP_02146810	zinc/manganese/iron ABC transporter*	34197	333	15	5	9	1.1	100	328	15	5	7	1.1
ZP_02145961	extracellular solute-binding protein, family 1	48328	322	15	4	6	0.5	125	366	20	5	7	0.8

[1]Detected with at least three different peptides. The values are from a representative experiment.
*Periplasmic component.
[$]EUF stands for Enzyme Unspecificity Factor, defined as the ratio number of unique peptides (no-enzyme MASCOT search) per number of unique peptides (trypsin MASCOT search).

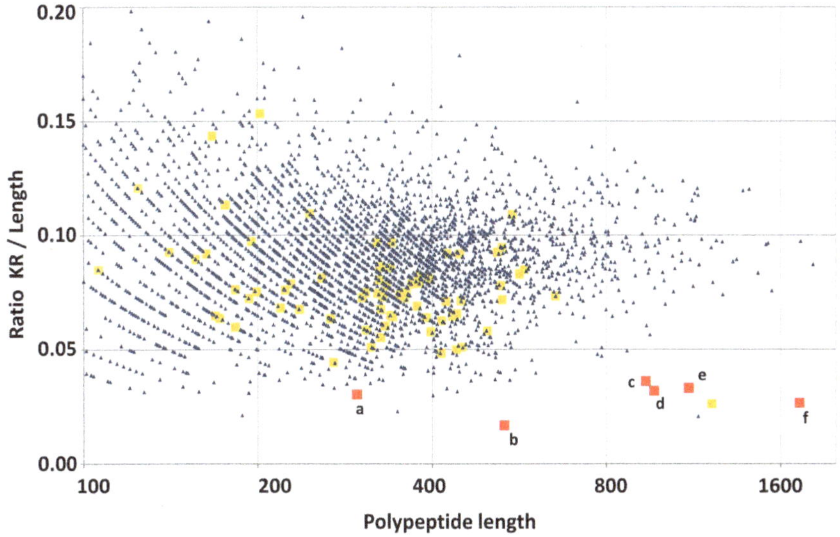

Figure 4. Occurrence of basic residues in proteins from *Phaeobacter strain DSM 17395.* The graph reports the ratio Lysine (K) and Arginine (R) residues per protein length of all the proteins with length above 100 residues encoded by *Phaeobacter strain DSM 17395*. Proteins are symbolized by a blue triangle. RTX-like proteins and the other exoproteins detected by tandem mass spectrometry are represented by red and yellow squares, respectively. The ZP_02143988.1 (a), ZP_02147451.1 (b), ZP_02144693.1 (c), ZP_02146960.1 (d), ZP_02144235.1 (e), and ZP_02146514.1 (f) RTX-toxins are indicated. Only the ZP_02143988.1 RTX-like protein has been detected by mass spectrometry.

mass spectrometry the proteins cleaved with a combination of different proteases (*e.g.* trypsin, chymotrypsin or EndoGluC proteases) in order to not underestimate, or simply miss, abundantly secreted proteins of interest. As an example, here the experimental proteolysis of the 55 kDa band with chymotrypsin allowed the identification of the ZP_02147451 RTX-toxin with 701 MS/MS spectra assigned to the protein out of 817 MS/MS spectra recorded, i.e. 85% % (**Table S5 in File S1**). Proteolysis with endo glu-C led to the identification of the same protein with 662 assigned MS/MS spectra out of 747 MS/MS spectra recorded, i.e. 89 (**Table S6 in File S1**). In this paper we outlined a strategy that could be systematically used to identify secreted proteins in environmental environments. It would be worth re-analyzing previous exoproteome datasets to check whether the lack of identification of an intense protein was common to several of these studies. However, the lack of the corresponding raw data (SDS-PAGE and MS/MS datasets) prevented such analysis. While RTX-proteins are found numerous in marine bacteria with more than a thousand members, some have been found involved in virulence of uropathogenic *E. coli* revealing strong medical interest [42]. Our results point at the specific structural characteristics of these proteins as basic residues have been counter-selected during evolution.

Supporting Information

Figure S1 Relative abundance of C-ter cleaved residues within the whole mass spectrometry-detected ZP_02147451 peptides.

File S1 Table S1, List of tryptic peptides identified in the whole exoproteome from *Phaeobacter strain DSM 17395* using a standard shotgun procedure. Table S2, List of proteins identified in the whole exoproteome from *Phaeobacter strain DSM 17395* using a standard shotgun procedure. Table S3, List of tryptic peptides identified in the 55 kDa SDS-PAGE band of the exoproteome from *Phaeobacter strain DSM 17395* using a standard shotgun procedure. Table S4, List of peptides identified in the 55 kDa band of the exoproteome from *Phaeobacter strain DSM 17395* using the FT/FT procedure and with "no-enzyme" as MASCOT search criteria. Table S5, List of peptides identified in the 55-kDa band of the exoproteome from *Phaeobacter strain DSM 17395* after chymotrypsin digestion using the FT/FT procedure. Table S6, List of peptides identified in the 55-kDa band of the exoproteome from *Phaeobacter strain DSM 17395* after endo-GluC digestion using the FT/FT procedure.

Author Contributions

Conceived and designed the experiments: ED JACO JA. Performed the experiments: ED JACO JA. Analyzed the data: ED JACO JA. Contributed reagents/materials/analysis tools: ED JACO JA. Wrote the paper: ED JACO JA.

References

1. Venter JC, Remington K, Heidelberg JF, Halpern AL, Rusch D, et al. (2004) Environmental genome shotgun sequencing of the Sargasso Sea. Science 304: 66–74.
2. Giebel HA, Kalhoefer D, Lemke A, Thole S, Gahl-Janssen R, et al. (2011) Distribution of Roseobacter RCA and SAR11 lineages in the North Sea and characteristics of an abundant RCA isolate. ISME J 5: 8–19.
3. Wagner-Dobler I, Biebl H (2006) Environmental biology of the marine Roseobacter lineage. Annu Rev Microbiol 60: 255–280.
4. Geng H, Belas R (2010) Molecular mechanisms underlying roseobacter-phytoplankton symbioses. Curr Opin Biotechnol 21: 332–338.
5. Armengaud J (2013) Microbiology and proteomics, getting the best of both worlds! Environ Microbiol 15: 12–23.
6. Armengaud J, Hartmann EM, Bland C (2013) Proteogenomics for environmental microbiology. Proteomics 13: 2731–42.

7. Christie-Oleza JA, Fernandez B, Nogales B, Bosch R, Armengaud J (2012) Proteomic insights into the lifestyle of an environmentally relevant marine bacterium. ISME J 6: 124–135.
8. Clair G, Armengaud J, Duport C (2012) Restricting fermentative potential by proteome remodeling: an adaptive strategy evidenced in Bacillus cereus. Mol Cell Proteomics 11: M111 013102.
9. Armengaud J, Christie-Oleza JA, Clair G, Malard V, Duport C (2012) Exoproteomics: exploring the world around biological systems. Expert Rev Proteomics 9: 561–575.
10. Christie-Oleza JA, Armengaud J (2010) In-depth analysis of exoproteomes from marine bacteria by shotgun liquid chromatography-tandem mass spectrometry: the *Ruegeria pomeroyi* DSS-3 case-study. Mar Drugs 8: 2223–2239.
11. Clair G, Roussi S, Armengaud J, Duport C (2010) Expanding the known repertoire of virulence factors produced by *Bacillus cereus* through early secretome profiling in three redox conditions. Mol Cell Proteomics 9: 1486–1498.
12. Christie-Oleza JA, Pina-Villalonga JM, Bosch R, Nogales B, Armengaud J (2012) Comparative proteogenomics of twelve roseobacter exoproteomes reveals different adaptive strategies among these marine bacteria. Mol Cell Proteomics 11: M111 013110.
13. Chagnot C, Zorgani MA, Astruc T, Desvaux M (2013) Proteinaceous determinants of surface colonization in bacteria: bacterial adhesion and biofilm formation from a protein secretion perspective. Front Microbiol 4: 303.
14. Wohlbrand L, Trautwein K, Rabus R (2013) Proteomic tools for environmental microbiology-a roadmap from sample preparation to protein identification and quantification. Proteomics 13: 2700–2730.
15. Martens T, Heidorn T, Pukall R, Simon M, Tindall BJ, et al. (2006) Reclassification of Roseobacter gallaeciensis Ruiz-Ponte et al. 1998 as Phaeobacter gallaeciensis gen. nov., comb. nov., description of Phaeobacter inhibens sp. nov., reclassification of Ruegeria algicola (Lafay et al. 1995) Uchino et al. 1999 as Marinovum algicola gen. nov., comb. nov., and emended descriptions of the genera Roseobacter, Ruegeria and Leisingera. Int J Syst Evol Microbiol 56: 1293–1304.
16. Ruiz-Ponte C, Cilia V, Lambert C, Nicolas JL (1998) Roseobacter gallaeciensis sp. nov., a new marine bacterium isolated from rearings and collectors of the scallop Pecten maximus. Int J Syst Bacteriol 48 Pt 2: 537–542.
17. Yoon JH, Kang SJ, Lee SY, Oh TK (2007) Phaeobacter daeponensis sp. nov., isolated from a tidal flat of the Yellow Sea in Korea. Int J Syst Evol Microbiol 57: 856–861.
18. Zhang DC, Li HR, Xin YH, Liu HC, Chi ZM, et al. (2008) Phaeobacter arcticus sp. nov., a psychrophilic bacterium isolated from the Arctic. Int J Syst Evol Microbiol 58: 1384–1387.
19. Druppel K, Hensler M, Trautwein K, Kossmehl S, Wohlbrand L, et al. (2014) Pathways and substrate-specific regulation of amino acid degradation in Phaeobacter inhibens DSM 17395 (archetype of the marine Roseobacter clade). Environ Microbiol. 16: 218–38.
20. Kossmehl S, Wohlbrand L, Druppel K, Feenders C, Blasius B, et al. (2013) Subcellular protein localization (cell envelope) in Phaeobacter inhibens DSM 17395. Proteomics 13: 2743–2760.
21. Zech H, Hensler M, Kossmehl S, Druppel K, Wohlbrand L, et al. (2013) Dynamics of amino acid utilization in Phaeobacter inhibens DSM 17395. Proteomics 13: 2869–2885.
22. Zech H, Hensler M, Kossmehl S, Druppel K, Wohlbrand L, et al. (2013) Adaptation of Phaeobacter inhibens DSM 17395 to growth with complex nutrients. Proteomics 13: 2851–2868.
23. Berger M, Neumann A, Schulz S, Simon M, Brinkhoff T (2011) Tropodithietic acid production in Phaeobacter gallaeciensis is regulated by N-acyl homoserine lactone-mediated quorum sensing. J Bacteriol 193: 6576–6585.
24. Brinkhoff T, Giebel HA, Simon M (2008) Diversity, ecology, and genomics of the *Roseobacter* clade: a short overview. Arch Microbiol 189: 531–539.
25. Seyedsayamdost MR, Carr G, Kolter R, Clardy J (2011) Roseobacticides: small molecule modulators of an algal-bacterial symbiosis. J Am Chem Soc 133: 18343–18349.
26. Seyedsayamdost MR, Case RJ, Kolter R, Clardy J (2011) The Jekyll-and-Hyde chemistry of Phaeobacter gallaeciensis. Nat Chem 3: 331–335.
27. Buddruhs N, Pradella S, Goker M, Pauker O, Pukall R, et al. (2013) Molecular and phenotypic analyses reveal the non-identity of the Phaeobacter gallaeciensis type strain deposits CIP 105210T and DSM 17395. Int J Syst Evol Microbiol 63: 4340–4349.
28. Dedieu A, Gaillard JC, Pourcher T, Darrouzet E, Armengaud J (2011) Revisiting iodination sites in thyroglobulin with an organ-oriented shotgun strategy. J Biol Chem 286: 259–269.
29. de Groot A, Dulermo R, Ortet P, Blanchard L, Guerin P, et al. (2009) Alliance of proteomics and genomics to unravel the specificities of Sahara bacterium *Deinococcus deserti*. PLoS Genet 5: e1000434.
30. Dupierris V, Masselon C, Court M, Kieffer-Jaquinod S, Bruley C (2009) A toolbox for validation of mass spectrometry peptides identification and generation of database: IRMa. Bioinformatics 25: 1980–1981.
31. Liu H, Sadygov RG, Yates JR 3rd (2004) A model for random sampling and estimation of relative protein abundance in shotgun proteomics. Anal Chem 76: 4193–4201.
32. Zivanovic Y, Armengaud J, Lagorce A, Leplat C, Guerin P, et al. (2009) Genome analysis and genome-wide proteomics of *Thermococcus gammatolerans*, the most radioresistant organism known amongst the *Archaea*. Genome Biol 10: R70.
33. Paoletti AC, Parmely TJ, Tomomori-Sato C, Sato S, Zhu D, et al. (2006) Quantitative proteomic analysis of distinct mammalian Mediator complexes using normalized spectral abundance factors. Proc Natl Acad Sci U S A 103: 18928–18933.
34. Baudet M, Ortet P, Gaillard JC, Fernandez B, Guerin P, et al. (2010) Proteomics-based refinement of *Deinococcus deserti* genome annotation reveals an unwonted use of non-canonical translation initiation codons. Mol Cell Proteomics 9: 415–426.
35. Bland C, Bellanger L, Armengaud J (2014) Magnetic Immunoaffinity Enrichment for Selective Capture and MS/MS Analysis of N-Terminal-TMPP-Labeled Peptides. J Proteome Res. in press.
36. Preston GM, Studholme DJ, Caldelari I (2005) Profiling the secretomes of plant pathogenic Proteobacteria. FEMS Microbiol Rev 29: 331–360.
37. Clair G, Lorphelin A, Armengaud J, Duport C (2013) OhrRA functions as a redox-responsive system controlling toxinogenesis in Bacillus cereus. J Proteomics 94: 527–539.
38. Linhartova I, Bumba L, Masin J, Basler M, Osicka R, et al. (2010) RTX proteins: a highly diverse family secreted by a common mechanism. FEMS Microbiol Rev 34: 1076–1112.
39. Moran MA, Belas R, Schell MA, Gonzalez JM, Sun F, et al. (2007) Ecological genomics of marine *Roseobacters*. Appl Environ Microbiol 73: 4559–4569.
40. Satchell KJ (2011) Structure and function of MARTX toxins and other large repetitive RTX proteins. Annu Rev Microbiol 65: 71–90.
41. Sheahan KL, Cordero CL, Satchell KJ (2007) Autoprocessing of the Vibrio cholerae RTX toxin by the cysteine protease domain. EMBO J 26: 2552–2561.
42. Vigil PD, Alteri CJ, Mobley HL (2011) Identification of in vivo-induced antigens including an RTX family exoprotein required for uropathogenic Escherichia coli virulence. Infect Immun 79: 2335–2344.

13

Seabird Nutrient Subsidies Benefit Non-Nitrogen Fixing Trees and Alter Species Composition in South American Coastal Dry Forests

Gilles Havik[1], Alessandro Catenazzi[2]*, Milena Holmgren[1]

1 Resource Ecology Group, Wageningen University, Wageningen, The Netherlands, **2** Department of Zoology, Southern Illinois University Carbondale, Carbondale, Illinois, United States of America

Abstract

Marine-derived nutrients can increase primary productivity and change species composition of terrestrial plant communities in coastal and riverine ecosystems. We hypothesized that sea nutrient subsidies have a positive effect on nitrogen assimilation and seedling survival of non-nitrogen fixing species, increasing the relative abundance of non-nitrogen fixing species close to seashore. Moreover, we proposed that herbivores can alter the effects of nutrient supplementation by preferentially feeding on high nutrient plants. We studied the effects of nutrient fertilization by seabird guano on tree recruitment and how these effects can be modulated by herbivorous lizards in the coastal dry forests of northwestern Peru. We combined field studies, experiments and stable isotope analysis to study the response of the two most common tree species in these forests, the nitrogen-fixing *Prosopis pallida* and the non-nitrogen-fixing *Capparis scabrida.* We did not find differences in herbivore pressure along the sea-inland gradient. We found that the non-nitrogen fixing *C. scabrida* assimilates marine-derived nitrogen and is more abundant than *P. pallida* closer to guano-rich soil. We conclude that the input of marine-derived nitrogen through guano deposited by seabirds feeding in the Pacific Ocean affects the two dominant tree species of the coastal dry forests of northern Peru in contrasting ways. The non-nitrogen fixing species, *C. scabrida* may benefit from sea nutrient subsidies by incorporating guano-derived nitrogen into its foliar tissues, whereas *P. pallida*, capable of atmospheric fixation, does not.

Editor: Paul V. A. Fine, University of California, Berkeley, United States of America

Funding: This study was partially funded by an Alberta Mennega Stichting grant to GH, a NWO Meervoud grant to MH (836.05.021) and a Swiss National Science Foundation Postdoctoral fellowship to AC (116305). The funders had no role in study design, data collection and analysis, decision to publish, or preparation of the manuscript.

Competing Interests: The authors have declared that no competing interests exist.

* E-mail: acatenazzi@siu.edu

Introduction

The composition and dynamics of ecological communities in coastal marine and terrestrial ecosystems are partially driven by nutrient fluxes that link sea and adjacent land [1–5]. Some of the most striking examples of this interaction are found in the unproductive arid west coasts of North and South America where sea nutrient subsidies from the productive Pacific Ocean permeate through whole food webs affecting terrestrial plants [4,6–8] and consumers [1,7,9–11], in various ways [12–14]. Marine nutrients reach land primarily through colonies of seabirds and marine mammals, or through organic debris, carcasses and algae deposited by tidal action [15,16], and are further dispersed through trophic interactions [11,14,17,18] and wind [4].

Seabird guano is a major nutrient path from the surrounding marine environments into terrestrial roosting and nesting sites worldwide [19]. It is well established that seabird guano increases soil nitrogen and phosphorus, although it can also affect soil pH, salinity and moisture [19]. These changes in soil characteristics generally translate into enhanced foliar nutrient levels and plant growth [20], especially when sufficient rainfall is available [6,11,12,21]. However, at very high guano concentrations, plant cover can be reduced [21] or completely depleted [22], which has been attributed to ammonium toxicity [23] or acidic conditions [19]. Plant communities in seabird colonies are often dominated by cosmopolitan and ruderal species [19,24] because the combined effects of seabirds include nutrient fertilization, physical disturbance and seed dispersal [19,25]. Experimental manipulations of seabird guano have found contrasting responses across tree species, suggesting that differences in plant functional traits could help explain changes in plant composition in seabird-associated areas [26].

The overall effects of marine subsidies on arid coastal woody plant communities likely depend on how herbivores respond to plant nutrient enrichment and on how they are distributed along the sea-land gradient of nutrient fertilization. In pulsed systems, herbivores can limit plant growth [27,28], but their effect on tree recruitment in arid coastal systems has not been directly assessed. Since herbivores usually respond positively to increasing plant nutrient content [29], we hypothesize that they can potentially offset the positive effects of sea nutrient subsides on tree seedling establishment.

In this paper, we assess the effects of seabird nutrient input and herbivory on tree seedling establishment in an arid coastal system. We hypothesised that non nitrogen-fixing tree species benefit more strongly from marine-derived nitrogen than nitrogen-fixing species since non-fixers depend on soil nitrogen sources and are not able to uptake atmospheric nitrogen through associations with rhizobial bacteria. We combined stable isotope analysis, field observations and experiments to test the following predictions: (1) marine nutrient fertilization has a positive effect on nitrogen assimilation of non-nitrogen fixing trees; (2) abundance of non-nitrogen fixing trees increases near the sea shore; (3) herbivory by lizards can offset the positive effects of nitrogen fertilization on tree seedling survival.

Materials and Methods

Study System

We conducted our work in the coastal dry forests of north-western Peru, a region where the rich marine waters of the Peru-Chile current face the arid Sechura desert [1,30]. Dry forests with an estimated surface of over 3 million ha expand through northern Peru and southern Ecuador [31]. These dry forests are dominated by *Prosopis pallida*, a nitrogen-fixing leguminous tree species, associated with other tree and shrub species such as the non-nitrogen fixing *Capparis scabrida* [32]. Vegetation becomes sparser closer to the coast (Fig. 1). We worked along the coast of Bayovar (5° 47′ S, 81° 04′ W; 8 m. a. s. l.; Fig. 1), near the oil pipeline terminal of PetroPerú and protected from human disturbance. This area is controlled by PetroPerú, and PetroPerú authorized our work in the area. We worked along the rocky seashore, rising about 7 m above the water and separated from the beach by cliffs. Large colonies of blue-footed boobies (*Sula nebouxii*) create abundant deposits of guano up to 50 cm thick (Fig. 1).

Rainfall is concentrated in the summer months (December-May) and is strongly influenced by El Niño Southern Oscillation (ENSO) events. Rainy ENSO events play a critical role in the regeneration of these forests, facilitating tree seedling establishment [27,33,34] and fast growth [35] that contributes to plants reaching a critical size at which they are less sensitive to herbivore damage [36]. Mean annual precipitation is ca. 50 mm (1961–1983; [37]), with high interannual variation (e.g. ranging from 2.8 mm in 1996 to 1639 mm in 1998). Mean annual temperature is 24.0°C, ranging between 21.1°C (August) and 27.9°C (February; [37]).

Study Species

The two tree species are *Prosopis pallida* (Fabaceae) and *Capparis scabrida* (Capparaceae). *Prosopis pallida* is able to grow under arid conditions with at least 50 to 250 mm rainfall annually [32]. It can grow in nutrient-poor soils where its roots fix nitrogen in association with rhizobial bacteria [32,38]. It is a fast growing tree with roots able to reach deep water sources up to 25 m below the surface [32]. Seed dispersal and germination are facilitated by a variety of animals [39]. Nevertheless, a high percentage of seeds remain dormant soon after falling from the tree [32]. Seedling establishment is greatly enhanced during rainy ENSO events [27,35]. *Capparis scabrida* grows under a wide range of conditions, often in association with *P. pallida* [40]. Seeds are dispersed by foxes and birds and have high germination rates [40].

The most important herbivore in this region is *Dicrodon guttulatum* (Teiidae), a lizard endemic to north-western Peru that feeds preferentially on *P. pallida* [41] and strongly limits tree seedling establishment [27]. This is the most abundant herbivore at our site because livestock are not allowed within the PetroPerú concession.

Soil Nutrients

We assessed soil nutrient concentrations along five transects perpendicular to the sea at increasing distances (10, 50, 100, 200 m) from the seashore. At each sampling location, we collected approximately 1 kg of soil from the top 30 cm of soil, and removed the rocks. Concentrations of NO_3, NH_4, PO_4 and K were measured at the Resource Ecology Laboratory, University of Wageningen. Dried samples were extracted using a 0.01*M* $CaCl_2$ solution previous to spectrophotometric determination of NO_3

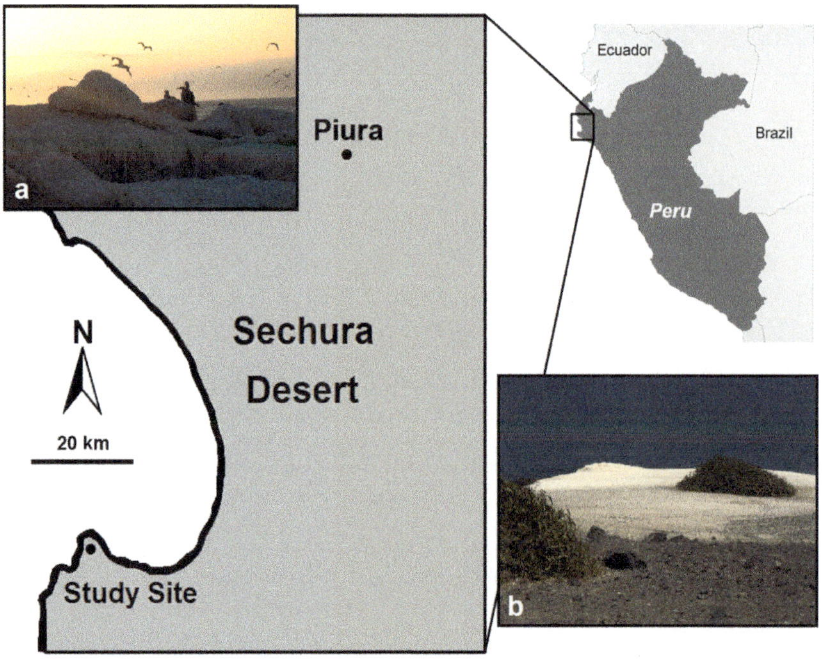

Figure 1. Location of the study site in Bayovar, north-western Peru. a) Guano accumulates in places where *Sula nebouxii* roost along shore (photograph G. Havik). b) *Capparis scabrida* growing in guano-rich soil (photograph A. Catenazzi).

and NH_4 by a Segmented Flow Analyzer (Skalar, Breda, The Netherlands). PO_4 concentration was measured spectrophotometrically after digestion with H_2O_2. Concentrations of total nitrogen (N) and carbon (C) were measured in an Elemental Analyzer (EA 1108 CHN-O, Fisons Instruments) based on the dynamic combustion method. Potassium concentration was assessed using the ammonium acetate pH 7.0 protocol.

Marine-derived Nitrogen in Plants and Soils

We measured nitrogen stable isotope ratios ($\delta^{15}N$) in soils and leaves of *P. pallida* and *C. scabrida* to assess the contribution of seabird-derived nitrogen in soil and foliar tissues. Values of $\delta^{15}N$ have been used successfully to assess the contribution of seabird-derived nitrogen to coastal and insular food webs [6,42,43] because seabird guano typically has high $\delta^{15}N$ values that can be over 30‰ [6,44]. When terrestrial plants assimilate nitrogen from guano, their $\delta^{15}N$ increases and approaches the $\delta^{15}N$ of guano: for example in the Gulf of California, $\delta^{15}N$ of the saltbush *Atriplex barclayana* was 5.26±0.85‰ on an island without seabirds, but 36.36±2.72‰ on a seabird island [6]. In the Sechura desert, $\delta^{15}N$ values were −0.50±0.20‰ for *P. pallida* and 2.82±0.98‰ for *C. scabrida* in places that lack guano [33]. Thus, we expected foliar tissues of *P. pallida* and *C. scabrida* near seabird roosting sites to have $\delta^{15}N$ values much higher than those measured in places lacking guano. For both tree species, we collected pooled samples of 5 leaves from 3 trees (at least 1.5 m tall, n = 15 leaves) growing at the same 4 distances from shore where other measurements were taken (10, 50, 100, 200 m) along six sea-inland transects (including the same transects used for sampling soil). We measured soil $\delta^{15}N$ in the samples used for measurements of nutrient concentrations. All samples were oven dried at 55°C for 72 hours and pulverized in a sterilized mill prior to isotope analysis. Analysis of N concentration and $\delta^{15}N$ was conducted on a Finnigan MAT Delta Plus continuous flow isotope ratio mass spectrometer at the Center for Stable Isotope Biogeochemistry, University of California, Berkeley. We analyzed 3–4 mg of leaves, and 10–75 mg of soil. Values of $\delta^{15}N$ are expressed relative to its international standard, atmospheric air. We analyzed 18% of our samples in duplicate. The average standard deviation for replicate samples was 0.12‰.

Spatial Tree Distribution

We evaluated the density of seedlings and trees of *P. pallida* and *C. scabrida* along the first 200 m from sea inland. We counted all seedlings and trees found in eight (300 m long × 60 m wide) transects perpendicular to shore. We divided each transect in six (50 m long × 60 m wide) plots.

Seedling Establishment along a Natural Sea-inland Gradient of Soil Nitrogen Concentration

We designed a field experiment to assess the effects of distance from shore and herbivory pressure on early seedling survival of *P. pallida* and *C. scabrida*. At four distances from shore along a natural sea-inland gradient (10, 50, 100, 200 m), we planted 20 seedlings (10 of each tree species). Half of the seedlings (five of each species) were protected against herbivores. At each distance location, seedlings were planted at 25 cm distance from each other [35]. These treatments were replicated five times, along the same sea-inland transects where we measured soil nutrients and $\delta^{15}N$ as described earlier. In summary, we planted 200 seedlings of each species for a total of 400 seedlings. Seedlings were protected against lizard browsing using plastic bottles (Ø: 10 cm, bottle top open). This method is commonly used in restoration projects in Peru because it is effective against herbivores [34] and experimental trials indicate that soil moisture content remains unaffected (File S1). Seedlings germinated from seeds collected locally and were transplanted when they were three weeks old in February 2009. We watered seedlings twice a week for the first seven weeks for a total of 270 mm, roughly equivalent to the rainfall during moderate ENSO events such as those in 1992 (235 mm, [34]) and 2002 (268 mm; [35]). No water was added during weeks with rainfall events of at least 5 mm. Total rainfall during the experimental time was 15 mm. Seedling survival was monitored regularly during three months.

Statistical Analyses

We compared the soil nutrient concentrations (log-transformed) and the nitrogen stable isotope ratios ($\delta^{15}N$) in soils along the sea-inland nutrient gradient using general linear models with distance as a fixed factor and transect as a random factor. We tested for differences in leaf N concentrations and leaf $\delta^{15}N$ for each tree species separately using general linear models including distance from shore as a fixed factor and transect as a random factor. We used generalized linear models to assess the effect of distance from shore on the number of naturally established plants for each species separately defining a Poisson distribution and using the factor distance from shore nested within transects.

To analyse the survival patterns of experimental seedlings, we used Cox proportional-hazards regression models, using the R-library survival [45]. We tested for differences in seedling survival for each tree species separately including distance from shore (log-transformed) and herbivory (protected or not) as covariables and cluster as a random factor to account for the grouping of 20 seedlings per location.

Results

Soil Nutrient Concentration along the Sea-inland Gradient

Soil nutrient concentrations were higher closer to shore (Table 1). Statistical tests indicate a significant effect of distance on [NO_3] ($F_{3,12} = 39.24$; $p<0.001$), [NH_4] ($F_{3,12} = 6.60$; $p = 0.007$), [PO_4] ($F_{3,12} = 7.84$; $p = 0.004$), and on [K] ($F_{3,12} = 18.23$; $p<0.001$; see File S2 for details). Transects had no significant effects in any of the models.

Nutrient Concentrations and $\delta^{15}N$ of Soil and Tree Leaves

Leaves of *P. pallida* had higher N concentrations (3.41% ±0.31) than leaves of *C. scabrida* (1.50% ±0.30; Table 1). The two species differed in their leaf N concentration pattern along the sea shore-inland gradient. Whereas N concentrations of *P. pallida* did not change among sea-distance classes ($F_{3,11} = 2.94$; $p = 0.08$), leaves of *C. scabrida* contained more N in plants growing near shore ($F_{3,12} = 15.75$; $p<0.001$).

The $\delta^{15}N$ of soil decreased with distance from shore (Fig. 2; $F_{3,10} = 8.35$; $p = 0.004$). Values of leaf $\delta^{15}N$ for *C. scabrida* were more enriched than $\delta^{15}N$ values for *P. pallida* (T- test = −14.71; $p<0.001$). Also the sea shore-inland pattern for leaf $\delta^{15}N$ differed between the two species (Fig. 2). Whereas the $\delta^{15}N$ for *P. pallida* did not vary among distance classes (Fig. 2; $F_{3,10} = 0.67$; $p = 0.59$), the $\delta^{15}N$ for *C. scabrida* decreased with distance from shore ($F_{3,12} = 3.54$; $p = 0.048$). At a distance of 10 m from shore, soil was mostly composed of guano, and $\delta^{15}N$ values averaged 22.24±2.75‰ for guano-rich soil, 19.94±1.66‰ for leaves of *C. scabrida*, and 0.87±0.82‰ for leaves of *P. pallida*.

Table 1. Nutrient concentrations in soil and leaves.

Nutrient Concentrations		Distance to shore							
		10 m		50 m		100 m		200 m	
Soil	Nitrate (μg/g)	94.49[a]	(40.56–887.75)	6.93[b]	(4.09–13.90)	1.84[c]	(1.44–2.04)	0.73[d]	(0.45–0.32)
	Ammonium (μg/g)	267.35[a]	(46.10–4993.15)	42.93[b]	(32.71–61.96)	66.61[ab]	(23.12–94.82)	50.64[ab]	(37.22–58.26)
	Phosphate (μg/g)	79.20[a]	(2.09–1114.80)	1.94[b]	(0.69–5.52)	1.30[b]	(0.30–3.47)	1.00[b]	(0.59–5.45)
	Potassium (μg/g)	425.00[a]	(233.00–1158.00)	95.60[b]	(58.40–98.00)	104.20[b]	(80.40–152.80)	93.00[b]	(76.20–152.80)
	%N	0.37[a]	(0.12–2.00)	0.12[ab]	(0.09–0.25)	0.12[b]	(0.10–0.18)	0.16[ab]	(0.12–0.26)
	%C	2.27[a]	(2.24–5.68)	1.45[ab]	(0.30–3.12)	1.03[b]	(0.49–1.75)	0.30[c]	(0.14–0.36)
Leaves	*C. scabrida* %N	1.80[a]	(1.58–2.25)	1.43[ab]	(1.27–2.05)	1.17[b]	(1.12–1.64)	1.43[ab]	(1.31–1.95)
	C. scabrida %C	41.73[a]	(40.90–42.76)	42.37[a]	(39.93–43.76)	41.18[a]	(39.55–42.57)	41.51[a]	(39.45–43.75)
	P. pallida %N	3.22[a]	(2.97–3.32)	3.23[a]	(2.91–3.91)	3.66[a]	(3.12–3.81)	3.48[a]	(3.13–3.97)
	P. pallida %C	44.58[a]	(42.59–45.73)	45.28[a]	(44.59–46.98)	46.09[a]	(45.35–47.28)	46.03[a]	(44.69–47.30)

Values are median and range (in parenthesis). Columns with same letters were not significantly different following post-hoc analysis ($p>0.05$).

Spatial Distribution of Naturally Established Trees

Overall, the density of *C. scabrida* in our study plots was higher than the density of *P. pallida* (Fig. 3). We found differences in tree abundance among sea-distance classes for both *P. pallida* (Wald $\chi^2 = 41.37$; $p = 0.028$) and *C. scabrida* (Wald $\chi^2 = 121.79$; $p<0.001$). However the distribution pattern was different. Whereas the abundance of *C. scabrida* was higher close to shore, the abundance of *P. pallida* was lower. Within the first 50 m from shore, the density of *C. scabrida* was higher than the density of *P. pallida* (T-test; $p = 0.01$; Fig. 3).

Survival of Planted Seedlings

The survival rate of *P. pallida* seedlings was lower than the survival of *C. scabrida* seedlings; Fig. 4). Protection from lizards had no significant effect on the survival of the two species (Robust-test; $p = 0.085$ and $p = 0.106$ for *P. pallida* and *C. scabrida* respectively). Since the effects of herbivores were not significant, we modelled seedling survival as a function of species and distance from shore including a species *distance interaction. We found that only the interaction was significant ($p<0.05$) indicating that the species responded differently to the sea distance gradient which is also reflected in the models for each species separately. Distance from shore had a positive significant effect on the survival rate of *C. scabrida* (Robust-test; $p = 0.0004$; Fig. 4) but no effect on the survival of *P. pallida* seedlings (Robust-test; $p = 0.826$; Fig. 4).

Discussion

Our study did not find a significant effect of herbivory on tree seedling survival in a tropical dry forest subsidized by seabird

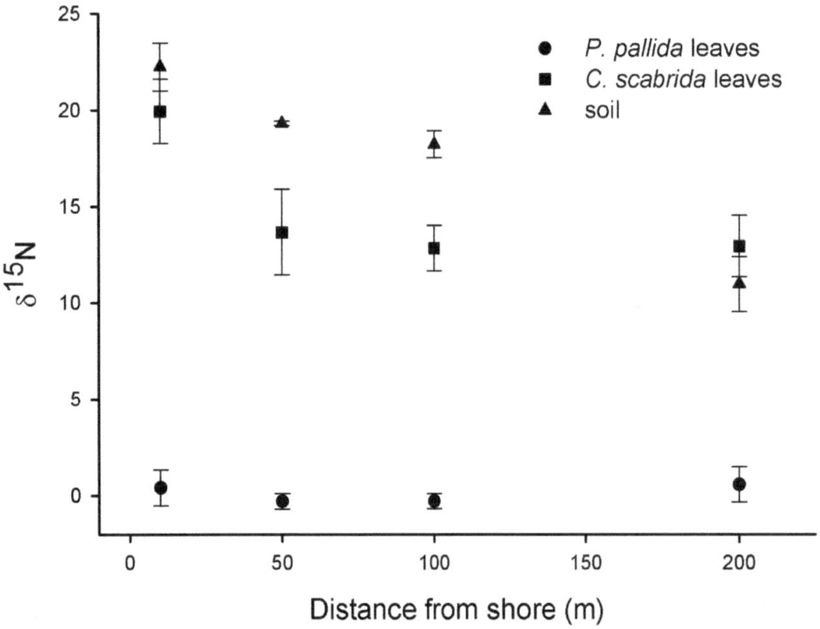

Figure 2. Nitrogen stable isotope ratios ($\delta^{15}N$; mean ± SE) for leaves of *Prosopis pallida* (circles), *Capparis scabrida* (squares), and soil samples (triangles) collected at 10, 50, 100 and 200 m from shore.

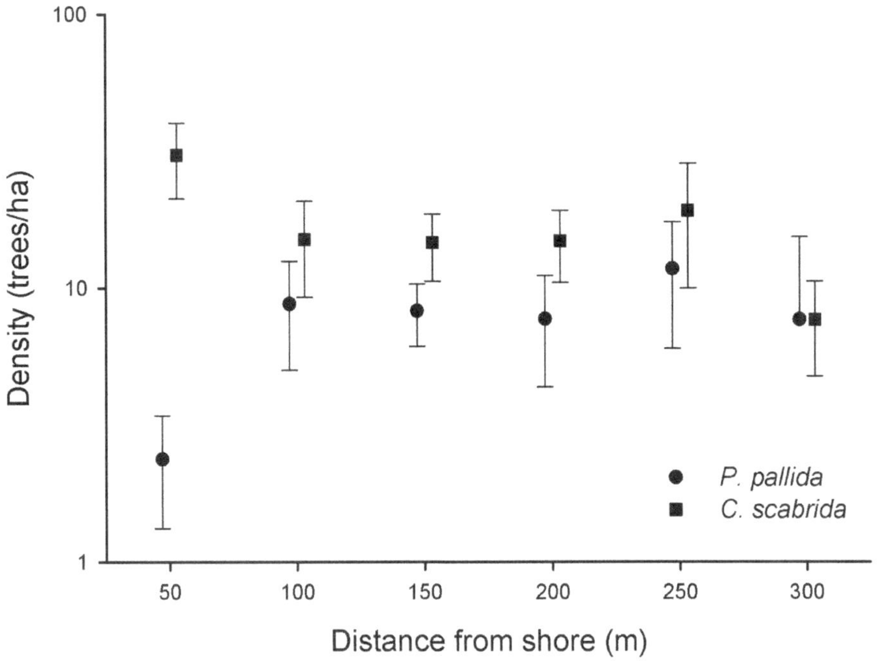

Figure 3. Density (mean ± SE; ha-1) of *Prosopis pallida* (circles) and *Capparis scabrida* (squares) seedlings and trees found along sea-inland transects.

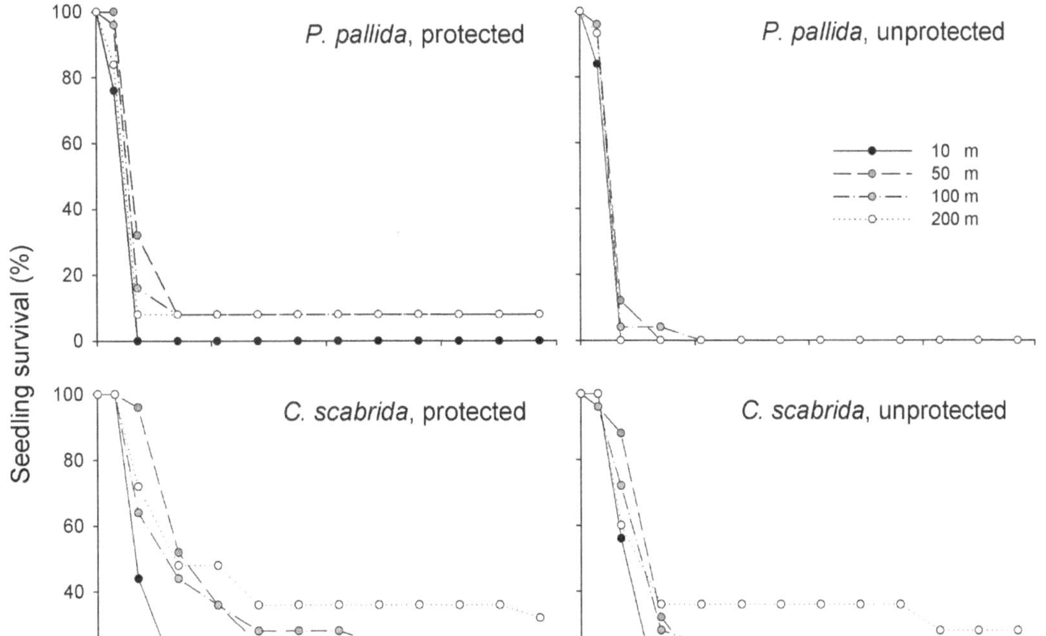

Figure 4. Survival (% individuals alive, n = 25) of seedlings of *Prosopis pallida* and *Capparis scabrida* in protected and unprotected experimental plots. Plots were located at increasing distance from shore (10, 50, 100 and 200 m). Overall survival (instead of mean and S.E.) illustrated here to facilitate comparison among distance classes.

guano. Woody plant species are common in tropical ecosystems receiving large amounts of seabird guano [26,43,46]. Previous studies have documented high densities of lizards in systems subsidized by marine-derived resources [9,13,47]. Therefore, we anticipated a positive effect of seabird-derived subsidies on the strength of top-down interactions at our study site. Our negative results confirm that it is difficult to predict the effect of seabird-derived resources on species interactions, or to generalize their effects on food web composition and functioning [48].

Our focal lizard species *Dicrodon guttulatum* is a widespread lizard in the Sechura desert [41] and has been identified as a major limiting factor for the regeneration of inland dry forests in northern Peru [27,34]. At our coastal site experimental protection from lizard herbivory did not affect survival of the two tree species. Therefore, top-down control by herbivorous lizards seems to be weaker in coastal dry forests than it is in inland dry forests. This difference might be explained by lower lizard density in coastal sites (unpublished data). *Dicrodon guttulatum* feeds primarily on *P. pallida* [41], and the abundance of *P. pallida* can be much higher at inland sites than it is at our coastal site, presumably supporting larger populations of lizards.

Although herbivory did not play a role in seedling survival, we observed differences in the distribution of the two tree species at our coastal site. Nutrient and stable isotopic analyses suggest alternative explanations for the observed pattern of increased dominance of the non-fixing species close to shore. The observed patterns of $\delta^{15}N$ and N concentrations in foliar tissue (Fig. 2) suggest a positive response to increased soil N concentrations near shore for the non-fixing *C. scabrida* but not for the fixing *P. pallida*. This result is in line with our hypothesis that sea-nutrient fertilization has a positive effect on N assimilation of the non-N fixer. The higher concentrations of NH_4 and NO_3 close to shore indicate that more soil N is available to plants near shore. Leaf tissues of *C. scabrida* trees growing near shore had elevated $\delta^{15}N$ values, suggesting assimilation of marine-derived N because $\delta^{15}N$ in guano-rich soil had similarly high values (Fig. 2). Higher N concentrations in *C. scabrida* tissue near shore support this result. In this coastal ecosystem, *C. scabrida* is the most abundant plant species suggesting that it can be an important vector of marine-derived nutrients to the terrestrial trophic food web through diverse types of consumers feeding on its leaves or fruits (e.g. foxes, goats, birds and insects).

The $\delta^{15}N$ of *P. pallida* was always close to zero, suggesting that atmospheric N was the source. Moreover, the N concentrations in foliar tissue varied little between *P. pallida* plants growing in N-rich soils near shore and plants growing further inland where N concentrations were lower. These results are particularly interesting compared to previous studies. Elevated soil N concentrations could inhibit N fixation [50], for example *Prosopis juliflora* had lower N-fixation rates and higher $\delta^{15}N$ under saline conditions [49]. Our data suggest that *P. pallida* assimilates atmospheric-fixed N regardless of soil N concentration in the soil and distance from shore. An alternative hypothesis, differences in fractionation during N assimilation, is unlikely to explain these patterns because our aggregated samples pooled leaves from different parts of the plants and from several individuals at each distance class, and because we collected replicated samples along a linear gradient of guano concentration. Therefore, given the lack of correspondence between the spatial patterns of $\delta^{15}N$ in soil and stability of $\delta^{15}N$ in leaves of *P. pallida*, the most parsimonious explanation is that *P. pallida* does not assimilate N from the soil even in places where NH_4 and NO_3 concentrations are high.

The density of naturally established trees of both species suggests a response to the sea-inland nutrient patterns consistent with results from the stable isotope analysis. Near shore, the density of *C. scabrida* was higher than the density of *P. pallida*. Density of *P. pallida* was low near shore, increased between 50 and 100 m, and remained relatively constant and similar to *C. scabrida* further away. These data suggest that *C. scabrida* growing near shore benefits more from high nutrient concentrations than *P. pallida*. High nutrient concentration may affect the competitive ability of the two tree species by promoting growth or survival of the non-fixing species. Our seedling transplant experiment did not detect an effect of distance from shore on survival. Therefore, high nutrient concentrations are likely to enhance growth rate of the non-fixer without affecting the growth rate of the fixer, as suggested by patterns of increased tissue N concentrations in *C. scabrida* near shore, and no change in tissue N concentration in *P. pallida*. The positive effects of marine nutrient subsidies on *C. scabrida* were restricted to areas very close to the sea guano deposits. Remarkably localized effects have also been found in other seabird studies worldwide [51], as well as in the transfer of marine-derived nitrogen accumulated and transported by salmon to riverine forests of North America [52,53].

An alternative hypothesis to explain increased dominance of the non-fixer near shore is that the fixer is suppressed by the guano nitrogen due to toxicity. This hypothesis relies on the assumption that *P. pallida* are getting their nitrogen from fixation rather than from the guano-rich soil near shore, as opposed to *C. scabrida* incorporating nitrogen from guano-rich soil. This assumption is supported by constant $\delta^{15}N$ values in the fixer and decreasing $\delta^{15}N$ in the non-fixer with distance from shore. Moreover, in a small laboratory experiment on the germination of *P. pallida* ($n = 50$ seeds), we found that seeds germinated better in soils coming from ≥50 m from shore than in guano-rich soils collected near shore [54]. However, our field experiment results show that seedling establishment close to shore is difficult for both tree species. The planted seedlings of *C. scabrida* did not grow well and ultimately died near shore, whereas the survival of *P. pallida* was very low throughout the 200 m sea-inland gradient (Fig. 4).

The toxic effect of N for both species' seedlings could be explained either by the acidifying effect of ammonia on the soil [55], by N toxicity, or by a combination of both [19]. The effects of P on nitrogen and non-nitrogen fixers are complex. Nitrogen fixing trees require relatively high supplies of P for the process of fixation [56]. In natural populations, the rates of N fixation increase with increasing soil P availability [57]. However, under non-limiting N conditions, experimental evidence shows that N acquisition strategies change with P supply, and fixers rely less on N fixation [58].

It has been suggested that the toxicity effects of guano may be more common in hyperarid regions [19]. We speculate that nutrient leaching during rainy years could drop nutrient concentrations below toxic levels. This would explain the apparent discrepancy between the natural distribution patterns and the outcome of the field experiment in our study system. Indeed the benefits of marine nutrient fertilization on arid plant communities can be particularly strong during rainy years, like those associated to ENSO, when water availability is not limiting [11,19]. It has been suggested that this positive effect can build up organic matter that in turn improves soil water retention and, as a consequence, long-term guano accumulation can have a positive effect on plant productivity even in dry years [8]. However, the relationship between guano accumulation and soil moisture is not always positive [19].

Salinity is another environmental factor that could change along the sea-inland gradient. Earlier work indicates that both *P. pallida* [32] and *C. scabrida* [40] can tolerate high soil salinity, but future manipulative experiments may contribute to further separate the role of nutrients and salinity.

We conclude that the input of marine-derived N through guano deposited by seabirds feeding in the Pacific Ocean affects the two dominant tree species of the coastal dry forests of northern Peru in contrasting ways. The non-N fixing species, *C. scabrida* may benefit from sea nutrient subsidies by incorporating marine-derived N into its foliar tissues, whereas *P. pallida*, capable of atmospheric N fixation, does not.

Acknowledgments

We thank Petroperú, Universidad de Piura and R. Rodríguez for logistic support. L. Urbina and N. Atarama assisted in the field and in the laboratory. E. Guerra shared her expertise on N fixation. A.M. van den Driessche conducted the soil nutrient analyses. P. Jansen supervised the survival analysis. E. N. Broadbent, R. J. Naiman and E. Chaneton commented on a previous version of this manuscript.

Author Contributions

Conceived and designed the experiments: GH AC MH. Performed the experiments: GH AC. Analyzed the data: GH AC MH. Contributed reagents/materials/analysis tools: GH AC MH. Wrote the paper: GH AC MH.

References

1. Koepcke HW, Koepcke M (1952) Sobre el proceso de transformación de la materia orgánica en las playas arenosas marinas del Perú. Publ Mus Hist Nat Javier Prado Ser A 8: 1–24.
2. Mellbrand K, Lavery PS, Hyndes G, Hambäck PA (2011) Linking land and sea: different pathways for marine subsidies. Ecosystems 14: 732–744.
3. Piovia-Scott J, Spiller DA, Schoener TW (2011) Effects of experimental seaweed deposition on lizard and ant predation in an island food web. Science 331: 461–463.
4. Polis GA, Anderson WB, Holt RD (1997) Toward an integration of landscape and food web ecology: the dynamics of spatially subsidized food webs. Annu Rev Ecol Syst 28: 289–316.
5. Spiller DA, Piovia-Scott J, Wright AN, Yang LH, Takimoto G, et al. (2010) Marine subsidies have multiple effects on coastal food webs. Ecology 91: 1424–1434.
6. Anderson WB, Polis GA (1999) Nutrient fluxes from water to land: seabirds affect plant nutrient status on Gulf of California islands. Oecologia 118: 324–332.
7. Polis GA, Hurd SD, Jackson TD, Sánchez-Piñero F (1997) El Niño effects on the dynamics and control of an island ecosystem in the Gulf of California. Ecology 78: 1884–1897.
8. Wait DA, Aubrey DP, Anderson WB (2005) Seabird guano influences on desert islands: Soil chemistry and herbaceous species richness and productivity. J Arid Environ 60: 681–695.
9. Catenazzi A, Donnelly MA (2007) The *Ulva* connection: marine algae subsidize terrestrial predators in coastal Peru. Oikos 116: 75–86.
10. Polis GA, Hurd SD, Jackson TD, Sánchez-Piñero F (1998) Multifactor population limitation: variable spatial and temporal control of spiders on Gulf of California islands. Ecology 79: 490–502.
11. Stapp P, Polis GA, Piñero FS (1999) Stable isotopes reveal strong marine and El Niño effects on island food webs. Nature 401: 467–469.
12. Anderson WB, Wait DA, Stapp P (2008) Resources form another place and time: responses to pulses in a spacially subsidized system. Ecology 89: 660–670.
13. Fariña JM, Sepulveda M, Reyna MV, Wallem KP, Ossa-Zazzali PG (2008) Geographical variation in the use of intertidal rocky shores by the lizard *Microlophus atacamensis* in relation to changes in terrestrial productivity along the Atacama Desert coast. J Anim Ecol 77: 458–468.
14. Polis GA (1998) Stability is woven by complex webs. Nature 395: 744–745.
15. Catenazzi A, Donnelly MA (2007) Role of supratidal invertebrates in the decomposition of beach-cast green algae (*Ulva* sp.). Mar Ecol Prog Ser 349: 33–42.
16. Polis GA, Hurd SD (1996) Linking marine and terrestrial food webs: allochthonous input from the ocean supports high secondary productivity on small islands and coastal land communities. Am Nat 147: 396–423.
17. Catenazzi A, Donnelly MA (2008) Sea lions (*Otaria flavescens*) as hosts of the common vampire bat (*Desmodus rotundus*). Mar Ecol Prog Ser 360: 285–289.
18. Rose MD, Polis GA (1998) The distribution and abundance of coyotes: The effects of allochthonous food subsidies from the sea. Ecology 79: 998–1007.
19. Ellis JC (2005) Marine birds on land: a review of plant biomass, species richness and community composition in seabird colonies. Plant Ecol 181: 227–241.
20. Schmidt S, Dennison WC, Moss GJ, Stewart GR (2004) Nitrogen ecophysiology of Heron Island, a subtropical coral cay of the Great Barrier Reef, Australia. Funct Plant Biol 31: 517–528.
21. Sánchez-Piñero F, Polis GA (2000) Bottom-up dynamics of alloctonous input: direct and indirect effects of seabirds on islands. Ecology 81: 3117–3132.
22. Murphy RC (1925) Bird islands of Peru. New York and London: GP Putnam's Sons.
23. Nesbit J, C. (1856) On agricultural chemistry, and the nature and properties of Peruvian guano. London and New York: Longman and Co. 128 p.
24. Baumberger T, Affre L, Torre F, Vidal E, Dumas P-J, et al. (2012) Plant community changes as ecological indicator of seabird colonies's impacts on Mediterranean Islands. Ecol Indic 15: 76–84.
25. Mulder CPH, Anderson WB, Towns DR, Bellingham PJ (2011) Seabird islands: Ecology, invasion, and restoration. New York: Oxford University Press.
26. Young HS, McCauley DJ, Dirzo R (2011) Differential responses to guano fertilization among tropical tree species with varying functional traits. Am J Bot 98: 207–214.
27. Holmgren M, López BC, Gutiérrez JR, Squeo FA (2006) Herbivory and plant growth rate determine the success of El Niño Southern Oscillation-driven tree establishment in semiarid South America. Global Change Biol 12: 2263–2271.
28. Illius AW, O'Connor TG (2000) Resource heterogeneity and ungulate population dynamics. Oikos 89: 283–294.
29. Hopcraft JGC, Olff H, Sinclair ARE (2009) Herbivores, resources and risks: alternating regulation along primary environmental gradients in savannas. Trends Ecol Evol 25: 119–128.
30. Mohtadi M, Hebbeln D, Marchant M (2005) Upwelling and productivity along the Peru-Chile Current derived from faunal and isotopic compositions of planktic foraminifera in surface sediments. Mar Geol 216: 107–126.
31. INRENA (1998) Mapa de Bosques Secos del Departamento de Piura. Memoria Descriptiva. Lima: Ministerio De Agricultura, Instituto Nacional de Recursos Naturales, Proyecto Algarrobo y Gobierno de los Países Bajos.
32. Pasiecznik NM, Felker P, Harris PJC, Harsh LN, Cruz G, et al. (2001) The *Prosopis juliflora-Prosopis pallida* complex: A monograph. Coventry, UK: HDRA.
33. Catenazzi A, Donnelly MA (2007) Distribution of geckos in northern Peru: long-term effect of strong ENSO events? J Arid Environ 71: 327–332.
34. Sitters J, Holmgren M, Stoorvogel JJ, López BC (2012) Rainfall-tuned management facilitates dry forest recovery. Restor Ecol 20: 33–42.
35. Squeo FA, Holmgren M, Miagros J, Albán L, Reyes J, et al. (2007) Tree establishment along an ENSO experimental gradient in the Atacama desert. J Veg Sci 18: 195–202.
36. Scheffer M, Van Nes EH, Holmgren M, Hughes T (2008) Pulse-driven loss of top-down control: the critical-rate hypothesis. Ecosystems 11: 226–237.
37. Bernex de Falen N, Reves B (1988) Atlas regional de Piura. Lima: Centro de Investigación y Promoción del Campesinado (CIPCA) y Pontificia Universidad Católica del Perú.
38. Shearer G, Kohl DH, Virginia RA, Bryan BA, Skeeters JL, et al. (1983) Estimates of N_2-fixation from variation in the natural abundance of ^{15}N in Sonoran desert ecosystems. Oecologia 56: 356–373.
39. Perevolotsky A (1991) Goats or scapegoats - the overgrazing controversy in Piura, Peru. Small Ruminant Res 6: 199–215.
40. Rodríguez-Rodríguez EF, Bussmann RW, Arroyo-Alfaro SJ, López-Medina SE, Briceño-Rosario J (2007) *Capparis scabrida* (Capparaceae) a species from Perú and Ecuador in urgent need of conservation plans. Arnaldoa 14: 269–282.
41. Van Leeuwen J, Catenazzi A, Holmgren M (2011) Spatial, ontogenetic and sexual effects on the diet of a Teiid lizard in arid South America. J Herpetol 45: 472–477.
42. Mituzani H, Wada E (1988) Nitrogen and carbon isotope ratios in seabird rookeries and their ecological implications. Ecology 69: 340–349.
43. Young HS, McCauley DJ, Dunbar RB, Dirzo R (2010) Plants cause ecosystem nutrient depletion via interruption of bird-derived spatial subsidies. Proc Natl Acad Sci U S A 107: 2072–2077.
44. Wada E, Kadonaga T, Matsuo S, Shibata R, Torii T (1975) ^{15}N abundance in nitrogen of naturally occurring substances and global assessment of denitrification from isotopic viewpoint. Geochem J 9: 139–148.
45. Therneau T (2012) A Package for survival analysis in S. R package version 2. 36–14.

46. Burger AE (2005) Dispersal and germination of seeds of *Pisonia grandis*, an Indo-Pacific tropical tree associated with insular seabird colonies. J Trop Ecol 21: 263–271.
47. Barrett K, Anderson WB, Wait DA, Grismer LL, Polis GA, et al. (2005) Marine subsidies alter the diet and abundance of insular and coastal lizard populations. Oikos 109: 145–153.
48. Kolb GS, Young HS, Anderson WB (2011) Effects of seabirds on island consumers. In: Mulder CPH, Anderson WB, Towns DR, Bellingham PJ, editors. Seabird islands: Ecology, invasion, and restoration. New York: Oxford University Press. 212–241.
49. Kurdali F, Al-Shamma'a M (2009) Natural abundances of ^{15}N and ^{13}C in leaves of some N_2-fixing and non-N_2 fixing trees and shrubs in Syria. Isotopes Environ Health Stud 45: 198–207.
50. Vitousek PM, Howarth RW (1991) Nitrogen limitation on land and in the sea: how can it occur? Biogeochemistry 13: 87–115.
51. Caut S, Angulo E, Pisanu B, Ruffino L, Faulquier L, et al. (2012) Seabird modulations of isotopic nitrogen on islands. PLoS One 7: e39125.
52. Helfield JM, Naiman RJ (2001) Effects of salmon-derived nitrogen on riparian forest growth and implications for stream productivity. Ecology 82: 2403–2409.
53. Hilderbrand GV, Hanley TA, Robbins CT, Schwartz CC (1999) Role of brown bears (*Ursus arctos*) in the flow of marine nitrogen into a terrestrial ecosystem. Oecologia 121: 546–550.
54. Havik G (2010) The role of seabird nitrogen input and lizard herbivores on the recruitment of dominant tree species in coastal dry forests of north Peru. Master thesis. Wageningen, The Netherhlands: Wageningen University.
55. Britto DT, Kronzucker HJ (2002) NH_4^+ toxicity in higher plants: a critical review. J Plant Physiol 159: 567–584.
56. Vitousek PM, Cassman K, Cleveland C, Crews T, Field CB, et al. (2002) Towards an ecological understanding of biological nitrogen fixation. Biogeochemistry 57: 1–45.
57. Isaac ME, Harmand JM, Lesueur D, Lelon J (2011) Tree age and soil phosphorus conditions influence N-fixation rates and soil N dynamics in natural populations of *Acacia senegal*. Forest Ecol Manage 261: 582–588.
58. Isaac ME, Harmand JM, Drevon JJ (2011) Growth and nitrogen acquisition strategies of *Acacia senegal* seedlings under exponential phosphorus additions. J Plant Physiol 167: 776–781.

14

Nitric Oxide Mediates Root K^+/Na^+ Balance in a Mangrove Plant, *Kandelia obovata*, by Enhancing the Expression of AKT1-Type K^+ Channel and Na^+/H^+ Antiporter under High Salinity

Juan Chen[1], Duan-Ye Xiong[1], Wen-Hua Wang[1], Wen-Jun Hu[1], Martin Simon[1], Qiang Xiao[2], Juan Chen[1], Ting-Wu Liu[3], Xiang Liu[1], Hai-Lei Zheng[1]*

1 Key Laboratory for Subtropical Wetland Ecosystem Research of MOE, College of the Environment and Ecology, Xiamen University, Xiamen, Fujian, China, **2** Laboratory of Biological Resources Protection and Utilization of Hubei Province, Hubei Institutes for Nationalities, Enshi, Hubei, China, **3** Department of Biology, Huaiyin Normal University, Huaian, Jiangsu, P.R. China

Abstract

It is well known that nitric oxide (NO) enhances salt tolerance of glycophytes. However, the effect of NO on modulating ionic balance in halophytes is not very clear. This study focuses on the role of NO in mediating K^+/Na^+ balance in a mangrove species, *Kandelia obovata* Sheue, Liu and Yong. We first analyzed the effects of sodium nitroprusside (SNP), an NO donor, on ion content and ion flux in the roots of *K. obovata* under high salinity. The results showed that 100 μM SNP significantly increased K^+ content and Na^+ efflux, but decreased Na^+ content and K^+ efflux. These effects of NO were reversed by specific NO synthesis inhibitor and scavenger, which confirmed the role of NO in retaining K^+ and reducing Na^+ in *K. obovata* roots. Using western-blot analysis, we found that NO increased the protein expression of plasma membrane (PM) H^+-ATPase and vacuolar Na^+/H^+ antiporter, which were crucial proteins for ionic balance. To further clarify the molecular mechanism of NO-modulated K^+/Na^+ balance, partial cDNA fragments of inward-rectifying K^+ channel, PM Na^+/H^+ antiporter, PM H^+-ATPase, vacuolar Na^+/H^+ antiporter and vacuolar H^+-ATPase subunit c were isolated. Results of quantitative real-time PCR showed that NO increased the relative expression levels of these genes, while this increase was blocked by NO synthesis inhibitors and scavenger. Above results indicate that NO greatly contribute to K^+/Na^+ balance in high salinity-treated *K. obovata* roots, by activating AKT1-type K^+ channel and Na^+/H^+ antiporter, which are the critical components in K^+/Na^+ transport system.

Editor: Shoba Ranganathan, Macquarie University, Australia

Funding: This study was financially supported by the Natural Science Foundation of China (NSFC) (30930076, 31260057, 30770192, 30670317), China Postdoctoral Science Foundation (2012M521278), and the Scholarship Award for Excellent Doctoral Student granted by the Chinese Ministry of Education. The funders had no role in study design, data collection and analysis, decision to publish, or preparation of the manuscript.

Competing Interests: The authors have declared that no competing interests exist.

* E-mail: zhenghl@xmu.edu.cn

These authors contributed equally to this work.

Introduction

Intracellular K^+/Na^+ balance is fundamental to the physiology of living cells and is crucial for plant normal growth [1,2]. Optimal K^+/Na^+ ratio is very important not only for the activities of many cytosolic enzymes, but also for maintaining the ideal osmoticum and membrane potential for cell volume regulation [3]. Nevertheless, high salinity condition disturbs intracellular K^+/Na^+ balance and causes ion toxicity and osmotic stress in plants [3].

In order to maintain the optimal cytosolic K^+/Na^+ balance and avoid the adverse effects of high salinity on plant growth and development, halophytes have developed different strategies to avoid excessive Na^+ accumulation and to maintain osmotic balance in plants. A common strategy involves the transport restriction of excess Na^+ *via* inhibiting non-selective cation channels (NSCCs) in the root cells [4]. Moreover, halophytes can elevate the Na^+ extrusion from the cytosol to external medium and/or Na^+ compartmentation into the vacuoles through trans-membrane transport proteins like plasma membrane (PM)-located Na^+/H^+ antiporter (SOS1) and tonoplast-located Na^+/H^+ antiporter (NHX1) [5,6]. The process of Na^+/H^+ antiporter-mediated Na^+ extrusion and Na^+ compartmentation is energy-dependent, and this energy is supplied by the proton-motive force, which can be generated by H^+-translocating pumps (e.g., H^+-ATPase and H^+-PPiase) [7,8].

As well known, maintaining a optimal K^+/Na^+ ratio in the cytoplasm is more important than simply maintaining a low Na^+ concentration in many plant species under high salinity [2]. Because Na^+ competes with K^+ for uptake into roots [3], NaCl-induced K^+ loss is an important plant response to high salinity [9]. The transcript levels of several K^+ transport-related genes, such as the shaker K^+ channel gene and the high affinity K^+ transport/K^+ uptake transporter-type gene, are either down- or up-regulated by salt treatment, which probably reflects the different capacities of plants to modulate K^+ uptake from the roots [3]. Noticeably, the inward-rectifying potassium channels (AKT1), a major route for

K^+ uptake from external environment by root epidermis, exhibited the high K^+/Na^+ selectivity at physiological K^+ and Na^+ concentrations [10]. Since the first *AKT1* from *Arabidopsis* was cloned in 1992 [11], *AKT1* genes have been identified in many other species, such as *TaAKT1* from *Triticum aestivum* [12], *OsAKT1* from *Oryza sative* [13], *CaAKT1* from *Capsicum annuum* [14] and *PutAKT1* from *Puccinellia tenuiora* [15]. Besides regulating K^+ uptake, AKT1 is also indirectly involved in SOS pathway [3]. Qi and Spalding [16] found that the defect in Na^+ efflux in the *sos* mutant could lead to excessive Na^+ in the cytoplasm that was inhibitory to AKT1, resulting in poor growth due to the impaired K^+ uptake. Mutant analyses showed that *akt1* mutant was sensitive to salt during early seedling development, indicating that AKT1 played a critical role in maintaining cytoplasmic K^+/Na^+ balance in salt-treated plants [17].

Nitric oxide (NO), an important signaling molecule, plays a critical role in wide range of physiological and developmental processes in plants including root formation, seed germination, stomatal closure, pollen tube growth and flowering [18]. Moreover, NO has been demonstrated to be involved in mediating the responses to biotic and abiotic stresses in plants, such as drought, salt, heat stress and disease resistance [19]. It was reported that exogenous NO significantly enhanced salt tolerance in maize seedlings through increasing the activities of H^+-ATPase and Na^+/H^+ antiporter in the tonoplast [20]. NO was also found to serve as a signal for inducing salt resistance by decreasing Na^+ content in reed callus [21]. Until now, studies were more focused on the effects of NO on avoiding excess Na^+ accumulation in cytoplast [6,20,21]. However, the precise mechanism of K^+ uptake and cytosolic K^+/Na^+ balance modulated by NO is not very clear yet, and the possible pathways for NO signaling that can regulate the expressions and activities of ion transporters to maintain a optimal K^+/Na^+ ratio in cytoplasm have not been investigated.

Kandelia obovata Sheue, Liu and Yong is the dominant mangrove species in East Asia and is mostly distributed in the regions of estuaries, deltas and riverbanks or along the coastlines of tropical and subtropical areas [22]. In contrast to some mangrove species with special morphological structure, such as salt glands and salt hairs, *K. obovata* is a typical non-secretor mangrove species [22]. *K. obovata* evolved many mechanisms to resist high salinity such as ultrafiltration by roots, osmotic adjustment and regulation of reactive oxygen species [17]. The most striking feature of *K. obovata* is the maintenance of cell turgor by accumulating inorganic ions in the vacuole and organic compounds in the cytoplasm [17]. A previous study found that *K. obovata* could exclude 99% of the salt in surrounding seawater by ultrafiltration [23]. Li et al [24] reported that *K. obovata* exhibited a higher capacity to restrict Na^+ uptake and accumulation under increasing salt treatments. However, the existing studies were mostly done at physiological and biochemical levels, the molecular mechanisms of salt tolerance in mangrove plants are still not well known. Recently, Lu et al [25] found that NO contributed to maintain ionic homeostasis by increasing Na^+ efflux and inhibiting K^+ loss in salt-treated *K. obovata* roots. Our previous study also found that NO could enhance salt secretion from salt glands and promote Na^+ sequestration into vacuole of another mangrove plant, *Avicennia marina* [6]. Although increasing evidence suggests that NO functions as an important intermediate for regulating ionic homeostasis under salt stress, further investigations are needed to clarify the precise molecular mechanism by which NO regulates K^+ uptake and K^+/Na^+ balance in the roots of mangrove plants.

In the light of the above context, in this study, we first characterized the effects of various concentrations of sodium nitroprusside (SNP, an NO donor) on K^+ and Na^+ content, K^+/Na^+ ratio, and net K^+ and Na^+ fluxes in NaCl-treated *K. obovata* roots using inductively coupled plasma mass spectrometry (ICP-MS), X-ray microanalysis and the scanning ion-selective electrode technique (SIET). Secondly, due to the fact that Na^+ extrusion from the cytoplast depend on PM H^+-ATPase as well as Na^+/H^+ antiporter, we investigated the protein expressions of H^+-ATPase and Na^+/H^+ antiporter using western blot analysis. Moreover, to further clarify the precise molecular mechanism by which NO regulates K^+ uptake and K^+/Na^+ balance in *K. obovata*, some critical genes related to K^+/Na^+ transport system (e.g., *AKT1*, *SOS1*, *NHX1*, etc.) were isolated and the transcriptional expressions of these genes were analyzed using quantitative real-time PCR (qRT-PCR). Based on above experiments, we attempted to clarify the detailed mechanism of NO in modulating cytosolic K^+/Na^+ balance, which might be closely related to the activity and expression of K^+/Na^+ transport system in salt-treated *K. obovata* roots.

Results

Effects of NO on element ratios and distribution in *K. obovata* roots

After *K. obovata* seedlings were treated with 400 mM NaCl and various concentrations of SNP for 15 days, the appropriate concentration of exogenous NO (100 μM SNP) significantly increased K^+ content and decreased Na^+ content, resulting in the increased K^+/Na^+ ratio compared with the solely NaCl-treated seedlings (Figures 1A, C and E). According to this result, 100 μM SNP, which could provide an optimal NO concentration for maintaining ionic balance in *K. obovata* seedlings, was used in the following experiments. To attribute the role of NO in enhancing K^+/Na^+ ratio in *K. obovata* roots exposed to high salinity, specific NO synthesis inhibitor (L-NNA) and NO scavenger (cPTIO) were used in the next experiments. As shown in Figures 1B, D and F, both L-NNA and cPTIO abolished the effects of NO and caused a decrease in K^+ content and K^+/Na^+ ratio, but an increase in Na^+ content in salt-treated *K. obovata* roots. Moreover, the results of X-ray microanalysis showed that 100 μM SNP significantly increased K^+ percentage, meanwhile decreased Na^+ percentage, resulting in the higher K^+/Na^+ ratio in both cortex and stele cells of *K. obovata* roots (Table 1). Remarkably, K^+/Na^+ ratio in stele cells was markedly higher than that in the cortex under various SNP concentrations (Table 1).

Effects of NO, specific inhibitor and NO scavenger on K^+ and Na^+ fluxes in *K. obovata* roots

As shown in Figures 2A and C, the control plants grown in Hoagland solution alone exhibited the slightly inward K^+ flux and outward Na^+ flux in *K. obovata* roots, although the flux rates oscillated during the period of recording. Compared with the control plants, 100 μM SNP did not significantly change K^+ and Na^+ flux profiles in *K. obovata* roots under no-salt condition (Figure 2). However, after *K. obovata* seedlings were treated with 400 mM NaCl for 15 days, the rectifications of K^+ flux in the root tips exhibited the evident efflux and reached the maximal mean K^+ efflux of 58 pmol $cm^{-2}s^{-1}$ (Figure 2B). It is noteworthy that the K^+ efflux in *K. obovata* roots under 100 μM SNP and 400 mM NaCl treatment for 15 days were inhibited, which was about 72% lower than that under solely NaCl-treated plants (Figures 2A and B).In contrast to the results of net K^+ flux, net Na^+ efflux in NaCl-treated *K. obovata* roots significantly enhanced and reached a mean rate of 737 pmol $cm^{-2}s^{-1}$, which was about 5-fold higher than that of the control (Figures 2C and D). After the plants were treated with 100 μM SNP and 400 mM NaCl for 15 days, net Na^+ efflux

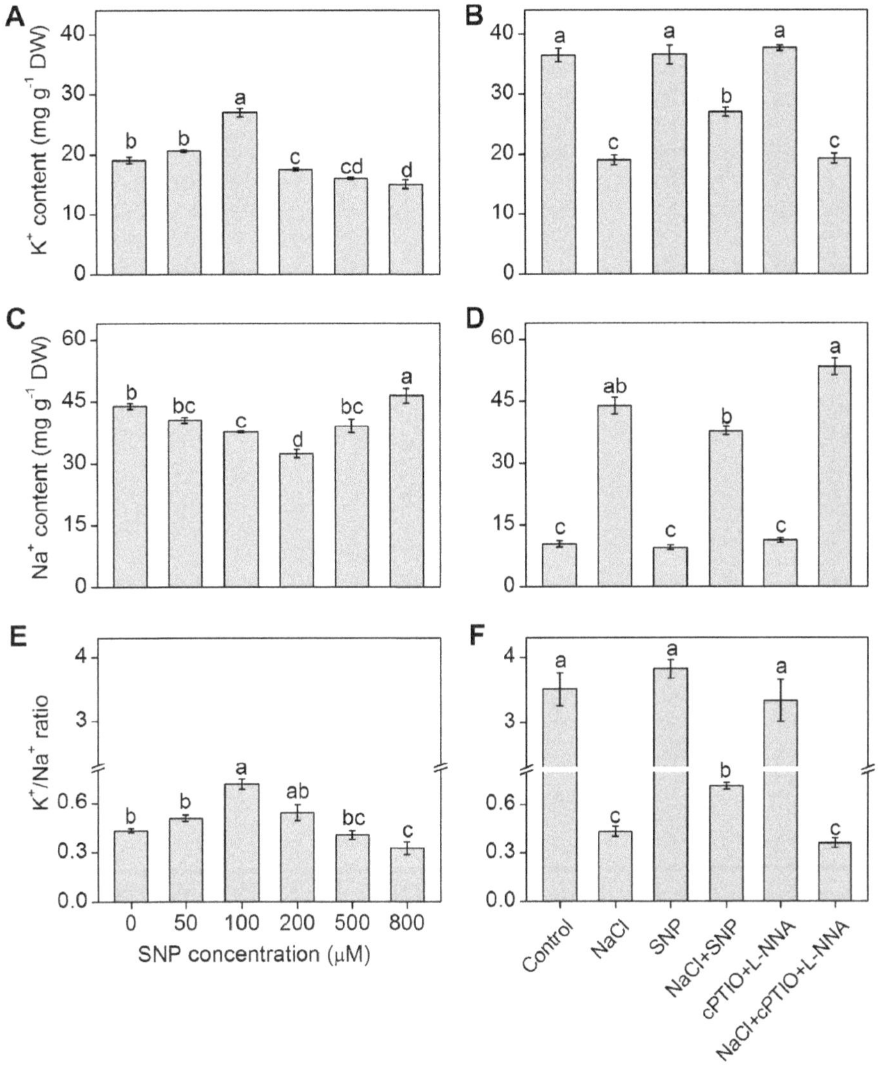

Figure 1. Effects of NO on K^+, Na^+ content and K^+/Na^+ ratio in *K. obovata* roots. K^+ content (A), Na^+ content (C) and K^+/Na^+ ratio (E) in *K. obovata* roots after 15 days of 400 mM NaCl and different concentrations of SNP treatment. Shown in (B), (D) and (F), respectively, show K^+ content, Na^+ content and K^+/Na^+ ratio in roots of *K. obovata* seedlings treated for 15 days with the following Hoagland solution-based treatments: Control (Hoagland solution), NaCl (400 mM), SNP (100 μM), NaCl + SNP (400 mM NaCl+100 μM SNP), cPTIO+L-NNA (200 μM cPTIO+100 μM L-NNA), NaCl+cPTIO+L-NNA (400 mM NaCl+200 μM cPTIO+100 μM L-NNA). Each column is the mean value of at least three independent seedlings and bars represent the SE of the mean. For each parameter, bars with different letters indicate significantly different at the 0.05 level.

in *K. obovata* roots increased by a maximal mean rate of 1300 pmol $cm^{-2}s^{-1}$ (about 10-fold higher than that with control condition) (Figures 2C and D).

Due to the fact that intracellular ionic balance is closely related to the activity of H^+-ATPase and Na^+/H^+ antiporter, H^+-ATPase inhibitor (vanadate) and Na^+/H^+ antiporter inhibitor (amiloride) were used for clarifying the critical role of H^+-ATPase and Na^+/H^+ antiporter in regulating K^+ and Na^+ fluxes. Figures 3A and C show that compared with the mean value of Na^+ flux before inhibitor addition, transient application of both 100 μM amiloride and 500 μM vanadate increased net K^+ efflux in the root tips of *K. obovata* seedlings under 400 mM NaCl treatment for 15 days. Unlike K^+ efflux, the addition of specific inhibitors decreased net Na^+ efflux (Figures 3B and D). Additionally, cPTIO, a specific NO scavenger, was applied to test the effects of NO on K^+ and Na^+ fluxes in NaCl-treated *K. obovata* roots. Similar to the patterns of ion flux affected by amiloride and vanadate, cPTIO also accelerated K^+ efflux and inhibited Na^+ efflux, as compared to the mean value of ion flux before cPTIO addition (Figures 3E and F).

Effects of NO on the protein expression of H^+-ATPase and Na^+/H^+ antiporter

Western-blot analysis exhibited that the protein expression level of H^+-ATPase and Na^+/H^+ antiporter reached the maximum value in NaCl-treated *K. obovata* roots in the presence of 100 μM SNP (Figures 4A and C). However, compared with NaCl treatment, the increased protein expression of H^+-ATPase and Na^+/H^+ antiporter induced by SNP was completely reversed by cPTIO and L-NNA addition (Figures 4B and D).

Effects of NO on transcriptional expression of several K^+/Na^+ balance related genes

The qRT-PCR analysis showed that 100 μM SNP remarkably up-regulated the relative expression levels of *AKT1* gene, which

Table 1. Effects of NO on the percentage of K^+, Na^+ and K^+/Na^+ ratio in cortex and stele section of *K. obovata* roots under 400 mM NaCl treatment for 15 days.

Measuring region		SNP concentration (μM)		
		0	100	800
	K^+	20.2±2.3[a]	21.9±1.2[a]	15.7±3.1[b]
Cortex	Na^+	30.5±1.7[ab]	25.0±4.2[b]	42.3±2.6[a]
	K^+/Na^+	0.7±0.1[b]	0.9±0.1[c]	0.4±0.1[a]
	K^+	23.1±2.2[c]	40.1±4.5[a]	31.3±4.7[b]
Stele	Na^+	32.3±1.6[b]	22.8±1.7[c]	42.0±3.5[a]
	K^+/Na^+	0.7±0.1[a]	1.8±0.1[c]	0.8±0.1[b]

The ratio of elements in roots was measured using the X-ray microanalysis. The results were calculated by expressing the percentage of atomic number for a particular element (K^+ or Na^+) in the total atomic number for all the elements (K^+, Na^+, Ca^{2+}, Mg^{2+}, Al^{3+} and Mn^{2+}) detected in the given region of the root samples. Each data are the means of 5–8 measurements. Values followed by the same letter in a line are not significantly different ($P>0.05$) as described by one-way ANOVA.

was about 4-fold higher than those of the control (0 μM SNP) (Figure 5A). Similarly, the relative expression levels of *SOS1*, *NHX1*, *HA1* and *VHA-c1* gene in 100 μM SNP-treated plants were also significantly higher than those in other treatments. In agreement with this, NO synthesis inhibitor (L-NNA) and NO scavenger (cPTIO) down-regulated the expression levels of *AKT1*, *SOS1*, *NHX1*, *HA1* and *VHA-c1* gene and reversed the effects of NO (Figure 5).

Discussion

NO modulates K^+/Na^+ balance by decreasing K^+ loss and increasing the expression of AKT1

There is no doubt that K^+ plays a vital role in salt adaptation of plants [3]. Under normal physiological conditions, the cytosolic K^+ is maintained at a steady state level [26]. However, under high salinity condition, due to physicochemical similarities between K^+ and Na^+, excess Na^+ tends to substitute K^+ and hence impairs cellular biochemistry [27]. Previous study reported that a strong correlation existed between the ability of plants to retain K^+ and salt tolerance [9]. Carden et al [28] also found that the barley variety with strong salt tolerance could better maintain root cytosolic K^+ under high-NaCl treatment. It should be noted that maintaining the steady K^+ level is also critical in halophytes [29]. For instance, Hariadi et al [29] found that shoot sap K^+ content increased along with the increased salt content in *Chenopodium quinoa* leaves. Lu et al [25] reported that salinity caused an evident K^+ efflux in the roots of two mangrove species, *Bruguiera gymnorrhiza* and *K. obovata*, which was similar to our study showing that K^+ efflux remarkably increased in NaCl-treated *K. obovata* roots (Figures 2C and D), indicating that K^+ plays an important role in osmotic adjustment under high salinity. Thus, minimizing NaCl-induced K^+ loss appears to be critical for enhancing salt tolerance of plants. NO, as a signaling molecule, is involved in the resistant response of plants to high salinity [19]. Zhao et al [21] found that NO enhanced salt resistance by increasing K^+/Na^+ ratio in the reed callus. Consistent with the previous study, application of SNP at medium concentrations (100 μM) significantly increased K^+ content and K^+/Na^+ ratio (Figures 1A and E; Table 1), while NO synthesis inhibitor and NO scavenger reversed the effects of NO, indicating that NO plays an important role in maintaining K^+ homeostasis in high salinity-treated *K. obovata* roots.

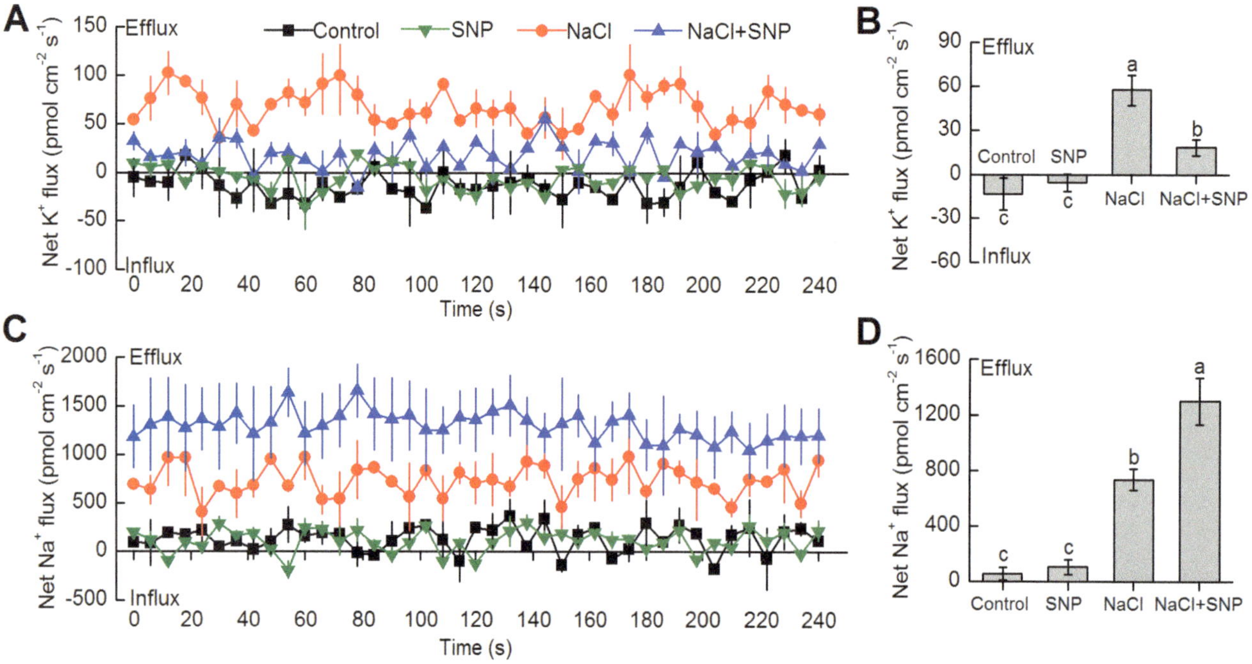

Figure 2. Effects of NaCl and SNP on K^+ and Na^+ fluxes in *K. obovata* roots. After 15 days of exposure to NaCl (400 mM), SNP (100 μM) and NaCl (400 mM)+SNP (100 μM), the net K^+ (A) and Na^+ (C) fluxes from the mature root zone (20 mm from the tip) of *K. obovata* seedlings. The positive value in the figures represents the net influx and negative value represents the net efflux. Each point represents the mean value of four individual roots and bars represent the SE of the mean. The mean fluxes of K^+ (B) and Na^+ (D) within the measuring periods are shown. Columns labeled with different letters indicates significantly different at $P<0.05$.

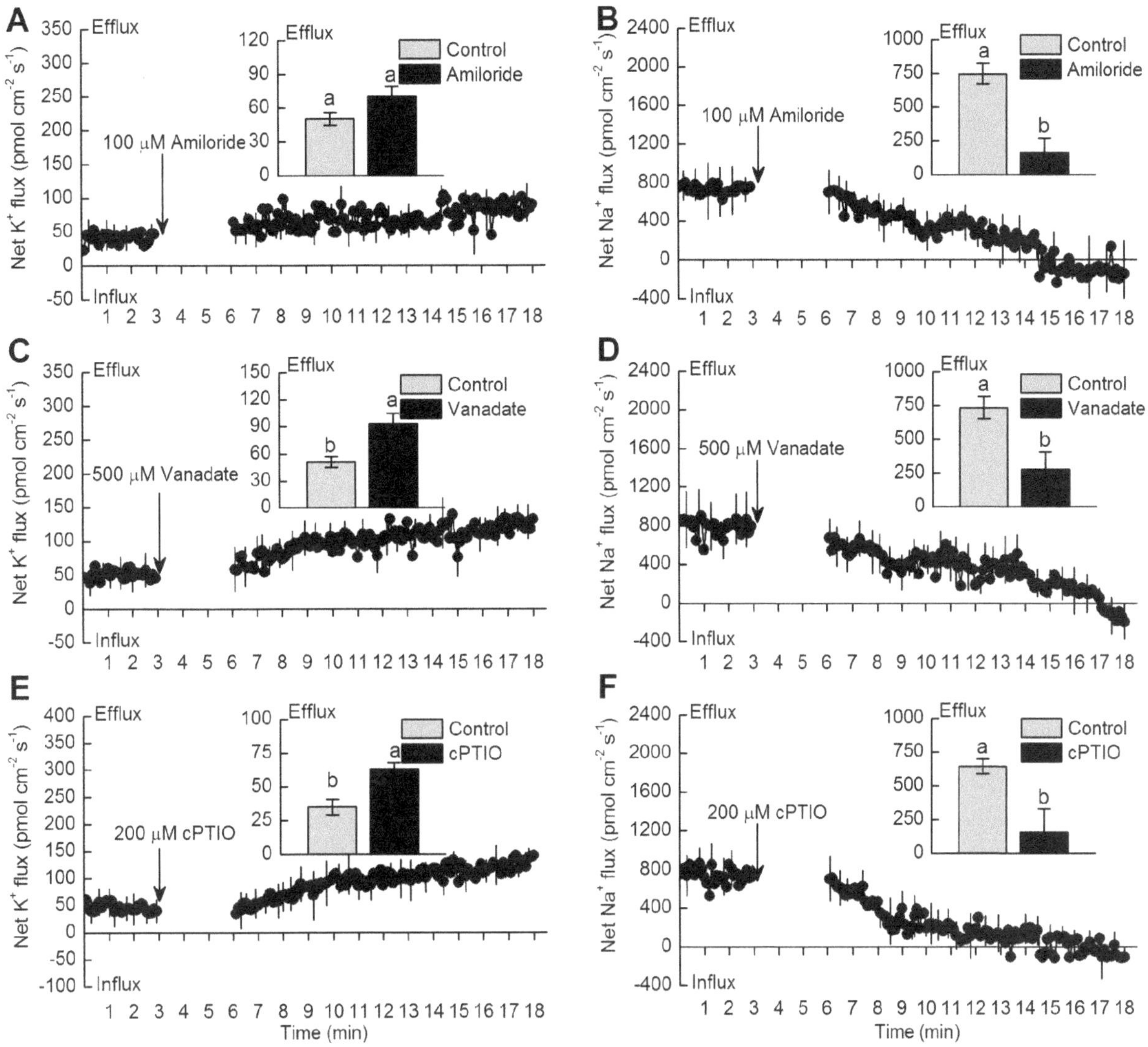

Figure 3. Changes in net K^+ and Na^+ fluxes after specific inhibitors and NO scavenger addition. Effects of specific Na^+/H^+ antiporter inhibitor (100 μM amiloride, A and B), H^+-ATPase inhibitor (500 μM vanadate, C and D) and NO scavenger (200 μM cPTIO, E and F) on the net K^+ and Na^+ fluxes in the mature root zone (20 mm from the tip) of *K. obovata* seedlings with 400 mM NaCl treatment for 15 days. The positive value in the figures represents the net influx and negative value represents the net efflux. The mean K^+ or Na^+ flux within the measuring periods is shown in the inserted section. Each point represents the mean value of four individual roots and bars indicate the SE of the mean. Before the addition of specific inhibitor or NO scavenger, steady K^+ or Na^+ flux was measured for about 3 min. Column labeled with different letters are significantly different at $P<0.05$.

It is widely established that under high salinity condition, the PM is strongly depolarized, which causes the remarkable K^+ loss *via* depolarization-activated (DA) channels, e.g., NSCCs and KORCs (outward rectifying K^+ channels) in some plants (e.g., barley, *Arabidopsis*, etc.) [1]. The higher H^+-ATPase activity can lead to hyperpolarization or repolarization of the PM, thus limit the K^+ efflux through DA-NSCC and DA-KORC [30]. In this study, salt shock-induced K^+ efflux in *K. obovata* roots was markedly enhanced by vanadate, an inhibitor of PM H^+-ATPase (Figure 3C). This finding is similar to recent reports indicating that the capacity to maintain low K^+ efflux and retain K^+ is highly dependent on an active PM H^+-ATPase in *Populus euphratica* and *K. obovata* [8,25]. It should be noted that NO could increase K^+ percentage and K^+/Na^+ ratio in NaCl-treated *Arabidopsis*, which was dependent on the increased PM H^+-ATPase activity [21]. For *K. obovata* under salt condition, previous study found that the decreased K^+ efflux induced by NO likely accounted for the less depolarized membrane potential maintained by high activity of PM H^+-ATPase [25]. Similarly, NaCl-induced transient K^+ efflux in the root tips of *K. obovata* was remarkably inhibited by PM H^+-ATPase inhibitor (Figure 3C), the protein and gene expressions of PM H^+-ATPase were significantly up-regulated *K. obovata* roots due to NO application in this study (Figures 4 and 5). These results suggested that the increased H^+-ATPase activity, which was induced by NO, might be necessary for decreasing the salt-induced K^+ efflux and for restricting K^+ loss in *K. obovata* grown under high salinity [1,8].

AKT1, the most important inward-rectifying shaker K^+ channel, whose gene is mainly expressed in root epidermal cells, is involved in K^+ acquisition by plants [15]. It's known that this type of channel plays an important role in K^+ uptake from the soil solution and greatly attributes to maintain and restore K^+ homeostasis in high salinity-treated plants [31]. The expression

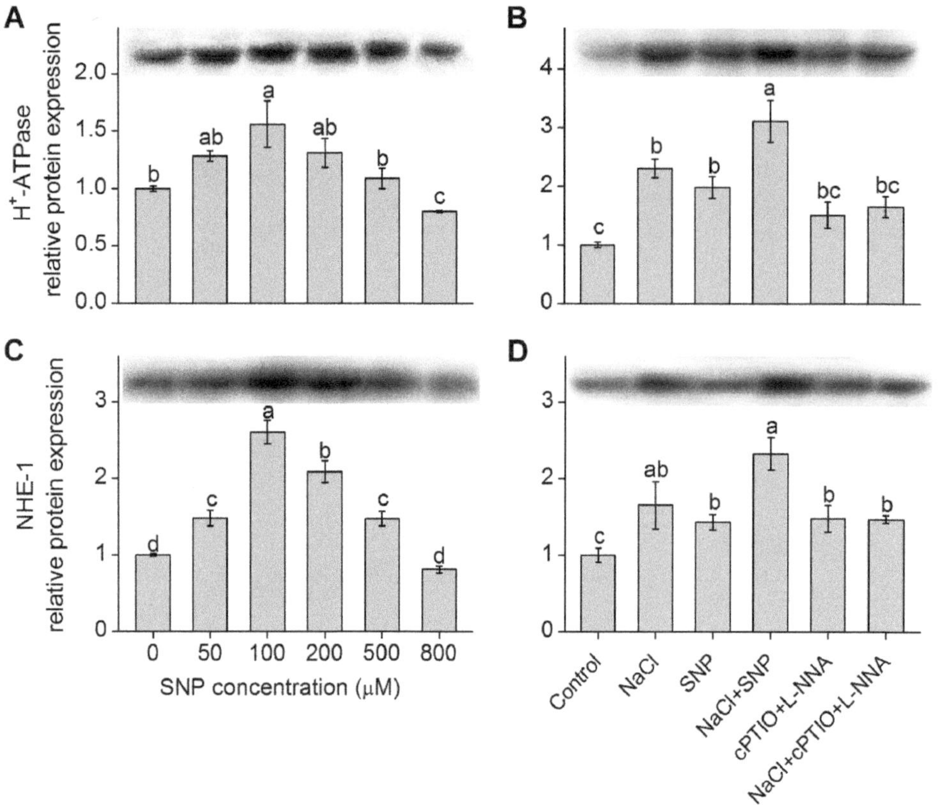

Figure 4. Western blot analysis of protein expression of plasma membrane (PM) H^+-ATPase and Na^+/H^+ antiporter (NHE-1). Shown in (A) and (C) are the relative protein expression level of *K. obovata* seedlings grown under 400 mM NaCl and various concentrations of SNP for 15 days. Shown in (B) and (D) represent the relative protein expression level of *K. obovata* seedlings treated with the following Hoagland solution-based solutions for 15 days: Control (only Hoagland solution), NaCl (400 mM), SNP (100 μM), NaCl+SNP (400 mM NaCl+100 μM SNP), cPTIO+L-NNA (200 μM cPTIO+100 μM L-NNA), NaCl+cPTIO+L-NNA (400 mM NaCl+200 μM cPTIO+100 μM L-NNA). Three independent experiments were performed. The relative intensity of bands (column charts) was quantified by densitometric analysis with Quantity One software and expressed relative to the intensity of the control band defined as 1. For each parameter, bars with different letters indicate significant difference at $P<0.05$.

of *AKT1* gene showed different responses to salinity in plants with different salt tolerance. Fuchs et al [32] reported that salt stress strongly down-regulated the transcriptional expression of *OsAKT1* gene in *Oryza sativa*, and a similar response was also found in *Arabidopsis* [33]. In contrast, the expression levels of the AKT1-type K^+ channel gene from *Puccinellia tenuiflora*, *PutAKT1*, seem to be unaffected by excessive external NaCl, and overexpression of *PutAKT1* significantly enhanced the resistence to high salinity in transgenic *Arabidopsis* [15]. Interestingly, previous study found that the *Arabidopsis sos* mutant had a growth defect under K^+-limiting conditions [3]. They speculated that the decreased Na^+ efflux in the *sos* mutant may result in high Na^+ accumulation in cytoplast, which may indirectly inhibit the activity of K^+ transporter such as AKT1 and thus regulate K^+ uptake in the roots [3]. In the present work, the transcriptional expression of *AKT1* was up-regulated in roots of NaCl-treated *K. obovata* (Figure 5B). The likely reason is that glycophytes (e.g., *Arabidopsis* and *O. sativa*) mainly grows in the soils with low salinity, so their mechanisms developed for K^+ uptake and transport may not operate as efficiently as in halophytes under high salinity condition. Most importantly, we found that SNP significantly up-regulated the transcriptional expression of *AKT1* gene in the roots of NaCl-treated *K. obovata* (Figures 5A and B), indicating that NO might play an important role in increasing K^+ uptake by inducing the expression of *AKT1* gene in salt-treated *K. obovata* roots. AKT also plays an important role in K^+ circulation through the xylem and phloem [34]. In this study, SNP significantly increased K^+ percentage in the stele cells (Table 1), indicating that NO-modulated K^+ accumulation and distribution in the roots may be related to the activity and expression of AKT1-type K^+ channel.

NO modulates K^+/Na^+ balance by increasing Na^+ extrusion

Maintaining a low Na^+ content in the cytoplasm is a key factor for cellular adaptation to high salinity condition [35]. Zhao et al [21] reported that NO served as a signal that induced salt resistance by decreasing Na^+ content and increasing K^+/Na^+ ratio in reed callus, which is dependent on the increased PM H^+-ATPase activity. In this study, we also confirmed that NO could decrease intracellular Na^+ accumulations and K^+/Na^+ ratio, whereas the treatment with NO synthesis inhibitor (L-NNA) and specific scavenger (cPTIO) completely blocked the effects of NO on element ratio (Figure 1 and Table 1), which were in accordance with previous results from Zhao et al [21]. SIET data showed that 100 μM SNP significantly enhanced net Na^+ efflux in salt-treated *K. obovata* roots (Figure 2A), while the inhibitors of PM Na^+/H^+ antiporter, H^+-ATPase and NO scavenger significantly decreased net Na^+ efflux (Figure 3A, C and E). These findings are similar to previous results showing that Na^+ efflux were enhanced by NO in *B. gymnorrhiza* and *K. obovata* through regulation of Na^+/H^+ antiporter and H^+-ATPase in the PM [25]. It is clear that NO functions in maintaining lower Na^+ content in *K. obovata* roots by

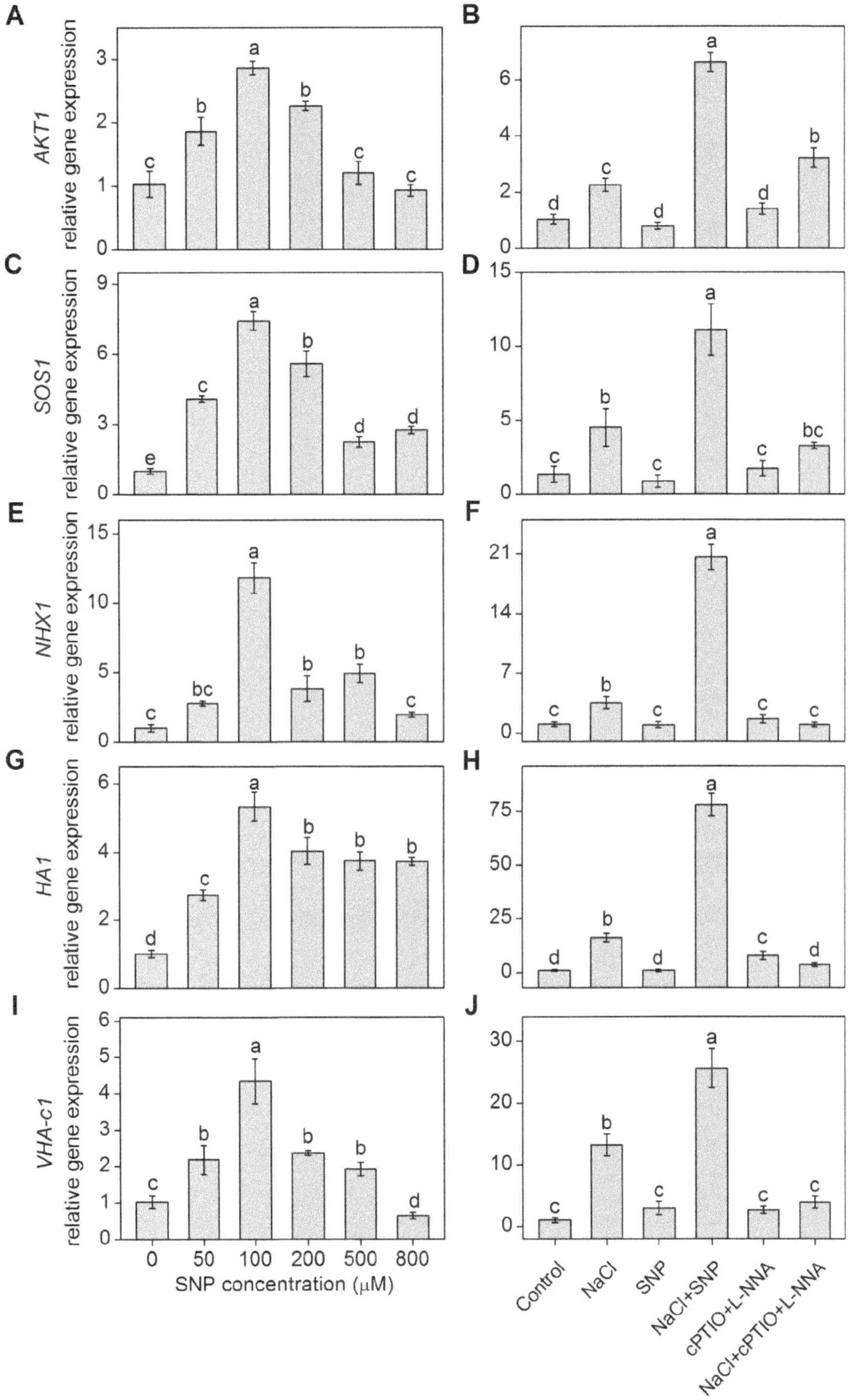

Figure 5. Effects of NO on transcriptional expression of several K^+/Na^+ balance related genes. Expressions of *AKT1* (A), *SOS1* (C), *NHX1* (E), *HA1* (G) and *VHA-c1* (I) genes in *K. obovata* roots in response to 400 mM NaCl and various concentrations of SNP for 15 days. Shown in B, D, F, H and J, respectively, stand for the gene expressions of *AKT1*, *SOS1*, *NHX1*, *HA1*, *VHA-c1* in the roots of *K. obovata* seedlings grown under the following Hoagland solution-based solutions for 15 days: Control (only Hoagland solution), NaCl (400 mM), SNP (100 μM), NaCl + SNP (400 mM NaCl+100 μM SNP), cPTIO+L-NNA (200 μM cPTIO+100 μM L-NNA), NaCl+cPTIO+L-NNA (400 mM NaCl+200 μM cPTIO+100 μM L-NNA). The relative transcriptional expression of each gene was performed by normalization with *18S* rRNA, a reference gene of *K. obovata*. Mean values and SE were calculated from three independent experiments. For each parameter, bars with different letters indicate significant difference at $P<0.05$.

increasing Na^+ extrusion to the external environment. However, NO is thought to be a "double-edged sword", because it is a reactive nitrogen species and readily reacts with the superoxide anion-radical (O_2^-). The formation of peroxynitrite ion ($ONOO^-$) could cause serious damage to cell structures [36]. Therefore, the protection or toxicity effect of NO depends on its concentration

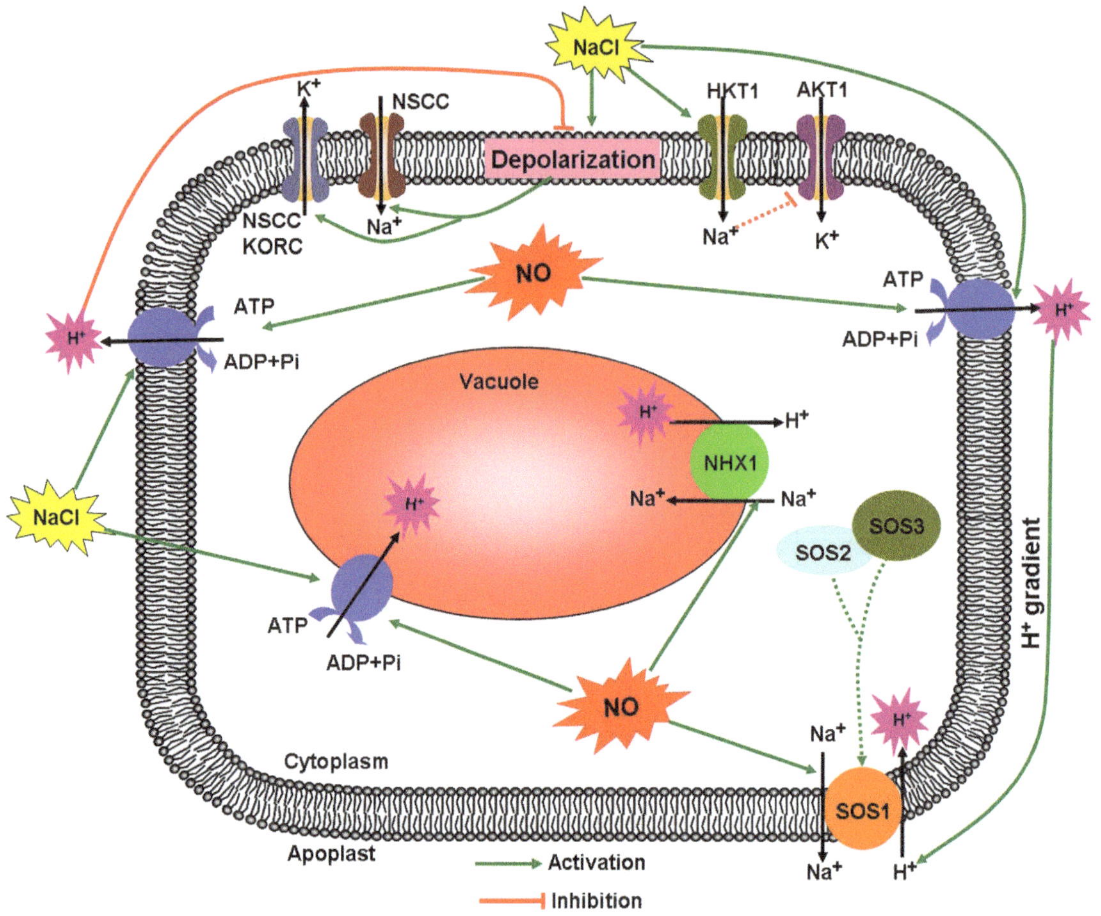

Figure 6. A schematic model for NO-modulated cellular K$^+$/Na$^+$ balance in *K. obovata* under high salinity condition. The arrows in solid lines denote the identified regulation. The hypothetical regulation is indicated by broken lines ending in t-bars.

and the plants developmental stages [6,20]. For instance, high SNP concentration (800 μM) induced the obvious increase in K$^+$/Na$^+$ ratio (Figure 1E), which may exaggerate the high salinity-induced damage in *K. obovata*.

At the cellular level, in order to maintain ion homeostasis, PM Na$^+$/H$^+$ antiporter, which is driven by PM H$^+$-ATPase, is important for removing excess Na$^+$ from the cytoplasm to the apoplast [3,37]. Recent research reported that up-regulation of PM H$^+$-ATPase activity facilitated Na$^+$ efux into the apoplast [20,21], contributing to alleviation of Na$^+$ toxicity in plants under salt stress [38]. The PM Na$^+$/H$^+$ antiporter encoded by *SOS1* has been suggested to be related to the long distance Na$^+$ transport and to Na$^+$ extrusion from meristem zone of the roots to external medium under high salinity [3,37,39]. Shi et al [37] confirmed that the *sos1* mutants are extremely sensitive to high concentration of external Na$^+$, and over-expression of Na$^+$/H$^+$ antiporter gene (*AtSOS1*) markedly improved salt tolerance in *Arabidopsis*. Importantly, previous studies have reported that NO alleviated salt toxicity by increasing the activities of H$^+$-ATPase [20,21]. Our previous results also suggested that the salt secretion in the salt glands of *A. marina*, which was enhanced by NO, was closely related to the increased expression levels of *HA1* and *SOS1* genes [6]. Similarly, in the present study, 100 μM SNP remarkably increased the expression of Na$^+$/H$^+$ antiporter and H$^+$-ATPase at both protein and gene levels in high salinity-treated *K. obovata* roots (Figures 4 and 5), which may greatly contribute to maintain lower Na$^+$ level in the cytoplast.

Moreover, Na$^+$ compartmentation is another efficient strategy for plant cells to cope with high salinity conditions [40]. The tonoplast Na$^+$/H$^+$ antiporter (NHX) and H$^+$-ATPase (VHA-c) function in mediating Na$^+$ sequestration into the vacuoles [41]. In accordance with the results of our previous study on *A. marina* [6], this study found that SNP significantly up-regulated the transcriptional level of *NHX1* and *VHA-c1* genes in salt-treated *K. obovata* roots (Figures 4 and 5). This result implies that NO might contribute to modulation of Na$^+$ compartmentation into the vacuoles, the detailed mechanism for this role of NO deserves further attention.

Pathway of NO-regulated K$^+$/Na$^+$ balance in salt-treated *K. obovata* root

Based on our results and current knowledge about the mechanisms of salt tolerance in plants, we propose a signaling pathway by which NO mediates K$^+$/Na$^+$ balance in salt-treated *K. obovata*. As shown in Figure 6, high salinity strongly depolarizes the PM and causes excessive Na$^+$ influx to the cytoplasm *via* NSCCs, simultaneously enhances K$^+$ efflux through depolarization-activated channels such as KORCs and NSCCs. As a result, the excessive Na$^+$ restricts K$^+$ absorption by competing with the site of K$^+$ uptake. Compared with no-salt condition, both high salinity and NO can enhance H$^+$ extrusion by activating PM H$^+$-ATPase.

The roles of up-regulated PM H^+-ATPase in regulating ionic homeostasis are manifold (Figure 6). Firstly, the increased H^+ extrusion by H^+-ATPase contributes to inhibit salt-induced PM depolarization, thus restricting K^+ efflux by KORCs and NSCCs [4]. Secondly, the enhanced H^+ pumping can sustain H^+ gradient, which is necessary for Na^+ extrusion out of the cytosol by Na^+/H^+ antiporter (SOS1) across the PM [42]. In addition, NO can also directly activate PM Na^+/H^+ antiport system and contribute to Na^+ efflux. Interestingly, the intracellular Na^+ participated in the potential negative regulation of AKT1 activity, implying that the lower cytoplasmic Na^+ level induced by NO may indirectly enhance K^+ uptake by some PM K^+ transporters such as AKT1, which can further increase the cytoplasmic K^+ level and maintain the K^+/Na^+ balance in salt-treated *K. obovata*.

In conclusion, this study provides the evidence that NO functions in decreasing the cytosolic K^+ loss and increasing the ability of K^+ uptake in salt-treated *K. obovata* through increasing the expression of PM H^+-ATPase and AKT1. Moreover, NO can also increase the protein and gene expression of H^+-ATPase and Na^+/H^+ antiporter, which further enhance Na^+ extrusion in salt-treated *K. obovata* roots. It is clear that NO modulates K^+/Na^+ balance in cytoplasm and further enhances salt tolerance of *K. obovata* through increasing the expression of some critical components in K^+/Na^+ transport system including AKT1-type K^+ channel, H^+-ATPase and Na^+/H^+ antiporter.

Materials and Methods

Plant materials and culture conditions

In May 2011, the mature propagules of *K. obovata* were collected from Mangrove Nature Reserve at Jiulongjiang River (latitude 24°24′N and longitude 117°23′E) in Fujian Province, China. Four healthy propagules of similar size were randomly planted in individual pots (15 cm in diameter and 18 cm in height) containing clean sand and were cultured in a greenhouse. The photosynthetically active radiation (PAR) was 800–1000 μmol m^{-2} s^{-1}, relative humidity was maintained at 60–70%, and daily temperature was 25–28°C. Seedlings were irrigated with tap water every 3 days and the pots were fertilized with a full-strength Hoagland solution biweekly as described by Chen et al [6]. NaCl and SNP treatments were started when the 4th pair of leaves came out from the apex of the growing shoot.

Treatments

Seedlings of uniform size were chosen and divided into two groups. As described by Kao et al [43] and Li et al [24], 400 mM NaCl, which significantly elevated Na^+ contents in the root and shoot tissues and inhibited the photosynthesis and growth of *K. obovata* seedlings, was considered as a high salinity for *K. obovata*. Therefore, 400 mM NaCl was used for salinity treatment in this study. In order to select a suitable concentration of SNP to use in the following experiments, different amounts of SNP (0, 50, 100, 200, 500, 800 μM) were added to the Hoagland solution containing 400 mM NaCl in the first group. NO synthesis inhibitor (*N*-nitro-L-arginine, L-NNA) and NO scavenger (2-(4-carboxyphenyl)-4,4,5,5- tetramethylimidazoline-1-oxyl-3-oxide, cPTIO) were used in the second group to clarify the role of NO. The second group consisted of a control (only Hoagland solution without NaCl and SNP) and five Hoagland solution-based treatments as follows: (1) NaCl, 400 mM NaCl; (2) SNP, 100 μM SNP; (3) NaCl+SNP, 400 mM NaCl+100 μM SNP; (4) cPTIO+L-NNA, 200 μM cPTIO+100 μM L-NNA; (5) NaCl+cPTIO+L-NNA, 400 mM NaCl+200 μM cPTIO+100 μM L-NNA. The treatment solutions were renewed three times a week to maintain the constant concentrations. After various treatments for 15 days, some fresh roots of *K. obovata* were collected and immediately used for K^+ and Na^+ content analysis, X-ray microanalysis and net K^+ and Na^+ efflux measurements. Other plant materials were immediately frozen in liquid nitrogen and stored at −80°C for further analyses such as western-blot and qRT-PCR.

Measurement of Na^+ and K^+ contents

The roots of *K. obovata* were collected and oven-dried for 48 h at 75°C. The dried samples (200 mg) were ashed for 12 h at 450°C in the muffle furnace and then dissolved in 1 ml HNO_3. After digestion, the acid solution was diluted to an appropriate volume with the deionized water. K^+ and Na^+ concentrations in the solution were determined by inductively coupled plasma mass spectrometry (ICP-MS, Perkin Elmer, Inc., Elan DRC-e).

X-ray microanalysis

Fresh roots of *K. obovata* were cut into 2–4 mm segments from tip with a razor blade and fixed immediately with 2.5% (w/v) glutaraldehyde for 5 h at 4°C. The samples were then rinsed with 0.1 M phosphate buffer solution (pH 7.0) before being dehydrated in a series of alcohol concentrations (50, 70, 80, 90, 95 and 100%) for 10 min. Then, standard procedures like freeze-drying, infiltration and polymerization as described by Fritz [44] were followed. Relative amount of elements in cortical and stellar cell of root tips were analyzed by scanning electron microscope (JSM6390, JEOL, Kyoto, Japan) equipped with an energy dispersive X-ray detector (Kenex, Valencia, CA, USA) [45]. Relative amount of K^+ and Na^+ levels was expressed as a percentage of the total atomic number for all the elements (K^+, Na^+, Ca^{2+}, Mn^{2+}, Mg^{2+} and Al^{3+}) that were detected from the root sections.

Measurement of net K^+ and Na^+ fluxes

Non-invasive Micro-test Technique (NMT) is a new method to measure non-invasively fluxes of specific ions and/or molecules inward or outward from cell without touching the sample by a micro-electrode. As one of the NMT, Scanning Ion-selective Electrode Technique (SIET) has been widely used in plant research [46]. In this study, net K^+ and Na^+ fluxes in the root tips of *K. obovata* were measured using SIET system (BIO-001A, Younger USA Sci. and Tech. Corp., Amherst, MA, USA) as described previously by Sun et al [8] and Chen et al [6]. Details concerning the principles of this method and preparation of ion-selective electrodes were available in previous publications [6,8]. In order to avoid the mechanical disturbance by the electrode movement, the K^+ and Na^+ fluxes must be measured slowly at approximately 1–2 s per point [8]. Ion-selective electrodes were calibrated before and after flux measurements in the following solutions containing different concentration of K^+ and Na^+: (1) 0.05, 0.1 and 0.5 mM K^+; (2) 0.5 and 5 mM Na^+. The final K^+ and Na^+ fluxes were calculated by Fick's law of diffusion as described by Xu et al [47]. According to Sun et al [8], the net ion influx is represented with positive value and net ion efflux with negative value. At least four individual roots were measured in an independent experiment. Each point in Figure 3 represents the mean ion flux of four individual roots. Data and image acquisition, preliminary processing, control of the electrode positioner and stepper-motor-controlled fine focus of the microscope stage were performed using Mage Flux software (http://www.youngerusa.com/mageux or http://xuyue.net/mageux) [4,8].

Steady-state SIET measurement of ion flux

The root segments with ca. 10 mm apices were sampled from the control, NaCl− and NaCl+SNP-treated plants at day 15. To avoid the effects of preloaded ion diffused from the surface of high salinity-treated *K. obovata* roots on ion flux measurement, according to the methods of Sun et al [8], the roots were rinsed with double-distilled water and immediately incubated in the measuring solution (0.1 mM KCl, 0.1 mM $CaCl_2$, 0.1 mM $MgCl_2$, 0.5 mM NaCl, 0.2 mM Na_2SO_4, 0.3 mM MES, pH 6.0) to equilibrate for 30 min. Afterwards, equilibrated samples were immobilized at the bottom of the measuring chamber, a 3 cm diameter plastic dish containing 2–3 ml of fresh measuring solution [6]. Net K^+ and Na^+ fluxes were measured in steady conditions for 10 to 15 min to insure that no fluctuation was present.

Transient K^+ and Na^+ Flux kinetics

The root tips of *K. obovata* seedlings grown in 400 mM NaCl were sampled and equilibrated in measuring solution for 30 min. After steady K^+ and Na^+ fluxes were recorded for 5–6 min, the kinetics K^+ and Na^+ fluxes affected by the inhibitors of PM H^+-ATPase and Na^+/H^+ antiporter and specific NO scavenger were examined by transiently adding sodium orthovanadate (100 μM), amiloride (500 μM) and cPTIO (200 μM), respectively. Steady fluxes of K^+ and Na^+ were recorded at least 5 min before addition of the inhibitor or scavenger. Then net K^+ and Na^+ fluxes were measured for another 20 min after the inhibitor or scavenger was added to the measuring solution. The data for the first 2–3 min were discarded due to diffusion effects after addition of SNP or inhibitor. All of the K^+ and Na^+ fluxes measurements were accomplished at Xuyue Science and Technology Company (Beijing, China).

SDS-PAGE and western-blot analysis

The root tips (1 g) of *K. obovata* were used for proteins extraction according to the method of Chen et al [6]. As described by Bradford [48], the protein concentrations were quantified using bovine serum albumin as a standard. For western-blot analysis, total proteins (150 μg from each sample) were separated by sodium dodecyl sulfate-polyacrylamide gel electrophoresis (SDS-PAGE) using 12% (w/v) acrylamide gels. After electrophoresis, the separated proteins were transferred to polyvinylidene difluoride (PVDF) membrane for 50 min. The membrane was blocked overnight at 37°C in Western Blocking Buffer (TIANGEN, China). The protein blot was probed with primary antibody of H^+-ATPase (AS07260, Agrisera, Sweden) or Na^+/H^+ antiporter (AS09484, Agrisera, Sweden), each diluted at the proportion of 1:2000 and agitated at room temperature for 3 h. Blots were then washed three times in phosphate buffered saline with Tween-20 solution (PBST) containing 50 mM Tris-HCl (pH 8.0), 150 mM NaCl, 0.05% Tween-20 (v/v), and followed by incubation with the secondary antibody (anti-rabbit IgG horse radish peroxidase-conjugated; Abcam, UK, 1:5000 dilution) for 2 h at room temperature. Finally, the blots were developed with ECL substrate detection solution (TIANGEN, China). Images of the blots were obtained using a CCD imager (FluorSMax Bio-Rad, Hercules, CA, USA). The Quantity One software (Bio-Rad, Hercules, CA, USA) was used to quantify the optical density value of the protein signals.

Total RNA extraction and gene cloning

Total RNA was isolated from the root tips of *K. obovata* with RNA purification reagents (Invitrogen Inc., CA, USA) according to the manufacturer's instructions. The RNA integrity and quality were confirmed by 1% agarose gel electrophoresis and ultraviolet spectrophotometer (Cary 50, Varian, USA), respectively. Total RNA (5 μg) was used for the first-strand cDNA synthesis using M-MLV reverse transcriptase (TaKaRa, Dalian, China). For the genes of inward-rectifying potassium channel (*AKT1*), PM Na^+/H^+ antiporter (*SOS1*), vacuolar Na^+/H^+ antiporer (*NHX1*), PM H^+-ATPase (*HA1*) and vacuolar H^+-ATPase subunit c (*VHA-c1*), the reverse transcription products were partially amplified by reverse transcription PCR using the degenerate primers and optimize reaction conditions as shown in Table S1. According to the methods of Chen et al [6], after analyzing the sequence of amplified fragments, partial sequences of *AKT1*, *SOS1*, *NHX1*, *HA1* and *VHA-c1* gene were acquired according to NCBI information (http://www.ncbi.nlm.nih.gov/).

Gene expression analysis by quantitative real-time PCR

Quantitative real-time PCR (qRT-PCR) assays were performed using the Rotor-Gene^{TM} 6000 real-time PCR instrument (Corbett Research, Mortlake, Australia). All reactions were carried out in a 10 μl final volume containing 0.6 μl of forward and reverse primers, 1 μl of cDNA (about 10 wng of mRNA), and 5 μl Faststart Universal SYBR Green Master (ROX, Mannheim, Germany). Amplifications were performed according to the following conditions: 10 min initial polymerase activation at 95°C, then 40 cycles of denaturation at 95°C, 30 s, annealing at appropriate temperature (Table S1) for 30 s and extension at 72°C for 30 s. Relative quantification values for each target gene were calculated by the $2^{-\Delta\Delta Ct}$ method [49]. Gene expression levels were normalized using *18S* rRNA gene of *K. obovata* (acquired from NCBI, Gene Bank number AY289625.1) as the internal standard. The gene-specific primers and optimized reaction conditions used for qRT-PCR are listed in Table S1.

Statistical analysis

Each experiment was repeated at least three times. All data were subjected to one-way analysis of variance (ANOVA) using the SPSS 13.0 (SPSS Chicago, IL, USA), and results were expressed as the mean ± SE. Difference at $P<0.05$ was considered significant between treatments.

Ethics statement

State Administration of Jiulongjiang River Mangrove Nature Reserve in Fujian Province, China, gave permission to collect the mature propagules of *K. obovata* in this site for investigating the salt tolerance mechanism of the mangrove plants, and they confirmed that our studies did not involve endangered or protected species.

Acknowledgments

We wish to thank Dr. Fei-Hua Wu for his assistance in element determination and Dr. Zhong-Lin Zhang and Dr. Isabwe Alain for critically editing the manuscript.

Author Contributions

Conceived and designed the experiments: JC DYX HLZ. Performed the experiments: JC DYX WHW JC QX. Analyzed the data: JC DYX MS QX TWL. Contributed reagents/materials/analysis tools: JC DYX WJH XL. Wrote the paper: JC DYX HLZ.

References

1. Chen Z, Pottosin II, Cuin TA, Fuglsang AT, Tester M, et al. (2007) Root plasma membrane transporters controlling K^+/Na^+ homeostasis in salt-stressed barley. Plant Physiol 145: 1714–1725.
2. Shabala S, Cuin TA (2008) Potassium transport and plant salt tolerance. Physiol Plantarum 133: 651–669.
3. Zhu JK (2003) Regulation of ion homeostasis under salt stress. Curr Opin Plant Biol 6: 441–445.
4. Sun J, Wang MJ, Ding MQ, Deng SR, Liu MQ, et al. (2010) H_2O_2 and cytosolic Ca^{2+} signals triggered by the PM H^+-coupled transport system mediate K^+/Na^+ homeostasis in NaCl-stressed *Populus euphratica* cells. Plant Cell Environ 33: 943–958.
5. Wu C, Gao X, Kong X, Zhao Y, Zhang H (2009) Molecular cloning and functional analysis of a Na^+/H^+ antiporter gene ThNHX1 from a halophytic plant *Thellungiella halophila*. Plant Mol Biol Rep 27: 1–12.
6. Chen J, Xiao Q, Wu F, Dong X, He J, et al. (2010) Nitric oxide enhances salt secretion and Na^+ sequestration in a mangrove plant, *Avicennia marina*, through increasing the expression of H^+-ATPase and Na^+/H^+ antiporter under high salinity. Tree Physiol 30: 1570–1586.
7. Silva P, Façanha AR, Tavares RM, Gerós H (2010) Role of tonoplast proton pumps and Na^+/H^+ antiport system in salt tolerance of *Populus euphratica* Oliv. J Plant Growth Regul 29: 23–34.
8. Sun J, Chen SL, Dai SX, Wang RG, Li NY, et al. (2009) NaCl-induced alternations of cellular and tissue ion fluxes in roots of salt-resistant and salt-sensitive poplar species. Plant Physiol 149: 1141–1153.
9. Chen Z, Newman I, Zhou M, Mendham N, Zhang G, et al. (2005) Screening plants for salt tolerance by measuring K^+ flux: a case study for barley. Plant Cell Environ 28: 1230–1246.
10. Serrano R, Rodriguez PL (2002) Plants, genes and ions. EMBO Rep 3: 116–119.
11. Sentenac H, Bonneaud N, Minet M, Lacroute F, Salmon JM, et al. (1992) Cloning and expression in yeast of a plant potassium ion transport system. Science 256: 663–665.
12. Buschmann PH, Vaidyanathan R, Gassmann W, Schroeder JI (2000) Enhancement of Na^+ uptake currents, time-dependent inward-rectifying K^+ channel currents, and K^+ channel transcripts by K^+ starvation in wheat root cells. Plant Physiol 122: 1387–1398.
13. Golldack D, Quigley F, Michalowski CB, Kamasani UR, Bohnert HJ (2003) Salinity stress-tolerant and-sensitive rice (*Oryza sativa* L.) regulate AKT1-type potassium channel transcripts differently. Plant Mol Biol 51: 71–81.
14. Martínez-Cordero M, Martínez V, Rubio F (2005) High-affinity K^+ uptake in pepper plants. J Exp Bot 56: 1553–1562.
15. Ardie SW, Liu S, Takano T (2010) Expression of the AKT1-type K^+ channel gene from *Puccinellia tenuiflora*, *PutAKT1*, enhances salt tolerance in *Arabidopsis*. Plant Cell Reports 29: 865–874.
16. Qi Z, Spalding EP (2004) Protection of plasma membrane K^+ transport by the salt overly sensitive1 Na^+-H^+ antiporter during salinity stress. Plant Physiol 136: 2548–2555.
17. Parida AK, Jha B (2010) Salt tolerance mechanisms in mangroves: a review. Trees-Struct Funct 24: 199–217.
18. Wilson ID, Neill SJ, Hancock JT (2008) Nitric oxide synthesis and signalling in plants. Plant Cell Environ 31: 622–631.
19. Wendehenne D, Durner J, Klessig DF (2004) Nitric oxide: a new player in plant signalling and defence responses. Curr Opin Plant Biol 7: 449–455.
20. Zhang YY, Wang LL, Liu YL, Zhang Q, Wei QP, et al. (2006) Nitric oxide enhances salt tolerance in maize seedlings through increasing activities of proton-pump and Na^+/H^+ antiport in the tonoplast. Planta 224: 545–555.
21. Zhao LQ, Zhang F, Guo JK, Yang YL, Li BB, et al. (2004) Nitric oxide functions as a signal in salt resistance in the calluses from two ecotypes of reed. Plant Physiol 134: 849–857.
22. Sheue CR, Liu HY, Yang WH (2003) *Kandelia obovata* (Rhizophoraceae), a new mangrove species from Eastern Asia. TAXON 52: 287–294.
23. Scholander PF (1968) How mangroves desalinate seawater. Physiol Plantarum 21: 251–261.
24. Li NY, Chen SL, Zhou XY, Li CY, Shao J, et al. (2008) Effect of NaCl on photosynthesis, salt accumulation and ion compartmentation in two mangrove species, *Kandelia candel* and *Bruguiera gymnorhiza*. Aquat Bot 88: 303–310.
25. Lu Y, Li N, Sun J, Hou P, Jing X, et al. (2013) Exogenous hydrogen peroxide, nitric oxide and calcium mediate root ion fluxes in two non-secretor mangrove species subjected to NaCl stress. Tree Physiol 33: 81–95.
26. Leigh RA (2001) Potassium homeostasis and membrane transport. J Plant Nutr Soil Sci 164: 193–198.
27. Maathuis FJM (2006) The role of monovalent cation transporters in plant responses to salinity. J Exp Bot 57: 1137–1147.
28. Carden DE, Walker DJ, Flowers TJ, Miller AJ (2003) Single-cell measurements of the contributions of cytosolic Na^+ and K^+ to salt tolerance. Plant Physiol 131: 676–683.
29. Hariadi Y, Marandon K, Tian Y, Jacobsen SE, Shabala S (2011) Ionic and osmotic relations in quinoa (*Chenopodium quinoa* Willd.) plants grown at various salinity levels. J Exp Bot 62: 185.
30. Sun J, Dai S, Wang R, Chen S, Li N, et al. (2009) Calcium mediates root K^+/Na^+ homeostasis in poplar species differing in salt tolerance. Tree Physiol 29: 1175–1186.
31. Alemán F, Nieves-Cordones M, Martínez V, Rubio F (2011) Root K^+ acquisition in plants: the *Arabidopsis thaliana* model. Plant Cell Physiol 52: 1603–1612.
32. Fuchs I, Stölzle S, Ivashikina N, Hedrich R (2005) Rice K^+ uptake channel OsAKT1 is sensitive to salt stress. Planta 221: 212–221.
33. Kaddour R, Nasri N, M'rah S, Berthomieu P, Lachaâl M (2009) Comparative effect of potassium on K and Na uptake and transport in two accessions of *Arabidopsis thaliana* during salinity stress. CR Biol 332: 784–794.
34. Berthomieu P, Conejero G, Nublat A, Brackenbury WJ, Lambert C, et al. (2003) Functional analysis of *AtHKT1* in *Arabidopsis* shows that Na^+ recirculation by the phloem is crucial for salt tolerance. EMBO J 22: 2004–2014.
35. Niu X, Bressan RA, Hasegawa PM, Pardo JM (1995) Ion homeostasis in NaCl stress environments. Plant Physiol 109: 735–742.
36. Arasimowicz M, Floryszak-Wieczorek J (2007) Nitric oxide as a bioactive signalling molecule in plant stress responses. Plant Sci 172: 876–887.
37. Shi H, Lee B, Wu SJ, Zhu JK (2002) Overexpression of a plasma membrane Na^+/H^+ antiporter gene improves salt tolerance in *Arabidopsis thaliana*. Nat Biotechnol 21: 81–85.
38. Wang CM, Zhang JL, Liu XS, Li Z, Wu GQ, et al. (2009) *Puccinellia tenuiflora* maintains a low Na^+ level under salinity by limiting unidirectional Na^+ influx resulting in a high selectivity for K^+ over Na^+. Plant Cell Environ 32: 486–496.
39. Qiu QS, Barkla BJ, Vera-Estrella R, Zhu JK, Schumaker KS (2003) Na^+/H^+ exchange activity in the plasma membrane of *Arabidopsis*. Plant Physiol 132: 1041–1052.
40. Tester M, Davenport R (2003) Na^+ tolerance and Na^+ transport in higher plants. Ann Bot 91: 503–527.
41. Fukuda A, Nakamura A, Hara N, Toki S, Tanaka Y (2011) Molecular and functional analyses of rice NHX-type Na^+/H^+ antiporter genes. Planta 233: 175–188.
42. Chen S, Polle A (2010) Salinity tolerance of Populus. Plant Biol 12: 317–333.
43. Kao WY, Tsai HC, Tsai TT (2001) Effect of NaCl and nitrogen availability on growth and photosynthesis of seedlings of a mangrove species, *Kandelia candel* (L.) Druce. J Plant Physiol 158: 841–846.
44. Fritz E (1989) X-ray microanalysis of diffusible elements in plant cells after freeze-drying, pressure-infiltration with ether and embedding in plastic. Scanning Microsc 3: 517–526.
45. Vázquez MD, Poschenrieder C, Corrales I, Barceló J (1999) Change in apoplastic aluminum during the initial growth response to aluminum by roots of a tolerant maize variety. Plant Physiol 119: 435–444.
46. Xu Y, Sun T, Yin LP (2006) Application of non-invasive microsensing system to simultaneously measure both H^+ and O_2 fluxes around the pollen tube. J Integr Plant Biol 48: 823–831.
47. Xu RR, Qi SD, Lu LT, Chen CT, Wu CA, et al. (2011) A DExD/H box RNA helicase is important for K^+ deprivation responses and tolerance in *Arabidopsis thaliana*. FEBS J 278: 2296–2306.
48. Bradford MM (1976) A rapid and sensitive method for the quantitation of microgram quantities of protein utilizing the principle of protein-dye binding. Anal Biochem 72: 248–254.
49. Livak KJ, Schmittgen TD (2001) Analysis of relative gene expression data using real-time quantitative PCR and the 2-[Delta][Delta] CT method. Methods 25: 402–408.

15

Plastic and the Nest Entanglement of Urban and Agricultural Crows

Andrea K. Townsend[1]*, Christopher M. Barker[2]

1 Department of Wildlife, Fish, and Conservation Biology, University of California Davis, Davis, California, United States of America, 2 Department of Pathology, Microbiology, and Immunology, School of Veterinary Medicine, University of California Davis, Davis, California, United States of America

Abstract

Much attention has been paid to the impacts of plastics and other debris on marine organisms, but the effects of plastic on terrestrial organisms have been largely ignored. Detrimental effects of terrestrial plastic could be most pronounced in intensively human-modified landscapes (e.g., urban and agricultural areas), which are a source of much anthropogenic debris. Here, we examine the occurrence, types, landscape associations, and consequences of anthropogenic nest material in the American crow (*Corvus brachyrhynchos*), a North American species that breeds in both urban and agricultural landscapes. We monitored 195 nestlings in 106 nests across an urban and agricultural gradient in the Sacramento Valley, California, USA. We found that 85.2% of crow nests contained anthropogenic material, and 11 of 195 nestlings (5.6%) were entangled in their nests. The length of the material was greater in nests in agricultural territories than in urban territories, and the odds of entanglement increased 7.55 times for each meter of anthropogenic material in the nest. Fledging success was significantly lower for entangled than for unentangled nestlings. In all environments, particularly urban, agricultural, and marine, careful disposal of potential hazards (string, packing and hay bale twine, balloon ribbon, wire, fishing line) could reduce the occurrence of entanglement of nestling birds.

Editor: Gabriele Sorci, CNRS, Université de Bourgogne, France

Funding: This work was funded by the University of California at Davis. The funders had no role in study design, data collection and analysis, decision to publish, or preparation of the manuscript.

Competing Interests: The authors have declared that no competing interests exist.

* E-mail: aktownsend@ucdavis.edu

Introduction

The consequences of plastic accumulation for marine organisms are well-documented and have received wide attention [1]. Impacts of marine debris include transport of alien species [2] and pollutants to new locations [3], smothering of sea floor biota [4], and sorption and toxicity of contaminants [5]. Reports of entanglement and choking of marine wildlife are voluminous: more than 250 species are known to ingest or have been entangled in marine debris [6] including seabirds, turtles, fish, crustaceans, and cetaceans, sometimes with documented consequences for fitness [7] or population size [4].

In contrast, impacts of debris on terrestrial organisms are poorly documented and have been largely ignored, perhaps because terrestrial debris is less conspicuous [8] and more difficult to measure [9] than marine debris. As was initially the case with marine wildlife [7], entangled and choking terrestrial wildlife might be difficult to detect without deliberate effort, and opportunistic observations may go unreported [10]. Available data suggest that the impacts of debris on terrestrial organisms are similar to those on marine organisms, including sorption of chemicals [11] and mortality linked to ingestion [12]. Likewise, a number of studies report entanglement of terrestrial organisms, including snakes in beer can tabs [13], tortoises in balloon ribbon [10], and birds in anthropogenic nest material [14–16].

Potential effects of debris on terrestrial organisms could be most substantial in intensively human-modified landscapes (e.g., urban and agricultural areas). One of the few previous studies of nest entanglement in terrestrial birds, for example, reported that 4.6% of Osprey (*Pandion haliaetus* [15]) nestlings in an agricultural setting were entangled in anthropogenic nest material (twine). Likewise, urban settings are a primary source of trash entering marine [1] and terrestrial protected areas [17]. Historically, urban biotas have received very little attention [18], although interest has intensified in recent years in response to the rapid spread of urban landscapes across the earth's surface [19]. Nevertheless, the effects of plastic debris on urban organisms have been largely ignored. Impacts of debris could be magnified in urban adapted species, which are often characterized by their ability to utilize anthropogenic resources [20,21]. A recent study of Chinese bulbuls *(Pycnonotus sinensis*), for example, showed that the proportion of anthropogenic material in nests increased with degree of urbanization [22]. The potential fitness consequences and adaptive significance of this adjustment, however, were not examined.

Herein, we examine the occurrence, types, landscape associations, and consequences of anthropogenic nest material in the American crow (*Corvus brachyrhynchos*; "crow" hereafter), a North American species that breeds in both urban and agricultural landscapes. This study is the first to explicitly document the links between terrestrial land cover, anthropogenic nest material, and nest entanglement in a terrestrial songbird.

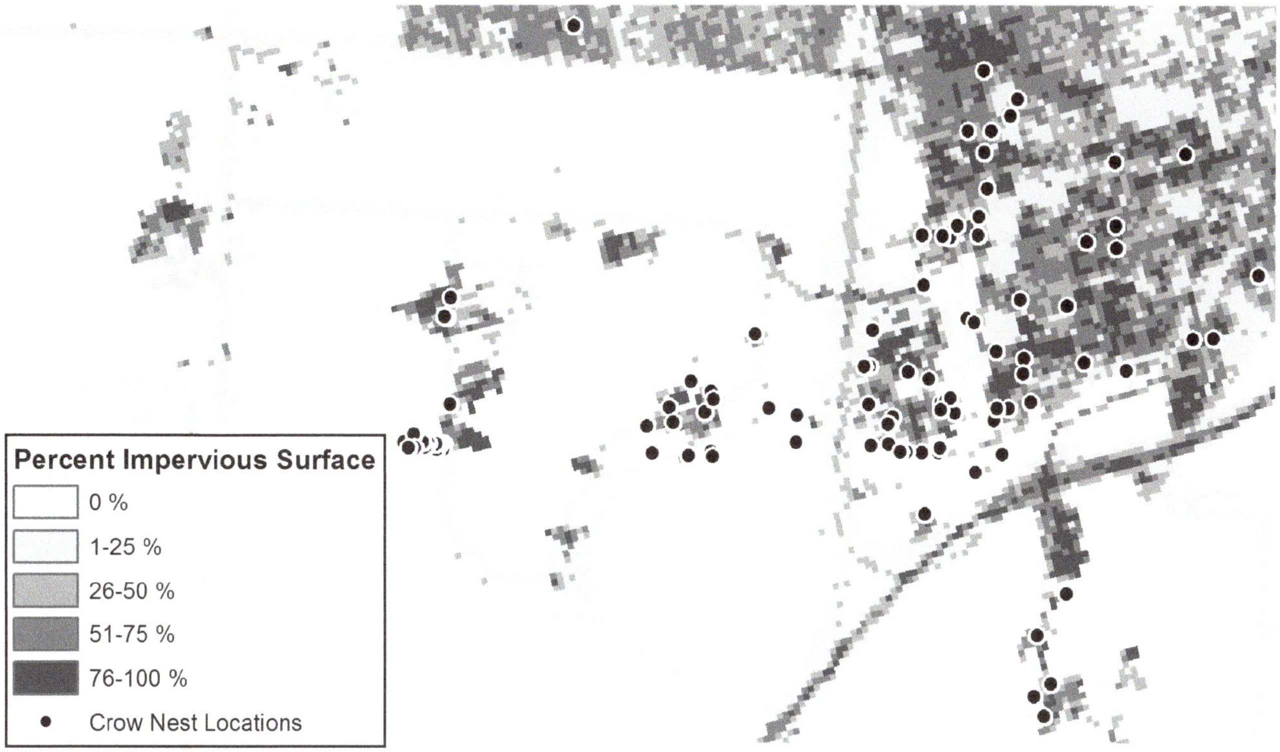

Figure 1. Map of study area in Davis, California, USA. All detected nests ($n = 106$) within this site were monitored for fledging success.

Methods

Entanglement, fledging success, and nest analysis

In 2012 and 2013, we monitored success of 106 crow nests across an urban to agricultural gradient in Davis, California. The study site spanned the urban campus of the University of California, Davis into the surrounding campus-owned agricultural areas (e.g., vineyards, pasture, and row crops; Fig. 1). Nests were situated on lateral tree branches (mean nest height ± SE: 9±0.5 m; $n = 106$ nests) and accessed by boom lift. Nestlings were checked for entanglement in the nest either once (approximately day 25 after hatching; $n = 45$ nests) or twice (<17 days after hatching and again on approximately day 25 after hatching; $n = 39$ nests). Entanglements were removed from live nestlings. Surviving nestlings were marked 25 days after hatching with a unique combination of color bands and a USGS band. After banding, nests were checked daily to monitor success or failure. The encounter rate of entangled nestlings represents a minimum entanglement rate, because dead, entangled nestlings could have been removed by parents or predators prior to nest checks.

A subset of 54 randomly selected nests was collected after the nestlings fledged or failed. We identified and measured the length (to the nearest 0.5 cm) of each piece of anthropogenic material longer than 5 cm from the lining of each nest. Corresponding

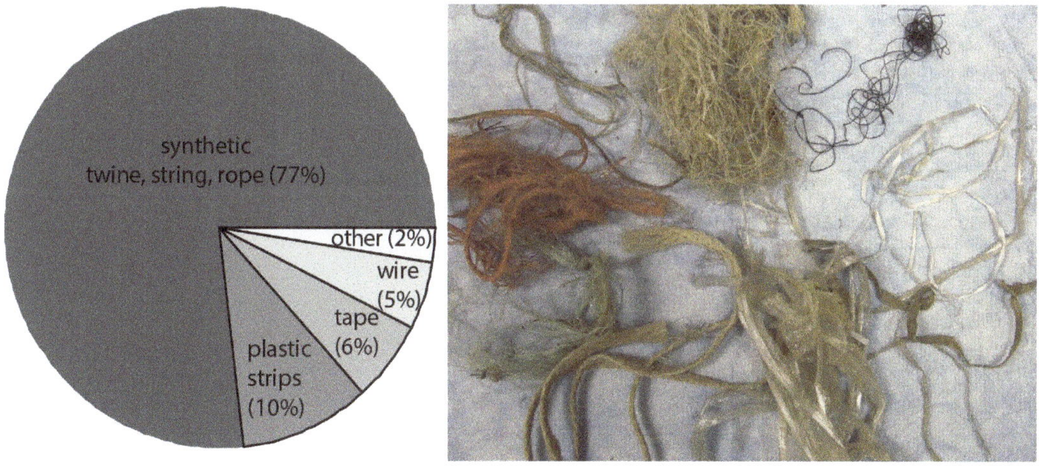

Figure 2. Percentage of each type of anthropogenic material in nests. Examples from nests are shown to right ($n = 678$ total items; 54 nests).

nestling entanglement data were available from 32 of these nests. We had no corresponding entanglement data from 22 of these nests because the broods failed (i.e., they were depredated or failed from other causes, which could have included entanglement) prior to the first nestling check.

To assess how the probability of entanglement varied with amount of anthropogenic nest material, we analyzed entanglement of individual nestlings (0/1) as the response in a generalized linear mixed model (GLMM; binomial error, penalized quasilikelihood method), with total length of anthropogenic nest material in the nest as the predictor. We specified nest as a random effect to account for non-independence among nestlings within a nest. We included in this analysis the subset of nestlings for which we had corresponding nest material data ($n = 64$ nestlings from 32 nests).

Landscape characterization

To characterize territories as primarily urban or agricultural, we created 1 ha buffers surrounding each nest site (Fig. 1). We estimated the average percent impervious surface within each buffer using the 2006 National Land Cover Database, which was the most recent version and representative of the study area [23]. We defined urban territories as those covered by an average of more than 50% impervious surface and agricultural territories as having 50% or less impervious surface. To assess how the amount of anthropogenic material varied with territory type, we examined (1) the total summed length of all anthropogenic nest material in each nest, and (2) the longest individual piece of anthropogenic nest material in each nest, in separate linear models with territory type (urban, agricultural) as the predictor. Length of anthropogenic material was square-root transformed to meet the assumption of normality. All statistical analyses were run in R v.3.0.1 [24], including the sp and raster packages for spatial analysis. All means and model parameter estimates are given ± SE.

Ethics statements

This work was performed under protocols approved by the Institutional Animal Care and Use Committee of the University of California, Davis (Permit Number: 16897). All work was conducted on the private property of the University of California, Davis. No protected species were sampled.

Results

Anthropogenic material greater than 10 cm was detected in 85.2% (46/54) of dissected nests. Only three of 36 nests in primarily agricultural territories (8.3%) and five of 18 nests in urban territories (27.8%) contained no anthropogenic material. The mean total length of anthropogenic material in these 54 nests was 292.0±49.3 cm (range: 0–1858.3 cm). Material included synthetic string, twine, or rope; plastic tape; strips of plastic or cloth (including elastic, ribbon, gauze bandages, fabric straps, and unraveled woven sacks), fishing line, balloon string, unraveled nets, and wire mesh (Fig. 2). Amount of anthropogenic material in the nest varied with landscape: total length of anthropogenic material was significantly longer in nests in agricultural areas than in urban areas (mean total length agricultural = 363.70±68.04 cm; $n = 36$ nests; mean total length urban = 163.60±43.64 cm; $n = 18$ nests; β (urban) = −6.03±2.81; t (−2.15); $p = 0.04$), and the longest piece of anthropogenic material was significantly longer, on average, in agricultural nests than in urban nests (mean longest piece agricultural = 64.30±8.23 cm; mean longest piece urban = 38.04±8.32 cm; β (urban) = −2.18±1.02; t (−2.15); $p = 0.04$).

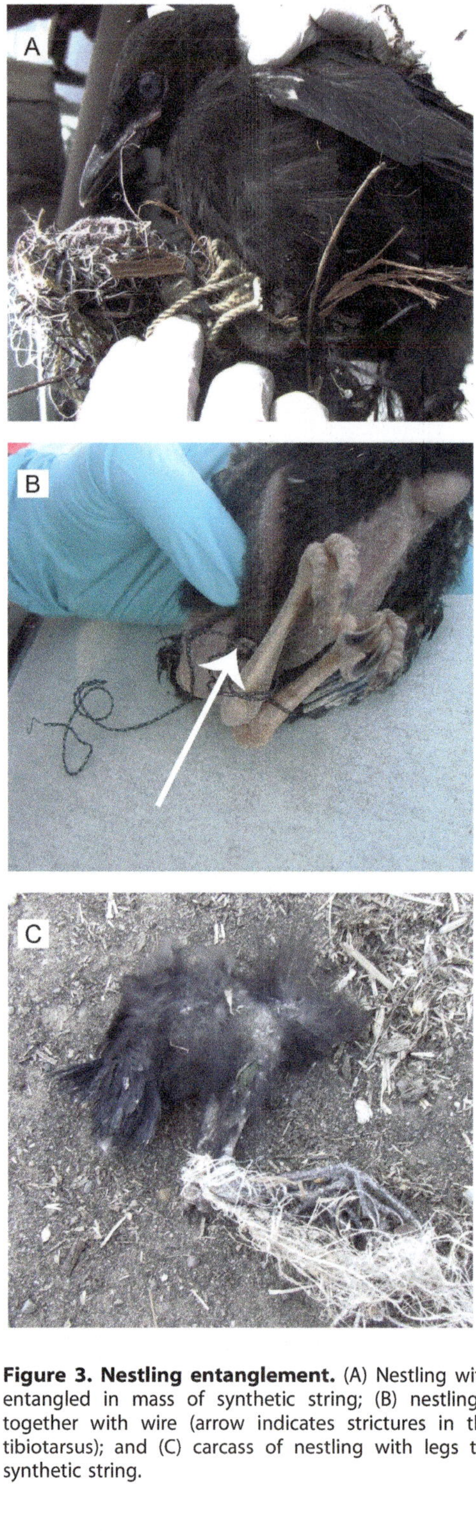

Figure 3. Nestling entanglement. (A) Nestling with tarsometarsus entangled in mass of synthetic string; (B) nestling with legs tied together with wire (arrow indicates strictures in the bone of the tibiotarsus); and (C) carcass of nestling with legs tied together by synthetic string.

Eleven of 195 nestlings (5.6%) were entangled (Fig. 3). The likelihood of entanglement increased with the total length of anthropogenic material in the nest (GLMM, 2.10±0.82 m; t (2.53); $p = 0.017$): the odds of entanglement increased 7.55 times for each meter of anthropogenic material in the nest. Nine nestlings were discovered entangled while still in the nest, and two were discovered dead below their nest tree. Entanglements were removed from all live nestlings, but in most cases, birds showed evidence of entanglement-related injury (e.g., strictures in the bone of the tibiotarsus (Fig. 3b); malformed toes). All nestlings that had been entangled (100%; 11/11 nestlings) failed to fledge, and the likelihood of fledging was significantly lower for birds that had been entangled than nestlings that had not been entangled (54.9%; 101/184 unentangled nestlings; χ^2 (6.8) = 0.009; 2-sample test for equality of proportions with continuity correction).

Discussion

In an early review of entanglement and ingestion of plastic debris by marine organisms, Laist (1987) listed three reasons why the significance of marine debris had been disregarded, all of which could apply to terrestrial organisms today: (1) the mechanics of entanglement were so straight-forward that they lacked "hidden mysteries;" (2) encounters between debris and wildlife could have been rare; and (3) the paucity of published reports appeared to confirm this overall rarity. Since that review, however, awareness and concern about the prevalence and problems associated with plastics in the marine environments have grown rapidly, and entire journal issues have been devoted to the topic [8]. Mechanical impacts of plastics on terrestrial organisms, however, are still largely disregarded.

We have shown that 85.2% of crow nests along an urban to agricultural gradient contained anthropogenic material, that the amount of material was higher in nests in agricultural areas, and that the likelihood of entanglement increased with length of anthropogenic material. All entangled nestlings (5.6% of the nestlings marked in this study) failed to fledge. Potential for entanglement and associated mortality could be widespread in birds in highly human-dominated (urban and intensive agricultural) landscapes. Anecdotal descriptions of anthropogenic nest material, including string, balloon ribbon, fishing line, plastic bags, paper, and dental floss, have been reported for many of the North American avian species defined as farmland species [25] (e.g., *Charadrius vociferous* [26], *Zenaida macroura* [16], *Cardinalis cardinalis* [27], *Sturnus vulgaris* [28], *Passer domesticus* [29]) and/or urban exploiters and adapters [20] (e.g., *Aeronautes saxatalis* [30], *Poecile rufescens* [31], *Psaltriparus minimus* [32], *Mimus polyglottos* [33], *Calypte anna* [34], *Haemorhous mexicanus* [35], and *Melozone crissalis* [36]). In some of these species, anthropogenic material has been anecdotally linked to entanglement or poor nest success [16,32,37].

Birds in urban and agricultural settings may use hazardous anthropogenic materials because they resemble their preferred, natural nest material (e.g., vines, grasses, strips of bark), analogous to marine turtles ingesting plastic bags because they resemble their jellyfish prey [38]. Some authors have suggested that highly modified environments, in general, could be ecological traps: animals are attracted to settle on the basis of historically adaptive cues, but cannot sustain a viable population because of low habitat quality [39,40]. Entanglement in anthropogenic nest material can be added to the suite of documented stressors of urban and intensive agricultural landscapes, including toxins [41], novel predators [42], pesticide usage [43], tillage [44], roads [45], and disease [46].

In some situations, anthropogenic nest material could be a beneficial resource, enabling nest construction in places where natural materials are limited [47]. Anthropogenic nest material could have other benefits: for example, cigarette butts incorporated into nests reduced the ectoparasite load of some urban birds [48]. How these potential benefits balance the entanglement hazards of some anthropogenic nest materials is unknown.

We found that the amount of anthropogenic nest material was greater in the agricultural landscape than the urban landscape, likely due to the ready availability of twine and shade cloth wire in the agricultural settings. The majority of crow nests across this urban and agricultural gradient contained some anthropogenic material, however, and nestling entanglement occurred across this land use gradient. In all environments, therefore, particularly urban, agricultural, and marine, careful disposal potential hazards (string, packing and hay bale twine, balloon ribbon, wire, fishing line) could reduce the occurrence of nestling entanglement.

Acknowledgments

Thanks to M. Raabe, M. Hinton, Y. Ha Suh, J. Hurd, and M. Saberi for nest dissection and to S. Wheeler and W. Boyce for advice and support.

Author Contributions

Conceived and designed the experiments: AKT CMB. Performed the experiments: AKT CMB. Analyzed the data: AKT CMB. Contributed reagents/materials/analysis tools: AKT CMB. Wrote the paper: AKT CMB.

References

1. Barnes DKA, Galgani F, Thompson RC, Barlaz M (2009) Accumulation and fragmentation of plastic debris in global environments. Philos Trans R Soc Lond B Biol Sci 364: 1985–1998.
2. Gregory MR (2009) Environmental implications of plastic debris in marine settings—entanglement, ingestion, smothering, hangers-on, hitch-hiking and alien invasions. Philos Trans R Soc Lond B Biol Sci 364: 2013–2025.
3. Mato Y, Isobe T, Takada H, Kanehiro H, Ohtake C, et al. (2000) Plastic resin pellets as a transport medium for toxic chemicals in the marine environment. Environ Sci Technol 35: 318–324.
4. Derraik JGB (2002) The pollution of the marine environment by plastic debris: a review. Mar Pollut Bull 44: 842–852.
5. Teuten EL, Saquing JM, Knappe DRU, Morton AB, Jonsson S, et al. (2009) Transport and release of chemicals from plastics to the environment and to wildlife. Philos Trans R Soc Lond B Biol Sci 364: 2027–2045.
6. Laist DW (1997) Impacts of marine debris: entanglement of marine life in marine debris including a comprehensive list of species with entanglement and ingestion records. In: Coe JM, Rogers DB, editors. Marine Debris, Sources, Impacts, and Solutions. New York, NY: Springer-Verlag. pp. 99–139.
7. Laist DW (1987) Overview of the biological effects of lost and discarded plastic debris in the marine environment. Mar Pollut Bull 18: 319–326.
8. Thompson RC, Swan SH, Moore CJ, Saal FS (2009) Introduction: our plastic age. Philos Trans R Soc Lond B Biol Sci 364: 1973–1976.
9. Zylstra ER (2013) Accumulation of wind-dispersed trash in desert environments. J Arid Environ 89: 13–15.
10. Walde AD, Harless ML, Delaney DK, Pater LL (2007) Anthropogenic threat to the desert tortoise (*Gopherus agassizii*): litter in the Mojave Desert. Western North American Naturalist 67: 147–149.
11. Gaylor MO, Harvey E, Hale RC (2012) House crickets can accumulate polybrominated diphenyl ethers (PBDEs) directly from polyurethane foam common in consumer products. Chemosphere 86: 500–505.
12. Mee A, Rideout BA, Hamber JA, Todd JN, Austin G, et al. (2007) Junk ingestion and nestling mortality in a reintroduced population of California condors *Gymnogyps californianus*. Bird Conserv Int 17: 119–130.
13. Herrington B (1985) Another reason for herpetologists to pick up their beer cans. Herpetol Rev 16: 113.
14. Montevecchi WA (1991) Incidence and types of plastic in gannets nests in the Northwest Atlantic. Can J Zool 69: 295–297.
15. Blem CR, Blem LB, Harmata PJ (2002) Twine causes significant mortality in nestling ospreys. Wilson Bulletin 114: 528–529.
16. Parker GH, Blomme CG (2007) Fish-line entanglement of nesting mourning dove, *Zenaida macroura*. Canadian Field Naturalist 121: 436–437.
17. McDonald RI, Forman RTT, Kareiva P, Neugarten R, Salzer D, et al. (2009) Urban effects, distance, and protected areas in an urbanizing world. Landsc Urban Plan 93: 63–75.
18. Gaston KJ (2010) Urbanisation. In: Gaston KJ, editor. Urban Ecology. New York: Cambridge University Press. pp. 10–34.
19. Chace JF, Walsh JJ (2006) Urban effects on native avifauna: a review. Landsc Urban Plan 74: 46–69.
20. Blair RB (1996) Land use and avian species diversity along an urban gradient. Ecol Appl 6: 506–519.
21. McKinney ML (2002) Urbanization, biodiversity, and conservation. Bioscience 52: 883–890.

22. Wang YP, Chen SH, Blair RB, Jiang PP, Ding P (2009) Nest composition adjustments by Chinese bulbuls *Pycnonotus sinensis* in an urbanized landscape of Hangzhou (E China). Acta Ornithologica 44: 185–192.
23. Fry JA, Xian G, Jin SM, Dewitz JA, Homer CG, et al. (2011) National land cover database for the conterminous United Sates. Photogramm Eng Remote Sensing 77: 859–864.
24. R Core Team (2013) R: A language and environment for statistical computing. Vienna, Austria: R Foundation for Statistical Computing.
25. Boutin C, Freemark KE, Kirk DA (1999) Farmland birds in southern Ontario: field use, activity patterns and vulnerability to pesticide use. Agric Ecosyst Environ 72: 239–254.
26. Jackson B, Jackson JA (2000) Killdeer (*Charadrius vociferus*). In: Poole A, editor. The Birds of North America Online. Ithaca: Cornell Lab of Ornithology.
27. Halkin SL, Linville SU (1999) Northern Cardinal (*Cardinalis cardinalis*). In: Poole A, editor. The Birds of North America Online Ithaca: Cornell Lab of Ornithology.
28. Cabe PR (1993) European Starling (*Sturnus vulgaris*). In: Poole A, editor. The Birds of North America Online. Ithaca: Cornell Lab of Ornithology.
29. Lowther PE, Cink CL (2006) House Sparrow (*Passer domesticus*). In: Poole A, editor. The Birds of North America Online. Ithaca: Cornell Lab of Ornithology.
30. Ryan TP, Collins CT (2000) White-throated Swift (*Aeronautes saxatalis*). In: Poole A, editor. The Birds of North America Online. Ithaca: Cornell Lab of Ornithology.
31. Dahlsten DL, Brennan LA, Mccallum DA, Gaunt SL (2002) Chestnut-backed Chickadee (*Poecile rufescens*). In: Poole A, editor. The Birds of North America Online. Ithaca: Cornell Lab of Ornithology.
32. Sloane SA (2001) Bushtit (*Psaltriparus minimus*). In: Poole A, editor. The Birds of North America Online. Ithaca: Cornell Lab of Ornithology.
33. Farnsworth G, Londono GA, Martin JU, Derrickson KC, Breitwisch R (2011) Northern Mockingbird (*Mimus polyglottos*). In: Poole A, editor. The Birds of North America Online. Ithaca: Cornell Lab of Ornithology.
34. Clark CJ, Russell SM (2012) Anna's Hummingbird (*Calypte anna*). In: Poole A, editor. The Birds of North America Online. Ithaca: Cornell Lab of Ornithology.
35. Badyaev AV, Belloni V, Hill GE (2012) House Finch (*Haemorhous mexicanus*). In: Poole A, editor. The Birds of North America Online. Ithaca: Cornell Lab of Ornithology.
36. Benedict L, Kunzmann MR, Ellison K, Purcell KL, Johnson RR, et al. (2011) California Towhee (*Melozone crissalis*). In: Poole A, editor. The Birds of North America Online. Ithaca: Cornell Lab of Ornithology.
37. Collins CT, Johnson EV (1982) Further records of White-throated Swifts utilizing man-made structures. Western Birds 13: 25–28.
38. Schuyler Q, Hardesty BD, Wilcox C, Townsend K (2012) To eat or not to eat? Debris selectivity by marine turtles. PLoS ONE 7: e40884.
39. Schlaepfer MA, Runge MC, Sherman PW (2002) Ecological and evolutionary traps. Trends Ecol Evol 17: 474–480.
40. Kokko H, Sutherland WJ (2001) Ecological traps in changing environments: ecological and evolutionary consequences of a behaviourally mediated Allee effect. Evol Ecol Res 3: 537–551.
41. Roux KE, Marra PP (2007) The presence and impact of environmental lead in passerine birds along an urban to rural land use gradient. Arch Environ Contam Toxicol 53: 261–268.
42. Leston LFV, Rodewald AD (2006) Are urban forests ecological traps for understory birds? An examination using Northern cardinals. Biol Conserv 131: 566–574.
43. Benton TG, Bryant DM, Cole L, Crick HQP (2002) Linking agricultural practice to insect and bird populations: a historical study over three decades. J Appl Ecol 39: 673–687.
44. Best LB (1986) Conservation tillage: ecological traps for nesting birds? Wildl Soc Bull 14: 308–317.
45. Coelho IP, Teixeira FZ, Colombo P, Coelho AVP, Kindel A (2012) Anuran road-kills neighboring a pen-urban reserve in the Atlantic Forest, Brazil. J Environ Manage 112: 17–26.
46. Boal CW, Mannan RW (1999) Comparative breeding ecology of Cooper's hawks in urban and exurban areas of southeastern Arizona. J Wildl Manage 63: 77–84.
47. Adams CE, Lindsey KJ (2010) Urban Wildlife Management. Boca Raton, FL.: CRC Press.
48. Suárez-Rodríguez M, López-Rull I, Garcia CM (2012) Incorporation of cigarette butts into nests reduces nest ectoparasite load in urban birds: new ingredients for an old recipe? Biol Lett 9.

16

The Use of a Predictive Habitat Model and a Fuzzy Logic Approach for Marine Management and Planning

Tarek Hattab[1,2]*, Frida Ben Rais Lasram[1], Camille Albouy[3], Chérif Sammari[4], Mohamed Salah Romdhane[1], Philippe Cury[2], Fabien Leprieur[5], François Le Loc'h[2]

1 UR 03AGRO1 Ecosystèmes et Ressources Aquatiques, INAT (Institut National Agronomique de Tunisie), Tunis, Tunisia, **2** UMR 212 Ecosystèmes Marins Exploités, IRD (Institut de Recherche pour le Développement), Sète, France, **3** Département de biologie, Chimie et Géographie, UQAR (Université du Québec à Rimouski), Québec, Canada, **4** Laboratoire du milieu marin, INSTM (Institut National des Sciences et Technologies de la Mer), Salammbô, Tunisia, **5** Laboratoire Ecologie des Systèmes Marins Côtiers UMR 5119, UM2 (Université de Montpellier 2), Montpellier, France

Abstract

Bottom trawl survey data are commonly used as a sampling technique to assess the spatial distribution of commercial species. However, this sampling technique does not always correctly detect a species even when it is present, and this can create significant limitations when fitting species distribution models. In this study, we aim to test the relevance of a mixed methodological approach that combines presence-only and presence-absence distribution models. We illustrate this approach using bottom trawl survey data to model the spatial distributions of 27 commercially targeted marine species. We use an environmentally- and geographically-weighted method to simulate pseudo-absence data. The species distributions are modelled using regression kriging, a technique that explicitly incorporates spatial dependence into predictions. Model outputs are then used to identify areas that met the conservation targets for the deployment of artificial anti-trawling reefs. To achieve this, we propose the use of a fuzzy logic framework that accounts for the uncertainty associated with different model predictions. For each species, the predictive accuracy of the model is classified as 'high'. A better result is observed when a large number of occurrences are used to develop the model. The map resulting from the fuzzy overlay shows that three main areas have a high level of agreement with the conservation criteria. These results align with expert opinion, confirming the relevance of the proposed methodology in this study.

Editor: Joshua S. Madin, Macquarie University, Australia

Funding: This work was partly funded by the Institut de Recherche pour le Développement(http://www.ird.fr/) and by the 'Fondation pour la Recherche sur la Biodiversité' (http://www.fondationbiodiversite.fr/). The funders had no role in study design, data collection and analysis, decision to publish, or preparation of the manuscript.

Competing Interests: The authors have declared that no competing interests exist.

* E-mail: hattab.tarek@gmail.com

Introduction

Understanding species distributions is essential for conservation planning and forecasting [1], particularly in the present context of stock depletion and species extinction. Species distribution maps play an important role in developing spatial management measures such as the identification of Essential Fish Habitats or the establishment Marine Protected Areas (MPAs) [2,3], which in turn contribute to sustainable ecosystem-based marine management [3–5]. However, information on the true distribution of many marine organisms remains limited, particularly for species that are difficult to detect [6]. Modelling species distributions based on data samples is one solution to address this lack of knowledge. For instance, Species Distribution Models (SDMs) relate species' distributions based on data samples with the associated environmental and geographical characteristics of the surveyed locations [7].

Nevertheless, the perception remains that the distributions of marine species are uncertain and dependent on the sampling process used to generate the models. Bottom trawls are commonly used as a sampling technique to assess the spatial distribution of commercial species and to obtain fisheries-independent abundance data [8]. However, there are some limitations in the species detections associated with this technique due to a range of factors, e.g., catchability, gear efficiency, and gear-specific selectivity [8]. A common problem in recording species' distribution results from a false absence, which occurs when a species is not available for capture despite occupying the site, or when a species occurs at a site but is simply not captured. Such data can severely limit the fit of many SDMs [9] and can decrease the reliability of prediction models (see [10,11]). The issue of false absences is further complicated by locations with favourable environmental conditions but where species are absent due to biotic interactions, dispersal limitations or fishing pressure. This latter case is particularly critical when modelling commercial exploited species [12]. Yet the problem of imperfect detection when modelling marine species distributions has been rarely mentioned in the literature [13].

Confirmed absences are very difficult to obtain, especially from bottom trawl survey data and for mobile species. A much higher level of sampling effort is required to ensure their reliability relative to presence data [14]. To cope with the lack of confirmed absence data, presence-only models, or profile techniques, have often been used [12]. These models differ from the group discrimination approaches that require presence–absence or abundance data [15]. Some well-known examples of profile-type models are the Ecological Niche Factor Analysis (ENFA; [16]), the Genetic Algorithm for Rule-Set Prediction [17], and the maximum entropy method [18]. Comparisons between the various SDMs reveal that group discrimination approaches tend to perform better than the profile-type, or presence-only models [19–21]. A common problem of the profile-type models is that their predictions are often overly optimistic, i.e., they predict the species occurring at too many locations [22].

In order to use group discrimination approaches, artificial absence data are increasingly being used in situations where no confirmed absence data are available. Typically called pseudo-absences, artificial absences are generated and inserted into the selected model in lieu of confirmed absence data. The method selected to generate the pseudo-absences is particularly important because it can influence the final quality of the model [21,22]. Several approaches have been suggested for the generation of pseudo-absences: *(i)* a random selection of absence points across the entire available area (e.g., [21,23]; *(ii)* a random selection with a geographically-weighted exclusion (e.g., [24]; and *(iii)* a selection of sampled locations (i.e., occurrence locations for other species) at which the target species has not been recorded (e.g., [5]). However, there are several issues with these approaches. The first two may produce false absences, even in environmentally-favourable areas for the species [25,26]. The latter approach is unsuitable for bottom trawl survey data because it is unlikely that an entire target group of species caught by a particular type of fishing gear will share a similar sampling bias.

To deal with presence-only records, Engler *et al.* [22] suggested an intermediate methodology between presence-only and presence-absence distribution models. This approach proposed the use of a habitat suitability map as a way of selecting weighted pseudo-absence data points. These points are added to the original presence-only data and used to improve the logistic regression procedure. This mixed method could be more suitable for bottom trawl survey data than the other approaches outlined above.

Distribution models often result in a predicted probability surface that is then translated into a presence-absence classification map for use in different conservation applications [27]. For instance, these presence-absence maps are commonly aggregated (the 'predict first, assemble later' strategy from [28]) to identify areas that will experience the greatest changes in species composition due to climate change [29]. To convert a probability surface into a binary map, a number of threshold selection methods have been proposed [27]. Given the variety of approaches available to generate a dataset, the method chosen can have a dramatic effect on a model's accuracy and its predictions [27,30], as well as the subsequent conservation planning decisions and outcomes [31]. To address the uncertainty in the presence-absence classification map, we propose the use of a fuzzy logic approach that can be applied in situations where vagueness and uncertainty exist. The fuzzy logic approach has several advantages in situations where: *(i)* there are no clear cut definitions; and *(ii)* results cannot be categorised as either 0 or 1 [32]. Given its current application in a range of scientific disciplines [33], it is logical to extend the use of fuzzy logic to the threshold selection procedure in the context of SDMs.

Predicting the distributions of commercially targeted marine species is particularly urgent in over-exploited and damaged ecosystems, such as the Gulf of Gabes. With a soft bottom, shallow slope and a high diversity of fishes, the Gulf of Gabes is the most important fishing ground in Tunisia [34,35]. This coastal area supports 60% of the national fishing fleet and contributes 42% of the national annual fish and crustacean production [36]. This intense fishing activity began in the early 1980s, after which fisheries experienced large fluctuations in landings until the late 1990s when catches began to progressively decline [35]. This was mainly due to the decreasing of demersal stocks caused by intense bottom trawling activities. Depletion rates are so alarming that in the near future the Gulf of Gabes will be subject to a habitat conservation management plan that excludes trawling activities.

In this study, we initially assess the reliability of the mixed approach proposed by Hengl *et al.* [37]. We combine ENFA predictions for generating pseudo-absences and regression-kriging (RK) for modelling the spatial distribution of 27 commercially-targeted species in the Gulf of Gabes based on bottom trawls survey data. We then propose a fuzzy logic framework to transform modelled probabilities of occurrence into binary predictions of species presence and absence. We illustrate this framework by applying it to the task of identifying areas that meet the conservation targets for the deployment of artificial anti-trawling reefs (AARs) in the Gulf of Gabes.

Materials and Methods

Study Area

The Gulf of Gabes is located in the southern Mediterranean Sea and covers the second widest continental shelf area of this semi-enclosed sea. The Gulf of Gabes has high fisheries productivity and it serves as a feeding and reproduction area for numerous populations of fishes and crustaceans [34]. Indeed, this ecosystem supports one of the most extensive biocenosis of seagrass (*Posidonia oceanica*; [38]), which constitutes a major nursery site for several marine species [39]. Accordingly, the Gulf of Gabes is one of the most productive ecosystems in the Mediterranean Sea and has great economic and ecological importance [38].

Bottom trawling is the predominant fishing activity in this area and the gear type that has the largest impact on the target demersal fishes [34]. The regular incursions of trawlers into areas that are shallower than their regulated depth have led to the extensive degradation of *P. oceanica* meadows [40]. Due to the lack of monitoring and surveillance activities carried out by the marine police and fishery authorities, illegal fishing still takes place. Consequently, effective management measures are required to prevent illegal fishing activities.

The proposed fisheries management plan for the Gulf of Gabes includes a perimeter of AARs that combine anti-trawling structures with artificial reefs. The major functions of these structures will be to protect the coastal zone marine ecosystems and species from the mechanical impacts of trawling. This measure will be especially important for high diversity communities, such as the *P. oceanica* seagrass beds and associated fauna of biological interest [41]. In addition, AARs aim to reduce the fishing mortality of commercial species, to protect nursery areas and juvenile fish, to create new fishing grounds and/or improve existing grounds, and to increase natural productivity [42]. The AARs will be deployed in areas that are chosen by considering both the presence of favourable habitats of commercial species and areas of high-density *P. oceanica* seagrass beds.

Bottom Trawl Survey Data and Model Variables

Species occurrence data were collected from the Tunisian bottom trawl survey database gathered by the National Institute of Marine Sciences and Technologies (INSTM, Tunisia) onboard the R/V Hannibal. The sampling net used (vertical opening trawl, GOV: 42/55) had a 20 mm diamond stretched mesh at the cod end and a 15 m horizontal opening. The trawl was towed at a speed of 2.9 knots for one hour and sampling areas were fixed

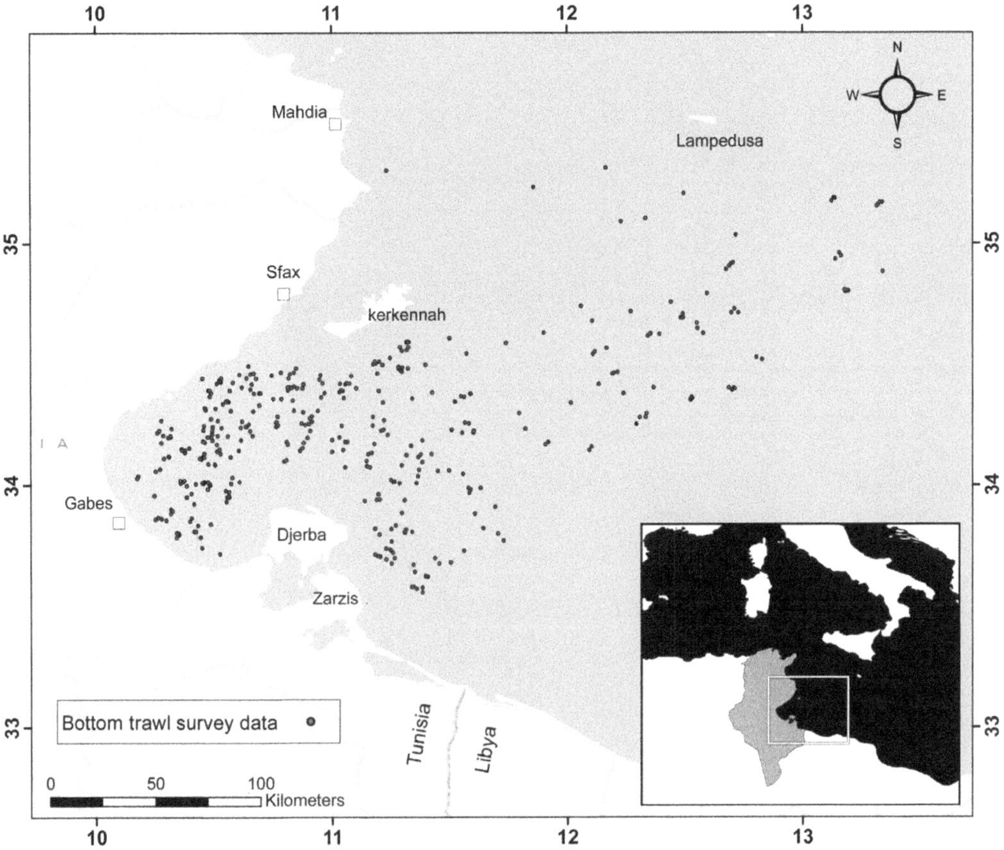

Figure 1. The Gulf of Gabes study area with black points indicating the location of the benthic trawls used in this study.

according to a stratified random sampling based on three depth strata (0–25 m, 25–65 m, and 65–200 m). A total of 360 trawl hauls were completed between 1998 and 2005. Central point geo-referenced position data for each trawl haul and the associated catch contents were extracted from the database (Figure 1).

To build the SDMs, we selected four local habitat variables (depth, slope, aspect, and seafloor type) and one spatial predictor (distance to shore). Previous studies undertaken at a similar spatial extent showed that these variables had a strong influence on species distributions in coastal environments (Figure 2) (e.g., slope [43,44], depth [5,45], aspect [43,44,46], distance to shore [47,48], and seafloor type [5]).

The bathymetry of the Gulf of Gabes was extracted as a digital gridded depth data set from a digital elevation model with a 90 m resolution. The bathymetric elevation data was derived from source soundings collected by the INSTM and referenced to the local tidal datum. The seafloor bathymetric slope and aspect were derived from the bathymetric base map using a 3×3 cell neighbourhood window around the processing cell. Respectively, these represent the rate of change in bathymetry and the azimuthal direction of the steepest slope over the analysis window. The aspect was transformed into two derived variables: Eastness (values close to 1 represent an eastward aspect, while values close to −1 represent a westward aspect) and Northness (values close to 1 represent a northward aspect, while values close to −1 represent a southward aspect).

To develop the map of the distance to shore, Euclidean distances were calculated from the shorelines and islands throughout the study area based on a gridded map with a 90 m resolution. Release and recapture locations were then sampled using ArcGIS 10 to obtain distance values. Digital seafloor type data were obtained from the INSTM. The original 15 seafloor types were grouped into eight broader seafloor types that represent the relatively distinct physical environments thought to influence the distributions of demersal marine species. We mapped the seafloor type by attributing a seafloor category to the center of each 0.0081 km^2 grid cell.

Selected Species

Since the advent of bottom trawl surveys in the Gulf of Gabes, 152 different species of fishes, cephalopods, and crustaceans have been identified. As small sample sizes pose challenges to any statistical analyses and result in decreased predictive potential, we decided to concentrate on the relatively common species (defined here as species present in >10% of trawl hauls). Having applied this criterion we retained a total of 27 species: 20 fishes, four cephalopods, two decapods, and one stomatopod (Table 1). Collectively, these species constituted 60% of total biomass of the landings and 98% of bottom trawl landings in the Gulf of Gabes [36].

The Modelling Framework

Confirmed absence data were not available for the 27 species selected for this study. As group discrimination approaches usually perform better than presence-only models (as discussed above), we selected a hybrid approach. This approach combines group discrimination and profile techniques to model species distributions (Figure 3) [22,25,37,49].

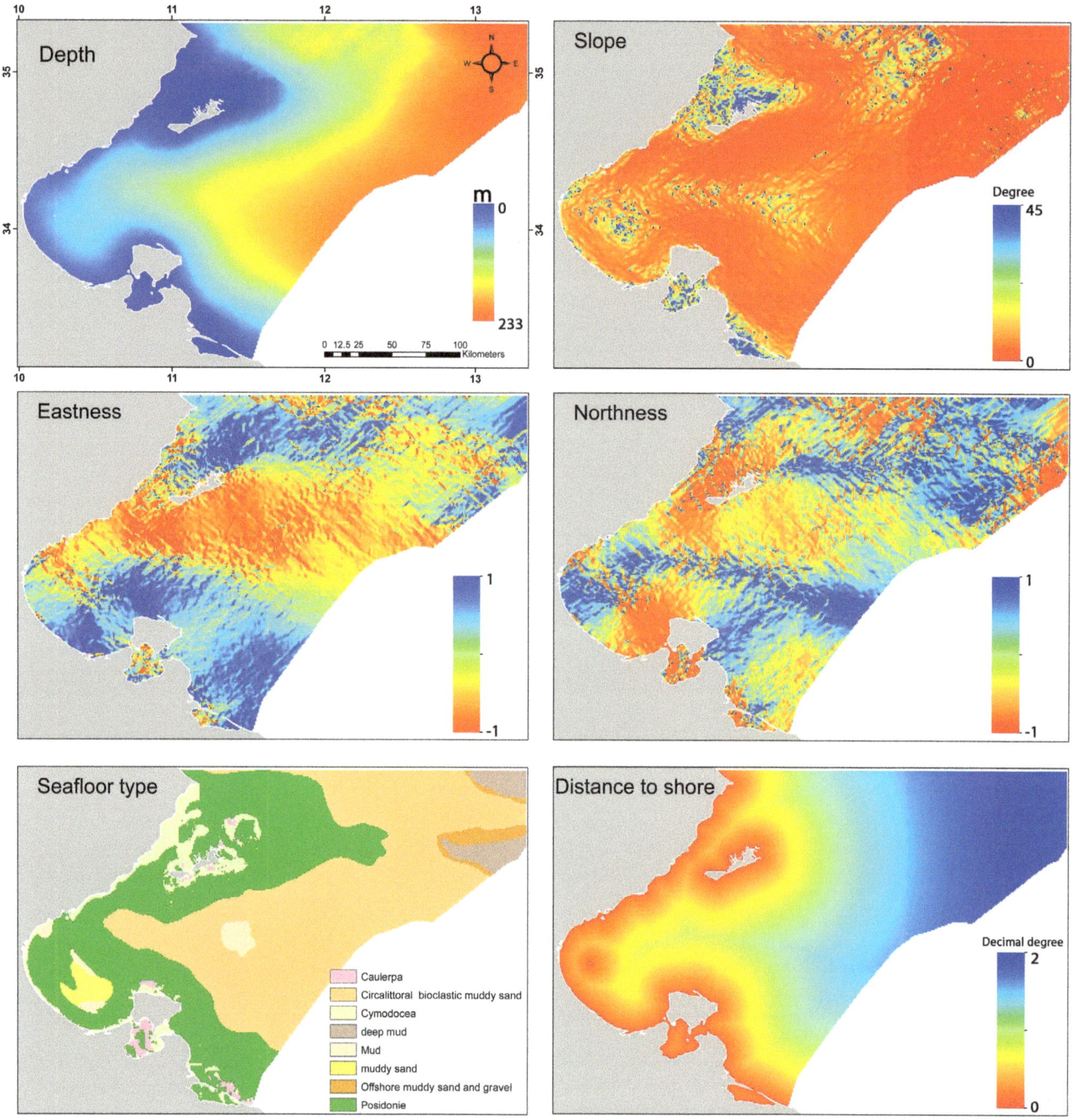

Figure 2. Maps of the six habitat variables (including two derived variables) used to generate the predicted distribution maps.

Presence-only models. ENFA [16] was used to create an habitat suitability map that depicts areas where species are unlikely to occur. ENFA is a specific ordination technique that compares a species' environmental niche and the environmental characteristics of the study area and assigns a degree of suitability to each point on a map (typically from 0 to 100). Thus, it quantifies the dissimilarity between an ecological niche and the ecological space. The first component of the technique, called the marginality factor, is defined as the standardised difference between the centroids of the ecological space and the ecological niche. The second component, the specialisation factors, are successively extracted from the n–1 residual dimensions and represent the narrowness of the ecological niche relative to the ecological space [16]. ENFA was preferred to another widely used presence-only model, namely maximum entropy method, as several studies recently showed that this technique may lead to spurious inferences (e.g., [50]). Then, ENFA is recognized as one of the best presence-only methods to model habitat suitability for marine species (see [6,51]).

To evaluate the accuracy of the ENFA, we performed a Monte-Carlo randomisation test with 100 permutations. This test for the significance of the marginality factor by randomising the locations of selected species within the study area. At each permutation the ENFA was performed on the random locations and the results

Table 1. Code, scientific name and the number of occurrences of each of the 27 species modelled in this study.

Code	Scientific name	Number of occurrences	Code	Scientific name	Number of occurrences
BALICAR	*Balistes carolinensi*	87	OCTOVUL	*Octopus vulgaris*	77
BOOPBOO	*Boops boops*	299	PAGEERY	*Pagellus erythrinus*	223
DIPLANN	*Diplodus annularis*	190	PAGRCAE	*Pagrus caeruleosticutus*	85
ELEDMOS	*Eledone moschata*	64	PENAKER	*Penaeus kerathurus*	219
ENGRENC	*Engraulis encrasicolus*	47	POMTSAL	*Pomatomus saltator*	51
GOBINIG	*Gobius niger*	61	SCOMSCO	*Scomber scombrus*	95
LITHMOR	*Lithognathus mormyrus*	111	SEPIOFF	*Sepia officinalis*	270
LOLIVUL	*Loligo vulgaris*	134	SERAHEP	*Serranus hepatus*	34
MERLMER	*Merluccius merluccius*	164	SOLEAEG	*Solea aegyptica*	104
METAMON	*Metapenaus monoceros*	46	SPARAUR	*Sparus aurata*	34
MUGICEP	*Mugil cephalus*	38	SPICMAE	*Spicara maena*	62
MULLBAR	*Mullus barbatus*	191	SQUIMAN	*Squilla mantis*	43
MULLSUR	*Mullus surmuletus*	69	TRACTRA	*Trachurus trachurus*	169
MUSTMUS	*Mustelus mustelus*	62			

were then evaluated against the observed locations. We ran the ENFA using the *R* adehabitat package [52,53].

Simulation of pseudo-absences. To generate pseudo-absence data, two methods were used to choose geographic coordinates: (*i*) at random across the Gulf of Gabes; and (*ii*) weighted by ENFA predictions and the geographical location of presence-only records [37]. We undertook both methods with the aim of assessing the level of improvement gained by using the environmentally- and geographically-weighted method as compared with the random method. The weighted method proposed by Hengl *et al.* [37] is based on both the Habitat Suitability Index (HSI, derived through ENFA) and the distance from the observations that are subsequently used to weight pseudo-absence points (Figure 3). Given that HSI values are scaled between 0 and 100, Hengl *et al.* [37] defined the probability distribution (τ) used to generate the pseudo-absence locations as:

$$\tau(x*) = \left[\frac{d_R(x) + (100 - HSI(x))}{2}\right]^2 \qquad (1)$$

where d_R is the normalised distance in the range [0, 100%], i.e., the distance from the observation points divided by the maximum distance. The square term is used to ensure that there are progressively more pseudo-absences at the edge of low HSI and large distances will approximately follow a Poisson distribution. In this way, pseudo-absences are located both in areas of low HSI (unsuitable habitat) and further away from the occurrence locations.

Based on Equation 1, the HSI and the map with buffers around the occurrences were combined to create a weighted map. We then performed the random generation of points with a probability density proportional to the values of the weighted map. To account for the variability arising from this weighted selection, 10 groups of pseudo-absences were generated. This allowed us to assess the stability of the final predictions for different simulations. For each set, the number of simulated pseudo-absences was equal to the number of presences. This is supported by the statistical theory of model-based designs, also known as "D-designs" [37]. According to this theory, the optimal design to minimise prediction variance is when an equal number of observations are at opposite value extremes [37,54] and there is a higher spreading in the feature space.

A total of 11 groups of pseudo-absences were obtained. One group was generated entirely at random and the remaining 10 groups were weighted by ENFA predictions and the geographical locations of the presence-only records (Figure 3).

Regression modelling. Once the pseudo-absences were simulated, they could then be combined with the occurrence locations to build a regression model to predict the probability distribution of occurrences. Prior to running the regression analysis, the six original habitat predictors were converted to principal components (to reduce their dimensions and the multicollinearity effect) using the Hill-Smith ordination method [55] that deals with mixed variable types (i.e., quantitative and factor).

We used a generalised linear model (GLM) [56] for the regression analysis, assuming a binomial error in the response variable. Model residuals were then analysed by fitting a variogram to assess their level of spatial dependence (Figure 3). Model residuals exhibited no spatial dependence for six species (i.e., the squid, *Loligo vulgaris*; the stomatopod, *Squilla mantis*; the demersal fishes, *Merlucius merlucius*, *Mugil cephalus*, *Mullus barbatus*, and *Mullus surmuletus*; and the pelagic fish, *Trachurus trachurus*). For the remaining species (n = 21), we used logistic Regression Kriging (RK) models that explicitly incorporate spatial dependence into predictions (Figure 3). This method assumes that the model residuals have a spatial structure resulting from either 'model' factors such as incorrectly specified or inadequate predictor variables, or 'real' factors such as biotic processes that cause spatial patterns [57]. It combines the predictions from a regression model along with the resulting kriged residuals [58]. Specifically, the regression modelling was supplemented with the use of variograms to assess the level of spatial dependence among residuals (Figure 3). Regression residuals were then interpolated and added back to the regression estimate (see [58] for more details). Finally, to select the most parsimonious model for each of the selected species, we applied an automatic stepwise model selection using the Akaike Information Criterion [59].

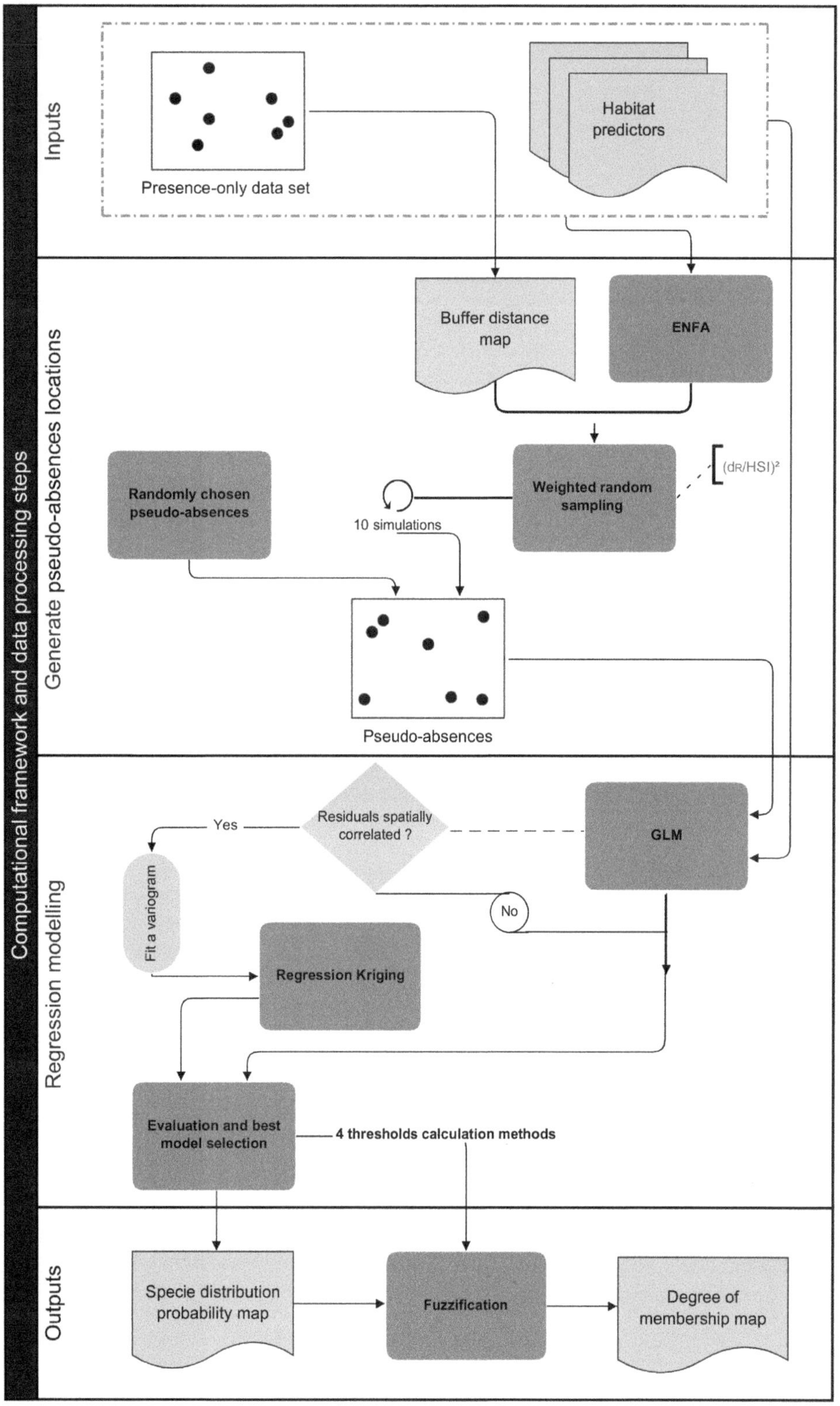

Figure 3. Computational framework and data processing steps.

Model accuracy. For every species, the predictive accuracy of the model was evaluated by a 10-fold cross validation [56]. The receiver operating characteristic curve method was then applied to derive the area under the curve (AUC) index [60] to measure the model's performance.

The simulation of pseudo-absences may generate absences far from the environmental conditions of presences, which may artificially increases the rate of well-predicted absences and hence the AUC scores [24,25,61]. In addition, the AUC test statistic may not always reflect a model's ability to prioritise areas in terms of their habitat suitability relative to alternative models (e.g., [61,62]). Model assessment was therefore supplemented with the Point Biserial Coefficient (PBC) [20,24,63], the sensitivity (presences correctly predicted as presences), and the specificity metrics (absences correctly predicted as absences) [60]. The PBC was calculated as a Pearson's correlation coefficient between the observation in the occurrence dataset (presence (1) or pseudoabsence (0)) and the prediction and therefore takes into account how far the prediction varies from the observation. An independent examination of the percentage of presence and absence errors was recommended by Lobo *et al* [61] to help in the model selection process according to the researcher's goals, rather than the use of a synthetic measure such as the AUC. Predictions were further inspected visually and compared to plotted occurrence data in order to assess their plausibility.

Finally, we used the Pearson's correlation coefficient to calculate the pairwise correlation between the final predictions maps derived from the 10 groups of pseudo-absences. This allowed us to assess the stability of predictions for each of the different weighted simulations of pseudo-absences.

Each data processing step was completed in *R*, drawing on code developed by Hengl *et al.* [37], and automating the calculation for several species simultaneously. For each modelled species, the regression models, 10-fold cross validation, and evaluation procedures were carried out for the 11 simulated groups of pseudo-absences.

Conservation Planning Procedure

Probability of occurrence maps can be generated using the most accurate model selected for each species from the 11 pseudo-absences groups. Once these maps have been converted to presence-absence data, they can be used to identify areas that meet the conservation targets for selection of AAR deployment sites. These targets specify the inclusion of both favourable habitats for commercially-targeted species and areas of high density *P. oceanica* seagrass beds.

Threshold approaches and fuzzy modelling. The most common method used to convert probabilities of occurrence to presence-absence data is the use of an optimum probability threshold [64]. Different methods have been proposed to select a probability threshold [27]. Among these, the most widely used threshold optimisation criteria are:

(1) *Sens = Spec* - The threshold where sensitivity equals specificity (i.e., where positive observations are equally as likely to be wrong as negative observations).
(2) *Max (Sens+Spec)* - The threshold that maximises the sum of sensitivity and specificity (i.e., it minimises the mean error rates for both positive and negative observations). This threshold is equivalent to finding the point on the receiver operating characteristic curve whose tangent has a slope of one.
(3) *MaxKappa* - The threshold that results in the maximum value of Kappa statistic.
(4) *MaxPCC* - The threshold that results in the maximum percent of correctly classified observations.

For each species, a probability threshold was determined using each of the four optimisation criteria outlined above. As these methods do not provide similar threshold values, the use of one method over another can influence conservation planning outcomes, e.g., modifying areas that are expected to be suitable for a given species [27]. To avoid the subjective selection of a particular threshold, predicted distributions of selected species were defined by fuzzy sets theory [65]. Fuzzy logic is useful in circumstances that involve uncertainty, imprecision, and vagueness by replacing the sharp boundary between the suitable and non-suitable classes with the concept of a degree of truth (membership). In the fuzzification process, crisp attribute values (probability of presence) are transformed linearly into a common suitability scale (0 to 1) using the fuzzy linear membership function. A membership value of 0 is assigned to the lowest probability threshold value (as calculated using the four criteria outlined above) and a value of 1 to the highest threshold value (see example in Figure 4). Each map cell was assigned a fuzzy membership value resulting from the fuzzy linear membership function.

Fuzzy overlay of conservation criteria. For each species, the favourability value of an area is defined as the degree of membership of that area to the fuzzy set of favourable areas for the species. Fuzzified inputs can be combined together to identify the most favourable area for the majority of commercial species by using a fuzzy operator. The fuzzy AND operator was applied to return the minimum of the fuzzy memberships from the fuzzy input maps. The result of this aggregation is a final fuzzy set expressing the site suitability for all key species (see example of the method in Figure 4). To identify areas with a high density of seagrass beds, a recent map showing seagrass recovery rates was obtained from the INSTM. This map was fuzzified using a fuzzy linear membership function that assigned a membership value of 0 for recovery rates less than 30% and a value of 1 for recovery rates greater than 60%. Finally, the maps that express the fuzzy memberships of site suitability for all key species and seagrass recovery rates were combined together using the fuzzy AND operator. For example, "IF the favourability value of an area for the species 1 IS high, AND the favourability value of an area for the species 2 IS high, etc..., AND the recovery rate of seagrass IS high THEN the area has high conservation criteria".

Results and Discussion

The Monte-Carlo randomisation tests show that the ENFA marginality factor is highly significant for each modelled species (all $p<0.001$). This implies that the habitat occupied by the species modelled differ unequivocally from the average environmental conditions found in the broader study area. This indicates that a species-specific habitat selection process takes place. The ENFA also show that the potential distributions of species are much larger than their realised distributions, based on sampling locations (see examples of three species in Figure 5). In contrast, incorporating pseudo-absence data into logistic RK models results in predicted distributions that are closer to the realised distributions (Figure 5).

When the pseudo-absence data are randomly selected, the potential distribution of the species are less extensive in comparison with those obtained using a weighted selection. This difference can be explained by the fact that a random simulation can select absences between observed occurrences points, therefore generating pseudo-absences in favourable areas. This can subsequently lead to an underestimation in the realised distributions [25,66].

For each method generating pseudo-absences, the distribution of the AUC, PBC, the sensitivity and the specificity metrics is shown in Figure 6. For each species, model accuracy differs according to the method used to generate pseudo-absences. All the models based on the environmentally- and geographically-

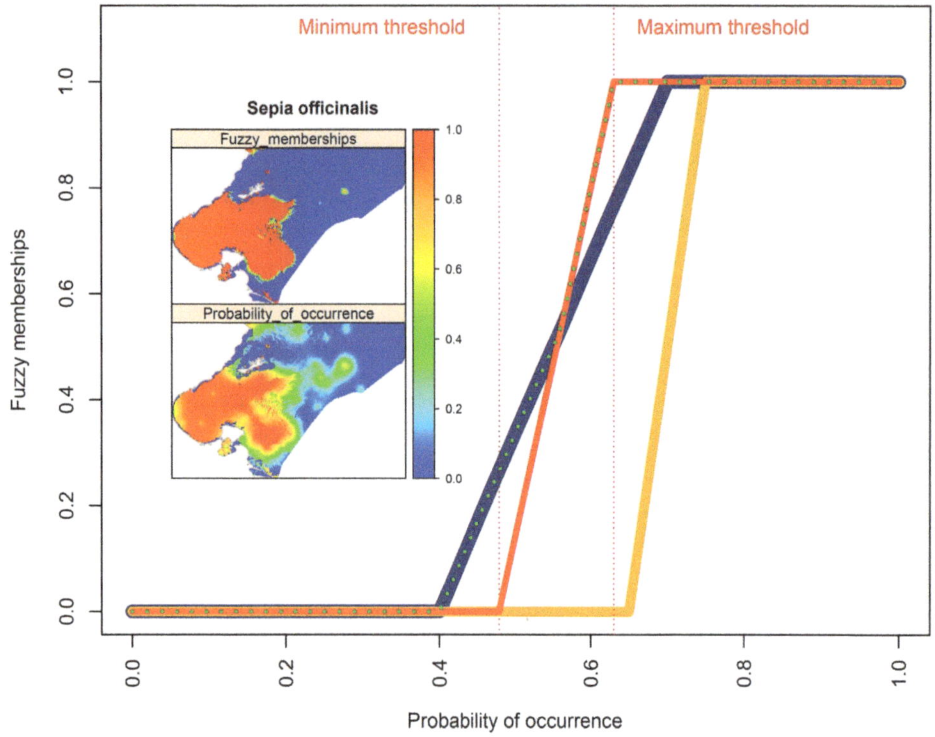

Figure 4. Threshold approaches and fuzzy modelling. An illustration of the fuzzification process performed using a fuzzy linear membership function. The red, orange, and blue curves represent the fuzzy membership sets of three different species. The dotted green line represents the fuzzy AND overlay outcomes. The inset maps display an example of the fuzzification process for the cuttlefish *(Sepia officinalis)* which corresponds with the red curve.

weighted method achieved a high level of accuracy as indicated by the four measures of model accuracy (Mean±standard deviation: 0.9±0.07, 0.65±0.23, 0.87±0.08 and 0.85±0.09 respectively for the AUC, PBC, sensitivity and specificity) (Figure 6). These values were significantly higher (Wilcoxon rank test: $p<0.001$) than those obtained by models based on the random selection method (Mean±standard deviation: 0.78±0.1, 0.36±0.26, 0.77±0.11 and 0.69±0.1 for the AUC, PBC, sensitivity and specificity, respectively).

Since all the resulting SDMs are based on pseudo-absences, both specificity and AUC scores estimate the degree of accuracy of the absence information used in the model training process. Thus, a high specificity score only implies that most of the data considered as absence data are correctly predicted and does not imply a high performance in the prediction of the unknown true absences. However, the sensitivity and PBC values were higher when the weighted method of generating pseudo-absence was used, implying a high performance of this method in the predictions of the known true presences as compared with the random method.

Models based on the weighted method of generating pseudo-absence data provide significantly better results on average, which aligns with the results of Engler *et al.,* Chefaoui & Lobo and Hengl *et al.* [22,25,37]. It contrasts, however, with the findings of Wisz & Guisan and Barbet-Massin *et al.* [19,26]. Using virtual species, these authors found that randomly-selected pseudo-absences yielded the most reliable species distribution models. However, this may be explained by the fact that both studies used a large number of pseudo-absences (e.g., 10.000 data points), whereas this study used the same number of simulated pseudo-absences as the number of occurrences. Currently, there is no consensus on the number of pseudo-absences that are required to optimise model predictions. Some authors suggest a ratio of 10:1 (e.g., [25,49]) while others recommend using large numbers of pseudo-absences when they are randomly selected (e.g., [20,26]). Intuitively, it makes sense to generate an equal number of pseudo-absence data points as occurrence data points [58,67] to avoid the bias caused by a presence-absence ratio that is too low [68,69]. Indeed this was the outcome in McPherson *et al.* [70]who found SDMs had the best predictive accuracy when prevalence values (the proportion of data points representing a species' presence) were around 0.5. Therefore, several authors recommend resampling the training data to balance presence and absence data points [70,71].

For each species, the range of pairwise correlations between the final habitat prediction maps (derived from each of the 10 groups of pseudo-absences) varied according to the number of occurrences used to develop the models (Figure 7). The lowest correlation occurs when the number of occurrences is less than 61. The array of evaluation measures, based on all the replicated runs, does not show a clear trend in relation to the number of occurrences used to fit the models. Even with a small number of occurrences, values of AUC and PBC indicate excellent predictive accuracy of the models. However, large fluctuations in the predictive accuracy and the models' subsequent predictions are observed when the models are based on a low number of occurrences. Indeed, the selection of pseudo-absences can induce variability in model predictions when several runs are made with a small set of occurrence data, each run having its own dataset for calibration.

Williams [72] found that the predictive ability of some ecological modelling approaches varies with a species' detectability. While presence-absence approaches generally have higher

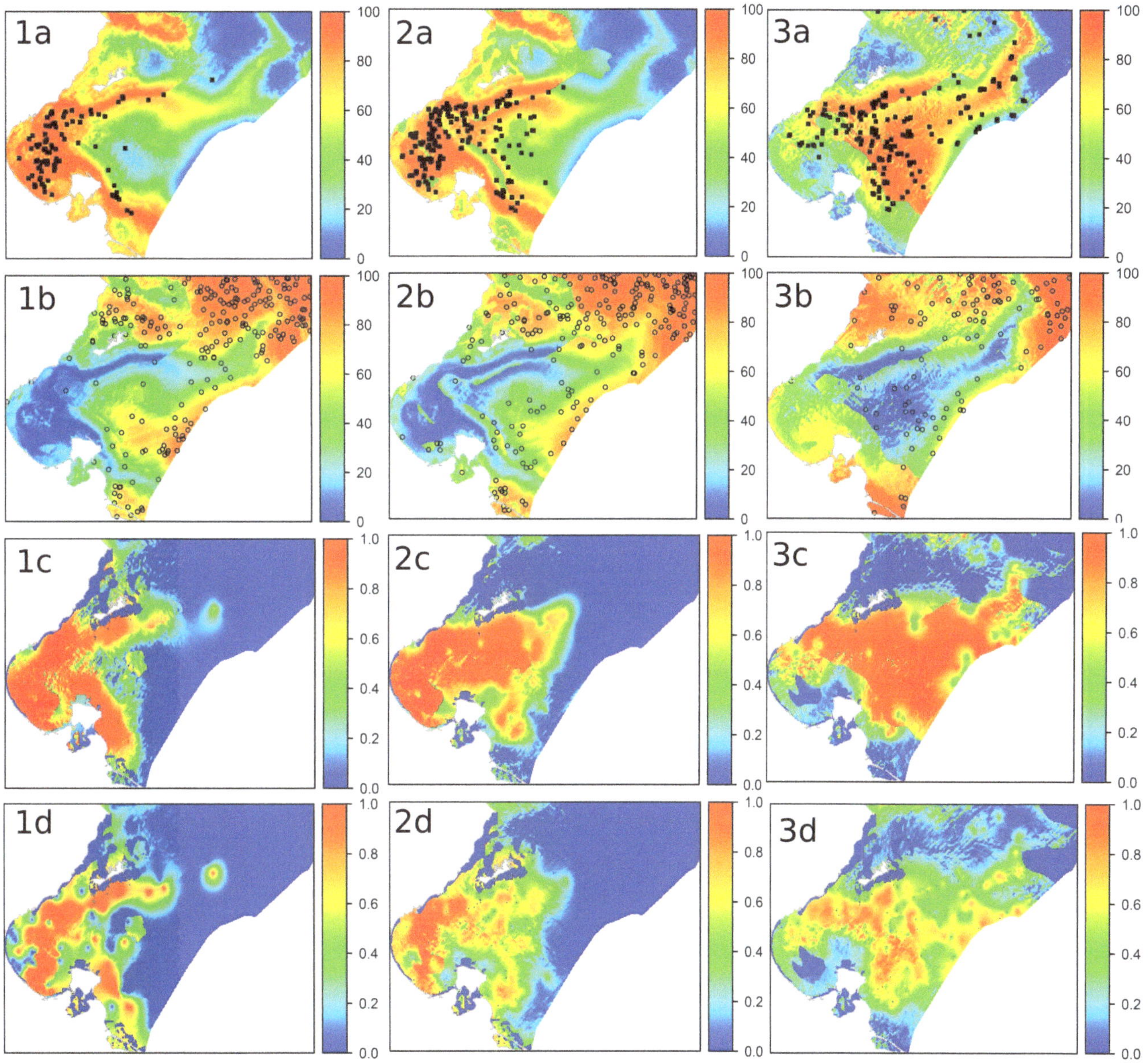

Figure 5. Spatial prediction maps in the Gulf of Gabes for three example species. (1) *Lithognathus mormyrus,* (2) *Penaeus kerathurus,* (3) *Pagellus erythrinus* (a) the habitat suitability index map with presence-only data; (b) the weighted map and the randomly-generated pseudo-absences using the Equation 1; (c) probabilities predicted using the binomial regression-kriging (RK) with a weighted selection of pseudo-absences; and (d) probabilities predicted using the binomial RK with a random selection of pseudo-absences.

predictive abilities for species with a high detectability, they do not perform as well as presence-only approaches when detectability is low. However, for some species such as the pelagic fish species, *Trachurus trachurus*, the fitted model is found to be moderately accurate though it is based on a relatively high number of data points (338 presences/pseudo-absences). This result may be related to the fact that the selected predictors (e.g., depth, seafloor type) have greater ecological significance to model the distribution of demersal and benthic species. [66]. This implies that the performance of SDMs strongly depends on the environmental predictors, which in turn depend of the type of organism that is studied [62]. In addition, the realism and the robustness of models may have been influenced by our automatic variable selection procedure [73]. Indeed, ecologically important variables may have been excluded from the stepwise models and/or non-meaningful variables may have been incorporated into models. This was particularly the case for some of the studied pelagic species (*Trachurus trachurus, Spicara maena, Scomber scombrus*). These species were therefore excluded from the conservation planning procedure.

Overall, our results suggest that simulating pseudo-absences with an environmentally and geographically weighted method rather than a purely heuristic approach enhances the accuracy of predictions. This method provides a robust result when a relatively large number of occurrence data points with good spatial coverage are used. The RK method also shows great potential as an approach to incorporate spatial dependence in SDMs, by combining information on species-habitat relationships (i.e., through the deterministic model) and error components. We believe that the combined ENFA and RK method has several

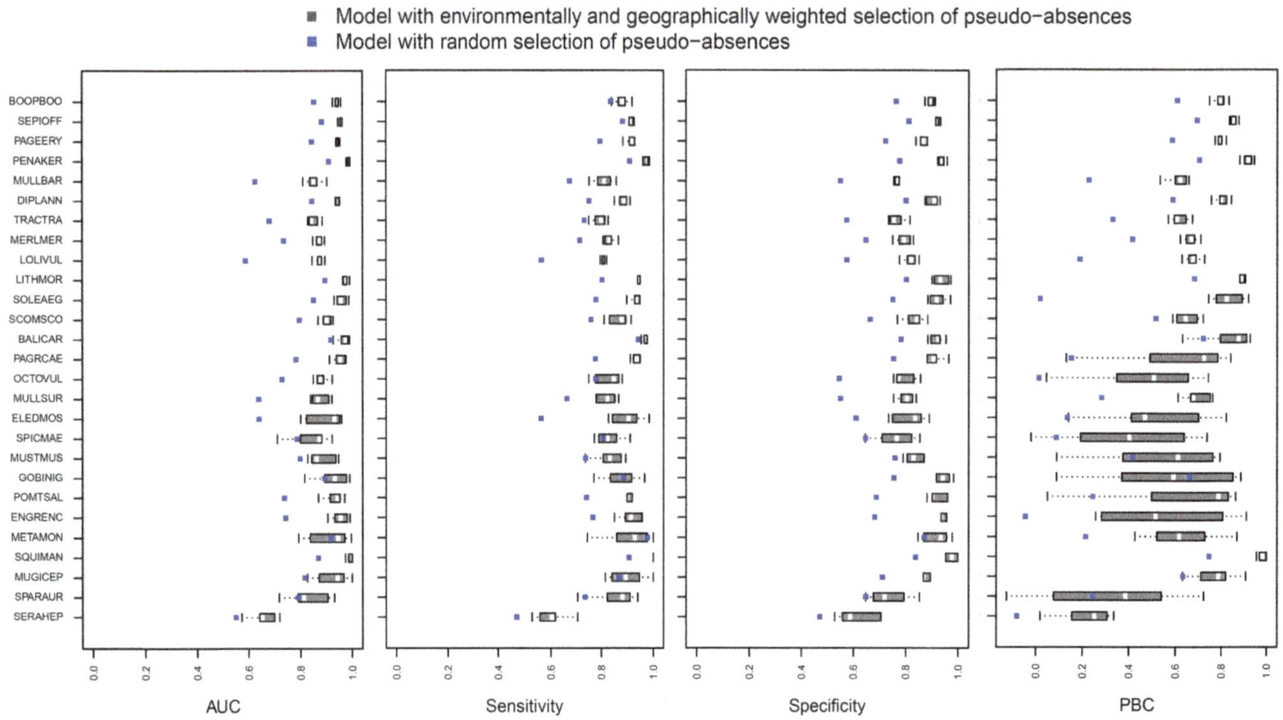

Figure 6. Distributions of the area under the curve (AUC), PBC, sensitivity and specificity values. Calculated AUC, PBC, sensitivity and specificity values for the 27 modelled species (species codes are listed in Table 1). Species are sorted (top to bottom) by decreasing number of occurrences (used to develop the models).

advantages when applied to trawl survey data, especially when addressing their imperfect ability to detect a species. Furthermore, this method applies both the spatial auto-correlation structure and the trend component of the spatial variation to make spatial predictions of species' distributions. This method can be applied to

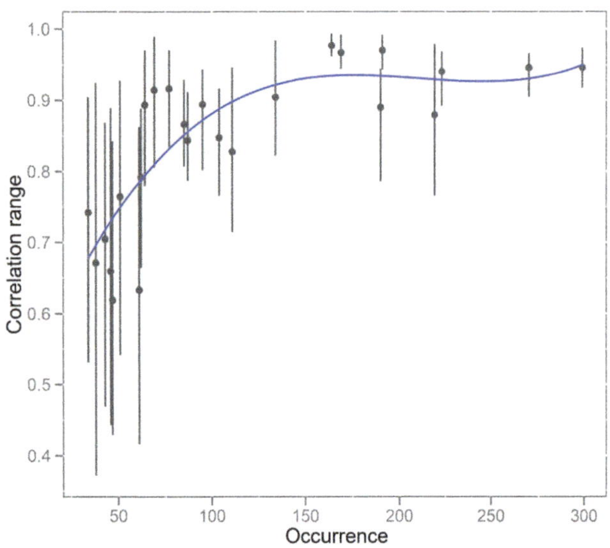

Figure 7. Stability of predictions. The range of Pearson's pairwise correlation coefficients between the final prediction maps according to the number of occurrences used to develop the models (red line: smooth curve fitted by Loess function).

other areas where survey data are available, such as the Medits (International bottom trawl surveys in the Mediterranean) dataset.

The second purpose of this study is to identify the areas required to meet the conservation targets of AARs based on probabilistic predicted maps. In this study we propose a novel method to develop these maps that uses fuzzy sets to address the uncertainty associated with the selection of probability threshold optimisation criteria. Fuzzy logic is interested in capturing partial truths, that is, how to reason about things that are not wholly true or false; while probability is concerned with making predictions about events based on a partial state of knowledge [65]. The fuzzy sets theory is used to transform the probability of presence into a membership degree using not only a single threshold value but several values obtained with different cut-off threshold optimization criteria.

Of the 27 species modelled in this study, an excellent or high predictive accuracy is only found for 12 benthic and demersal species. It was these species that were selected for the fuzzy overlay of conservation criteria. Of these, the 2004–2005 stock assessment results for the Gulf of Gabes report that *Solea aegyptiaca, Octopus vulgaris*, and *Sepia officinalis* were fully exploited (Othman Jarboui, personal communication: INSTM) and *Mullus barbatus, Pagrus caeruleostictis*, and *Pagellus erythrinus* were overexploited.

The resultant fuzzy overlay map (Figure 8) highlights three main areas that meet the conservation criterion to a high level. The largest area is located south of Kerkennah Island while the remaining two areas are in the coastal region off Mahres Harbour and north of Jerba Island. As well as meeting the AAR conservation criteria, the Kerkennah Island area has already been reported as a biodiversity hotspot for megabenthic fauna [74] and is currently proposed as a potential MPA [75]. The area north of Jerba Island was also proposed as suitable for MPA establishment by Ben Mustapha & Afli [75]. In addition to its dense seagrass

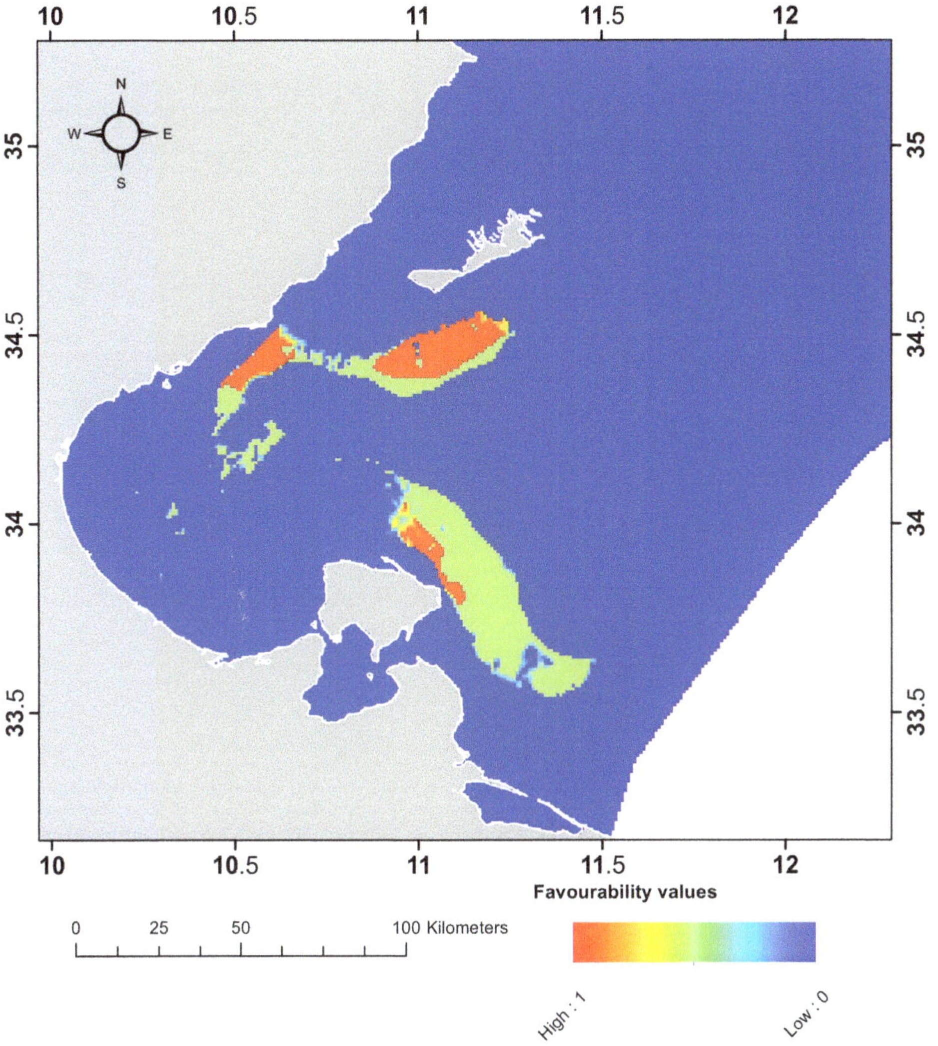

Figure 8. Map of favourability values for the deployment of AARs.

beds, the area is also characterised by coralligenous assemblages and is a recognised nursery site for several juvenile of commercially-targeted species [75]. These expert opinions corroborate our findings and confirm the relevance of the established methodology for the selection of AAR deployment sites.

Conclusions

The comparison between the two pseudo-absence data generation methods reveals that the environmentally- and geographically-weighted method has significant potential to reduce prediction errors in SDMs. When confirmed absence data are not available, we recommend using the method proposed by Hengl *et al.* [37], which combines ENFA and RK predictions to deal with an imperfect ability to detect a species and incorporate spatial dependence in predictions. This study proposes a novel method to developing predicted maps that uses fuzzy sets to address the uncertainty associated with selecting probability threshold criteria. In the context of conservation planning, we illustrate the advantages of this novel method, using it to identify areas within the Gulf of Gabes that meet the conservation targets for AARs. Three key areas that met the conservation criteria to a high level were identified, and these areas are recommended as deployment sites for AARs. The location and spatial arrangement of reef units must be carefully planned. Reef characteristics such as the number of modules, the distance between modules as a function of trawl parameters, and the weight of modules as a function of exposure to currents should all be informed by scientific studies. Additional factors that should be considered in the deployment of AARs are economic costs and socio-economic effects.

The areas with a nonzero favourability value for AAR deployment cover 1578 km^2 and encompass 30% of seagrass beds in good condition. However, these areas represent biodiversity targets without regard to cost. In this context this study can contribute to effective conservation planning in a broader prioritization process, which must be optimized by accounting for the cost of conservation. For instance, future works should focus on systematic conservation planning [76,77] that attempts to solve a cost-effectiveness problem (i.e. how to achieve the most conservation given limited resources). This is particularly important in the Gulf of Gabes where the diverse stakeholder group often holds conflicting values and opinions (e.g., conflicts between professional and local artisanal fishers).

Acknowledgments

The authors would like to acknowledge the National Institute of Marine Sciences and Technologies (INSTM, Tunisia) that contributed unpublished data. We are very grateful to Jane Alpine for English editing. We also thank the two anonymous referees for their helpful comments on the manuscript.

Author Contributions

Conceived and designed the experiments: TH. Performed the experiments: TH. Analyzed the data: TH FLL FBRL. Contributed reagents/materials/analysis tools: TH CS MSR PC. Wrote the paper: TH FLL FBRL FL CA PC CS MSR.

References

1. Harris PT, Whiteway T (2009) High seas marine protected areas: Benthic environmental conservation priorities from a GIS analysis of global ocean biophysical data. Ocean & Coastal Management 52: 22–38. doi:10.1016/j.ocecoaman.2008.09.009.
2. Valavanis VD, Pierce GJ, Zuur AF, Palialexis A, et al (2008) Modelling of essential fish habitat based on remote sensing, spatial analysis and GIS. Hydrobiologia 612: 5–20. doi:10.1007/s10750-008-9493-y.
3. Johnson AF, Jenkins SR, Hiddink JG, Hinz H (2012) Linking temperate demersal fish species to habitat: scales, patterns and future directions. Fish and Fisheries. doi:10.1111/j.1467-2979.2012.00466.x.
4. Degraer S, Verfaillie E, Willems W, Adriaens E, Vincx M, et al (2008) Habitat suitability modelling as a mapping tool for macrobenthic communities: An example from the Belgian part of the North Sea. Continental Shelf Research 28: 369–379. doi:10.1016/j.csr.2007.09.001.
5. Maxwell DL, Stelzenmüller V, Eastwood PD, Rogers SI (2009) Modelling the spatial distribution of plaice (Pleuronectes platessa), sole (Solea solea) and thornback ray (Raja clavata) in UK waters for marine management and planning. Journal of Sea Research 61: 258–267. doi:10.1016/j.seares.2008.11.008.
6. MacLeod CD, Mandleberg L, Schweder C, Bannon SM, Pierce GJ (2008) A comparison of approaches for modelling the occurrence of marine animals. Hydrobiologia 612: 21–32. doi:10.1007/s10750-008-9491-0.
7. Elith J, Leathwick JR (2009) Species Distribution Models: Ecological explanation and prediction across space and time. Annual Review of Ecology, Evolution, and Systematics 40: 677–697. doi:10.1146/annurev.ecolsys.110308.120159.
8. Hoffman JC, Bonzek CF, Latour RJ (2009) Estimation of Bottom Trawl Catch Efficiency for Two Demersal Fishes, the Atlantic Croaker and White Perch, in Chesapeake Bay. Marine and Coastal Fisheries 1: 255–269. doi:10.1577/C08-048.1.
9. Guisan A, Edwards TC, Hastie T (2002) Generalized linear and generalized additive models in studies of species distributions: setting the scene. Ecological Modelling 157: 89–100.
10. Anderson R. (2003) Real vs. artefactual absences in species distributions: tests for Oryzomys albigularis (Rodentia: Muridae) in Venezuela. Journal of Biogeography 30: 591–605.
11. Loiselle B., Howell C., Graham C., Goerck J., Brooks T, et al (2003) Evitando Dificultades Resultantes del Uso de Modelos de Distribución de Especies en Planeación de Conservación. Conservation Biology 17: 1591–1600. doi:10.1111/j.1523-1739.2003.00233.x.
12. Robinson L, Elith J, Hobday A, Pearson R, Kendall B, et al (2011) Pushing the limits in marine species distribution modelling: lessons from the land present challenges and opportunities. Global Ecology and Biogeography 20: 789–802.
13. Monk J (2013) How long should we ignore imperfect detection of species in the marine environment when modelling their distribution? Fish and Fisheries. doi:10.1111/faf.12039.
14. Mackenzie DI, Royle JA (2005) Designing occupancy studies: general advice and allocating survey effort. Journal of Applied Ecology 42: 1105–1114. doi:10.1111/j.1365-2664.2005.01098.x.
15. Anderson R., Dudik M, Ferrier S, Guisan A, J Hijmans R, et al (2006) Novel methods improve prediction of species' distributions from occurrence data. Ecography 29: 129–151.
16. Hirzel AH, Hausser J, Chessel D, Perrin N (2002) Ecological-niche factor analysis: how to compute habitat-suitability maps without absence data? Ecology 83: 2027–2036.
17. Stockwell D (1999) The GARP modelling system: problems and solutions to automated spatial prediction. International Journal of Geographical Information Science 13: 143–158. doi:10.1080/136588199241391.
18. Phillips SJ, Anderson RP, Schapire RE (2006) Maximum entropy modeling of species geographic distributions. Ecological Modelling 190: 231–259. doi:10.1016/j.ecolmodel.2005.03.026.
19. Segurado P, Araújo MB (2004) An evaluation of methods for modelling species distributions. Journal of Biogeography 31: 1555–1568. doi:10.1111/j.1365-2699.2004.01076.x.
20. Elith J, H. Graham C, P. Anderson R, Dudík M, Ferrier S, et al (2006) Novel methods improve prediction of species' distributions from occurrence data. Ecography 29: 129–151. doi:10.1111/j.2006.0906-7590.04596.x.
21. Barbet-Massin M, Jiguet F, Albert CH, Thuiller W (2012) Selecting pseudo-absences for species distribution models: how, where and how many? Methods in Ecology and Evolution 3: 327–338. doi:10.1111/j.2041-210X.2011.00172.x.
22. Engler R, Guisan A, Rechsteiner L (2004) An improved approach for predicting the distribution of rare and endangered species from occurrence and pseudo-absence data. Journal of Applied Ecology 41: 263–274.
23. Torres LG, Read AJ, Halpin P (2008) Fine-scale habitat modeling of a top marine predator: do prey data improve predictive capacity. Ecological Applications 18: 1702–1717.
24. Jones MC, Dye SR, Pinnegar JK, Warren R, Cheung WW (2012) Modelling commercial fish distributions: Prediction and assessment using different approaches. Ecological Modelling 225: 133–145.
25. Chefaoui RM, Lobo JM (2008) Assessing the effects of pseudo-absences on predictive distribution model performance. Ecological Modelling 210: 478–486. doi:10.1016/j.ecolmodel.2007.08.010.
26. Wisz MS, Guisan A (2009) Do pseudo-absence selection strategies influence species distribution models and their predictions? An information-theoretic approach based on simulated data. BMC Ecology 9: 8. doi:10.1186/1472-6785-9-8.
27. Freeman EA, Moisen GG (2008) A comparison of the performance of threshold criteria for binary classification in terms of predicted prevalence and kappa. Ecological Modelling 217: 48–58. doi:10.1016/j.ecolmodel.2008.05.015.
28. Ferrier S, Guisan A (2006) Spatial modelling of biodiversity at the community level. Journal of Applied Ecology 43: 393–404. doi:10.1111/j.1365-2664.2006.01149.x.
29. Albouy C, Guilhaumon F, Araújo MB, Mouillot D, Leprieur F (2012) Combining projected changes in species richness and composition reveals climate change impacts on coastal Mediterranean fish assemblages. Global Change Biology 18: 2995–3003. doi:10.1111/j.1365-2486.2012.02772.x.
30. Nenzén HK, Araújo M (2011) Choice of threshold alters projections of species range shifts under climate change. Ecological Modelling 222: 3346–3354.
31. Wilson KA, Westphal MI, Possingham HP, Elith J (2005) Sensitivity of conservation planning to different approaches to using predicted species distribution data. Biological Conservation 122: 99–112. doi:10.1016/j.biocon.2004.07.004.
32. Teh LCL, Teh LSL (2011) A fuzzy logic approach to marine spatial management. Environmental Management 47: 536–545.
33. Montero J (2009) Fuzzy Logic and Science. In: Seising R, editor. Views on Fuzzy Sets and Systems from Different Perspectives. Studies in Fuzziness and Soft Computing. Springer Berlin Heidelberg, Vol. 243. 67–77.
34. Hattour A (1991) Le chalutage dans les eaux Tunisiennes réalités et considérations législatives particulièrement dans les Golfe de Tunis et de Gabès. Note de l'Institut National Scientifique et Technique d'Océanographie et de Pêche de Salammbô: 13p.
35. Najar B, Ben Mariem S, Hadj Ali M (2010) Évolution des profils des débarquements de poissons dans la région de Gabes, Tunisie. Commission International pour l'Exploration Scientifique de la Mer Méditerranée 39: 601p.
36. Direction Générale de la Pêche et de l'Aquaculture (DGPA) (2010) Annuaire des statistiques des pêches en Tunisie. Ministère de l'Agriculture, Tunisie.
37. Hengl T, Sierdsema H, Radović A, Dilo A (2009) Spatial prediction of species' distributions from occurrence-only records: combining point pattern analysis, ENFA and regression-kriging. Ecological Modelling 220: 3499–3511. doi:10.1016/j.ecolmodel.2009.06.038.
38. Batisse M, De Grissa AJ (1995) Marine Region 3: Mediterranean. In G Kelleher, C Bleakley and S Wells (eds) A Global Representative System of Marine Protected Areas Vol I, The Great Barrier Reef Marine Park Authority, The World Bank and IUCN, Washington, DC: 77–104.
39. Francour P (1997) Fish Assemblages of Posidonia oceanica Beds at Port-Cros (France, NW Mediterranean): Assessment of Composition and Long-Term Fluctuations by Visual Census. Marine Ecology 18: 157–173. doi:10.1111/j.1439-0485.1997.tb00434.x.
40. Ben Mustapha K, Hattour A, Mhetli M, El Abed A, Tritar B (1999) Etat de la Bionomie Benthique des Etages Infra et Circalittoral du Golfe de Gabès. Bulletin de l'Institut National des Sciences et Technologies de la Mer 26: 5–48.
41. Jensen A, Collins K, Lockwood AP (2000) Artificial Reefs in European Seas. Lockwood, A.P. (Eds.). Springer. 508 p.
42. Munoz-Perez JJ, Mas G, Jose M, Naranjo JM, Torres E, et al (2000) Position and monitoring of anti-trawling reefs in the Cape of Trafalgar (Gulf of Cadiz, SW Spain). Bulletin of Marine Science 67: 761–772.
43. Moore C., Harvey E., Niel K (2010) The application of predicted habitat models to investigate the spatial ecology of demersal fish assemblages. Marine Biology 157: 2717–2729. doi:10.1007/s00227-010-1531-4.
44. Monk J, Ierodiaconou D, Bellgrove A, Harvey E, Laurenson L (2011) Remotely sensed hydroacoustics and observation data for predicting fish habitat suitability. Continental Shelf Research 31: 17–27. doi:10.1016/j.csr.2010.02.012.
45. Katsanevakis S, Maravelias CD (2008) Bathymetric distribution of demersal fish in the Aegean and Ionian Seas based on generalized additive modeling. Fisheries Science 75: 13–23. doi:10.1007/s12562-008-0033-5.

46. Pittman SJ, Brown KA (2011) Multi-Scale Approach for Predicting Fish Species Distributions across Coral Reef Seascapes. PLoS ONE 6: e20583. doi:10.1371/journal.pone.0020583.
47. Palialexis A, Georgakarakos S, Karakassis I, Lika K, Valavanis VD (2011) Prediction of marine species distribution from presence–absence acoustic data: comparing the fitting efficiency and the predictive capacity of conventional and novel distribution models. Hydrobiologia 670: 241–266. doi:10.1007/s10750-011-0673-9.
48. Monk J, Ierodiaconou D, Harvey E, Rattray A, Versace VL (2012) Are We Predicting the Actual or Apparent Distribution of Temperate Marine Fishes? PLoS ONE 7: e34558. doi:10.1371/journal.pone.0034558.
49. Jiménez-Valverde A, Gómez JF, Lobo JM, Baselga A, Hortal J (2008) Challenging species distribution models: the case of Maculinea nausithous in the Iberian Peninsula. Annales Zoologici Fennici. Vol. 45. 200–210.
50. Yackulic CB, Chandler R, Zipkin EF, Royle JA, Nichols JD, Campbell Grant E, Veran S (2012) Presence-only modelling using MAXENT: when can we trust the inferences? Methods in Ecology and Evolution: 236–243. doi:10.1111/2041-210x12004.
51. Monk J, Ierodiaconou D, Versace VL, Bellgrove A, Harvey E, et al (2010) Habitat suitability for marine fishes using presence-only modelling and multibeam sonar. Marine Ecology Progress Series 420: 157–174.
52. Calenge C (2006) The package "adehabitat" for the R software: A tool for the analysis of space and habitat use by animals. Ecological Modelling 197: 516–519. doi:10.1016/j.ecolmodel.2006.03.017.
53. R Development Core Team (2011) R: A Language and Environment for Statistical Computing. Vienna, Austria. Available: http://www.R-project.org/.
54. Montgomery DC (2007) Design and Analysis of Experiments, 6th Edition Set. New York: Wiley. 752 p.
55. Hill M, Smith A (1976) Principal component analysis of taxonomic data with multi-state discrete characters. Taxon 25: 249–255.
56. McCullagh P, Nelder JA (1989) Generalized Linear Models, Second Edition. 2nd edition London. Chapman and Hall. 512 p.
57. Miller J, Franklin J, Aspinall R (2007) Incorporating spatial dependence in predictive vegetation models. Ecological Modelling 202: 225–242. doi:10.1016/j.ecolmodel.2006.12.012.
58. Hengl T, Heuvelink GBM, Rossiter DG (2007) About regression-kriging: From equations to case studies. Computers & Geosciences 33: 1301–1315. doi:10.1016/j.cageo.2007.05.001.
59. Chambers JM, Hastie T, editors (1991) Statistical models in S. London: Chapman & Hall. 608 p.
60. Fielding AH, Bell JF (1997) A review of methods for the assessment of prediction errors in conservation presence-absence models. Environmental Conservation 24: 38–49.
61. Lobo JM, Jiménez-Valverde A, Real R (2008) AUC: a misleading measure of the performance of predictive distribution models. Global Ecology and Biogeography 17: 145–151.
62. Austin M (2007) Species distribution models and ecological theory: A critical assessment and some possible new approaches. Ecological Modelling 200: 1–19. doi:10.1016/j.ecolmodel.2006.07.005.
63. Zheng B, Agresti A (2000) Summarizing the predictive power of a generalized linear model. Statistics in medicine 19: 1771–1781.
64. Manel S, Williams HC, Ormerod SJ (2001) Evaluating presence–absence models in ecology: the need to account for prevalence. Journal of Applied Ecology 38: 921–931.
65. Zadeh L (1965) Fuzzy sets. Information and Control 8: 338–353. doi:10.1016/S0019-9958(65)90241-X.
66. Hanberry B., He H., Palik BJ (2012) Pseudo-absence generation strategies for Species Distribution Models. PloS one 7: e44486.
67. Zarnetske PL, Edwards Jr TC, Moisen GG (2007) Habitat classification modeling with incomplete data: Pushing the habitat envelope. Ecological Applications 17: 1714–1726.
68. King G, Zeng L (2001) Logistic Regression in Rare Events Data. Political Analysis 9: 137–163.
69. Dixon PM, Ellison AM, Gotelli NJ (2005) Improving the precision of estimates of the frequency of rare events. Ecology 86: 1114–1123. doi:10.1890/04-0601.
70. McPherson JM, Jetz W, Rogers D (2004) The effects of species' range sizes on the accuracy of distribution models: ecological phenomenon or statistical artefact? Journal of Applied Ecology 41: 811–823. doi:10.1111/j.0021-8901.2004.00943.x.
71. Liu C, Berry PM, Dawson TP, Pearson RG (2005) Selecting thresholds of occurrence in the prediction of species distributions. Ecography 28: 385–393. doi:10.1111/j.0906-7590.2005.03957.x.
72. Williams AK (2003) The influence of probability of detection when modeling species occurrence using GIS and survey data [PhD thesis]. Virginia Polytechnic Institute and State University. Available: http://vtechworks.lib.vt.edu/handle/10919/11129. Accessed 1 October 2012.
73. Guisan A, Thuiller W (2005) Predicting species distribution: offering more than simple habitat models. Ecology letters 8: 993–1009.
74. El Lakhrach H, Hattour A, Jarboui O, Elhasni K, Ramos-Espla A (2012) Spatial distribution and abundance of the megabenthic fauna community in Gabes gulf (Tunisia, eastern Mediterranean Sea). Mediterranean Marine Science 13: 12–29.
75. Ben Mustapha K, Afli A (2007) Quelques traits de la biodiversité marine de Tunisie: Proposition d'aires de conservation et de gestion. Report of the MedSudMed Expert Consultation on Marine Protected Areas and Fisheries Management. MedSudMed Technical Documents. Rome (Italy). 32–55.
76. Margules CR, Pressey RL (2000) Systematic conservation planning. Nature 405: 243–253.
77. Watson JE, Grantham H, Wilson KA, Possingham HP (2011) Systematic conservation planning: Past, present and future. Wiley-Blackwell: 136–160.

17

Relationships between Meiofaunal Biodiversity and Prokaryotic Heterotrophic Production in Different Tropical Habitats and Oceanic Regions

Antonio Pusceddu[1]*, Cristina Gambi[1], Cinzia Corinaldesi[1], Mariaspina Scopa[1,2], Roberto Danovaro[1,3]

1 Dipartimento di Scienze della Vita e dell'Ambiente, Università Politecnica delle Marche, Ancona, Italy, **2** Istituto Zooprofilattico Sperimentale dell'Abruzzo e del Molise 'G. Caporale', Experimental Regional Centre for Fishery and Acquacolture, Termoli, Italy, **3** Stazione Zoologica Anton Dohrn, Napoli, Italy

Abstract

Tropical marine ecosystems are among the most diverse of the world oceans, so that assessing the linkages between biodiversity and ecosystem functions (BEF) is a crucial step to predict consequences of biodiversity loss. Most BEF studies in marine ecosystems have been carried out on macrobenthic diversity, whereas the influence of the meiofauna on ecosystem functioning has received much less attention. We compared meiofaunal and nematode biodiversity and prokaryotic heterotrophic production across seagrass, mangrove and reef sediments in the Caribbean, Celebes and Red Seas. For all variables we report the presence of differences among habitats within the same region, and among regions within the same habitat. In all regions, the richness of meiofaunal taxa in reef and seagrass sediments is higher than in mangrove sediments. The sediments of the Celebes Sea show the highest meiofaunal biodiversity. The composition of meiofaunal assemblages varies significantly among habitats in the same region. The nematode beta diversity among habitats within the same region is higher than the beta diversity among regions. Although one site per habitat was considered in each region, these results suggest that the composition of meiofaunal assemblages varies primarily among biogeographic regions, whereas the composition of nematode assemblages varies more considerably among habitats. Meiofauna and nematode biodiversity and prokaryotic heterotrophic production, even after the removal of covariate effects linked with longitude and the quantity and nutritional quality of organic matter, are positively and linearly linked both across regions and within each habitat type. Our results confirm that meiofauna and nematode biodiversity may influence benthic prokaryotic activity, which, in turn, implies that diversity loss could have negative impacts on ecosystem functioning in these systems.

Editor: Maura (Gee) Geraldine Chapman, University of Sydney, Australia

Funding: This study was carried out in the framework of the projects Ricerca di Ateneo 2004–2011 financed by the Polytechnic University of Marche. This study is a contribution to the FIRB Project EXPLODIVE financially supported by MIUR Italy. The funders had no role in study design, data collection and analysis, decision to publish, or preparation of the manuscript.

Competing Interests: The authors have declared that no competing interests exist.

* E-mail: a.pusceddu@univpm.it

Introduction

Marine coastal ecosystems provide crucially important goods and services to the human beings [1,2]. Marine coastal ecosystems are characterized by the presence of a complex and heterogeneous mosaic of habitats, such as sandy beaches, rocky shores, mangrove forests, seagrass meadows, coral reefs and transitional zones linking terrestrial and coastal ecosystems. Nevertheless, because of their proximity with the planet region most dense in human population, the coastal oceans are exposed to multiple anthropogenic stressors, including, among the others, aquaculture, dredging, mining, pollution, species invasion, over-harvesting and destructive fishing practices, large-scale oil and gas operations, watershed and offshore renewable energy development, coastal engineering habitat having strong and direct effects on the marine biodiversity [3,4].

The annual loss rates of five of the most important biogenic marine habitats (seagrass beds, salt marshes, coral and oyster reefs and mangrove forests) range between 1 and 9% [5]. Based on historical evidence, the total global loss of these habitats ranges from ca. 19% for coral reefs (in between 2004–2008) [6], to 29% (since 1879) for seagrass [7], to 35% (since 80ies) for mangrove forests [8] to >85% (in the last 20–130 years) for oyster reefs [9]. Many of these habitats, which are also hot spots of biodiversity, belong to tropical regions.

Ecological theory predicts that biodiversity can control ecosystems' functioning, although outputs of correlative and manipulative investigations have provided at times equivocal or contrasting results [10]. The relationships between biodiversity and functioning of marine ecosystems are most often positive [11], so that any biodiversity loss could result in a decrease of the ecosystem functioning [12] and, consequently, in a lower provision of goods and services to the humans [2,13]. A biodiversity loss can thus potentially impair the ecosystems' capacity to sustain humanity [14]. This is of particular concern in tropical and subtropical environments which host an important fraction of the biodiversity of the coastal oceans and are among the world regions that will experience the earliest emergence of historically unprecedented climates and changes in biodiversity [15].

Either correlative or manipulative approaches for assessing the shape and strength of the relationships between biodiversity and ecosystem functioning in marine ecosystems have been mostly

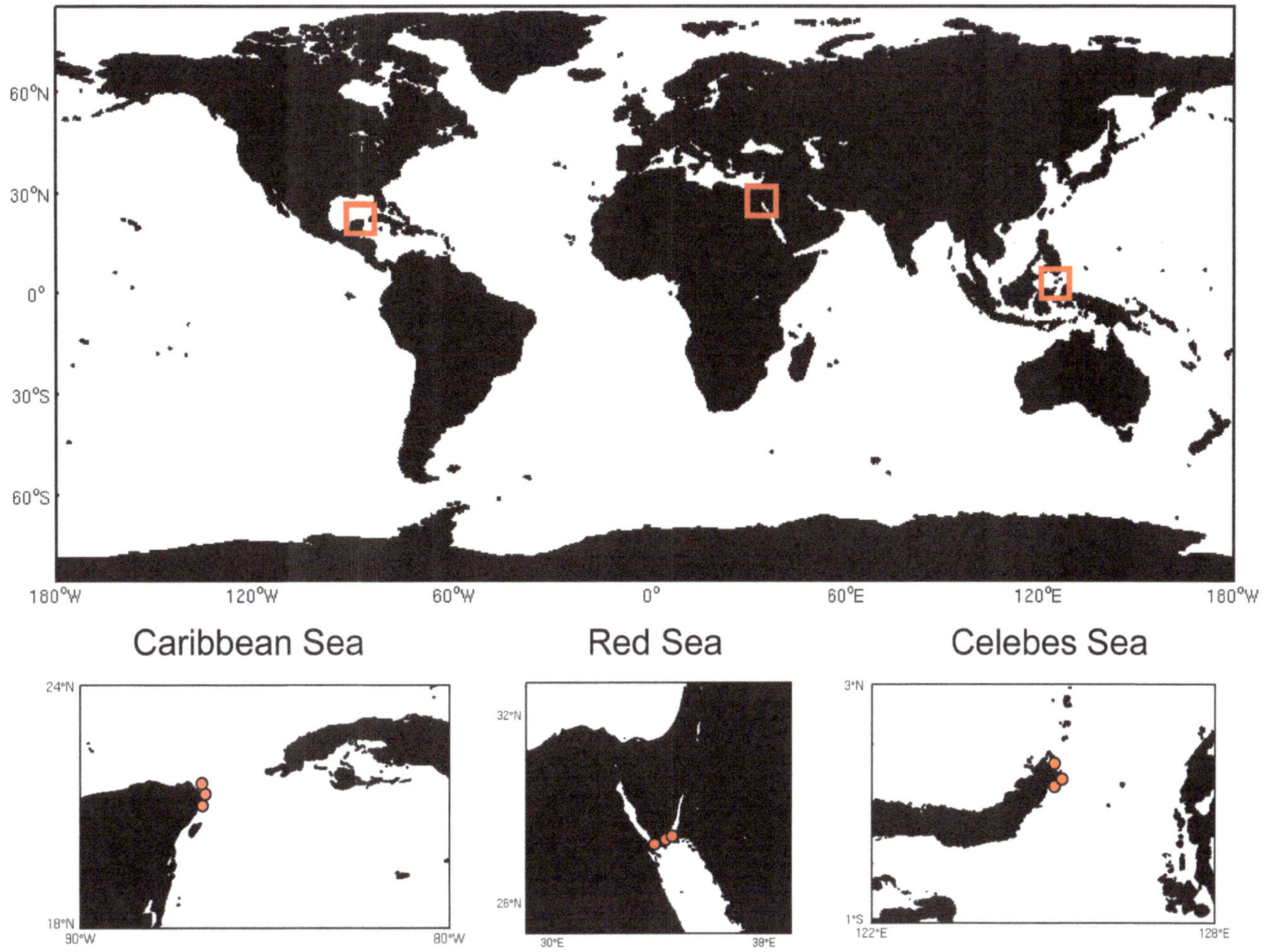

Figure 1. Sampling areas and location of the investigated regions with details of the positioning of each habitat type.

carried out using the diversity of macro-fauna or macro-algae as the independent variable [16]. More recently, meiofauna, the most abundant benthic group of metazoans in marine ecosystems [17], have been utilized for investigating, though only with a correlative approach, the linkages between biodiversity and ecosystem functioning in deep-sea sediments [18,19]. Among metazoan meiofauna, nematodes respond rapidly to many different sources of natural and anthropogenic disturbance affecting sedimentary environments; thus, also due to their high abundance, species and functional (trophic) diversity, nematodes are an ideal tool to investigate the relationships between biodiversity and ecosystem functioning [20]. Meiofauna play also a key ecological role in linking detrital (and prokaryotic) resources with higher trophic levels: in fact most of the meiofaunal taxa eat microalgae, prokaryotes and detritus and, at the same time, it is known that meiofauna are a food source for macrofauna and fishes [21–25]. Meiofauna and nematodes, based on laboratory and *in situ* experiments, are in fact able to influence microbial activities and to graze their production [17,26]. Meiofauna and nematodes are also very sensitive to the broad variations in natural environmental conditions (tidal influence, river inputs and local rainfall, food availability, sediment chemistry, bottom current regimes, habitat heterogeneity, among the others) [17,27] that characterize marine sediments across all spatial scales and water depths [28–31].

The functioning of marine sedimentary ecosystems relies on the rates of organic matter cycling, which is also related to the production of heterotrophic prokaryotes: in turn, these are related to the food quantity and food availability [32,33]. This holds true, in particular for highly detrital ecosystems like seagrass beds [34], mangroves [35] and coral reefs [36], where organic detritus, through the so-called microbial loop, is firstly incorporated into prokaryotic biomass, then enters higher trophic levels through bacterivorous, detritivorous and deposit feeders inhabiting the benthos [32]. Thus, it can be hypothesized that, in those ecosystems, the shape and strength of the relationship between biodiversity and ecosystem function could be influenced by the quantity and bioavailability of food resources.

To provide insights about the relationships between meiofaunal biodiversity and the ecosystem functioning in different tropical sedimentary habitats, we analyzed the richness of meiofaunal taxa and of nematode species (biodiversity) along with sedimentary organic matter quantity and nutritional quality, and prokaryotic heterotrophic production (functioning) across three habitats (i.e., seagrass, mangrove and reef sediments) in the Caribbean, Red and Celebes Seas. We also explored how some of the potential drivers of biodiversity considered in our study (longitude, quantity and quality of trophic sources) could influence the biodiversity-ecosystem functioning linkage. By comparing different habitats and regions we posed the following questions: i) how the different target variables vary among habitats and regions?; ii) whether and how changes the relationship between biodiversity and ecosystem functioning across different habitats?.

Materials and Methods

Ethical Statement

No special permits were requested at the time of sampling as in all regions sediment samples were collected out of protected areas and respecting local legislation. No protected species or taxa were sampled.

Study Areas and Sampling Activities

The sampling strategy included a total of nine sampling sites, from three tropical regions: the Caribbean, Red and Celebes Seas (Fig. 1, Table 1). In each region, three habitat types were visited once each: seagrass, mangrove and coral reef sediments.

In the Caribbean Sea, sediment samples were collected in the east coast of the Yucatan peninsula (Mexico) from a seagrass bed (Isla Mujeres), a coral reef (Xelha) and a mangrove forest (dominated by *Rizophora sp.*). In the Red Sea, sediment samples were collected in the Nabq lagoon from a seagrass bed, a mangrove forest dominated by *Avicennia marina* and a coral reef. In the Celebes Sea (Sulawesi Island, Indonesia), sediment samples were collected from a seagrass bed (dominated by *Enhalus acoroides*), a mangrove forest and a coral reef.

Sampling sites (each representing one type of habitat) within a region were 10–35 kilometers distance apart. At all sampling sites, replicate sediments within each habitat have been collected from three 0.5×0.5 m quadrates randomly selected from one 5×5 m area per habitat. The choice of this sampling area was constrained by the limited capability of operational movement in the three regions. Although this could have possibly reduced the ability to identify precisely the patterns of variability at the habitat scale, this design allowed the collection of sediments from the different habitats in each region within a reasonable time interval (hours-days).

Sediment samples were collected by scuba diving using Plexiglas manual corers (internal diameter 4.6 cm). The top first centimeter of three independent corers was collected for the determination of the quantity and biochemical composition of sedimentary organic matter, prokaryotic abundance, biomass and C production.

All samples for organic matter and biomass determinations were stored at −20°C until the analyses in the laboratory, whereas samples for the determination of prokaryotic heterotrophic production were immediately incubated as described below.

Three additional independent corers, collected for the analyses of meiofaunal abundance, biomass and biodiversity, were preserved in buffered formalin (4%), stained with Rose Bengal (0.5 gL^{-1}) and stored at 4°C until analysis.

Sedimentary Organic Matter

The sedimentary contents of phytopigments and biopolymeric C pools reflect the overall trophic conditions of marine coastal sediments, whereas the algal fraction of biopolymeric C pools reflects the food quality of sedimentary detritus [37]. Chloroplastic pigments (chlorophyll-*a* and phaeopigments) were analyzed fluorometrically [38]. Pigments were extracted from the top centimeter of each core with 90% acetone (24 h in the dark at 4°C). After centrifugation (800×g), the supernatant was used to determine the functional chlorophyll-*a* and acidified with 0.1 N HCl to estimate the amount of phaeopigments. Chlorophyll-*a* and phaeopigment concentrations were summed up and reported as total phytopigment concentrations. Total phytopigments, after

Table 1. Biochemical composition of sedimentary organic matter, prokaryotic biomass and C production in the investigated habitats and regions.

Region	Habitat	Latitude	Longitude	Depth	Chlorophyll-*a*	Phaeopigment	Protein	Carbohydrate	Lipid	Biopolymeric Carbon	Prokaryotic biomass	Prokaryotic heterotrophic production
		° N	°E	m	$\mu g\ g^{-1}$	$\mu g\ g^{-1}$	$mg\ g^{-1}$	$mg\ g^{-1}$	$mg\ g^{-1}$	$mgC\ g^{-1}$	$\mu gC\ g^{-1}$	$\mu gC\ g^{-1}\ d^{-1}$
Caribbean	Seagrass	21.6	−87.0	0.50	1.4±0.2	8.7±1.5	1.44±0.37	1.66±0.42	0.54±0.11	1.77±0.42	1.73±0.42	1.73±0.42
	Mangrove	21.3	−86.8	0.50	4.8±0.1	92.6±5.0	10.46±1.42	21.24±3.29	8.36±1.00	19.89±0.89	71.4±11.1	1.78±0.28
	Reef	20.5	−86.8	3.00	2.7±2.8	34.7±5.9	0.83±0.14	2.49±0.22	0.52±0.01	1.79±0.15	0.47±0.04	0.21±0.02
Red	Seagrass	28.2	34.4	0.50	0.2±0.1	16.4±4.7	0.61±0.15	0.53±0.03	0.03±0.00	0.54±0.06	1.04±0.34	1.04±0.34
	Mangrove	28.2	34.4	0.50	0.1±0.0	13.4±1.4	0.28±0.06	0.07±0.04	0.02±0.00	0.18±0.02	1.71±0.24	1.20±0.17
	Reef	28.2	34.4	2.00	0.4±0.1	49.9±8.0	0.67±0.09	0.10±0.04	0.05±0.03	0.41±0.07	0.36±0.08	0.16±0.03
Celebes	Seagrass	1.6	125.4	0.50	2.2±1.0	20.4±8.9	2.72±0.46	7.23±0.61	1.05±0.26	7.17±0.45	4.04±2.04	4.04±2.04
	Mangrove	1.5	125.2	0.50	0.7±0.3	4.0±1.9	1.78±0.21	1.63±0.22	0.63±0.31	2.35±0.19	6.38±0.53	2.23±0.18
	Reef	1.6	125.5	5.00	1.0±0.3	5.8±1.1	0.44±0.12	0.80±0.10	0.20±0.04	0.90±0.08	8.24±0.59	3.59±0.26

Table 2. Results of PERMANOVA and pairwise test for differences in sedimentary organic matter quantity and biochemical composition, prokaryotic biomass and heterotrophic production.

	Source	DF	MS	Pseudo-F	P	Pairwise tests: Region (Habitat)	Pairwise tests: Habitat (Region)
Phytopigments	**Region**	2	3.8	57.14	***	Car [M>R>S]	S [Car = Red = Cel]
	Habitat	2	0.4	6.07	*	Red [R>S = M]	M [Car>Red>Cel]
	Region × Habitat	4	3.1	46.71	***	Cel [S>M = R]	R [Red = Car>Cel]
	Residuals	18	0.1				
Proteins	**Region**	2	1.8	187.69	***	Cel [S>M>R]	S [Cel>Car>Red]
	Habitat	2	1.2	127.98	***	Car [M>S>R]	M [Car>Cel>Red]
	Region × Habitat	4	1.2	127.14	***	Red [R = S>M]	R [Car>Cel]
	Residuals	18	0.0				
Carbohydrates	**Region**	2	5.8	736.42	***	Car [M>R = S]	S [Cel>Car>Red],
	Habitat	2	1.3	163.86	***	Red [S>R = M]	M [Car>Cel>Red]
	Region × Habitat	4	2.4	301.53	***	Cel [S>M>R]	R [Car>Red>Cel]
	Residuals	18	0.0				
Lipids	**Region**	2	2.2	278.69	***	Car [M>S = R]	S [Cel>Car>Red]
	Habitat	2	1.2	145.26	***	Red [R>S>M]	M [Car>Cel>Red]
	Region × Habitat	4	1.2	144.26	***	Cel [S>M = R]	R [Car>Cel>Red]
	Residuals	18	0.0				
Biopolymeric C	**Region**	2	4.6	1076	***	Car [M>R = S]	S [Cel>Car>Red]
	Habitat	2	1.5	346.56	***	Red [R = S>M]	M [Car>Cel>Red]
	Region × Habitat	4	2.1	501.85	***	Cel [S>M>R]	R [Car>Cel>Red]
	Residuals	18	0.0				
Biochemical composition	**Region**	2	20.2	42.83	***	Car [M≠R = S]	S [Cel≠Car≠Red]
	Habitat	2	10.1	21.43	***	Red [R≠[M = S]	M [Cel≠Car≠Red]
	Region × Habitat	4	15.2	32.14	***	Cel [M≠S≠R]	R [Cel≠Red]
	Residuals	18	0.5				
Prokaryotic biomass	**Region**	2	9.2	138.3	***	Car [S = M>R]	S [Cel>Car = Red]
	Habitat	2	11	163.5	***	Red [S = M>R]	M [Car = Cel>Red]
	Region × Habitat	4	5.9	88.7	***	Cel [R>M = S]	R [Cel>Car = Red]
	Residuals	18	0				
Prokaryotic heterotrophic production	**Region**	2	2.0	64.7	***	Car [S = M>R]	S [Cel> Red = Car]
	Habitat	2	0.5	18.7	***	Red [S = M>R]	M [Car = Cel>Red]
	Region × Habitat	4	0.3	10.6	***	Cel [S = R>M]	R [Cel>Car = Red]
	Residuals	18	0.0				

DF = degrees of freedom; MS = mean squares; Pseudo-F = F statistic; P = probability level; *** = P<0.001; * = P<0.05; ns: not significant. Car = Caribbean Sea; Red = Red Sea; Cel = Celebes Sea; M = mangrove; R = reef; S = seagrass.

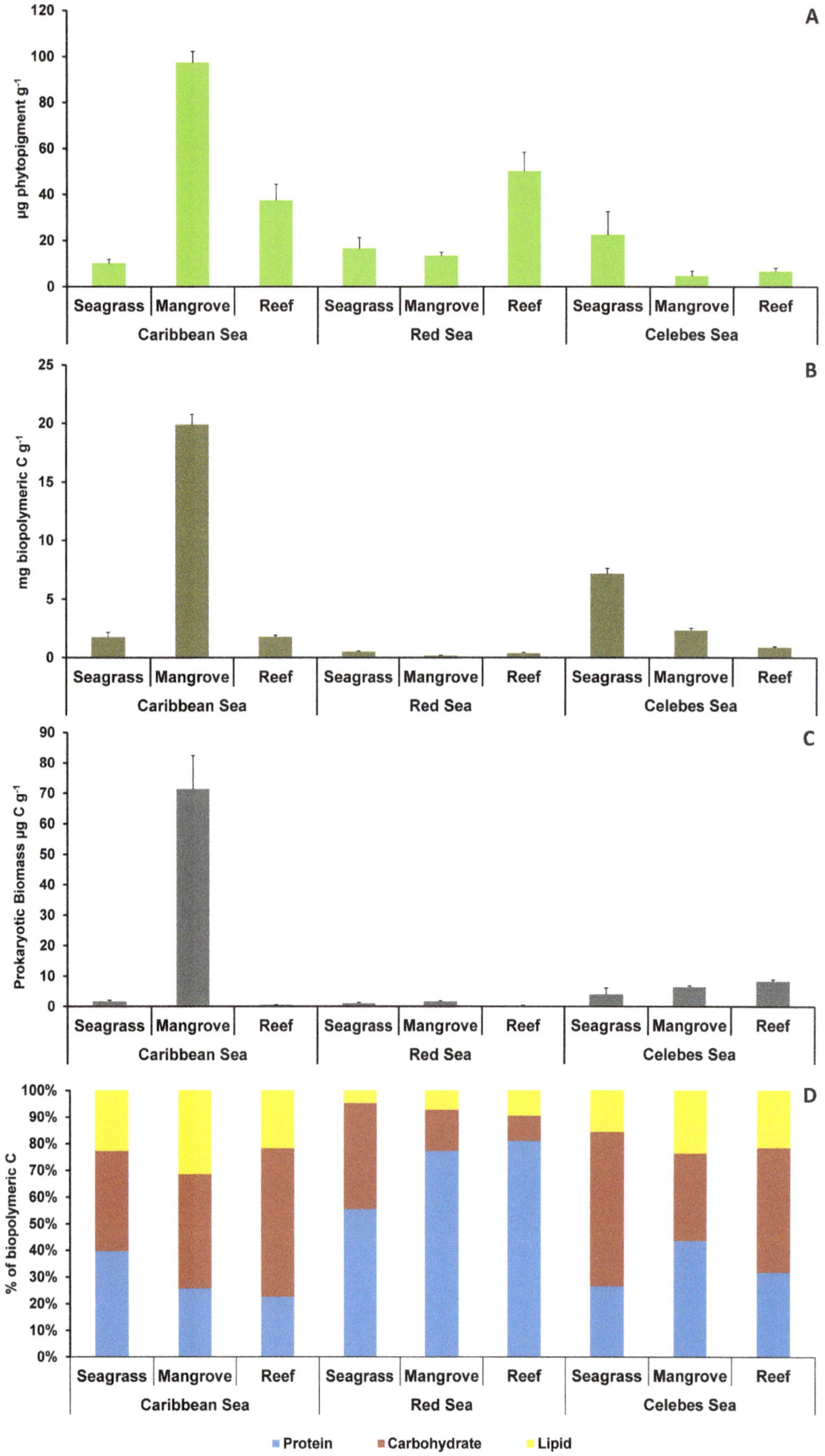

Figure 2. Biochemical composition of sedimentary organic matter. Reported are the concentrations of A) total phytopigments, B) biopolymeric carbon, C) prokaryotic biomass, and D) the biochemical composition (as percentage contribution of the biopolymeric carbon content)

of the sedimentary organic matter in seagrass, mangrove and reef habitats of the Caribbean, Red and Celebes Seas. Reported are mean values ± standard deviation.

transformation into C equivalents using 30 as a conversion factor [39], are thus used as an estimate of the organic material of algal origin including either the living (chlorophyll-*a*) and senescent/detrital (phaeopigment) fractions [40].

Protein, carbohydrate and lipid analyses were carried out on the top 1 cm of three independent corers from each site using photometric protocols [39]. Protein, carbohydrate and lipid sedimentary contents were converted to carbon equivalents by using the following conversion factors: 0.49, 0.40 and 0.75 mg C mg^{-1}, respectively and their sum referred as biopolymeric C (BPC) [37].

Prokaryotic Biomass and Heterotrophic Production

Prokaryotic biomass was determined from the prokaryotic cell biovolume and abundance by epifluorescence microscopy. To estimate the cell biovolume we examined at least 100 cells for sediment sample classifying prokaryotes into three different size classes: small (<0.065 μm^3), medium (0.065–0.320 μm^3) and large (0.320–0.574 μm^3) and the biovolume was converted into carbon content assuming 310 fg C μm^3 as a conversion factor [41]. The total prokaryotic abundance was determined using epifluorescence microscopy [39]. Briefly, sediment subsamples (0.25 g) were added with 1125 µl of 2% formalin and 125 µl of pyrophosphate (final concentration, 5 mM). Then, these were treated with ultrasounds (three times for 1 min; Branson sonifier 2200, 60 W) to increase the extraction efficiency and diluted 250–500 times using autoclaved and 0.2 ml pre-filtered seawater. One ml of the supernatant was stained with Acridine Orange (final concentration, 0.01%), for 5 min in the dark) and filtered under vacuum (< 100 mmHg) using 0.2 µm pore size black Nuclepore Polycarbonate filters. The filters were then washed twice with 3 ml of 0.2 µm pre-filtered and sterilized reagent grade water, mounted on microscope slides, and analyzed under an epifluorescence microscope (Zeiss Axioskop 2; magnification, ×1000). For each slide, at least 10 fields were observed, for a total of at least 400 cells counted per filter. Prokaryotic abundance is expressed as cells g^{-1} dry sediment (after desiccation at 60°C for 24 h).

Benthic prokaryotic heterotrophic production was measured by ^{3}H-leucine incorporation [42]. Sediment sub-samples (200 µl), added to a saturating aqueous solution of ^{3}H-leucine (6-µCi final concentration per sample), were incubated for 1 h in the dark at in situ temperature. After incubation, prokaryotic C incorporation was stopped with 1.7 ml of 80% ethanol before scintillation counting. Sediment blanks were made adding ethanol immediately before ^{3}H-leucine addition. Data were normalized to sediment dry weight after desiccation (60°, 24 h).

Meiofaunal Abundance and Biomass

For metazoan meiofaunal extraction, sediment samples were sieved through a 500 µm and a 30 µm mesh, respectively, to retain the smallest organisms. The fraction remaining on the latter sieve (including organisms with a size of 30–500 µm) was resuspended and centrifuged three times with Ludox HS 40 (diluted with tap water to arrange density to 1.18 g cm^{-3}) for muddy samples, whereas for sandy samples, meiofauna were extracted using decantation (repeated 10 times for 95% of efficiency) [39,43]. After the extraction the sediments have been carefully checked to search for remnant organisms. No organisms were observed in the residual sediments after the treatment with Ludox or decantation. All metazoan animals were counted and classified per taxon under a stereomicroscope using Delfuss cuvettes, after staining with Rose Bengal (0.5 g L^{-1}). For the determination of meiofaunal biomass, we calculated the individual biomass of all animals belonging to different taxa. Nematode biomass was calculated from biovolume (n = 100 per sample) using Andrassy's formula ($V = L \times W^2 \times 0.063 \times 10^{-5}$; body length L in µm and width W in µm). For all other taxa, the biovolume was measured for all of the specimens encountered. Body volume was derived from measurements of body length (L; in mm) and width (W; in mm) using the formula $V = L \times W^2 \times C$; where C is the approximate conversion factor for each meiofaunal taxon [44]. The body volume was multiplied by an average density (1.13 g cm^{-3}) to obtain the biomass (µg DW) assuming that the dry : wet weight ratio was 20–25%, and the C content was considered as 40% of the dry weight [44].

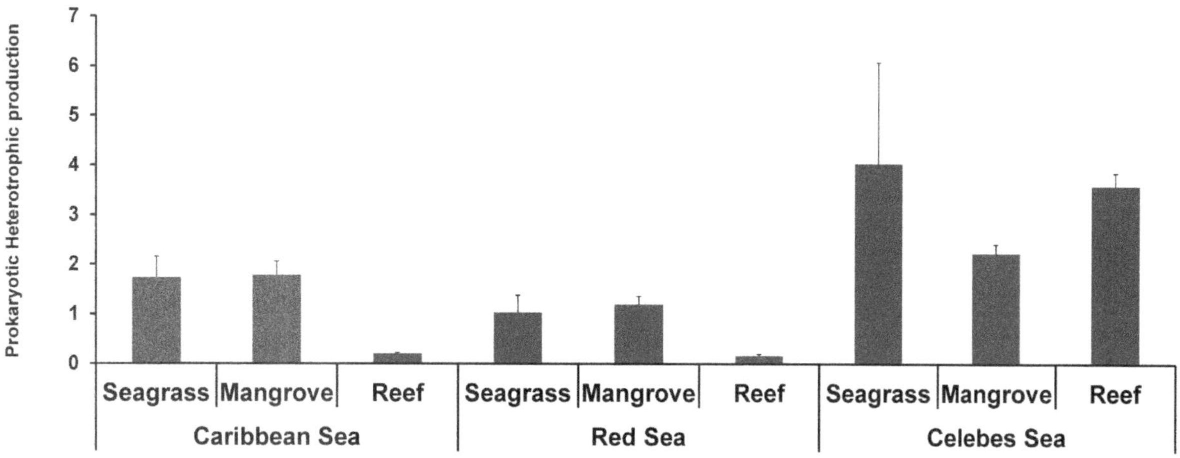

Figure 3. Ecosystem functioning as prokaryotic heterotrophic production (µgC g^{-1} d^{-1}) in seagrass, mangrove and reef habitats of the Caribbean, Red and Celebes Seas.

Table 3. Meiofaunal abundance and richness of taxa, nematode diversity as Species Richness, ES(51), evenness (Pielou's J) and index of trophic diversity in all habitats and regions.

Region	Habitat	Latitude	Longitude	Depth	Meiofaunal abundance	Richness of meiofaunal taxa	Species Richness	Expected species number	Pielou's evenness	Trophic diversity
		° N	°E	m	Ind. 10 cm^{-2}			ES(51)	J	ITD
Caribbean	Seagrass	21.6	−87.0	0.50	1268±731	11.3±4.2	33	14.1±4.3	0.82±0.05	0.49±0.04
	Mangrove	21.3	−86.8	0.50	2474±117	5.0±1.0	13	6.7±0.1	0.48±0.01	0.80±0.02
	Reef	20.5	−86.8	3.00	2871±1307	9.0±1.0	28	10.9±2.0	0.68±0.00	0.38±0.02
Red	Seagrass	28.2	34.4	0.50	1475±166	11.0±1.7	n/a	n/a	n/a	n/a
	Mangrove	28.2	34.4	0.50	343±41	4.0±1.0	n/a	n/a	n/a	n/a
	Reef	28.2	34.4	2.00	488±253	6.0±2.0	n/a	n/a	n/a	n/a
Celebes	Seagrass	1.6	125.4	0.50	1604±267	16.3±2.5	59	24.5±4.0	0.90±0.03	0.44±0.08
	Mangrove	1.5	125.2	0.50	627±192	9.7±1.5	40	16.1±2.9	0.79±0.04	0.43±0.08
	Reef	1.6	125.5	5.00	706±271	17.0±1.0	77	27.5±1.2	0.90±0.05	0.45±0.05

n/a = not analyzed.

Nematode Biodiversity

Nematode species richness and assemblage composition was determined only in the Celebes and Caribbean Seas. For the analysis of nematode diversity, 100 specimens (or all of the retrieved nematodes if <100) were randomly picked up from each of the three independent replicates at each sampling station. Nematodes were mounted on slides (following the formalin-ethanol-glycerol technique to prevent dehydration) [39] and identified to species level. Species identity was not considered in this study but, for the purposes of estimating species richness and other diversity indexes, the different morphotypes belonging to each genus were indicated as sp1, sp2, sp3 and considered as separate species [45]. Nematode species richness was estimated as the total number of species identified in each habitat. Since species richness is strongly influenced by the number of the individuals identified, to standardize the values of nematode diversity, the species-abundance data were converted into rarefaction diversity indices [46,47]. The expected number of species for a theoretical sample of 51 nematode specimens, ES(51), was calculated. Previous studies have shown that this approach enables the provision of robust data on species richness. Although it is far from being perfect [48], the expected species number is the density-independent index most commonly used for the comparison of areas with a non-standardized sample size [31,49]. The Pielou's index (J') was also calculated [50].

We measured also the beta diversity of meiofaunal and nematode assemblages among habitats within each region and among regions within each habitat using the SIMPER analysis based on Bray-Curtis matrixes, and expressed as percentage of dissimilarity [48]. Taxon and species-abundance data were presence/absence transformed prior to the analysis to search for specific assemblages in each habitat and region, irrespectively of the taxon and species relative abundance. Diversity indexes and the dissimilarity estimates were calculated using the routine DIVERSE and SIMPER, respectively included in the PRIMER6+ software [51].

Functional diversity of nematodes was estimated using the Index of Trophic Diversity (ITD) calculated as $ITD = g_1^2+g_2^2+g_3^2\ldots+g_n^2$, where g is the relative contribution of each trophic group to the total number of individuals and n is the number of trophic groups (with $n = 1–4$) [52]. Nematodes were divided into four groups as follows: (1A) no buccal cavity or a fine tubular one - selective (bacterial) feeders; (1B) large but unarmed buccal cavity-non-selective deposit feeders; (2A) buccal-cavity with scraping tooth or teeth-epistrate or epigrowth (diatom) feeders; (2B) buccal cavity with large jaws-predators/omnivores [52]. For four trophic guilds, the Index of Trophic Diversity ranges from 0.25 (highest trophic diversity; i.e., the four trophic guilds account for 25% each) to 1.0 (lowest diversity; i.e., one trophic guild accounts for 100% of nematode abundance) [43].

Ecosystem Functioning

Resembling one of the approach used for assessing biodiversity and ecosystem functioning relationships in the deep sea [18], we used prokaryotic heterotrophic production as proxy of the benthic ecosystem functioning and related it to biodiversity, estimated in terms of richness of meiofaunal taxa and nematode expected species number [ES(51)].

Statistical Analyses

To assess separately the differences in the quantity of sedimentary trophic resources, prokaryotic variables (biomass and heterotrophic production), total meiofaunal abundance, biomass, and richness of taxa among habitats within each region,

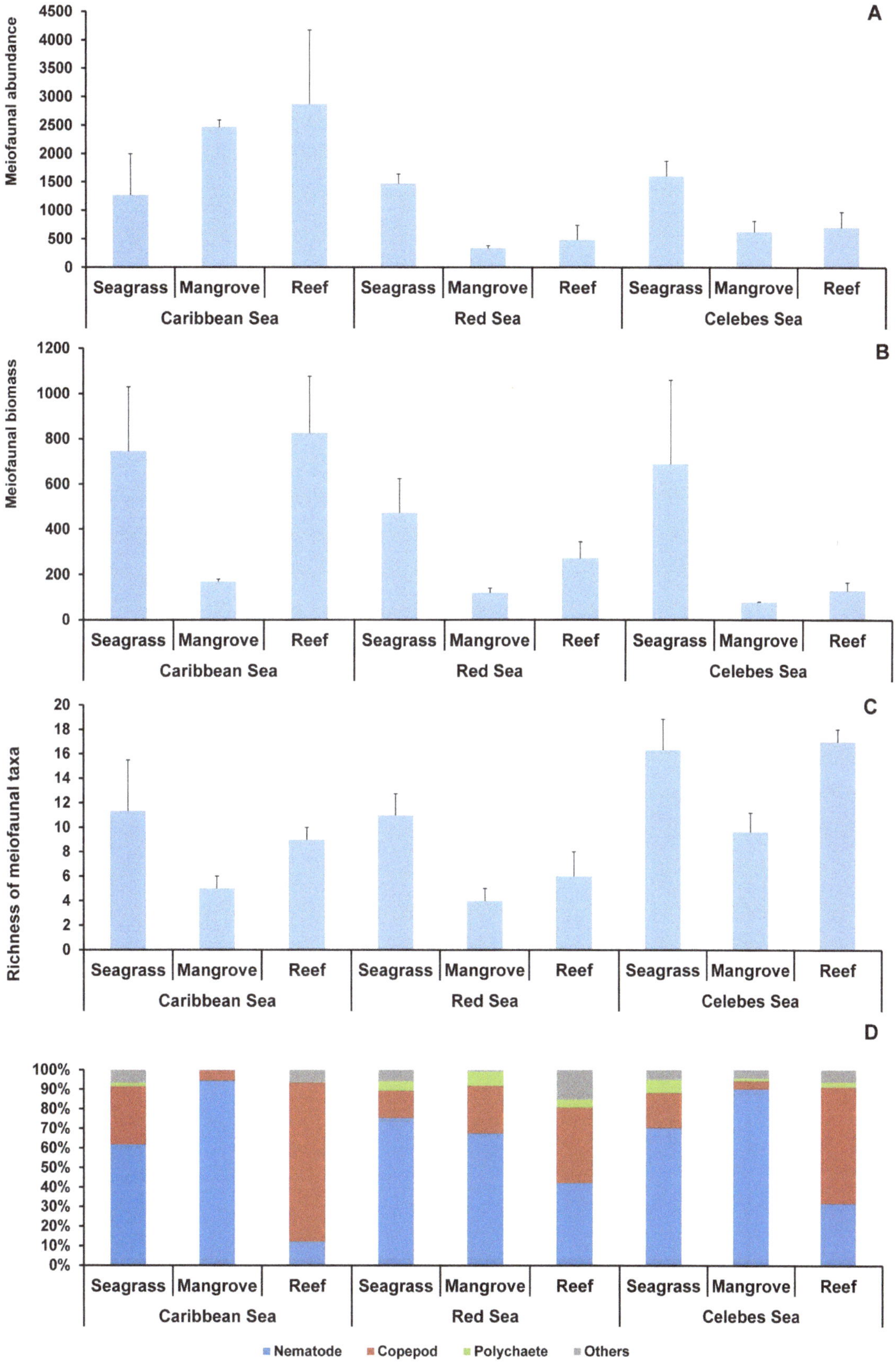

Figure 4. Meiofaunal assemblages. Illustrated are meiofaunal: A) abundance (ind 10 cm^{-2}), B) biomass (μgC 10 cm^{-2}), C) richness of taxa, and D) assemblage composition in seagrass, mangrove and reef habitats of the Caribbean, Red and Celebes Seas. Reported are average values ± standard deviation.

Table 4. Results of PERMANOVA testing for differences in meiofaunal abundance, biomass, richness of taxa and assemblages composition among habitats across regions and among regions across habitats.

						Pairwise tests	
	Source	**DF**	**MS**	**Pseudo-F**	**P**	**Region (Habitat)**	**Habitat (Region)**
Meiofaunal abundance	**Habitat**	2	0.74	5.54	**	Car [S = M = R]	S [Car = Red = Cel]
	Region	2	3.31	24.9	***	Red [S>M = R]	M [Car>Cel>Red]
	Habitat × Region	4	1.3	9.8	***	Cel [S>M = R]	R [Car>Cel = Red]
	Residuals	18	0.13				
Meiofaunal Biomass	**Habitat**	2	2.0	6.91	***	Car [R = S>M]	S [Car = Red = Cel]
	Region	2	5.0	17.77	***	Red [S = R>M]	M [Car>Red = Cel]
	Habitat × Region	4	0.6	2	ns	Cel [S>R = M]	R [Car>Red = Cel]
	Residuals	18	0.3				
Richness of meiofaunal taxa	**Habitat**	2	135.81	33.04	***	Car [R = S>M]	S [Cel>Red = Car]
	Region	2	103.70	25.23	***	Red [R = S>M]	M [Cel>Car = Red]
	Habitat × Region	4	7.70	1.87	ns	Cel [R = S>M]	R [Cel>Car = Red]
	Residuals	18	4.11				
Assemblages composition	**Habitat**	2	3430	5.67	***	Car [S = M = R]	na
	Region	2	2759	2.80	ns	Red [S≠M = R]	na
	Habitat × Region	4	985	1.63	ns	Cel [M≠S = R]	na
	Residuals	18	605				

Results of the pairwise tests among levels of the relevant factors are also reported. DF = degrees of freedom; MS = mean squares; Pseudo-F = F statistic; P = probability level; *** = $P<0.001$; ** = $P<0.01$; ns: not significant. Car = Caribbean Sea; Red = Red Sea; Cel = Celebes Sea; M = mangrove; R = reef; S = seagrass; na = not applicable.

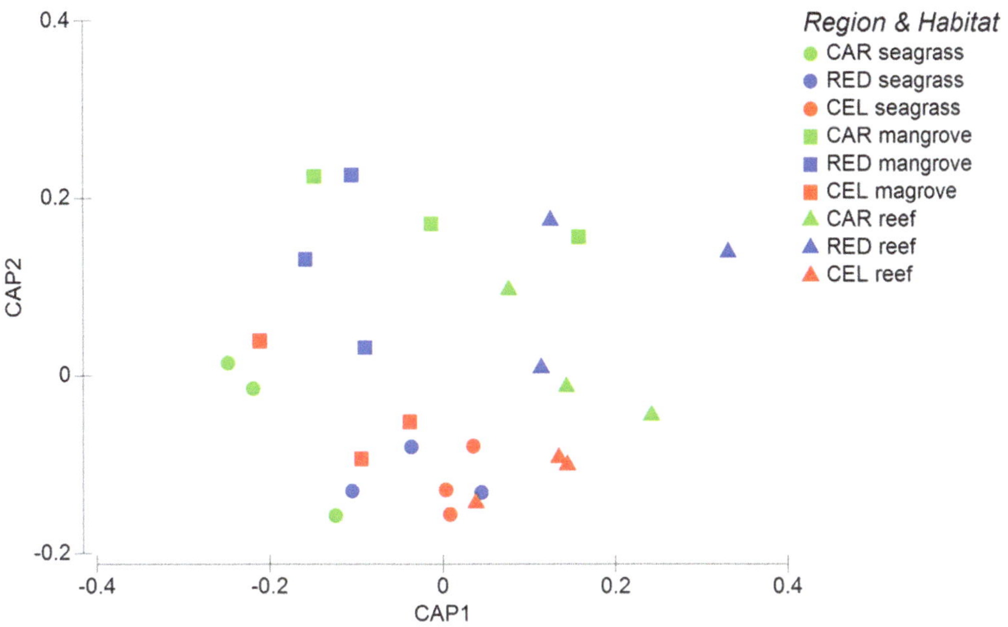

Figure 5. Bi-plots after the canonical analysis of principal coordinates illustrating differences in the composition of total meiofaunal taxa in the sediments of the investigated oceanic regions and habitats. CAR = Caribbean Sea (green); CEL = Celebes Sea (red); RED = Red Sea (blue).

and among regions within each habitat, we used two-way permutational analyses of variance (PERMANOVA) under a reduced model. The analyses were carried out using Habitat (3 fixed levels: mangrove, reef and seagrass; orthogonal to Region) and Region (3 fixed levels: Celebes, Red and Caribbean Seas) as main sources of variance. The same design was used for the analysis of nematode species diversity, but, in this case, the factor region included two levels only (Celebes and Caribbean Seas).

The same design has been used as the basis for a distance-based permutational multivariate analyses of variance (PERMANOVA) [53] to test variations among habitats and regions in the biochemical composition of sedimentary organic matter (based on Euclidean distance matrices), in the meiofaunal and the nematode assemblages (based on Bray Curtis similarity matrices after presence/absence transformations). The analyses were carried out using the permutation of residuals under a reduced model. In both uni- and multivariate tests, significant terms were investigated using a posteriori pair-wise comparisons with the PERMANOVA *t* statistic and 999 permutations.

The variations in meiofaunal communities and nematode assemblages among regions and habitats were illustrated using bi-plots produced after a canonical analysis of principle coordinates (CAP).

Biodiversity vs. ecosystem functioning relationships were assessed using distance-based analyses for a linear model using the routine DISTLM before and after the removal of covariates associated with the geographical location (longitude) and the availability of food resources (here evaluated in terms of biopolymeric C and protein to carbohydrate ratio), as synthetic descriptors of quantity and nutritional quality of food, respectively.

PERMANOVA, CAP and DISTLM tests were carried out using the homonymous routines included in the PRIMER6+ software [51].

Results

Sedimentary Organic Matter

The results of the two-way PERMANOVA carried out including the three habitat types in the Caribbean, Red and Celebes Seas revealed the presence of a significant effect of the interaction Habitat × Region for phytopigment, biopolymeric C and prokaryotic biomass (Table 2).

In the Caribbean Sea, the highest phytopigment and biopolymeric C contents occurred in mangrove sediments; in the Red Sea the highest total phytopigment contents occurred in reef sediments, whereas in the Celebes Sea the highest contents were observed in the seagrass sediments (Fig. 2A–C). Comparing seagrass habitats in different regions, phytopigment and biopolymeric C contents in the Celebes Sea were significantly higher than those in both the Caribbean and the Red Seas, whereas values in the Caribbean mangrove sediments were consistently higher than those in the Celebes and Red Seas.

Prokaryotic biomass in seagrass sediments did not vary significantly across regions (Table 2). In mangrove sediments prokaryotic biomass showed the highest values in the Caribbean Sea and the lowest in the Red Sea, whereas in reef sediments it was significantly higher in the Celebes Sea than in the Caribbean and Red Seas (Table 2).

The pairwise tests revealed also the presence of significant differences in the biochemical composition of sediments among regions (for mangrove and seagrass sediments) (Table 2). The reef sediments of the Celebes and Caribbean Seas exhibited a similar biochemical composition characterized by the dominance of carbohydrates, whereas the protein fraction dominated in the Red Sea reef sediments (Fig. 2D).

Prokaryotic Biomass and Heterotrophic Production

Prokaryotic biomass and heterotrophic production values were characterized by a significant effect of the Habitat × Region interaction (Table 2). In the Caribbean and Red Seas, prokaryotic

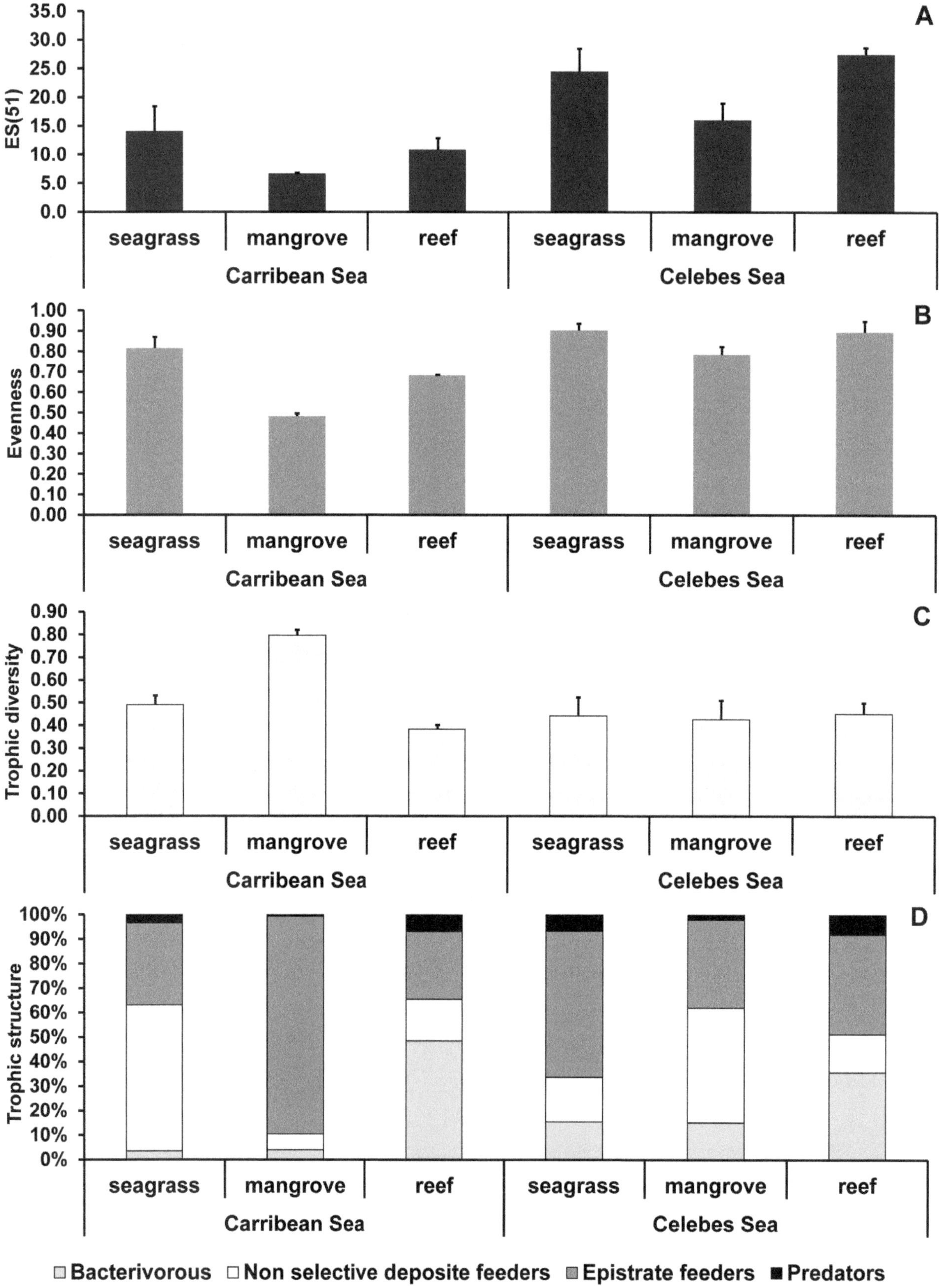

Figure 6. Nematode diversity. Illustrated are A) expected species number [ES(51)], B) evenness (Pielou's J), C) index of trophic diversity (ITD) and D) trophic composition of nematode assemblages in seagrass, mangrove and reef habitats of the Caribbean, Red and Celebes Seas.

Table 5. Results of PERMANOVA testing for differences in nematode biodiversity, as nematode Species Richness, ES(51), Pielou's evenness and index of trophic diversity (ITD), and nematode assemblage composition among habitats across regions and among regions across habitats.

						Pairwise test	
	Source	**DF**	**MS**	**Pseudo-F**	**P**	**Region (Habitat)**	**Habitat (Region)**
Species richness	**Region**	1	2.39	59.13	**	Car [S = R>M]	S [Cel>Car]
	Habitat	2	0.3	7.45	*	Cel [S = R>M]	M [Cel>Car]
	Region × Habitat	2	0.03	0.79	ns		R [Cel>Car]
	Residuals	12	0.04				
ES(51)	**Region**	1	2.44	93.8	***	Car [R = S>M]	S [Cel>Car]
	Habitat	2	0.5	19.36	***	Cel [R = S>M]	M [Cel>Car]
	Region × Habitat	2	0.05	1.79	ns		R [Cel>Car]
	Residuals	12	0.03				
Evenness	**Region**	1	0.06	148.78	***	Car [S>R>M]	S [Car = Cel]
	Habitat	2	0.03	67.18	***	Cel [R = S>M]	M [Cel>Car]
	Region × Habitat	2	0.01	17.29	***		R [Cel>Car]
	Residuals	12	0.00				
ITD	**Region**	1	0.02	16.37	***	Car [M>S>R]	S [Cel = Car]
	Habitat	2	0.02	16.27	***	Cel [M = R = S]	M [Car>Cel]
	Region × Habitat	2	0.03	21.33	***		R [Cel = Car]
	Residuals	12	0.00				
Nematode assemblage	**Region**	1	7074	4.62	**	Car [S≠M≠R]	S [Cel = Car]
	Habitat	2	7267	1.43	ns	Cel [S = M = R]	M [Cel≠Car]
	Region × Habitat	2	5071	3.31	***		R [Cel = Car]
	Residuals	12	1532				

Results of the pairwise tests among levels of the relevant factors are also reported. df = degrees of freedom; MS = mean squares; Pseudo-F = F statistic; *** = $P<0.001$; ** = $P<0.01$; * = $P<0.05$; ns: not significant. Car = Caribbean Sea; Red = Red Sea; Cel = Celebes Sea; M = mangrove; R = reef; S = seagrass.

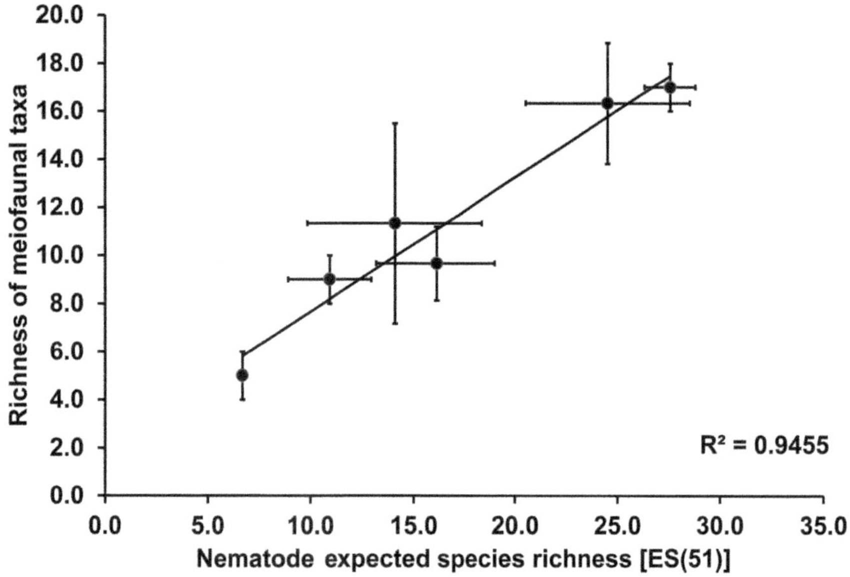

Figure 7. Relationship between richness of meiofaunal taxa and nematode biodiversity as ES(51) in tropical habitats from different oceanic regions. Illustrated are mean values ± standard deviation.

heterotrophic production values in seagrass and mangrove sediments were significantly higher than those in reef sediments, whereas in the Celebes Sea the mangrove sediments were characterized by values significantly lower than those in seagrass and reef sediments (Fig. 3). Overall, prokaryotic heterotrophic production in the sediments of the Celebes Sea were generally and significantly higher than those in the two other regions.

Meiofaunal Abundance, Biomass and Community Composition

Data on meiofaunal abundance, and richness of taxa, nematode diversity as species richness, ES(51), evenness (Pielou's J) and index of trophic diversity in all habitats and regions are reported in Table 3.

The results of the PERMANOVA on meiofaunal abundance revealed a significant effect of the interaction Habitat × Region (Table 4, Fig. 4A). In more detail, the pairwise tests revealed that, in both the Celebes and Red Seas, meiofaunal abundance was significantly higher in seagrass than in mangrove and reef sediments, whereas in the Caribbean Sea values did not vary significantly among habitats. The pairwise tests revealed also that the meiofaunal abundance in both mangrove and reef habitats was significantly higher in the Caribbean than in the Celebes and Red Seas. Meiofaunal abundance in seagrass sediments did not vary significantly across regions.

Meiofaunal biomass varied significantly among habitats (within the same region) and among regions (comparing the same habitat) (Table 4, Fig. 4B). Meiofaunal biomass was generally lower in mangrove sediments (range 79–170 μgC 10 cm^{-2}) than in the other two habitats (range 131–826 μgC 10 cm^{-2}). Significantly higher values of meiofaunal biomass were observed in the Caribbean followed by the Celebes and the Red Seas (Table 4).

The richness of meiofaunal taxa (Fig. 4C) ranged from 4 (in the mangrove sediments of the Red Sea) to 17 (in the reef sediments of the Celebes Sea) and varied significantly among habitats and among regions (Table 4). In more detail, the pairwise tests revealed that in all regions the richness of meiofaunal taxa in seagrass and reef sediments was consistently and significantly higher than in mangrove sediments. The richness of meiofaunal taxa in the Celebes Sea was significantly higher than that in the Caribbean and Red Seas (Table 4, Fig. 4C).

In all habitats and regions, with exception of reef sediments of the Celebes and Caribbean Seas, nematodes were the dominant taxon (52–97%), followed by copepods (3–25%) and polychaetes (0–13%) (Fig. 4D). Only reef sediments of the Caribbean and Celebes Seas were dominated by copepods (78 and 54% of total meiofaunal abundance, respectively). The contribution of all other identified taxa (acarians, amphipods, bivalves, cladocerans, cnidarians, cumaceans, gastrotrichs, gastropods, isopods, kinorinchs, decapods larvae, oligochaetes, ostracods, priapulians, tanaidaceans, tardigrades, termosbanaceans and turbellarians) varied from 0 to 25% of the total meiofaunal abundance (Fig. S1).

The PERMANOVA test showed that the composition of meiofaunal communities varied significantly only among habitats (Table 4). Moreover, the pairwise tests revealed that differences in the composition of meiofaunal communities were not consistent among habitats within each region (Table 4). Consequently, the results of the CAP analyses revealed that the differences among regions in the composition of meiofaunal assemblages were not well defined (Fig. 5).

In the Caribbean Sea, the highest meiofaunal beta diversity occurred between seagrass and mangrove sediments (60%). In the Red Sea, the highest meiofaunal beta diversity occurred between seagrass and reef sediments (44%), whereas in the Celebes Seas the highest beta diversity occurred between mangrove and reef sediments (29%) (Table S1). When contrasting the regions, the highest meiofaunal beta diversity among seagrass sediments is observed between the Red and the Celebes Sea (30%), whereas for mangrove sediments the highest beta diversity is observed between the Caribbean and Celebes Sea (61%). For reef sediments the highest meiofaunal beta diversity is observed between the Red and Celebes Seas (48%).

Nematode Biodiversity

The results of PERMANOVA conducted on nematode species richness and expected species number [ES(51)] revealed significant

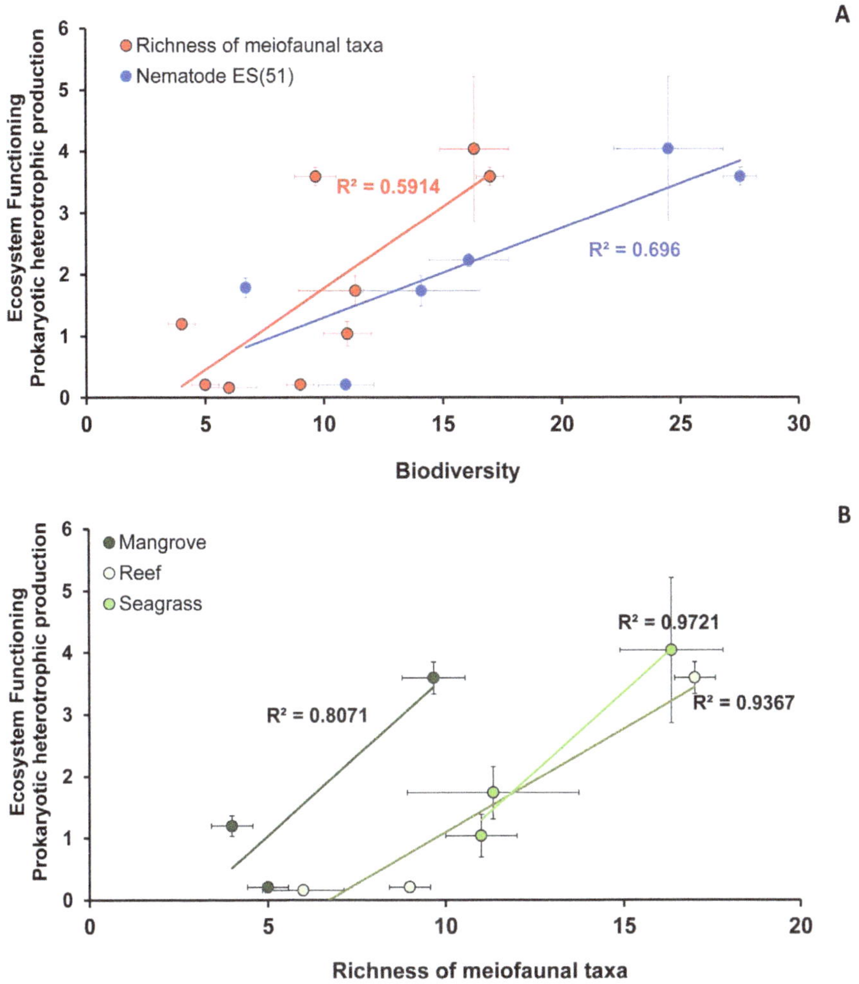

Figure 8. Relationship between biodiversity and ecosystem functioning. Illustrated are the relationships between: A) richness of meiofaunal taxa and prokaryotic heterotrophic production (μgC g^{-1} d^{-1}); nematode diversity as ES(51) and prokaryotic heterotrophic production (μgC g^{-1} d^{-1}); B) richness of meiofaunal taxa and prokaryotic heterotrophic production in different habitat types. Reported are R^2 values. $P<0.01$ for all linear regressions. Error bars indicate standard deviations among replicates (n = 3).

effects of the factors Region and Habitat, but no significant interaction effects (Table 5). Overall, in both the Celebes and Caribbean Seas the mangrove sediments displayed consistently and significantly lower values of nematode species richness, expected species number and evenness values (Table 5; Fig. 6). In all habitats, nematode expected species number and evenness

Table 6. Effects of biodiversity, measured as richness of meiofaunal taxa and nematode expected species number ES(51), on ecosystem functioning, as prokaryotic heterotrophic production before and after the removal of the covariables' effect: longitude, sediment biopolymeric organic C content and the protein to carbohydrate ratio.

Independent variable	R^2	SS	Pseudo-F	P
Richness of Meiofaunal Taxa	0.56	39.25	32.39	***
Covariates	0.63	43.52	12.82	***
Taxa Richness after removal of covariates	0.72	6.53	7.37	*
Nematode ES(51)	0.73	39.97	42.24	***
Covariates	0.77	42.46	15.67	***
Nematode ES(51) after the removal of covariates	0.88	5.78	10.94	***

In the regression analyses, all tests were based on Euclidean distances calculated among observations from untransformed data, using all data from different regions and habitats. The following abbreviations are used: R^2 = regression coefficient, SS = sum of squares, Pseudo-F = statistic F; P = probability level (*** = $P<0.001$; * = $P<0.05$).

were consistently higher in the Celebes than in the Caribbean Sea (with the only exception of evenness in seagrass sediments) (Table 5; Fig. 6A–B). Values of the index of trophic diversity did not vary among habitats in the Celebes Sea, whereas decreased from mangrove to seagrass and reef sediments in the Caribbean Sea. Values of the trophic diversity index varied significantly among regions only in mangrove sediments (Fig. 6C). In seagrass and reef sediments of the Caribbean Sea the selective (bacterial) and non-selective deposit feeders cumulatively represented more than 60% of the total nematode abundance, whereas in mangrove sediments dominated the epistrate feeders (more than 80% of the total abundance, Fig. 6D). In mangrove and reef sediments of the Celebes Sea the selective and non-selective deposit feeders cumulatively represented more than 50% of the total nematode abundance, whereas in seagrass sediments dominated the epistrate feeders (more than 60% of the total abundance). In both the Celebes and Caribbean Seas and in all habitats predator nematodes represented a minor fraction of the total nematode abundance (range 1–7%).

The richness of meiofaunal taxa was significantly and positively correlated with the number of nematode species as ES(51) (Fig. 7).

The results of PERMANOVA showed the presence of a significant effect of the interaction Habitat × Region on the composition of nematode assemblages (Table 5). The pairwise comparisons revealed that the composition of the nematode assemblages varied among regions only in mangrove sediments, whereas differences among habitats were consistently present only in the Caribbean Sea (Table 5). In the Celebes Sea, the nematode beta diversity among habitats was high and ranged from 64% (seagrass vs. reefs) to 79% (reefs vs. mangroves) (Table S1). In the Caribbean Sea the nematode beta diversity among habitats (74–93%) was slightly higher than that in the Celebes Sea (64–71%). Nematode beta diversity among regions was 68%, 77% and 83% in seagrass, mangrove and reef sediments, respectively. Overall the nematode beta diversity among the Caribbean and Celebes was 79%.

Relationship between Biodiversity and Ecosystem Functioning

The richness of meiofaunal taxa and the nematode expected species number [ES(51)] were significantly and positively related with prokaryotic heterotrophic production across all regions and in each habitat (Fig. 8A–B). The linear relationships between biodiversity and ecosystem functioning were not influenced by any of the tested covariates (i.e., longitude, biopolymeric C contents and the values of the protein to carbohydrate ratio) only when all data, irrespectively of region or habitat, were included in the analysis (Table 6).

Discussion

Meiofaunal and Nematode Biodiversity in Tropical Marine Sediments

Meiofaunal abundance and biomass values reported in this study fall within the range of those previously reported for the same regions (Caribbean Sea, Celebes Sea and Red Sea) or from other tropical and subtropical habitats [54–60]. Nonetheless, although our survey is limited to one replicate habitat per region, we report some differences in the structure and biodiversity of meiofaunal assemblages comparing both the same habitat from different regions and different habitats within the same region. For instance, in all investigated regions both seagrass and reef sediments generally host the highest number of meiofaunal taxa. These results suggest that, at both investigated spatial scales (i.e., oceanic region and habitat) the differences in meiofaunal diversity appear more evident than those in terms of abundance or biomass.

Although the use of correlations must be always considered with caution, the presence of a significant positive correlation between the number of higher meiofaunal taxa and the number of the nematode expected species number (Fig. 7) suggests that, at least for the investigated regions and habitats, the analysis of biodiversity based on nematodes could be a good proxy for the analysis of the patterns of the whole meiofaunal biodiversity. Such relationship, however, should be further explored on a much larger set of tropical sedimentary environments.

The analysis of structural nematode biodiversity (either expressed in terms of species richness, expected species number, or evenness) is significantly higher in the Coral triangle (Celebes, Indonesia) than in the Caribbean Sea, whatever the considered habitat is. Moreover, although a proper replication of habitats included in our survey would have shed more light on the actual differences among habitats in each region, we report here that in both the Celebes and Caribbean Seas, seagrass beds and reef sediments apparently host a significantly higher species number than mangrove sediments.

A comparative analysis of nematode diversity among different habitats, including reef sediments [60], temperate seagrass beds [61,62] and mangrove sediments in Australia, Africa, Asia and South America [59,63–65] is difficult as different estimates (indices) of diversity have been used in different studies. Moreover, such a comparison could be biased because of the different geographic locations and sampling efforts [66] as well as by the lack of temporal replication of the surveys. In presence of these potential biases, which altogether could weaken or reinforce the differences among habitats and regions, the results from our study can be nevertheless used at least in comparative terms, as, at all investigated areas, we adopted the same sampling strategy (and sampling effort), sample storage, methodology of extraction and determination/identification procedures.

The comparison of the biodiversity of the different habitats from each region suggests that each single habitat provided an important and significant contribution to the regional (i.e., gamma) nematode diversity only in the Caribbean Sea. In fact, the nematode beta diversity among seagrass, mangrove and reef sediments is >64% in both the Celebes and Caribbean Seas, but the post-hoc tests after the PERMANOVA revealed that the differences in the composition of nematode assemblages were significant only in the Caribbean Sea. This result would suggest that factors regulating nematode gamma diversity in the Caribbean and Celebes Sea are far different each other and leave open this issue to further investigations.

Soft bottoms are typically characterized by environmental variations that operate at the scale of a few centimeters on micro- and meiofaunal assemblages [33]. For instance, a recent study conducted from a single mangrove system from northwestern Brazil reported the existence of significant differences in nematode assemblage structure among micro-habitats [29]. Altogether our results and the evidences available from the literature pinpoint that tropical marine sediments are possibly characterized not only by high levels of nematode biodiversity, but also by high levels of beta diversity at different spatial scales, from the micro-scale to the regional one.

Relationships between Biodiversity and Ecosystem Functioning

Empirical and theoretical studies increasingly argue that biodiversity regulates the ecosystem functions that are responsible for the production of natural goods and services [2,67–70]. Many investigations relating biodiversity and ecosystem functioning have been performed using controlled field experiments that assemble model (and often non-natural) communities to measure the effects of changes in diversity on several ecosystem processes. Meta-analyses have also shown that species diversity generally has a positive effect on ecosystem processes and that this effect is remarkably consistent across trophic levels and different ecosystems [11–12,70].

In this study, the relationships between the richness of meiofaunal taxa and the nematodes species diversity vs. prokaryotic heterotrophic production are positive linear across all investigated regions and habitats (Fig. 8A–B). Our results are slightly different from those recently reported from coral reefs, where using fish as a model for biodiversity, the relationship appears to be positive exponential [71]. Nevertheless, this allows us to hypothesize that positive relationships could be a peculiar characteristic of tropical shallow habitats, but underpin also that the shape of the relationship (linear vs. exponential) could vary when different components of biodiversity (e.g., meiofauna, nematodes and fish) are taken into account. Differences in the shape of the relationships between biodiversity and ecosystem functioning from different marine environments also emerge when comparing habitats from different water depth. For instance, a positive exponential relationship has been observed in the deep Atlantic, Pacific and Antarctic Oceans and in the deep Mediterranean Sea (at depths ranging 200–8200 m) [18], whereas a recent study conducted along the upper slope off New Zealand (at depths ranging 264–1238 m) reported linear negative or null relationships [19]. Moreover, a recent manipulative experiment carried out on natural nematode assemblages response to thermal stress [72] led to hypothesize the same probability of (saturating) rivet-like [73] or idiosyncratic [74] relationships between nematode species richness and ecosystem functioning.

The large variability in the shape of the relationships between marine biodiversity and ecosystem functioning can be due to: i) the relatively limited number of regions and habitats considered so far, ii) the different approaches used (i.e., correlative from the real world and manipulative experiments in the laboratory), and iii) the different environmental characteristics of the investigated regions and habitats (e.g., coastal vs. deep-sea sediments). These discrepancies do not allow us making any robust speculation about the possible mechanisms explaining the different shapes in the relationship between biodiversity and ecosystem functioning observed in different marine ecosystems. Nevertheless, our results corroborate the presence of a pre-eminently positive effect of biodiversity on marine ecosystem functioning, and let us concluding that any loss in marine metazoan biodiversity (whatever the phylum considered) could result in a variably severe impairment of marine ecosystem functions.

Finally, it must be taken into account that the correlative approach used in this and other previous studies leaves open yet the conceptual possibility to interpret the reverse relationship, i.e. addressing whether and how ecosystem functions control biodiversity. This was not among our aims, but we must acknowledge that, in our study, the reverse relationship would remain positive linear and this would confirm that meiofaunal (and nematode) biodiversity are tightly dependent on the functioning of the microbial loop [26].

Acknowledgments

We acknowledge constructive criticism of two reviewers that helped us to significantly improve an early version of our manuscript.

Author Contributions

Conceived and designed the experiments: AP RD. Performed the experiments: AP RD CC. Analyzed the data: CG MS RD. Contributed reagents/materials/analysis tools: AP RD CC. Wrote the paper: AP CG CC MS RD. Species identification: CG MS. Statistical analysis: AP CG. Supported revision and intellectual contribution: AP CG RD.

References

1. Costanza R, d'Arge R, de Groot R, Farber S, Grasso M et al. (1997) The value of the world's ecosystem services and natural capital. Nature 387: 253–260.
2. Worm B, Barbier EB, Beaumont N, Duffy JE, Folke C, et al. (2006) Impacts of biodiversity loss on ocean ecosystem services. Science 314: 787–790.
3. Crain CM, Kroeker K, Halpern BS (2008) Interactive and cumulative effects of multiple human stressors in marine systems. Ecol Lett 11: 1304–1315.
4. Claudet J, Fraschetti S (2010) Human-driven impacts on marine habitats: A regional meta-analysis in the Mediterranean Sea. Biol Conserv 143: 2195–2206.
5. Duarte CM, Dennison WC, Orth RJW, Carruthers TJB (2008) The charisma of coastal ecosystems: addressing the imbalance. Estuar Coast 31: 233–238.
6. Wilkinson CR (2008) Status of the coral reefs of the world: 2008. Global Coral Reef Monitoring Network and Institute of Marine Science, Townsville, 8 pp.
7. Waycott M, Duarte CM, Carruthers TJB, Orth RJ, Dennison WC, et al. (2009) Accelerating loss of seagrasses across the globe threatens coastal ecosystems. Proc Natl Acad Sci USA 106: 12377–12381.
8. Valiela I, Bowen JL, York JK (2001) Mangrove forests: one of the world's most threatened major tropical environments. Bioscience 51: 807–815.
9. Beck MB, Brumbaugh RD, Airoldi L, Carranza A, Coen LD, et al. (2009) Shellfish reefs at risk: a global analysis of problems and solutions. The Nature Conservancy, Arlington, VA, 52 pp.
10. Loreau M (2010) Linking biodiversity and ecosystems: towards a unifying ecological theory. Proc Trans R Soc B 365: 49–60.
11. Cardinale BJ, Srivastava DS, Duffy JE, Wright JP, Downing AL, et al. (2006) Effects of biodiversity on the functioning of trophic groups and ecosystems. Nature 443: 989–992.
12. Loreau M (2008) Biodiversity and Ecosystem Functioning: The Mystery of the Deep Sea. Curr Biol 18: 126–128.
13. Bulling MT, Hicks N, Murray L, Paterson DM, Raffaelli D, et al. (2010) Marine biodiversity-ecosystem functions under uncertain environmental futures. Philos Trans R Soc Lond B Biol Sci. 365: 2107–2116.
14. Cardinale BJ, Duffy JE, Gonzalez A, Hooper DU, Perrings C, et al. (2012) Biodiversity loss and its impact on humanity. Nature 486: 59–67.
15. Solan M, Raffaelli DG, Paterson DM, White PCL, Pierce GJ (2006) Marine biodiversity and ecosystem function: empirical approaches and future research needs. Mar Ecol Prog Ser 311: 175–178.
16. Mora C, Frazier AG, Longman RJ, Dacks RS, Walton MM, et al. (2013) The projected timing of climate departure from recent variability. Nature 502: 183–187.
17. Giere O (2009) Meiobenthology: The Microscopic Motile Fauna of Aquatic Sediments. 2nd Edition, Springer, Berlin.
18. Danovaro R, Gambi C, Dell'Anno A, Corinaldesi C, Fraschetti S, et al. (2008) Exponential decline of deep-sea ecosystem functioning linked to benthic biodiversity loss. Curr Biol 18: 1–8.
19. Leduc D, Rowden AA, Pilditch CA, Maas EW, Probert PK (2013) Is there a link between deep-sea biodiversity and ecosystem function? Mar Ecol 34: 334–344.
20. Danovaro R, Gambi C, Höss S, Mirto S, Traunspurger W et al. (2009) Case studies using nematodes assemblage analysis in aquatic habitats. In: Wilson MJ, Kakouli-Duarte T (Eds) Nematodes as environmental indicators, CABI, Wallingford, UK, 146–171.

21. Watzin MC (1983) The effects of meiofauna on settling macrofauna: meiofauna may structure macrofaunal communities. Oecologia 59: 163–166.
22. Montagna PA (1984) In situ measurement of meiobenthic grazing rates on sediment bacteria and edaphic diatoms. Mar Ecol Progr Ser 18: 119–130.
23. De Morais LT, Bodiou JY (1984) Predation on meiofauna by juvenile fish in a western Mediterranean flatfish nursery ground. Mar Biol 82: 209–215.
24. Heip C, Vincx M, Vranken G (1985) The ecology of marine nematodes. Ocean Mar Biol Ann Rev 23: 399–489.
25. Danovaro R, Fraschetti S, Belgrano A, Curini-Galletti M, Vincx M, et al. (1995) The potential impact of meiofauna on the recruitment of macrobenthos in a subtidal coastal benthic community of the Ligurian Sea: a field result. In: Biology and Ecology of Shallow Coastal Waters, Eleftheriou A, Ansell AD, Smith CJ (eds), Olsen and Olsen, Fredensborg, Denmark, 115–122.
26. van Oevelen D, Middelburg JJ, Soetaert K, Moodley L (2006) The fate of bacterial carbon in an intertidal sediment: Modeling an in situ isotope tracer experiment. Limnol Oceanogr 51(3): 1302–1314.
27. Wilson MJ, Kakouli-Duarte T (2009) Nematodes as environmental indicators. CAB International, UK, 326 pp.
28. Barnes N, Bamber RN, Bennell G, Cornelius N, Glassom D, et al. (2011) Assessment of regional and local biodiversity in tropical and subtropical coastal habitats in the East African Marine Ecoregion. Biodivers Conserv 10: 2075–2109.
29. Pinto TK, Austen MCV, Warwick RM, Somerfield PJ, Esteves AM, et al. (2013) Nematode diversity in different microhabitats in a mangrove region. Mar Ecol 34(3): 257–268.
30. Vanreusel A, Fonseca G, Danovaro R, Da Silva MC, Esteves AM, et al. (2010) The contribution of deep-sea macrohabitat heterogeneity to global nematode diversity Mar Ecol 31(1): 6–20.
31. Gambi C, Pusceddu A, Benedetti-Cecchi L, Danovaro R (2014) Species richness, species turnover and functional diversity in nematodes of the deep Mediterranean Sea: searching for drivers at different spatial scales. Global Ecol Biogeogr 23: 24–39.
32. Pusceddu A, Danovaro R (2009) Exergy, ecosystem functioning and efficiency in a coastal lagoon: the role of auxiliary energy. Estuar Coast Shelf Sci 84: 227–236.
33. Danovaro R, Armeni M, Dell'Anno A, Fabiano M, Manini E, et al. (2001) Small-scale distribution of bacteria, enzymatic activities and organic matter in coastal sediments. Microb Ecol 42(2): 177–185.
34. Danovaro R (1996) Detritus-bacteria-meiofauna interactions in a seagrass bed (*Posidonia oceanica*) of the NW Mediterranean. Mar Biol 127: 1–13.
35. Nagelkerken I, Blaber SJM, Bouillon S, Green P, Haywood M, et al. (2008) The habitat function of mangroves for terrestrial and marine fauna: A review. Aquatic Bot 89: 155–185.
36. Max LM, Hamilton SL, Gaines SD, Warner RR (2013) Benthic processes and overlying fish assemblages drive the composition of benthic detritus on a central Pacific coral reef. Mar Ecol Progr Ser 482: 181–195.
37. Pusceddu A, Dell'Anno A, Fabiano M, Danovaro R (2009) Quantity and bioavailability of sediment organic matter as signatures of benthic trophic status. Mar Ecol Progr Ser 375: 41–52.
38. Lorenzen CJ, Jeffrey SW (1980) Determination of chlorophyll and phaeopigments spectrophotometric equations. Limnol Oceanogr 12: 343–346.
39. Danovaro R (2010) Methods for the study of deep-sea sediments, their functioning and biodiversity. CRC Press, Boca Raton, USA, 428 pp.
40. Stephens MP, Kadko DC, Smith CR, Latasa M (1997) Chlorophyll a and pheopigments as tracers of labile organic carbon at the central equatorial Pacific seafloor. Geochim Cosmochim Acta 61: 4605–4619.
41. Fry JC (1990) Determination of biomass. In: Methods in Aquatic Bacteriology (Austin B, Ed), John Wiley and Sons LTD, 27–72.
42. van Duyl FC, Kop AJ (1994) Bacterial variation in North Sea sediments: clues to seasonal and spatial variations. Mar Biol 120: 323–337.
43. Heip C, Vincx M, Vranken G (1985) The ecology of marine nematodes. Oceanogr Mar Biol Ann Rev 23: 399–489.
44. Feller RJ, Warwick RM (1988) Energetics. In Introduction to the study of meiofauna (Higgins RP, Thiel H Eds) Smithsonian Institution Press, London, 181–196.
45. De Mesel I, Lee HJ, Vanhove S, Vincx M, Vanreusel A (2006) Species diversity and distribution within the deep-sea nematode genus Acantholaimus on the continental shelf and slope in Antarctica. Polar Biol 29: 860–871.
46. Sanders HL (1968) Marine benthic diversity: a comparative study. Am Nat 102: 243–282.
47. Hurlbert SH (1971) The non-concept of species diversity: a critique and alternative parameters. Ecology 52: 577–586.
48. Gray JS (2000) The measurement of marine species diversity, with an application to the benthic fauna of the Norwegian continental shelf. J Exp Mar Biol Ecol 250: 23–49.
49. Soetaert K, Heip C (1990) Sample-size dependence of diversity indices and the determination of sufficient sample size in a high-diversity deep-sea environment. Mar Ecol Progr Ser 59: 305–307.
50. Pielou EC (1975) Ecological diversity, John Wiley, New York, 165 pp.
51. Anderson MJ, Gorley RN, Clarke KR (2008) PERMANOVA+ for PRIMER: Guide to Software and Statistical Methods. PRIMER-E, Plymouth, United Kingdom.
52. Wieser W (1953) Die Beziehung zwischen Mundhöhlengestalt, Ernährungsweise und Vorkommen bei freilebenden marinen Nematoden. Arkiv Zool 2–4: 439–484.
53. Anderson MJ (2001) A new method for non-parametric multivariate analysis of variance. Austral Ecol 26: 32–46.
54. De Troch M, Gurdebeke S, Fiers F, Vincx M (2001) Zonation and structuring factors of meiofauna communities in a tropical seagrass bed (Gazi Bay, Kenya). J Sea Res 45: 45–61.
55. Netto SA, Gallucci F (2003) Meiofauna and macrofauna communities in a mangrove from the Island of Santa Catarina, South Brazil. Hydrobiologia 505: 159–170.
56. Armenteros M, Martín I, Williams JP, Creagh B, González-Sansón G, et al. (2006) Spatial and temporal variations of meiofaunal communities from the western sector of the Gulf of Batabanó, Cuba. I. Mangrove systems. Estuar Coast 29: 124–132.
57. Chinnadurai G, Fernando OJ (2007) Meiofauna of mangroves of the southeast coast of India with special reference to the free-living marine nematode assemblage. Estuar, Coast Shelf Sci 72: 329–336.
58. Xuan QN, Vanreusel A, Vu Thanh N, Smol N (2007) Biodiversity of Meiofauna in the Intertidal Khe Nhan Mudflat, Can Gio Mangrove Forest, Vietnam with Special Emphasis on Free Living Nematodes. Ocean Sci J 42: 135–152.
59. Barnes N, Bamber RN, Bennell G, Cornelius N, Glassom D, et al. (2011) Assessment of regional and local biodiversity in tropical and subtropical coastal habitats in the East African Marine Ecoregion. Biodivers Conserv 10: 2075–2109.
60. Alongi DM (1988) Microbial-meiofaunal interrelationships in some tropical intertidal sediments. J Mar Res 46: 349–365.
61. Danovaro R, Gambi C (2002) Biodiversity and trophic structure of nematode assemblages in seagrass systems: evidence for a coupling with changes in food availability. Mar Biol 141: 667–677.
62. Danovaro R., Gambi C., Mirto S. (2002). Meiofaunal production and energy transfer efficiency in a seagrass *Posidonia oceanica* bed in the Western Mediterranean. Mar Ecol Progr Ser 234: 95–104.
63. Alongi DM (1987) Inter-estuary variation and intertidal zonation of free-living nematode communities in tropical mangrove systems. Mar Ecol Prog Ser 40: 103–114.
64. Ólafsson E (1995) Meiobenthos in mangrove areas in eastern Africa with emphasis on assemblages structure of free-living marine nematodes. Hydrobiologia 312: 47–57.
65. Somerfield PJ, Gee JM, Aryuthaka C (1998) Meiofaunal communities in a Malaysian mangrove forest. J Mar Biol Ass UK 78: 717–732.
66. Benedetti-Cecchi L, Maggi E (2012) The analysis of biodiversity-ecosystem function experiments: portioning richness and density-dependent effects. In (eds: Solan M, Aspden RJ, Paterson DM) Marine biodiversity and ecosystem functioning: frameworks, methodologies and integration, 73–195.
67. Naeem S, Thompson LJ, Lawler SP, Lawton JH, Woodfin RM (1994) Declining biodiversity can alter the performance of ecosystems. Nature 368: 734–736.
68. Tilman D, Reich PB, Knops J, Wedin D, Mielke T, et al. (2001) Diversity and productivity in a long-term grassland experiment. Science 294: 843–845.
69. Heemsbergen DA, Berg MP, Loreau M, van Hal JR, Faber JH, et al. (2004) Biodiversity effects on soil processes explained by interspecific functional dissimilarity. Science 306: 1019–1020.
70. Balvanera P, Pfisterer AB, Buchmann N, He JS, Nakashizuka T, et al. (2006) Quantifying the evidence for biodiversity effects on ecosystem functioning and services. Ecol Lett 9: 1146–1156.
71. Mora C, Aburto-Oropeza O, Bocos AA, Ayotte PM, Banks S, et al. (2011) Global Human Footprint on the Linkage between Biodiversity and Ecosystem Functioning in Reef Fishes. PloS Biology 9(4): e1000606.
72. Gingold R, Moens T, Rocha-Olivares A (2013) Assessing the response of nematode communities to climate change-driven warming: a microcosm experiment. PLoS ONE 8: e66653.
73. Ehrlich PR, Ehrlich A (1981) Extinction: The Causes and Consequences of the Disappearance of Species. Random House, New York. 305 pp.
74. Lawton JH (1994) What do species do in ecosystems? Oikos 71: 367–374.

18

Effects of Salinity and Nutrient Addition on Mangrove *Excoecaria agallocha*

Yaping Chen, Yong Ye*

Key Laboratory of the Ministry of Education for Coastal and Wetland Ecosystem, College of the Environment and Ecology, Xiamen University, Xiamen, Fujian, China

Abstract

Effects of salinity on seed germination and growth of young (1 month old) and old (2-year old) seedlings of *Excoecaria agallocha* were investigated. Combined effects of salinity and nutrient level were also examined on old seedlings. Seed germination was best at 0 and 5 psu salinity. 15 psu salinity significantly delayed root initiation and decreased final establishment rate. All seeds failed to establish at 25 psu salinity. Young seedlings performed best at 0 and 5 psu, but growth was stunned at 15 psu, and all seedlings died within 90 days at 25 psu. Old seedlings grew best at salinities below 5 psu and they survived the whole cultivation at 25 psu. This indicated that *E. agallocha* increased salt tolerance over time. Gas exchange was significantly compromised by salinities above 15 psu but evidently promoted by high nutrient. Proline accumulated considerably at high nutrient, and its contents increased from 0 to 15 psu but decreased at 25 psu salinity. Lipid peroxidation was aggravated by increasing salinity beyond 15 psu but markedly alleviated by nutrient addition. These responses indicated that *E. agallocha* was intolerant to high salinity but it can be greatly enhanced by nutrient addition.

Editor: Randall P. Niedz, United States Department of Agriculture, United States of America

Funding: This study was supported by projects funded by National Natural Science Foundation of China (Grant No. 41276077 and 41076049). The funders had no role in study design, data collection and analysis, decision to publish, or preparation of the manuscript.

Competing Interests: The authors have declared that no competing interests exist.

* E-mail: yeyong@xmu.edu.cn

Introduction

Mangrove forests are distributed along coastlines and periodically inundated by seawater. The particularity of their habitat makes salinity an important factor limiting propagule germination, seedling growth and reproduction of mangrove trees [1–3]. Many studies dealt with the effects of salinity on mangroves. A negative relationship between seedling emergence rate and salt content was obtained in *Avicennia marina* [4]. Increasing salinity delays root initiation of *Acanthus ilicifolius* and reduces final seedling establishment rates of *Aegiceras corniculatum* [5,6]. Moderate salinities stimulate growths of *Av. marina* [7] and *Ae. corniculatum* [8], but further increases in salinity inevitably decrease their growths. Biomass accumulation and stem elongation of *Sonneratia alba* and *S. lanceolata* show a tendency to be stimulated by low salinity and inhibited by further increasing salinity [9]. Medina and Francisco [10] obtained significant decreases in leaf number and area of *Av. germinans* as soil salinity increased. Under extreme salinity stress, accelerated leaf mortality rates of mangrove seedlings are often accompanied by decreases in leaf production rates, finally leading to the deaths of plants [9,11,12]. High salinity can cause osmotic stress and reduce the availability of water, resulting in stomatal closure and reduced supply of carbon dioxide [13,14]. In addition, salt stress can induce ion toxicities such as membrane disorganization, production of reactive oxygen species, and disturbance of nutrient balance [4,15,16]. Increases in salinity reduce nitrogen accumulation in *Kandelia obovata* [17] and inhibit the uptake of K^+ by *Av. marina*, resulting in damage to photosynthetic apparatus [4]. On the other hand, during long term of acclimation to saline conditions, mangroves evolve various strategies to cope with high salinity, including anatomical [18–20], physiological [21], and molecular [22,23] mechanisms. Patel et al. [4] showed that with increases in salinity, *Av. marina* increased contents of leaf proline to alleviate NaCl stress. In order to defend salt-induced oxidative damage, plants are equipped with oxygen radical detoxifying enzymes such as superoxide dismutase, peroxidase and catalase [6,24]. Accumulation of inorganic ions in vacuoles is common pattern observed in mangrove plants under saline conditions [25,26], which serves not only to increase cellular osmolarity to counter osmotic stress but also to avoid increases in ionic strength of the cytoplasm [21,27]. However, previous studies on this topic mostly focused on a certain growth stage of mangroves, how salinity influences mangroves in a dynamic developmental process is not well known.

Nutrient deficiency is another main problem limiting mangrove growth [28–30]. Different from most terrestrial soils, mangrove sediments are frequently waterlogged by seawater. Waterlogging results in anaerobic environment, which greatly restrains nitrification and consequently leads to low nutrient bioavailability of mangrove sediments [20,31]. Nutrient addition significantly accelerates shoot elongation of *Rhizophora mangle* [32], increases leaf and branch growths of *Av. germinans* [33] and remarkably enlarges leaf area of *Ceriops tagal* [34]. Increases in nutrient availability evidently increase stomatal conductance and photosynthetic rate of *K. obovata* seedlings [17]. For *Av. germinans* trees, nutrient addition increases nitrogen investment in the osmotically compatible solute, glycine betaine, which improves water status of tissues and enhanced photosynthesis [35,36]. Studies on plant anatomical mechanism demonstrated that nutrient addition might enhance water supply to leaves and increase hydraulic conductivity by stimulating root growth and/or improving some aspects of the water conducting pathway [1,37,38]. The applications of nutrient fertilizers were also reported to alleviate the phytotoxicity of heavy metals [20,39,40], contribute to osmoregulation [17,41] and stimulate antioxidant systems [42], all of which endow plants with stronger resistance to adverse environments including prolonged waterlogging and extreme salinities.

Although the ecological and economic importance of mangrove forests are widely acknowledged, these special coastal ecosystems have for a long time been intensively affected by both anthropogenic activities and natural disasters, and suffered from significant habitat loss on a world wide scale [43–45]. Thus, restoring these fragmented ecosystems becomes an imperative purpose for local government. In Southeast China, more than 50% mangrove forests are man-made [46], most of which are dominated by viviparous species, such as *Avicennia marina* and *Kandelia obovata.* Unlike naturally occurring forests characterized by high biodiversity, these artificial forests are often mono-species or merely have several dominant species, rendering the system rather vulnerable to both internal and external disturbance. In order to enrich the biodiversity of mangrove forests and further promote the recovery of their ecosystem service, application of multi-species in mangrove reforestation is of great importance.

Excoecaria agallocha, known as "milk mangrove", is an important medicinal plant and among the few non-viviparous mangrove species in Southeast China, while its restoration in nature is seldom reported. Continuous habitat loss or deterioration accompanied with little compensation has caused great challenges for survival and regeneration of non-viviparous mangrove species like *E. agallocha.* Previous researches on *E. agallocha* are mostly focused on its heredity gene and medicinal properties [47–50], but few can be found on its ecological adaptation to environments. Nandy et al. have ever recorded good growth of mature *E. agallocha* trees in fresh water [18], but its early response to saline conditions is still unknown. Therefore, the present study is designed to test the following three hypotheses about *E. agallocha*, expecting to fill in gaps about this species. Firstly, what is the optimal salinity for seed germination of *E. agallocha*? Our field observation shows that few *E. agallocha* seedlings can be found in forests although there are many seeds and mature *E. agallocha* trees frequently occur. Thus, we hypothesize that the saline condition in the field might be detrimental for natural regeneration of this species. Secondly, will *E. agallocha* change its salinity tolerance over time? Then, what is its tolerance limit at each developmental stage? Since juvenile stages of plants are generally more sensitive to environmental factors as compared to adult plants, we hypothesize that *E. agallocha* will increase its tolerance to salinity over time and the salinity level in its natural habitat probably approaches the talerance limit of early *E. agallocha* seedlings. Thirdly, will fertilization greatly enhance seedling growth or alleviate seedlings under stressed conditions? Since nutrient limitation is prevailing in mangrove forests, we hypothesize that nutrient addition would greatly promote growth and physiological performance of *E. agallocha* seedlings under saline conditions. To test these hypotheses, we devised a series of experiments, and the results were expected to provide instrumental information for field restoration of this species.

Materials and Methods

Ethics statement

No specific permits were required for the described field studies and the field studies did not involve endangered or protected species.

Experiment for seed germination

E. agallocha is non-viviparous and one mature fruit contains three propagules. Mature and healthy propagules of *E. agallocha* were collected from mangrove reserve in Futian (114°1′31.57″E, 22°31′15.40″N), Shenzhen of China, which is seldom affected by human activities. The mangrove forests have average soil salinity of 15.25 psu, organic matter of 4.09%, total nitrogen of 0.139%, total phosphorus of 0.086% and total potassium of 1.25% [51]. The mature propagules were yellowish-brown in color, each with fresh weight of 30±1 mg and diameter of 4.0±0.2 mm.

Six polyethylene pots (12 cm diameter and 18 cm height) for each of the four salinity treatments were each filled with 2.5 kg coastal sand washed twice with tap water. Then, 20 propagules were sown in each pot. The pots were irrigated with artificial seawater (obtained by dissolving raw marine salt in tap water) of salinities of 0, 5, 15 and 25 psu. The water level was kept 3 to 4 mm above the sand surface and tap water was added daily to compensate evapotranspiration loss. Seawater in each pot was renewed weekly to ensure that it would not become stale. All pots were kept in an uncontrolled greenhouse under natural temperature and light. The cultivation lasted for 30 days during which air temperature was 25°C±5°C and average sunshine time was about 7.2 h/d. The seeds were considered rooted when the radicle reached 3 mm, and the unfurling of the first pair of leaves was defined as seedling establishment. The numbers of rooted seeds and established seedlings in each pot was recorded daily during the first 15 days and then every two days during the last 15 days.

Experiment for young seedlings

Twelve polyethylene pots (21 cm diameter and 17 cm height), each containing 4 kg tap-water-washed coastal sand, were prepared. Tap water was added to each pot until the water level was 3 to 4 mm above the sand surface. Then, every three seedlings of uniform size (1.7±0.1 mm stem basal diameter and 3.5±0.2 cm stem height, at one true leaf stage) which had been cultivated under freshwater condition for 30 days from propagules were transplanted in each pot. After one week acclimatization, these seedlings were allowed to expose different salinity treatments (0, 5, 15 and 25 psu) each of which had three replicate pots. Seawater for each treatment was prepared by dissolving raw marine salt in 1.0 strength Hoagland solution [52] and was weekly renewed. Tap water was added to compensate for water loss through evapotranspiration. The treatments lasted for 120 days under the same greenhouse as above. Morphological characteristics including leaf number, maximum leaf area, stem height and stem basal diameter were recorded every 30 days.

Experiment for old seedlings

Seedlings used for this experiment had been incubated for two years under freshwater condition. Every three uniformly sized seedlings were transplanted in each of the 24 polyethylene pots with the same size and medium as those in the young seedling experiment. Treatments began after the seedlings succeeded in colonizing the new environment (one week after transplantation). The experimental design was a completely randomized split-plot, containing eight treatments with two nutrient levels (LN: low nutrient, seawater without nutrient addition; HN: high nutrient, seawater in 1.0 strength Hoagland solution) and four salinity levels (0, 5, 15 and 25 psu). Each treatment had three replicate pots. Treatment solutions were replaced once a week, and tap water was daily added to maintain the water level of 3 to 4 mm above the sand surface. After treatments began, these seedlings were allowed to grow for 120 days under the same greenhouse as above.

Stem height and basal diameter, leaf number and maximum leaf area were recorded every 30 days. After 120 days treatment, all seedlings were harvested, rinsed thoroughly with deionized water, dried at 105°C to constant weight for the measurements of the biomass and its partition (shoot to root biomass ratio, S/R).

Relative growth rates (RGRs) were estimated. Before treatments began, 20 transplanted seedlings were randomly selected to measure stem height (*H*), stem basal diameter (*D*) and total biomass (*B*). According to the logarithmic relation between *B* and D^2H, the equation was simulated as following:

$$\lg B = 0.5807 \lg (D^2H) + 0.18 \quad (p < 0.01)$$

The initial total biomass of each seedling was then estimated using the above equation and RGRs were calculated as:

$$\mathrm{RGR} = (\ln B_1 - \ln B_0)/(t_1 - t_0)$$

Here, B_0 and B_1 are total biomass at the beginning (t_0) and end (t_1) of the experimental period [53].

All physiological analyses were done during the last week of the cultivation. For each seedling, 3 to 4 fully expanded leaves were selected to measure net photosynthetic rate (P_n), transpiration rate (T_r) and stomatal conductance (g_s) by using a portable photosynthesis system (LI-6400XT, U.S.). Gas exchange measurements were conducted between 10 a.m. and 11 a.m. with ambient CO_2 concentration of 387±2 μmol ml^{-1}, illumination intensity of 1000 μmol m^{-2} s^{-1}, air temperature of 32.4±0.8°C and relative humidity of 63%±3%. Water use efficiency (WUE) was calculated through dividing P_n by T_r. These measured leaves were sampled for further chemical analyses.

Activities of catalase (CAT) and superoxide dismutase (SOD) as well as malondialdehyde (MDA) content were determined according to those described by Aebi [54] and Ye et al. [55]. Salt content in leaf tissues was measured following the method by Ye et al. [55] with some modification. One leaf was ground in double distilled water. The homogenate was centrifuged at 10000×*g* for 3 min and the conductivity of the supernatant was measured with a conductivity meter made by Orion. Conductivity values were then converted to salt contents and the final salt content (%) in leaf tissue solution was expressed as salt per gram of water in leaf tissue. Proline contents in leaf tissues were determined according to the method described by Bates et al. [56], using an extract of 0.3 g fresh leaf materials in aqueous sulphosalicylic acid. The extracted proline was reacted with ninhydrin to form a chromophore and then the absorbance at 520 nm was measured for final determination of proline content.

Statistical analysis

All values were expressed as mean ± standard deviation (S.D.) values of 6 (germination experiment) or 3 (seedling experiments) replicates. For either young or old seedlings cultivated with 1.0 strength Hoagland solution, differences in parameters obtained in each measurement event among four salinity treatments (0, 5, 15 and 25 psu) were analyzed by 1-way ANOVA. For old seedlings, differences among salinity treatments and between nutrient levels, and interactions of these two factors were analyzed by 2-way ANOVA. If any significant difference was found, the Student-Newman-Keuls multiple comparison method was involved.

Results

Responses to salinity of *E. agallocha* at different developmental stages

At 0 and 5 psu, propagules of *E. agallocha* began to emerge roots 1–2 d after being sowed (Fig. 1a). Root emergence continued for a maximum period of 15 and 14 d at 0 and 5 psu, and the final rooting rates were both over 90%. Compared with 0 or 5 psu treatments, salinity of 15 and 25 psu delayed root initiation by several days. Root emergence at 15 and 25 psu was firstly recorded on the 6th day, with final rooting rates of 70.0±1.2% and 60.0±2.3%, respectively. Propagules of *E. agallocha* had similar seedling establishment between 0 and 5 psu, with first pair leaves unfurled 9 days after sowing and final establishment rates over 85%. (Fig. 1b). Compared with low salinities, increasing salinity level to 15 psu postponed seedling establishment for 6 days and the final establishment rate decreased to 37.5%. At 25 psu, all seeds failed to establish though some of them had initiated roots.

In general, growth performance of young seedlings was favorable at salinity of 0 and 5 psu, followed by 15 psu and 25 psu (Fig. 2). Below 15 psu treatments, all seedlings survived throughout the 120 days' cultivation, whereas at 25 psu, plant

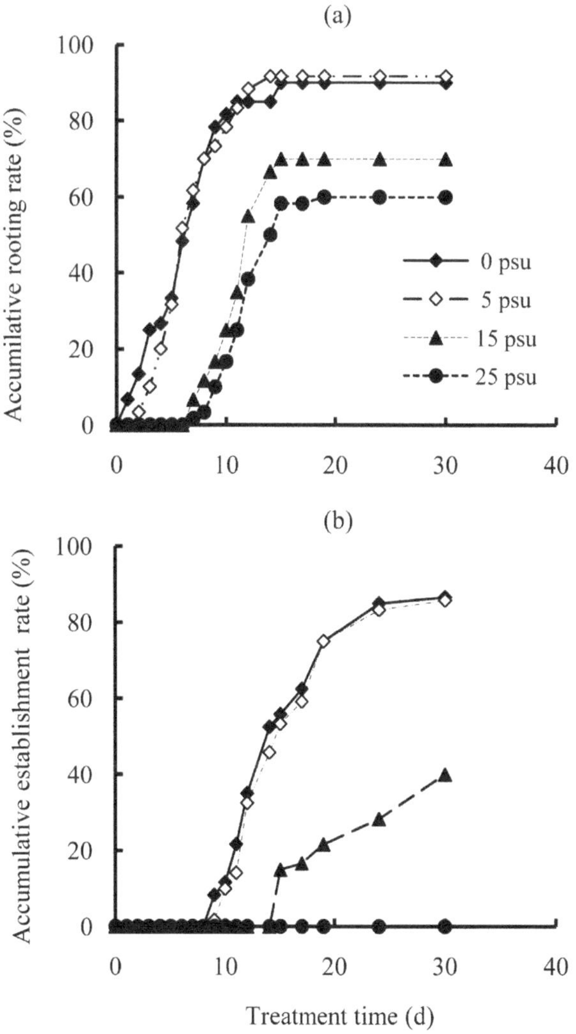

Figure 1. Effects of salinity on seed germination of *E. agallocha*.

deaths occurred after 60 days and all seedlings died within 30 days. Dynamics of all the measured morphologic parameters showed similar tendency at 0 and 5 psu. The stem basal diameter and stem height of seedlings at 0 or 5 psu increased gradually with time, while leaf number and leaf area increased considerably during the first 60 days and then changed little. Elevating salinity to 15 psu evidently compromised growth performance. Stem basal diameter and stem height increased little with time and leaf number even declined, whereas the maximum leaf area increased during the first 90 days and decreased afterwards. At 25 psu, all seedlings failed to survive the whole experiment period.

Old seedlings were different from young seedlings in response to salinity. All seedlings survived the whole experiment period. Growth performance declined in the order of 5 psu ≥0 psu >15 psu >25 psu (Fig. 3). Similar to young seedlings, at 0 and 5 psu, stem basal diameter, stem height and leaf number increased gradually over time, while increases in maximum leaf area were less visible. At the end of the cultivation, all growth parameters except leaf number showed no differences between 0 and 5 psu treatment. Patterns of stem basal diameter and stem height were similar between 15 and 25 psu, with no significant changes over time. At 15 psu, leaf number increased a little while maximum leaf area declined with time. At 25 psu, older leaves gradually detached from seedlings, but few new leaves were produced during the experiment, resulting in evident declines in leaf number and maximum leaf area.

Combined effects of salinity and nutrient on old seedlings of *E. agallocha*

Both salinity and nutrient had significant effects on morphologic parameters of old seedlings after 120 days cultivation, and their interactive effects on stem basal diameter, stem height and leaf number were very significant (Table 1). At any salinity, seedling growth was more vigorous at high than low nutrient, while the differences in morphologic parameters between the two nutrient levels gradually decreased as salinity increased (Fig. 4). At high nutrient, stem basal diameter and stem height varied little either between salinity of 0 and 5 psu or between 15 and 25 psu, while the values at salinities above 15 psu were significantly lower than those at salinities below 5 psu. The maximum value of leaf number occurred at 5 psu, about 1.3, 2.3 and 15.5 times those at 0, 15 and 25 psu, respectively. Similarly, maximum leaf area was highest at 5 psu, not significantly different from that at 0 psu, but 37.5% and 219.0% higher than those at 15 and 25 psu, respectively. Decreasing nutrient availability reduced the differences in stem basal diameter, stem height and leaf number among salinities. Similar to those at high nutrient, decreases in stem basal diameter, leaf number and maximum leaf area at low nutrient were significant with increasing salinity, while significant variations in stem height were not detected among salinities.

Salinities above 15 psu significantly depressed biomass accumulation of old *E. agallocha* seedlings, and the inhibitory effects of high salinity became more pronounced when coupled with low nutrient (Fig. 5). No significant interaction between salinity and nutrient was detected on total biomass (Table 1), while the value of total biomass at low nutrient only accounted for 43.5%, 41.6%, 53.5% and 83.5% of those at high nutrient at 0, 5, 15 and 25 psu, respectively. Similarly, in spite of significant higher values of S/R at high than low nutrient, patterns of S/R with salinity were similar between the two nutrient levels: significantly increasing with salinity increasing from 0 to 5 psu and then decreasing sharply with further increasing salinity from 5 to 25 psu. High

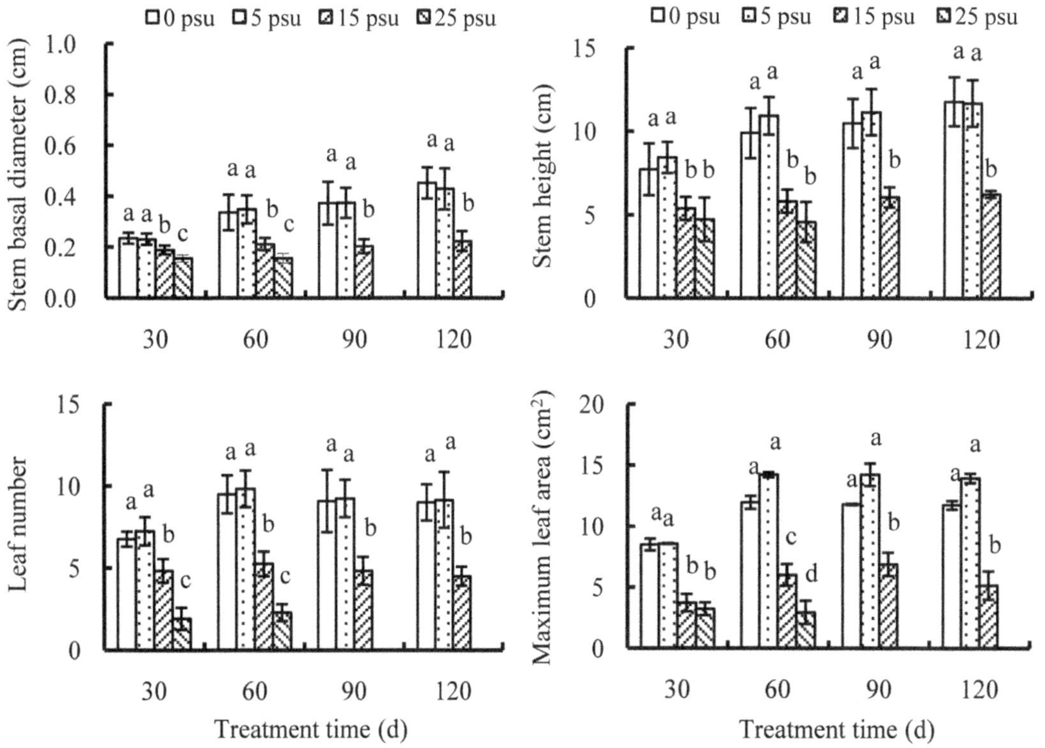

Figure 2. Effects of salinity on growth of young seedlings. Growth parameters include stem basal diameter, stem height, leaf number and maximum leaf area of young *E. agallocha* seedlings cultivated with 1.0 strength Hoagland solution. Mean and SD of three replicates are shown, and means at the same salinity with different letters are significantly different at 0.05 according to 1-way ANOVA test.

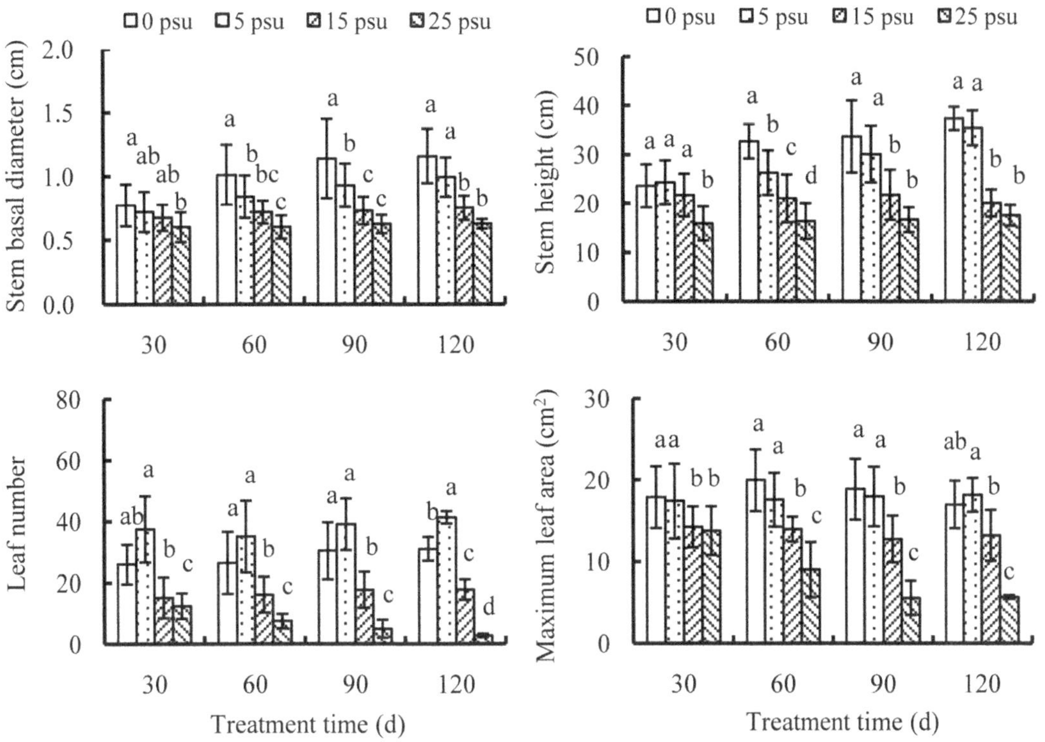

Figure 3. Effects of salinity on growth of old seedlings. Growth parameters include stem basal diameter, stem height, leaf number and maximum leaf area of old *E. agallocha* seedlings cultivated with 1.0 strength Hoagland solution. Mean and SD of three replicates are shown, and means at the same salinity with different letters are significantly different at 0.05 according to 1-way ANOVA test.

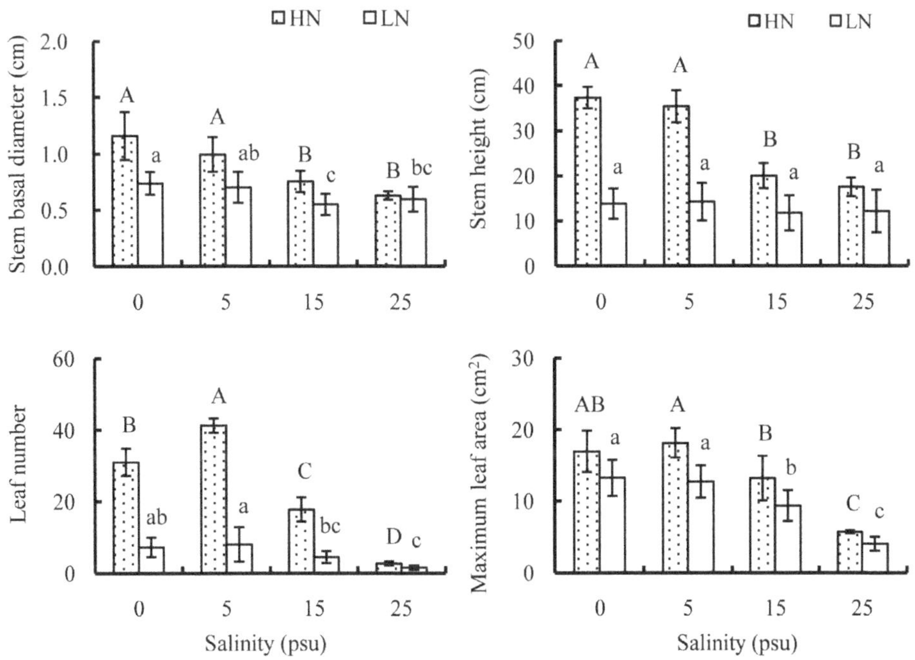

Figure 4. Effects of salinity and nutrient on growth of old seedlings. Growth parameters of stem basal diameter, stem height, leaf number and maximum leaf area were measured after 120 days cultivation. Mean and SD of three replicates are shown, and means at the same nutrient level with different letters are significantly different at 0.05 according to 1-way ANOVA test. HN: high nutrient, fertilized with 1.0 strength Hoagland nutrient solution; LN: low nutrient, unfertilized.

Table 1. Results of 2-way ANOVA on growth and physiological parameters of old *E. agallocha* seedlings showing interactions between salinity and nutrient.

Parameter	Factor		
	Salinity	Nutrient	Salinity × Nutrient
Stem basal diameter	20.284***	45.784***	4.568**
Stem height	28.050***	172.901***	17.173***
Leaf number	62.729***	215.333***	27.049***
Maximum leaf area	35.792***	24.819***	1.031
Total biomass	5.497**	12.453**	1.608
S/R	5.249**	17.128***	0.407
RGR	71.920***	54.150***	12.003***
P_n	135.704***	211.834***	19.325***
T_r	109.575***	274.099***	46.485***
g_s	140.511***	490.896***	85.538***
WUE	29.155***	16.489**	6.076**
Salt content	25.311***	0.501	1.162
Proline	16.828***	39.328***	3.206
MDA	29.697***	25.805**	2.633
CAT	2.821	926.322***	1.401
SOD	3.450*	4.047	2.532

F-values are given and significant effects are denoted as: *0.01<*P*<0.05, **0.001<*P*<0.01, ***P<0.001.

salinity or low nutrient significantly compromised RGR of *E. agallocha* seedlings, while the decline trend in RGR with salinity was more pronounced at high than low nutrient.

Effects of salinity were significant on gas exchange of *E. agallocha* leaves (Fig. 6). Gas exchange was also significantly enhanced by high nutrient, as shown by significantly increasing P_n, T_r, g_s and WUE at high compared to low nutrient. However, differences in P_n, T_r and g_s among the four salinity treatments seemed more pronounced at high than low nutrient. At high nutrient, gas exchange was greatly promoted by elevating salinity level from 0 to 5 psu, but significantly depressed as salinity increased to 15 psu. As increasing salinities from 15 to 25 psu, P_n and T_r continued to decline while g_s and WUE no more significantly changed. Differently, at low nutrient, gas exchange differed little between 0 and 5 psu salinities, but significantly inhibited at 15 psu, while gas exchange at 25 psu was comparable to that at 15 psu.

Salt contents in *E. agallocha* leaves were significantly affected by salinity but not by nutrient level, and there were no interactive effects between salinity and nutrient level on this parameter (Table 1). Increasing salinity from 0 to 5 psu had no significant effect on leaf salt contents, but elevating salinity level from 5 to 15 increased the values by 55.0% and 30.6% at high and low nutrient levels, respectively (Fig. 7). Salt contents varied little with further increasing salinity from 15 to 25 psu. Proline contents were significantly enhanced by high nutrient (Table 1, Fig. 7), with the values at high nutrient about 2.4, 1.3, 1.1 and 1.7 times those at low nutrient at salinity of 0, 5, 15 and 25 psu, respectively. However, there were no consistent responses in proline contents to salinity. Elevating salinity level from 0 to 15 psu greatly accelerated proline accumulation in leaves, but further increasing salinity from 15 to 25 psu imposed an opposite effect on proline accumulation.

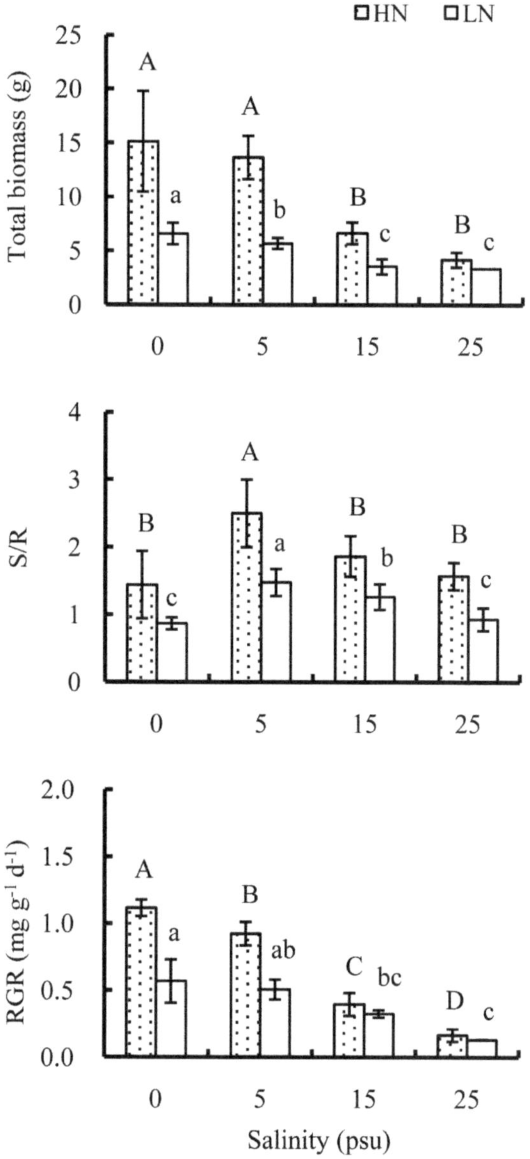

Figure 5. Effects of salinity and nutrient on biomass partition and RGR of old seedlings. Parameters of total biomass, S/R and RGR were measured after 120 days cultivation. Mean and SD of three replicates are shown, and means at the same nutrient level with different letters are significantly different at 0.05 according to 1-way ANOVA test. HN: high nutrient, fertilized with 1.0 strength Hoagland nutrient solution; LN: low nutrient, unfertilized.

MDA contents in leaves were evidently promoted either by elevating salinity level or decreasing nutrient level (Table 1). High nutrient decreased MDA contents by 14.8%, 35.8%, 10.3% and 28.1% at 0, 5, 15 and 25 psu, respectively (Fig. 8). There were no significant effects of salinity on CAT activity, but CAT activity was steeply stimulated by nutrient addition, with values at high nutrient about 6.7, 6.3, 5.0 and 5.4 times those at low nutrient at salinity of 0, 5, 15 and 25 psu, respectively. In contrast, SOD activity showed no significant differences between nutrient levels, but significantly activated by increasing salinity from 5 to 25 psu at high nutrient.

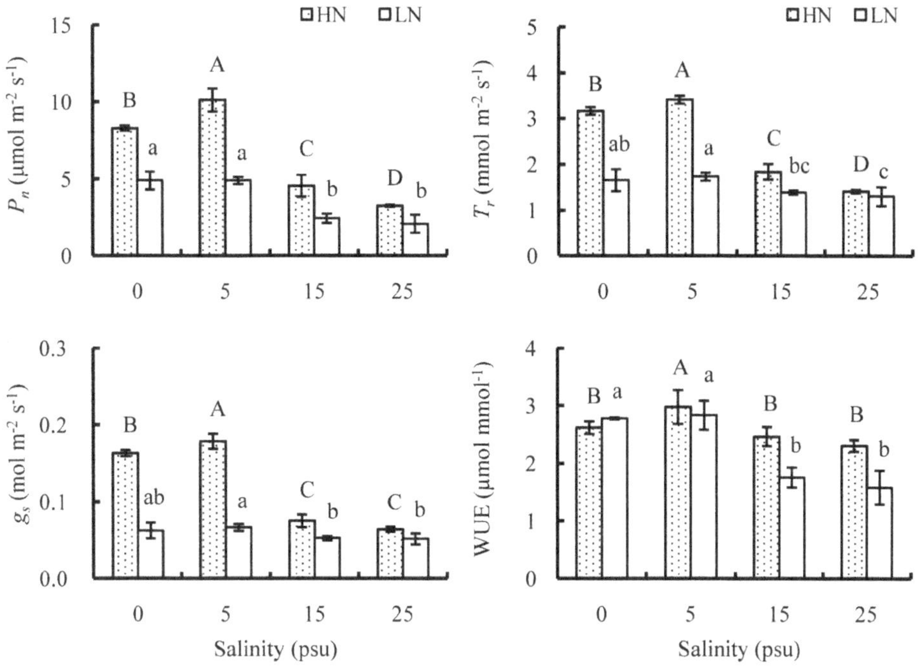

Figure 6. Effects of salinity and nutrient on leaf gas exchange of old seedlings. Physiological parameters include P_n, T_r, g_s and WUE. Mean and SD of three replicates are shown, and means at the same nutrient level with different letters are significantly different at 0.05 according to 1-way ANOVA test. HN: high nutrient, fertilized with 1.0 strength Hoagland nutrient solution; LN: low nutrient, unfertilized.

Discussion

Effects of salinity at different developmental stages

Seed germination is extensively reported to be reduced, retarded [4,57,58] or completely inhibited [59] at high salinities. Similarly, seeds of *Excoecaria agallocha* showed clear preference for low salinities from 0 to 5 psu, and salinity beyond 15 psu significantly delayed root initiation and seedling establishment. High salinities could inhibit seed germination either by impeding water uptake [60] or through facilitating the intake of toxic ions [61,62]. When compared to seed rooting, effects of high salinities seemed more severe on seedling establishment. Salinity of 25 psu was deadly to *E. agallocha* seeds, under which no seedlings succeeded to establish. Likewise, root initiation of *Ae. corniculatum* was not evidently depressed by increasing salinity, but the final establishment percentages decreased significantly at salinities over 25 psu [6]. One possible explanation is that when departing from mother trees, seeds often store a certain amount of water to help go through water deficit during early germination; when further development gradually depletes internal water storage, while water uptake from external environment is inhibited by high salinity, seedlings would be confronted with severe physiological drought, leading to the failure of establishment.

Although mangroves can survive under a wide range of salinity, their growth, stature and productivity varied dramatically along salinity gradients [1–3]. Salinity affects both leaf initiation and leaf area expansion [7,63], and high salinities generally result in dwarfed plants with small and thick leaves [4,19]. Similar to mature *E. agallocha* trees [18], young *E. agallocha* seedlings had vigorous growth at low salinity of 0 and 5 psu, with stems elongated and basal diameter expanded over time. Accelerated

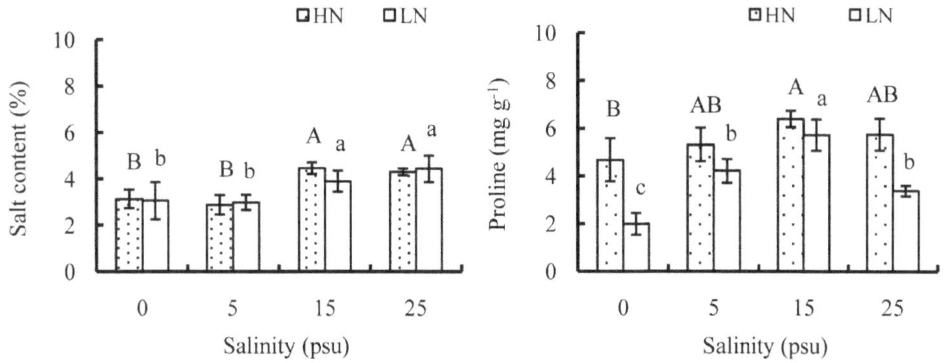

Figure 7. Effects of salinity and nutrient on salt content and proline accumulation of old seedlings. Mean and SD of three replicates are shown, and means at the same nutrient level with different letters are significantly different at 0.05 according to 1-way ANOVA test. HN: high nutrient, fertilized with 1.0 strength Hoagland nutrient solution; LN: low nutrient, unfertilized.

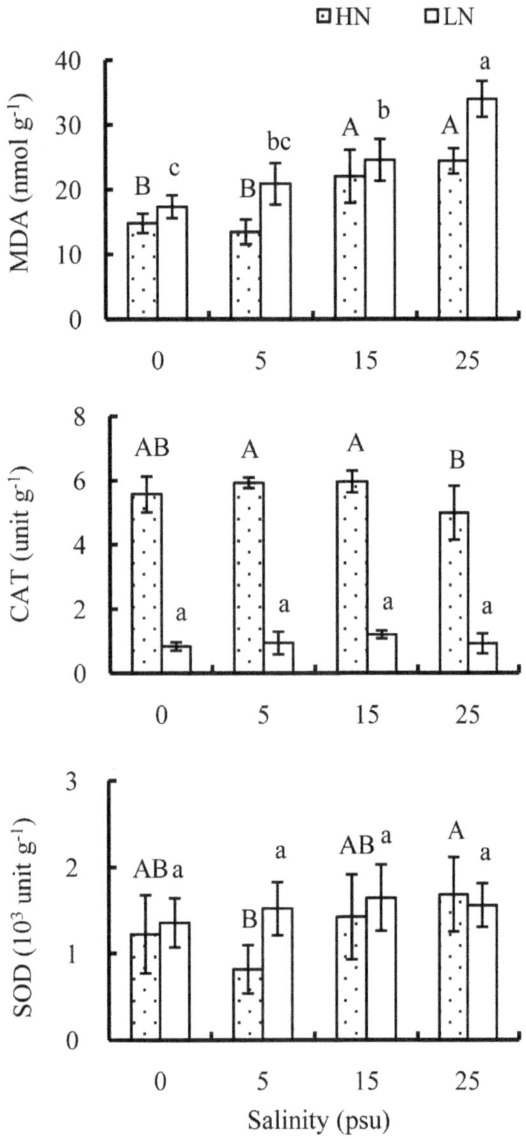

Figure 8. Effects of salinity and nutrient on MDA contents, CAT and SOD activities of old seedlings. Mean and SD of three replicates are shown, and means at the same nutrient level with different letters are significantly different at 0.05 according to 1-way ANOVA test. HN: high nutrient, fertilized with 1.0 strength Hoagland nutrient solution; LN: low nutrient, unfertilized.

senescence of old leaves coupled with decreases in leaf initiation at 15 psu resulted in negative leaf gain over time, implying that 15 psu approaches the limit of salt tolerance for young *E. agallocha* seedlings. Salinity beyond 25 psu was lethal to young seedlings since all plants died within 90 days.

Sensitivity of plants to salinity depends on their developmental stages. As compared to young seedlings of *E. agallocha*, old seedlings were more tolerant to high salinities. Although old seedlings were severely retarded at 15 psu with small and dark leaves, the stressed plants were healthy in appearance with no wilting symptoms observed. In addition, these seedlings slowly but continually produced new leaves, leading to a slight increase in leaf number over time, which implied that old seedlings of *E. agallocha* can tolerate salinity up to 15 psu. High salinity or prolonged salt stress can accelerate leaf mortality and depress leaf production, frequently leading to plant death in the long term [11,64,65]. Likewise, accelerated leaf senescence and mortality of *E. agallocha* seedlings at 25 psu was accompanied by little leaf production, resulting in negative foliage gain over time. At the end of the experiment, the leaf number approximately declined to zero, suggesting that although having lived through the whole experiment period, old seedlings of *E. agallocha* cannot survive salinities above 25 psu in the long run.

Combined effects of salinity and nutrient

At either nutrient level, seedling growths were most vigorous at salinities of 0 to 5 psu, and increasing salinity beyond 15 psu significantly depressed seedling growth. Fewer leaves along with smaller leaf area may be adaptations to salt stress, expecting to reduce water loss through evapotranspiration. Similar to the observation by Ball and Sobrado [66], nutrient addition to some extent alleviated salt-induced reduction of seedling growth. Salt tolerance often involves the synthesis and accumulation of N-based compounds which act in osmotic adjustment [41,67,68]. At high nutrient level, the improved salt tolerance of *E. agallocha* seedlings might be attributed to enhanced osmoregulation by additional nitrogen supply.

Nutrient addition promoted shoot growth and increased branch number of *Av. germinans* [33], enhanced leaf production and expansion of *Av. marina* [32,34] and promoted stem elongation of dwarfed mangroves [38]. Likewise, adding nutrient induced vigorous growth of *E. agallocha*, especially at low salinities. However, promoting effects of nutrient addition was compromised at high salinities, as shown by narrowed differences in morphologic parameters between the two nutrient levels as salinities increased from 5 to 25 psu. Similar results were also attained for other mangrove species like *K. candel* [17] and *Av. marina* [26]. When grown at favorable salinities, seedling growth might be primarily limited by nutrient availability, thus, adding nutrient can greatly enhance growth performance. On the contrary, when adverse saline conditions became the dominate factor over nutrient limit to depress seedling growths, high cost of osmotic adjustment [17] and ion toxicity [4,69] might prevent plants from taking advantage of improved nutrient supply. High contents of Na^+ in soils typically reduce the uptake of K^+ [25,70] and other nutrients like Ca^{2+} and Mg^{2+} [4], leading to ion imbalance in leaves [63,71], while the presence of excessive Cl^- salts could inhibit NO_3^- uptake [63], which interrupts protein synthesis mechanism [72].

In addition, high salinity and low nutrient significantly decreased biomass accumulation and RGR, and changed biomass partitioning of *E. agallocha* seedlings. Similar observation was also recorded in mangrove species of *Av. marina* and *B. parviflora* [4,63], *K. candel* [17] and *S. lanceolata* [9]. Reduction of leaf expansion rates was reported to be one main reason for growth decrease under salt stress [73]. In this study, decreases in total biomass and RGRs were accompanied by decreases in both leaf number and area. As salinity increased from 5 to 25 psu or as nutrient level decreased, a shift in biomass allocation from shoots to roots of *E. agallocha* seedlings occurred. One possible explanation might be that increasing biomass investments to roots could facilitate water and nutrient uptake under high salinities. In the same way, seedlings at low nutrient enhanced root allocation in order for nutrient acquisition.

Decreases in stomatal conductance (g_s) with increasing salinity are typical responses of mangroves [19,63]. Stomatal closure at high salinities minimizes water loss through transpiration, which is however, accompanied by reduced carbon gain, leading to suppression of photosynthesis [18,25,74,75]. In the present study, net photosynthetic rate (P_n) and transpiration rate (T_r) of *E.*

agallocha seedlings decreased in parallel with g_s as salinity increased from 5 to 25 psu. Moreover, in spite of a sharp decrease in T_r through stomatal regulation, *E. agallocha* seedlings failed to maintain high water use efficiency (WUE) at elevated salinities. Dramatic decreases in WUE coupled with significantly lowered P_n possibly accounted for the poor growth performance at salinities above 15 psu. In addition to stomatal limitation, nutrient imbalance might be another possible reason for decreased photosynthesis at high salinities, since salt induced ion imbalance, especially K and Mg deficiency, was extensively reported to suppress chlorophyll synthesis and photosynthetic capacity [18,70]. On the contrary, nutrient addition significantly enhanced P_n, T_r, g_s as well as WUE of *E. agallocha* seedlings, similar observation was also recorded on *K. candel* seedlings [17] and dwarfed *R. mangle* trees [38]. Furthermore, gas exchange in terms of P_n, T_r and g_s was interactively affected by nutrient and salinity, which is in conformity with growth responses in morphological parameters and RGR. Although P_n, T_r and g_s were promoted by high nutrient at any salinity, the differences in these parameters between nutrient levels decreased with increasing salinity.

When in salt stressed habitat, mangroves could accumulate inorganic ions or compatible solutes like betaine, proline, or sugar alcohol as osmotica to counter the toxic effects of salinity [6,21,76]. In this study, increasing salinities beyond 15 psu induced evident increases in salt content of *E. agallocha* seedlings, while nutrient addition had no significant effect on salt contents. Differently, proline contents increased significantly with salinity increasing from 0 to 15 psu, but decreased as it increased to 25 psu. Flowers et al. [41] indicated that the accumulation of proline is strongly affected by nitrogen availability, and the inhibition of nitrogen uptake at high salinities may be the reason for decreased proline contents. In the same way, high nutrient significantly enhanced proline accumulation in *E. agallocha* seedlings, possibly contributing to osmotic adjustment.

An inevitable consequence of salt stress is the excessive generation of reactive oxygen species which can severely permeabilize membranes by causing lipid peroxidation [16,77,78]. MDA, an effective indicator of oxidation damage, accumulated considerably with increasing salinities beyond 15 psu, and its contents were significantly higher at low than high nutrient level, which suggests that lipid peroxidation was aggravated under salt stress, while nutrient addition could greatly alleviate oxidation damage. In addition, antioxidative enzymes like CAT, SOD and peroxidase are often triggered under stressed conditions to defend against oxidation damage [23,79,80]. SOD activities of *E. agallocha* seedlings were stimulated by increasing salinities from 5 to 25 psu, especially at high nutrient. However, CAT was not activated by salt stress, but it was greatly promoted by increasing nutrient availability, which possibly explained the significantly lowered MDA contents at high nutrient.

Conclusion

E. agallocha is highly sensitive to salinity, especially at early developmental stages. The favorable salinity range for seed germination is below 5 psu, and salinity of 15 psu decreased seedling establishment rate to 37%, which partly explained why few seedlings can be found in the mangrove reserve which has soil salinity slightly higher than 15 psu. The adverse saline condition in the field might act as a primary obstacle for natural regeneration of this species. Thus, artificial breeding and culture should be adopted to ensure higher survival rate of *E. agallocha* seedlings.

Then when to replant this species in the field? Results from the present study indicate that seedlings as young as one-month old can hardly tolerant salinity above 15 psu, while for two-year old seedlings, they can adapt to salinity up to 15 psu although failing to surviving 25 psu salinity in the long run, which indicates that *E. agallocha* increased salinity tolerance over time and two-year old seedlings could be chosen for *E. agallocha* restoration in sites where soil salinity is up to 15 psu. In addition to salinity, nutritional limitation prevailing in mangrove forests is another challenge for successful reforestation. The current study shows that nutrient addition could not only greatly enhance plant growth, but also alleviate salt-induced damage to plant physiology. Therefore, applying nutrient rationally could be very helpful in promoting survival of *E. agallocha* seedlings in plantations with relatively high salinity.

On the other hand, we should be aware that due to the particularity of their natural habitat, field restoration of mangrove forests is confronted with many other challenges. Duration and frequency of seawater inundation, predation by crabs and gastropods, soil physiochemical properties, inter- and intra-species competition, and anthropogenic disturbance might be equally important in determining whether *E. agallocha* seedlings would sustainably develop in nature. And researches on these environmental factors as well as various other mangrove species are under way in order for the prosperity of mangrove ecosystems.

Acknowledgments

We thank Yang Zhang and Feilong Tang for their helps in the field work.

Author Contributions

Conceived and designed the experiments: YY. Performed the experiments: YC. Analyzed the data: YC. Contributed reagents/materials/analysis tools: YY. Wrote the paper: YC.

References

1. Lovelock CE, Feller IC, Mckee KL, Engelbrecht BMJ, Ball MC (2004) The effect of nutrient enrichment on growth, photosynthesis and hydraulic conductance of dwarf mangroves in Panamá. Funct Ecol 18: 25–33.
2. McKee KL, Feller IC, Popp M, Wanek W (2002) Mangrove isotopic (δ^{15}N and δ^{13}C) fractionation across a nitrogen vs. phosphorus limitation gradient. Ecology 83: 1065–1075.
3. Naidoo G (2006) Factors contributing to dwarfing in the mangrove *Avicennia marina*. Ann Bot-London 97: 1095–1101.
4. Patel NT, Gupta A, Pandey AN (2010) Salinity tolerance of *Avicennia marina* (Forssk.) Vierh. from Gujarat coasts of India. Aquat Bot 93: 9–16.
5. Ye Y, Lu CY, Wong YS, Tam NFY (2004) Diaspore Traits and Inter-tidal Zonation of Non-viviparous Mangrove Species. Acta Botanica Sinica, 46(8): 896–906.
6. Ye Y, Tam NFY, Lu CY, Wong YS (2005) Effects of salinity on germination, seedling growth and physiology of three salt-secreting mangrove species. Aquat Bot 83: 193–205.
7. Clough BF (1984) Growth and salt balance of the mangrove *Avicennia marina* (Forsk.) Vierh. and *Rhizophora stylosa* Griff. in relation to salinity. Aust J Plant Physiol 11: 419–430.
8. Ball MC, Anderson JM (1986) Sensitivity of photosystem II to NaCl in relation to salinity tolerance: comparative studies with thylakoids of the salt-tolerant mangrove, *Avicennia marina*, and the salt-sensitive pea, *Pisum sativum*. Aust J Plant Physiol 13: 689–698.
9. Ball MC, Pidsley SM (1995) Growth responses to salinity in relation to distribution of two mangrove species, *Sonneratia alba* and *S. lanceolata*, in northern Australia. Funct Ecol 9: 77–85.
10. Medina E, Francisco M (1997) Osmolality and δ^{13}C of leaf tissue of mangrove species from environments of contrasting rainfall and salinity. Estuar Coast Shelf S 45: 337–344.
11. Munns R, Termaat A (1986) Whole-plant responses to salinity. Aust J Plant Physiol 13: 143–160.
12. Suarez N, Medina E (2005) Salinity effect on plant growth and leaf demography of the mangrove, *Avicennia germinans* L. Trees 19: 721–727.

13. Tanaka Y, Hibino T, Hayashi Y, Tanaka A, Kishitani S, et al. (1999). Salt tolerance of transgenic rice over-expressing yeast mitochondrial Mn-SOD in chloroplasts. Plant Sci 148: 131–138.
14. Li N, Chen SL, Zhou XY, Li CY, Shao J, et al. (2008) Effect of NaCl on photosynthesis, salt accumulation and ion compartmentation in two mangrove species, *Kandelia candel* and *Bruguiera gymnorrhiza*. Aquat Bot 88: 303–310.
15. Khan MA, Ungar IA, Showalter AM (2000) The effect of salinity on the growth, water status, and ion content of a leaf succulent perennial halophyte, *Suaeda fruticosa* (L.) Forssk. J Arid Environ 45: 73–84.
16. Hasegawa PM, Bressan RA, Zhu JK, Bohnert HJ (2000) Plant cellular and molecular responses to high salinity. Annu Rev Plant Physiol Plant Mol Biol 51: 463–499.
17. Kao WY, Tsai HC, Tsai TT (2001). Effect of NaCl and nitrogen availability on growth and photosynthesis of seedlings of a mangrove species, *Kandelia candel* (L.) Druce. J Plant Physiol 158: 841–846.
18. Nandy P, Das S, Ghose M, Spooner-Hart R (2007) Effects of salinity on photosynthesis, leaf anatomy, ion accumulation and photosynthetic nitrogen use efficiency in five Indian mangroves. Wetl Ecol Manag 15: 247–357.
19. Naidoo G, Hiralal O, Naidoo Y (2011) Hypersalinity effects on leaf ultrastructure and physiology in the mangrove *Avicennia marina*. Flora 206: 814–820.
20. Cheng H, Wang YS, Ye ZH, Chen DT, Wang YT, et al. (2012) Influence of N deficiency and salinity on metal (Pb, Zn and Cu) accumulation and tolerance by *Rhizophora stylosa* in relation to root anatomy and permeability. Environ Pollut 164: 110–117.
21. Dasgupta N, Nandy P, Sengupta C, Das S (2012) Protein and enzymes regulations towards salt tolerance of some Indian mangroves in relation to adaptation. Trees 26: 377–391.
22. Nandy P, Dasgupta N, Das S (2009) Differential expression of physiological and biochemical defense mechanisms of some Indian mangroves towards salt tolerance. Physiol Mol Biol Plants 151(2): 151–160.
23. Dasgupta N, Nandy P, Tiwari C, Das S (2010) Salinity imposed changes of some isozymes and total leaf protein expression in five mangroves from two different habitats. J Plant Interact 5(3): 211–221.
24. Prochazkova D, Sairam RK, Srivastava GC, Singh DV, (2001) Oxidative stress and antioxidant activity as the basis of senescence in maize leaves. Plant Sci 161(4): 765–771.
25. Ball MC, Farquhar GD (1984) Photosynthetic and stomatal responses of two mangrove species, *Aegiceras corniculatum* and *Avicennia marina*, to long term salinity and humidity conditions. Plant Physiol 74: 1–6.
26. Naidoo G (1987) Effects of salinity and nitrogen on growth and plant water relations in the mangrove *Avicennia marina* (Forssk.) Vierh. New Phytol 107: 317–326.
27. Mimura T, Kura-Hotta M, Tsujimura T, Ohnishi M, Miura M, et al. (2003) Rapid increase of vacuolar volume in response to salt stress. Planta 216: 397–402.
28. Feller IC, Whigham DF, O'Neill JP, McKee KM (1999) Effects of nutrient enrichment on within-stand nutrient cycling in mangrove ecosystems in Belize. Ecology 80: 2193–2205.
29. Feller IC, McKee KL, Whigham DF, O'Neill JP (2003) Nitrogen vs. phosphorus limitation across an ecotonal gradient in a mangrove forest. Biogeochemistry 62: 145–175.
30. Reef R, Feller IC, Lovelock CE (2010) Nutrition of mangroves. Tree Physiol 30: 1148–1160.
31. Tam NFY, Wong YS (1998) Variations of soil nutrient and organic matter content in a subtropical mangrove ecosystem. Water Air Soil Poll 103: 245–261.
32. Feller LC, Mckee KL, Whigham DF, O'Neill JP (2002) Nitrogen vs. phosphorus limitation across an ecotonal gradient in a mangrove forest. Biogeochemistry 00: 1–31.
33. Whigham DF, Verhoeven JTA, Samarkin V, Megonigal PJ (2009) Responses of *Avicennia germinans* (Black Mangrove) and the Soil Microbial Community to Nitrogen Addition in a Hypersaline Wetland. Estuar Coast 32: 926–936.
34. Yates EJ, Ashwath N, Midmore DJ (2002) Responses to nitrogen, phosphorus, potassium and sodium chloride by three mangrove species in pot culture. Trees 16: 120–125.
35. Popp M, Lied W, Meyer AJ, Richter A, Schiller P, et al. (1996) Sample preservation for determination of organic compounds: microwave versus freeze-drying. J Exp Bot 47: 1469–1473.
36. Lovelock CE, Feller IC (2003) Photosynthetic performance and resource utilization of two mangrove species coexisting in a hypersaline scrub forest. Oecologia 134: 455–462.
37. Nadini A, Salleo S (2000) Limitation of stomatal conductance by hydraulic traits: sensing or preventing xylem cavitation. Trees 15: 14–24.
38. Lovelock CE, Ball MC, Choat B, Engelbrecht BMJ, Holbrook NM, et al. (2006) Linking physiological processes with mangrove forest structure: phosphorus deficiency limits canopy development, hydraulic conductivity and photosynthetic carbon gain in dwarf *Rhizophora mangle*. Plant Cell Environ 29: 793–802.
39. Lin CC, Zhu TC, Liu L, Wang DL (2010) Influences of major nutrient elements on Pb accumulation of two crops from a Pb-contaminated soil. J Hazard Mater 174: 202–208.
40. Leblebici Z, Aksoy A (2011) Growth and lead accumulation capacity of *Lemna minor* and *Spirodela polyrhiza* (Lemnaceae): interactions with nutrient enrichment. Water Air Soil Poll 214: 175–184.
41. Flowers TJ, Troke PF, Yeo AR (1977) The mechanism of salt tolerance in halophytes. Annu Rev Plant Physiol 28: 89–121.
42. Shi GR, Cai QS, Liu CF, Wu L (2010) Silicon alleviates cadmium toxicity in peanut plants in relation to cadmium distribution and stimulation of antioxidative enzymes. J Plant Growth Regul 61: 45–52.
43. Chen YP, Ye Y (2013) Growth and physiological responses of saplings of two mangrove species to intertidal elevation. Mar Ecol Prog Ser 482: 107–118.
44. Ellison AM, Farnsworth EJ (1997) Simulated sea level change alters anatomy, physiology, growth, and reproduction of red mangrove (Rhizophora mangle L). Oecologia 112: 435–446.
45. Krauss KW, Allen JA (2003) Factors influencing the regeneration of the mangrove Bruguiera gymnorrhiza (L.) Lamk. on a tropical Pacific island. For Ecol Manag 176: 49–60.
46. He FL (2004) Study on ecological restoration of mangrove ecosystem. Environmental Science and Technology 27: 81–83.
47. Zhang ZH, Zhou RC, Tang T, Huang YL, Zhong Y, et al. (2008) Genetic variation in central and peripheral populations of *Excoecaria agallocha* from Indo-West Pacific. Aquat Bot 89: 57–62.
48. Li MY, Xiao Q, Pan JY, Wu J (2009) Natural products from semi-mangrove flora: source, chemistry and bioactivities. Nat Prod Rep 26(2): 281–298.
49. Chandrasekaran M, Kannathasan K, Venkatesalu V, Prabhakaret K (2009) Antibacterial activity of some salt marsh halophytes and mangrove plants against methicillin resistant *Staphylococcus aureus*. World J Microb Biot 25(1): 155–160.
50. Satyan RS, Aveek N, Eganathan P, Parida A (2010). Comparative histochemical localization of secondary metabolites in seed-raised and in vitro propagated plants of *Excoecaria agallocha* Linn. (Euphorbiaceae), the milky mangrove tree of historical significance. Biotech Histochem 85(5): 285–293.
51. Zhang YB, Lin P, Wei XY, Zhuang TC (2008) Effects of salinity on microbial densities of soil in the dilution plate technique applied in mangrove areas. Acta Ecologoca Sinica 28: 1287–1295.
52. Hoagland DR, Arnon DI (1950) The water-culture method for growing plants without soil. California Agricultural Experiment Station. Circular 347:1–32.
53. Hunt R (1978) Plant Growth Analysis. The Institute of Biology's Studies in Biology no. 96. Edward Arnold, London.
54. Aebi H (1984) Catalase in vitro. Method Enzymol 105: 121–126.
55. Ye Y, Wong YS, Tam NFY (2005) Acclimation of a dominant mangrove plant (*Kandelia candel*) to soil texture and its response to canopy shade. Hydrobiologia 539: 109–119.
56. Bates LS, Waldren RP, Teare ID (1973) Rapid Determination of Free Proline for Water Stress Studies. Plant Soil 39: 205–207.
57. Abari AK, Nasr MH, Hojjati M, Bayat D (2011) Salt effects on seed germination and seedling emergency of two Acacia species. Afr J Plant Sci 5(1): 52–56.
58. Khan MA, Gulzar S (2003) Light, salinity, and temperature effects on the seed germination of perennial grasses. Am J Bot 90(1): 131–134.
59. Benabderrahim MA, Haddad M, Hamza H, Ferchichi A (2011) Germination and emergence variability of alfalfa (*Medicago sativa* L.) landraces collected in Southern Tunisia oases. Span J Agric Res 9(1): 135–143.
60. Werner JE, Finkelstein RR (1995) Arabidopsis mutants with reduced response to NaCl and osmostic stress. Physiol Plantarum 93: 659–666.
61. Rehman S, Harris PJC, Bourne WF, Wilkin J (1997) The effect of sodium chloride on germination and the potassium and calcium contents of Acacia seeds. Seed Sci Technol 25(1): 45–57.
62. Kaveh H, Nemati H, Farsi M, Jartoodeh SV (2011) How salinity affect germination and emergence of tomato lines. J Biol Environ Sci 5(15): 159–163.
63. Parida AK, Das AB, Mittra B (2004) Effects of salt on growth, ion accumulation, photosynthesis and leaf anatomy of the mangrove, *Bruguiera parviflora*. Trees 18: 167–174.
64. Srivastava DS, Jefferies RL (1995) The effect of salinity on the leaf and shoot demography of two arctic forage species. J Ecol 83: 421–430.
65. Warwick NWM, Bailey PCE (1998) The effect of time of exposure to NaCl on leaf demography and growth for two non-halophytic wetland macrophytes, Potamogeton tricarinatus F. Muell. and A. Benn. Ex A. Benn. and Triglochin procera R. Br. Aquat Bot 62: 19–31.
66. Ball MC, Sobrado MA (1999) Ecophysiology of mangroves: challenges in linking physiological processes with patterns in forest structure. In: Press MC, Scholes JD, Barker MG, (eds) Physiological plant ecology. The British Ecological Society, Blackwell Science, London 331–346.
67. Cavalieri AJ, Huang AHC (1981) Accumulation of proline and glycinebetaine in *Spartina alterniflora* Loisel. in response to NaCl and nitrogen in the marsh. Oecologia 49: 224–228.
68. Naidoo G, McKee KL, Mendelssoh IA (1992) Anatomical and metabolic responses to waterlogging and salinity in *Spartina alterniflora* and *S. patens* (Poaceae). Am J Bot 79: 765–770.
69. Feigin A (1985) Fertilization management of crops irrigated with saline water. Plant Soil 89: 285–299.
70. Chow WS, Ball MC, Anderson JM (1990) Growth and photosynthetic responses of spinach to salinity: implications of K^+ nutrition for salt tolerance. Aust J Plant Physiol 17: 563–567.
71. Popp M, Larher F, Weigel P (1985) Osmotic adaptation in Australian mangroves. Vegetatio 61: 247–253.
72. Parida AK, Das AB (2004) Effects of NaCl stress on nitrogen and phosphorous metabolism in a true mangrove *Bruguiera parviflora* grown under hydroponic culture. J Plant Physiol 161: 921–928.

73. Rawson HM, Munns R (1984) Leaf expansion in sunflower as influenced by salinity and short-term changes in carbon fixation. Plant Cell Environ 7: 207–213.
74. Clough BF, Sim RG (1989) Changes in gas exchange characteristics and water use efficiency of mangroves in response to salinity and vapour pressure deficit. Oecologia 79: 38–44.
75. Nandy P, Das S, Ghose M (2005) Relation of leaf micromorphology with photosynthesis and water efflux in some Indian mangroves. Acta Bot Croat 64(2): 331–340.
76. Popp M (1984) Chemical composition of Australian mangroves. I. Inorganic ions and organic acids. Zeitschr Pflanzenphysiol 113: 395–409.
77. Meloni DA, Oliva MA, Martinez CA, Cambraia J (2003) Photosynthesis and activity of superoxide dismutase, peroxidase and gluthathione reductase in cotton under salt stress. Environ Exp Bot 49: 69–76.
78. Prochazkova D, Wilhelmova N (2007) Leaf senescence and activities of the antioxidant enzymes. Biol Plantarum 51: 401–406.
79. Macfarlane GR, Burcrad MD (2001) Photosynthetic pigments and peroxidase activity as indicators of heavy metal stress in the grey mangrove, *Avicennia marina* (Forsk.) Vierh. Mar Pollut Bull 42(3): 233–240.
80. Dasgupta N, Nandy P, Das S (2011) Photosynthesis and antioxidative enzyme activities in five Indian mangroves with respect to their adaptability. Acta Physiol Plant 33: 803–810.

19

Global Patterns of Abundance, Diversity and Community Structure of the *Aminicenantes* (Candidate Phylum OP8)

Ibrahim F. Farag, James P. Davis¤, Noha H. Youssef, Mostafa S. Elshahed*

Department of Microbiology and Molecular Genetics, Oklahoma State University, Stillwater, Oklahoma, United States of America

Abstract

We investigated the global patterns of abundance, diversity, and community structure of members of the *Aminicenantes* (candidate phylum OP8). Our aim was to identify the putative ecological role(s) played by members of this poorly characterized bacterial lineages in various ecosystems. Analysis of near full-length 16S rRNA genes identified four classes and eight orders within the *Aminicenantes*. Within 3,134 datasets comprising ~1.8 billion high throughput-generated partial 16S rRNA genes, 47,351 *Aminicenantes*-affiliated sequences were identified in 913 datasets. The *Aminicenantes* exhibited the highest relative abundance in hydrocarbon-impacted environments, followed by marine habitats (especially hydrothermal vents and coral-associated microbiome samples), and aquatic, non-marine habitats (especially in terrestrial springs and groundwater samples). While the overall abundance of the *Aminicenantes* was higher in low oxygen tension as well as non-saline and low salinity habitats, it was encountered in a wide range of oxygen tension, salinities, and temperatures. Analysis of the community structure of the *Aminicenantes* showed distinct patterns across various datasets that appear to be, mostly, driven by habitat variations rather than prevalent environmental parameters. We argue that the detection of the *Aminicenantes* across environmental extremes and the observed distinct community structure patterns reflect a high level of intraphylum metabolic diversity and adaptive capabilities that enable its survival and growth in a wide range of habitats and environmental conditions.

Editor: Ramy K. Aziz, Cairo University, Egypt

Funding: This work was supported by the National Science Foundation Microbial Observatories Program (grant EF0801858). The funders had no role in study design, data collection and analysis, decision to publish, or preparation of the manuscript.

Competing Interests: The authors have declared that no competing interests exist.

* E-mail: Mostafa@okstate.edu

¤ Current address: Oklahoma Medical Research Foundation, Oklahoma City, Oklahoma, United States of America

Introduction

During the last quarter century, culture-independent diversity surveys have been extensively utilized to investigate bacterial diversity in almost all accessible habitats on earth [1–5]. These surveys have collectively demonstrated that the scope of bacterial diversity is much broader than previously expected based on culture-based assessments [6,7], with a large fraction of the 16S rRNA gene sequences encountered not belonging to known cultured bacterial phyla. The term candidate phylum (CP) was thus proposed to describe such lineages [2].

One of the most important challenges facing microbial ecologists is to elucidate the putative metabolic capabilities and ecological roles of these candidate phyla, as well as the underlying ecological factors controlling their observed patterns of abundance, diversity, and community structure on a global scale. Various environmental genomics approaches have been utilized to obtain genomic fragments and partial genome assemblies from these lineages. These include construction and screening of large insert (Fosmid and BAC) libraries [8–11], direct metagenomic surveys and subsequent implementation of novel binning approaches to reconstruct genomes from metagenomic sequence data [12–15], and single cell genomics [16–19]. Collectively, these efforts have yielded valuable insight regarding the genomic characteristics and putative metabolic capabilities of multiple novel candidate phyla. Further, in several incidents, these insights were successfully utilized as a stepping-stone for enrichment and isolation of some of these lineages [20–22].

Genomic approaches are extremely valuable for deciphering putative metabolic capabilities of uncultured bacterial lineages. However, information from genomic studies is derived from a single sampling event in a single environment, and often from a single cell within the sample [17,19]. Extrapolation of such information to imply similar capabilities and genomic features to all members and lineages within an entire bacterial phylum is hence inappropriate. This is especially true since a single bacterial phylum could exhibit a bewildering array of metabolic capabilities.

A complementary approach that has previously been utilized on an ecosystem level [23–27], but rarely utilized in a global phylocentric context, relies on using *in silico* database mining approaches to examine patterns of distribution of members of a specific candidate phyla in 16S rRNA gene diversity surveys. This approach could clarify the patterns of abundance, diversity, and community structure of the targeted lineage. This phylocentric strategy could greatly benefit from the dramatic increase in the number and size of publicly available 16S rRNA gene datasets; brought about by utilizing next generation sequencing technologies in recent ambitious initiatives to catalogue 16S rRNA gene diversity on a global scale [28–30].

Here, we describe a comprehensive examination of the global distribution of members of the *Aminicenantes* (candidate phylum

OP8) using *in silico* database mining approaches. Our aim was to understand the putative ecological role(s) played by members of this poorly characterized bacterial lineages in various ecosystems and to demonstrate the utility of in silico database mining approaches in extracting meaningful ecological patterns from high throughput 16S rRNA gene datasets. Candidate phylum OP8 was first identified in sediments from the Obsidian Pool in Yellowstone National Park [2]. Since then, it has subsequently been identified in a wide range of terrestrial and marine habitats [31–34]. A recent study has described two near candidate phylum OP8 genome assemblies from 38 partial single cell genomes obtained from deep sediments of a brackish lake (Sakinaw lake, British Columbia, Canada), and the name *Aminicenantes* was proposed for this candidate phylum to highlight the high proportion of genes encoding aminolytic enzymes identified in both assemblies [16]. Our results highlight the ubiquitous nature of the *Aminicenantes*, and identify various environmental conditions impacting its global abundance and distribution in various habitats. We argue that these observed patterns suggest that, collectively, members of the *Aminicenantes* exhibit a high level of intra-phylum metabolic and adaptive diversities, and are hence capable of survival, and growth in a wide range of environmental extremes.

Materials and Methods

1. A taxonomic outline of the *Aminicenantes*

While the candidate phylum *Aminicenantes* (CD-OP8) is recognized in several curated taxonomic outlines e.g. Greengenes [35] and SILVA [36], only a fairly low number of *Aminicenantes* sequences are deposited in these databases (109, and 12, in Greengenes and SILVA, respectively). The continuous deposition of new near full-length 16S rRNA gene sequences in GenBank database repository, coupled to the sporadic updates of curated taxonomic schemes, raises the prospect that additional *Aminicenantes* 16S rRNA sequences putatively representing novel high rank (class/ order) lineages have been deposited in GenBank but have yet to be included in taxonomic schemes. Therefore, as a preliminary step, we aimed to identify and classify all GenBank-deposited *Aminicenantes* 16S rRNA gene sequences and produce an updated and comprehensive taxonomic outline of this phylum. To this end, we queried GenBank NR database using BlastN [37], to identify the closest relatives of each of the 109 *Aminicenantes* sequences currently recognized in Greengenes and SILVA databases. The 500 closest relatives of each sequence were downloaded; and duplicates, sequences shorter than 800 bp, and chimeric sequences, identified using Galaxy [38], were removed. The remaining sequences (n = 2955) were aligned to a collection of reference sequences representing all *Aminicenantes* sequences, as well as sequences from a collection of 17 phyla, and 8 candidate phyla using ClustalX [39]. The phylogenetic positions of putative *Aminicenantes* sequences were evaluated using Distance, Parsimony, Maximum likelihood, and Bayesian approaches as previously described [40]. Sequences were deemed representative of a new class/order within the *Aminicenantes* if two or more distinct sequences remained reproducibly monophyletic and formed a bootstrap-supported independent clade upon varying the composition and size of the data set used for phylogenetic analysis [41].

2. Identification of *Aminicenantes* members in next Generation 16S rRNA gene datasets

Publicly available datasets generated using high throughput sequencing technologies (Pyrosequencing and Illumina) were downloaded from MG-RAST [42], VAMPS (http://vamps.mbl.edu/index.php), and GenBank SRA[43] (through the mirror web interface of DNA Databank of Japan http://www.ddbj.nig.ac.jp) in December 2012. Preliminary analysis indicated the absence of the *Aminicenantes* in human and metazoan microbiome samples and hence these datasets were excluded from further analysis (with the notable exception of rumen samples which were included). In total, 3,141 datasets from 110 different studies with 1,820,857,401 distinct 16S rRNA sequences were included in the analysis (Table S1 in File S1). All datasets were quality screened to filter all the sequences with lengths less than 50 base pairs, sequences with ambiguous nucleotides, and sequences with hompolymer stretches more than 8 bps. Sequences were classified using classify.seqs commands package in MOTHUR v.1.29.0 [44], using Silva alignment and Greengenes classification scheme. Sequences were identified as members of the *Aminicenantes* using a cutoff of 70% confidence threshold, as well as by confirmation of such assignment by sporadic manual insertion of putative *Aminicenantes* sequences into reference phylogenetic trees as described above. The subphylum level affiliation of all high throughput *Aminicenantes* sequences identified were determined using the updated taxonomic scheme produced in this study using near full length 16S rRNA gene sequences as described above. All analyses were conducted on a the HPC Cowboy super computer, a 252 compute nodes with dual six core CPUs and 32 GB RAMs server, 2 fat nodes with 256 GB RAM, GPU cards and 120 TB very fast disk storage at the OSU High Performance Computing Center at Oklahoma State University.

3. Classification of next-generation datasets according to habitat type and prevalent environmental conditions

All datasets included in this study were classified according to two different classification schemes: habitat type as well as prevalent environmental conditions. These classifications were used to determine the ecological prevalence and distribution patterns of various members of the *Aminicenantes*. Habitat-based classification scheme involved binning all 3,141 datasets into five major habitat types: Marine, aquatic non-marine, soil, hydrocarbon-impacted, and rumen/other (dust, animal-associated habitats and air). Due to the heterogeneity of geochemical and environmental conditions observed in marine, aquatic non-marine, and soil habitats, these three habitats were further sub-classified into multiple sub-habitat types, determined through the analysis of the projects' available metadata (Table 1). For classification of datasets according to prevalent environmental conditions, three different classification schemes using temperature, oxygen tension, and salinity were utilized (Table 2). Classification based on prevalent pH conditions was not feasible due to the frequent absence of accurate pH metadata in a large proportion of the datasets, as well as the exceedingly low number of datasets that appear to originate from environments with preeminently low (e.g. <3) or high (e.g. >9) pH. A detailed description of all habitats examined in this study is presented in Table S1 in File S1.

4. Deciphering ecological preferences and patterns of distribution of the *Aminicenantes*

The distribution and preferences of *Aminicenantes* were identified by correlating *Aminicenantes* relative abundance (% of sequences affiliated with *Aminicenantes* in the dataset), diversity, and community structure to its distribution in various habitats and sub-habitats, as well as across various environmental conditions.

Rarefaction curves were used to compare diversities of *Aminicenantes* community in different datasets as previously described [45]. We chose rarefaction curve analysis since it provides a sample size unbiased estimate of diversity and is hence

Table 1. Classification and overall patterns of *Aminicenantes* relative abundance in various habitats and sub-habitats[1].

Dataset type	Total datasets	Datasets with *Aminicenantes* (%)	Average *Aminicenantes* abundance (%)	Maximum relative abundance
Total datasets	3,141	918 (29.22%)	0.20%[2]	10.20%
Total 16S rRNA sequences	1,820,857,401	47,315	0.0026%	
Marine datasets	1,154	248 (21.50%)	0.28%	5.28%
Deep marine sediments	32	30	0.50%	2.89%
Coral associated microbiome	19	10	0.89%	4.67%
Pelagic	390	40	0.20%	2.46%
Hydrothermal vents	101	60	0.23%	5.28%
Coastal	612	107	0.20%	1.87%
Aquatic non-marine datasets	1,665	645 (38.74%)	0.15%	10.20%
Spring and ground water	25	10	2.80%	10.20%
Temperate freshwater	1569	587	0.11%	2.50%
Salt marshes	71	48	0.03%	0.67%
Soil datasets	276	14 (5.072%)	0.07%	0.80%
Agriculture	28	2	0.03%	0.06%
Grassland	140	10	0.00%	0.00%
Heavy metal/hydrocarbon contaminated	8	1	0.00%	0.01%
Arid and Semi-arid	46	0	0%	0%
Permafrost	54	1	0.01%	0.01%
Hydrocarbon-impacted datasets	14	10 (71.43%)	0.32%	0.95%
Herbivorous gut and other datasets[3]	32	1 (3.125%)	0.02%	0.02%

[1]A detailed description of every dataset is provided as supplementary material (Table S1 in File S1).
[2]Average abundance values in datasets where *Aminicenantes* sequences were identified.
[3]26 Datasets were designated "other"; these datasets originated from dust, air and animal associated habitat. See Supplementary Table S1 in File S1 for details.

useful in comparing datasets with wide variations in the numbers of sequences examined. In brief, rarefaction curves were constructed for all datasets with more than 50 sequences belonging to the *Aminicenantes*. Rarefaction curve plots were used to rank the datasets in order of diversity. Datasets with intersecting rarefaction curves were given the same rank. The datasets were ranked from one (least diverse) to 198 (most diverse) and subsequently binned into diversity categories as follows; "very low" (ranks 1–40), "low" (41–80), "medium" (81–120), "high" (121–160), and "very high" (161–198) categories. The ranks were then used to correlate *Aminicenantes* diversity to specific environmental factors using Spearman rank correlation and the significance of these correlations were tested in R [46].

The community structure profiles i.e. the proportion of various *Aminicenantes* lineages in various datasets were examined to reveal overall patterns of community structure in different habitats and under different environmental conditions. In addition, to zoom in on the patterns of *Aminicenantes* community structure in datasets where *Aminicenantes* represents a significant fraction of the overall bacterial community, the community structures in datasets with more than 50 *Aminicenantes* sequences (n = 198) were compared using principal-component analysis (PCA) and biplots were constructed using the R statistical package. In this analysis, the relative position of datasets is indicative of the level of their similarity, the directions of the class/subclass arrows are indicative of their respective maximal abundances, and the lengths of the arrows are proportional to the differential abundances of such lineages.

Results

1. A revised taxonomic outline of the *Aminicenantes*

A total of 142 near-full length 16S rRNA *Aminicenantes* gene sequences were identified in GenBank NR database (Table S2 in File S1). Detailed phylogenetic analysis grouped the *Aminicenantes* sequences into four candidate classes: OP8-1, OP8-2, OP8-3 and OP8-unclassified. Candidate class OP8-1 has the largest number of near full-length *Aminicenantes* sequences and is comprised of five distinct orders (OP8-1_HMMV, OP8-1_SHA-124, OP8-1_OPB95, OP8-1_unclassified, and OP8-1_YNP) (Figure 1, Table S2 in File S1). In contrast, classes OP8-2, OP8-3, and OP8-unclassified have a lower number of near full-length sequences and are not further sub-classified into candidate orders. This revision of *Aminicenantes* phylogeny hence increased the number of recognized near full-length 16S rRNA gene sequences by 30.3%, and added one candidate class (OP8-3) and one candidate order (OP8-1_YNP) to the Greengenes taxonomic outline, the most detailed *Aminicenantes* classification scheme in curated databases.

2. Identification of members of *Aminicenantes* in next generation 16S rRNA gene datasets

We used pyrosequencing- and Illumina-generated 16S rRNA gene datasets available in three publicly available gene repositories (VAMPS, GenBank, and MG-RAST) [42,43] to identify the patterns of relative abundance, diversity, and community structure of members of the *Aminicenantes*. Within 3,141 datasets comprising ~1.8 billion 16S rRNA gene sequences, 47,315 (0.0026%) from

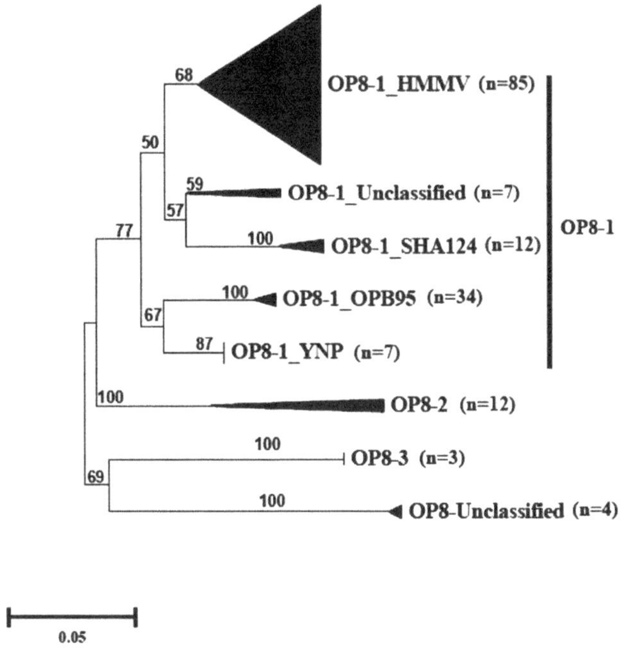

Figure 1. An updated taxonomic outline of the *Aminicenantes*. The Distance NJ tree was constructed using Jukes-Cantor corrections in MEGA5 [64]. Bootstrap values (in percent) are based on 1000 replicates and are shown for branches with more than 50% bootstrap support. Numbers in parentheses represent the number of sequences in each OP8 sub-phylum.

918 (29.2%) different datasets were affiliated with the *Aminicenantes* (Table 1, Table S1 in File S1).

3. Patterns of *Aminicenantes* abundance

Overall relative abundance of *Aminicenantes* varied widely between various datasets, and ranged between 0 and 10.2% (encountered in MG-RAST dataset number 4455892, obtained from groundwater heavily contaminated by arsenic in the Ganges-Brahmaputra Delta region of Bangladesh, [47] (Table 1, Figure S1 in File S1). Although *Aminicenantes* has been identified in a substantial fraction (29.2%) of examined datasets, it invariantly constituted a minor fraction of the bacterial community identified, and rarely exceeded 5% in all datasets (Table 1, Figure S1 in File S1).

Based on incidence of occurrence (i.e. percentages of datasets in which sequences affiliated with the *Aminicenantes* were identified), and relative abundance of *Aminicenantes* in various datasets (Table 1, Figure S1 in File S1), members of the *Aminicenantes* appear to be most abundant in hydrocarbon-impacted habitats, being identified in 71.4% of the datasets (10/14), with an average abundance of 0.321%. The *Aminicenantes* was also frequently identified in marine (21.5% of datasets) and aquatic non-marine (38.74% of datasets) habitats, with average relative abundances of 0.275%, 0.146%, respectively (Table 1, Figure S1 in File S1). On the other hand, the *Aminicenantes* were rarely identified in soils and rumen habitats (Table 1, Figure S1 in File S1).

Aminicenantes abundance also demonstrated distinct patterns in relation to oxygen tension, temperature, and salinity (Table 2, Figure S2 in File S1). The *Aminicenantes* were most abundant in anaerobic habitats (58% of datasets, average 0.46%) e.g. Mai Po mangrove marshes in Hong Kong, heavy metal contaminated ground water in Bangladesh [47], active hydrothermal vent sediments from the Mid-Atlantic Ridge [48], anoxic sulfide and sulfur-rich terrestrial spring in southwestern Oklahoma (Zodletone spring) [40], and anoxic sediments from the Guaymas [3] and Cariaco Basins [49] (Table S1 in File S1). However, the *Aminicenantes* were also identified in much lower abundance in few oxic habitats e.g. water and sediments from coastal and open ocean sites surveyed from South Atlantic to the Caribbean seabed (Table S1 in File S1), coastal water of western English channel [50], and soils and sediments of hypersaline lake, La Sal del Rey's in southern Texas, USA [51] (Table S1 in File S1). Temperature profile of *Aminicenantes* abundance indicated an extremely rare occurrence in low temperature terrestrial and marine habitats (e.g. in datasets from the Canadian, Alaskan and European tundra and arctic soils, as well as the Amundsen sea [50,52]), (Table S1 in File S1), and a slightly higher preference (based on incidence of occurrence) to habitats with temperate, medium, elevated, and extremely elevated temperatures (Table 2, Figure S2 in File S1). Salinity wise, *Aminicenantes* was present at all levels of salinities, with slightly higher relative abundances in non-saline, and low salinity habitats (Table 2, Figure S2 in File S1).

4. Patterns of *Aminicenantes* community structure

Examination of patterns of *Aminicenantes* community composition across habitats revealed several distinct patterns. For example, order OP8-1_HMMV appears to be prevalent in marine environments, where it represented 53.5% of the total *Aminicenantes* sequences identified in marine datasets (Figure 2a). Class OP8-1_unclassified appeared to be the prominent lineage in aquatic non-marine environments, where it represented 77% of the total number of *Aminicenantes* sequences (Figure 2a). Order OP8-2 was the prevalent *Aminicenantes* lineage in hydrocarbon-impacted environments where it represented 66% of the total number of sequences. Although extremely rare in the rumen, the *Aminicenantes* sequences identified in a single dataset from this habitat belonged to order OP8-1_OPB95. PCA analysis conducted on datasets with more than 50 *Aminicenantes* sequences (n = 198, Figure 2b) confirmed such patterns where most of the environments from marine origins clustered along the OP8-1_HMMV species arrow (circles in Figure 2b), most of the environments from aquatic non-marine origins clustered along the OP8-1_unclassified species arrow (stars in Figure 2b), and the majority of the hydrocarbon-impacted environments clustered in the direction of the OP8-2 species arrow (diamonds in Figure 2b).

Sub-classification of habitats (Figures 2c-h) further revealed additional patterns at the sub-habitat level, especially in marine, soil, and aquatic non-marine habitats systems. Within marine environments, the prevalence of OP8-1_HMMV was more pronounced in coral-associated, pelagic, and deep marine datasets (Figure 2c). Indeed, in marine datasets with >50 *Aminicenantes* sequences, OP8-1_HMMV represents the majority (more than 80%) of the total *Aminicenantes* sequences in all coral-associated and pelagic datasets, as well as in the majority (10 out of 13) of deep sediment datasets. OP8-1_HMMV also represented the majority of *Aminicenantes* sequences in a few of the coastal (three out of 15) and hydrothermal (one out of six) datasets. Accordingly, those samples clustered together along the OP8-1_HMMV species arrow in the PCA biplot (red circles representing one vent sample, green circles representing five pelagic samples, yellow circles representing ten deep sediment samples, black circles representing five coral-associated samples, and blue circles representing three coastal samples in Figure 2d). In the remaining marine samples, the majority of *Aminicenantes* datasets has a mixed community of OP8-1_HMMV and other lineages, and so had an intermediary position between species arrows in the PCA biplot. In rare cases,

Table 2. Patterns of *Aminicenantes* relative abundance in datasets classified by prevalent environmental conditions[1].

Dataset type	Total datasets	Datasets with *Aminicenantes* (%)	Average *Aminicenantes* abundance (%)[2]	Maximum relative abundance
Oxygen Tension				
Oxic	2,787	735(26.4%)	0.10%	2.50%
Hypoxic	101	35 (34.65%)	0.17%	2.90%
Anoxic	253	148 (58.5%)	0.46%	10.20%
Temperature[3]				
Low	317	4 (1.26%)	0.004%	0.01%
Temperate	2657	807 (30.372%)	0.19%	10.20%
Medium	53	48 (90.56%)	0.02%	0.06%
Elevated	11	6 (54.55%)	0.06%	0.20%
Extremely elevated	103	53 (51.46%)	0.24%	5.28%
Salinity[4]				
Non-Saline	1,863	575 (30.86%)	0.16%	10.20%
Low Salinity	1,179	274 (23.24%)	0.26%	5.30%
Moderate salinity	77	51 (66.23%)	0.02%	0.20%
Hypersaline	22	18 (81.81%)	0.07%	0.68%

[1]A detailed description of every dataset is provided as supplementary material (Supplementary Table S1 in File S1).
[2]Average abundance values in datasets where *Aminicenantes* sequences were identified.
[3]Temperature classifications: Low: Arctic, Antarctic, subarctic, and permafrost marine and terrestrial conducive to the growth of psychrophilic microorganisms; temperate: Habitats in temperate ecosystems e.g. lakes, soils in continental settings; Medium: Habitats with temperatures around 37°C e.g. rumen; Elevated: habitats with temperatures conducive to the growth of thermophiles (50–80°C) e.g. Alberta oil sands tailings pond; Extremely elevated: habitats conducive to the growth of hyperthermophiles (>80°C degrees) e.g. Hydrothermal vents.
[4]Salinity classifications: Non-saline: Environments with <1% salinity; Low salinity: Marine environments, and environments with comparable salinities; Moderate salinities: Environments with salinities around 5–15% e.g. Alberta oil sands tailings pond and Huabei Oilfield in China; Hypersaline: Environments with >15% salinity.

some datasets did not contain any OP8-1_HMMV sequences. For example, all *Aminicenantes*-affiliated sequences from three hydrothermal vent samples belonged to the newly proposed candidate class OP8-3, and hence clustered in the direction of OP8-3 species arrow in the PCA biplot (red circles in Figure 2d).

Within aquatic non-marine habitats, the overall majority of *Aminicenantes* sequences belonged to subclass OP8-1_unclassified (Figure 2e). The majority (85.1% of datasets originating from the two non-saline aquatic non-marine sub-habitats (temperate freshwater lakes, and spring and groundwater samples) showed >70% of *Aminicenantes*-affiliated sequences belonging to the order OP8-1_unclassified and were hence clustered along the OP8-1_unclassified arrow in the PCA biplot (black and red stars, Figure 2f). However, two notable exceptions to this pattern were observed: 1. In several datasets, a mixed community of OP8-1_unclassified with other lineages was observed (e.g. 11 freshwater lake samples had a mixed community of OP8-1_unclassified (53.6±2.4%), OP8-1_OPB95 (34.2±3.2%), and OP8-1_SHA-124 (11.3±1.9%), and one sample from a sinkhole had a mixed *Aminicenantes* community of OP8-1_OPB95 (43.7%), OP8-1_unclassified (32.1%), and OP8-2 (22.4%). 2. In few datasets, OP8-1_unclassified order was absent e.g. sewage samples with high abundance (> 90%) of OP8-1_OPB95 (Figure 2f).

While the *Aminicenantes* class OP8-1_unclassified was the prevalent lineage in the majority of aquatic non-marine habitats originating from temperate freshwater lakes, as well as spring and groundwater datasets; a distinct community structure was observed in aquatic non-marine habitats with low to moderate salinity (Figure 2e). Within these habitats, e.g. three samples from the Amazon-Guianas estuaries, and a salt marsh samples from Cabo Rojo, PR, the majority of *Aminicenantes*-affiliated sequences belonged to order OP8-1_HMMV (83.7±7.98%). Accordingly, those samples clustered along the HMMV species arrow in various PCA plot (Figure 2f).

Finally, a relatively small number of *Aminicenantes*-affiliated sequences were present in soil samples. Those were mainly affiliated with orders OP8-1_unclassified, OP8-1_OPB95, and OP8-1_SHA-124. Some unique patterns were observed at the sub-habitat level e.g. the prevalence of OP8-1_unclassified order in samples from permafrost soils (Figure 2g). However, it is important to note that the *Aminicenantes* exhibited an extremely rare distribution in all soil datasets examined, being only identified in 14 out of 276 datasets, with an extremely low average relative abundance (0.07%). Therefore, the significance of the observed patterns, given their extreme rarity, and doubtful ecological role in soil habitats, is questionable.

We also studied the effect of environmental conditions (O_2 tension, temperature, and salinity) on *Aminicenantes* community structure in various datasets. When environments were classified based on their salinity, we observed a shift in the prevalence of various *Aminicenantes* lineages, with order OP8-1_unclassified representing the majority of *Aminicenantes*-affiliated sequences in non-saline habitats, as opposed to order HMMV in low and moderate salinity environments, and class OP8-2 in hypersaline environments (Figure 3a). We also observed an effect of temperature on the pattern of *Aminicenantes* community structure changes, where order OP8-1_unclassified and class OP8_2 dominated in low temperature and psychrophilic habitats, as opposed to orders OP8-1_OPB95 and HMMV in thermophilic and hyperthermophilic habitats (Figure 3b). However, the uneven number of samples belonging to each category (Table 3) could possibly skew these results. Finally, no remarkable effect of O_2 tension on *Aminicenantes* community structure was observed (Figure 3c).

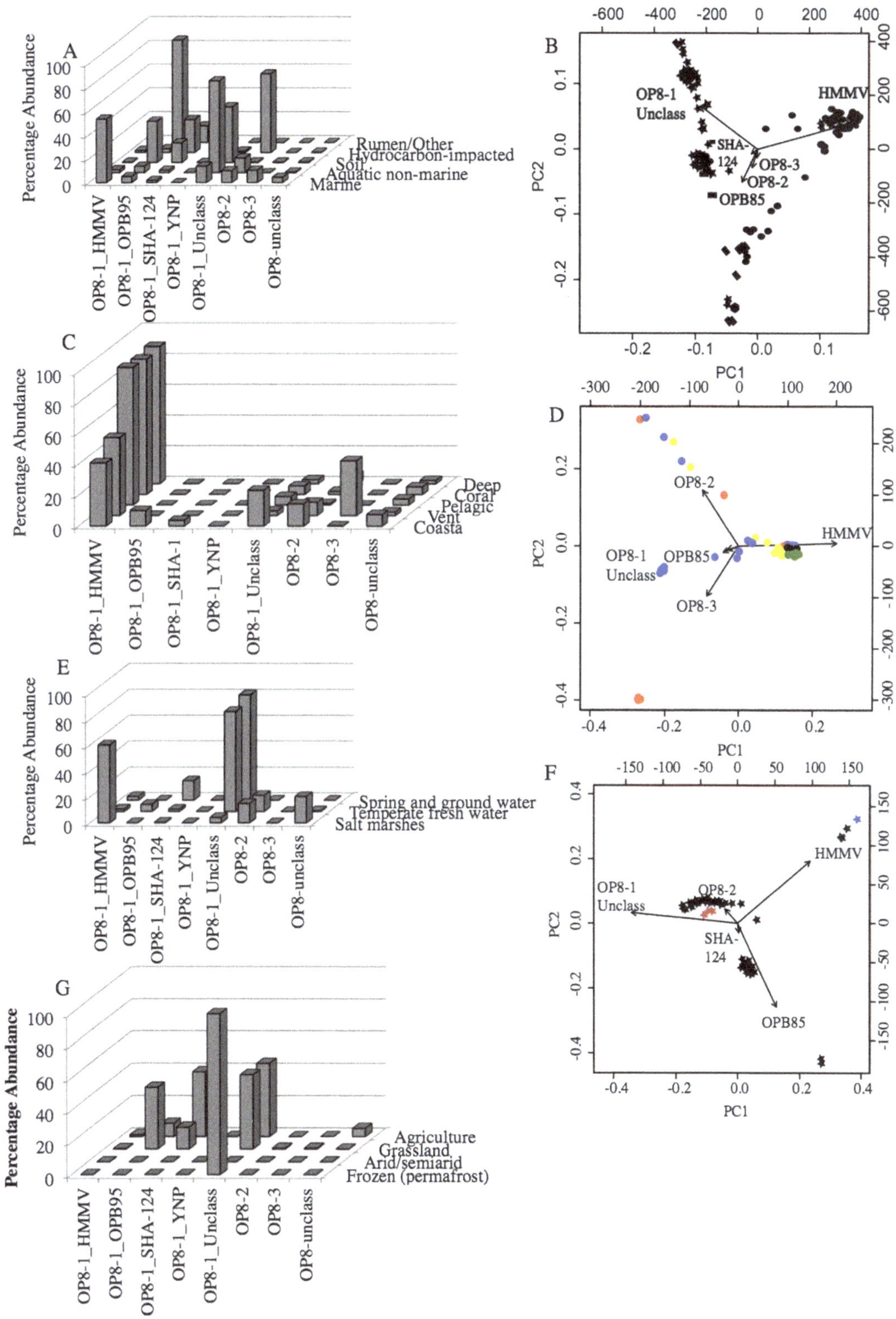

Figure 2. Aminicenantes relative abundance and community in various habitats. (A) Relative abundance of Aminicenantes-affiliated sequences in marine, aquatic non-marine, soil, hydrocarbon-impacted, and rumen/other habitats. (B) PCA biplot of the community structure of Aminicenantes in datasets belonging to marine (●), aquatic non-marine (★), soil (n), hydrocarbon-impacted (u), and rumen (⬢) with >50 Aminicenantes sequences. The biplot was generated in R using the prcomp and biplot functions in library labdsv. The first 2 axes explained 73% of the variance. There are two sets of axis scales on the biplot; the ones on the right and top correspond to the axis scores for samples, and the bottom and left axes correspond to the loadings of the variables (in this case, OP8 subphyla). (C) Relative abundance of *Aminicenantes*-affiliated sequences in various marine subhabitats. (D) PCA biplot of the community structure of *Aminicenantes* in marine datasets classified as coastal (blue), pelagic (green), hydrothermal vent (red), coral-associated (black), and deep sediment (yellow). There are two sets of axis scales on the biplot; the ones on the right and top correspond to the axis scores for samples, and the bottom and left axes correspond to the loadings of the variables (in this case, OP8 subphyla). (E) Relative abundance of *Aminicenantes*-affiliated sequences in environments originating from aquatic non-marine habitats. (F) PCA biplot

of the community structure of *Aminicenantes* in aquatic non-marine datasets classified as freshwater (black), spring and groundwater (red), and salt marshes (blue). There are two sets of axis scales on the biplot; the ones on the right and top correspond to the axis scores for samples, and the bottom and left axes correspond to the loadings of the variables (in this case, OP8 subphyla). (G) Relative abundance of *Aminicenantes*-affiliated sequences in environments originating from soil habitats. Since only one soil dataset contained >50 *Aminicenantes* sequence, a PCA soil biplot is not feasible.

5. Patterns of *Aminicenantes* diversity

One hundred and ninety-eight datasets with more than 50 *Aminicenantes*-affiliated sequences were included in the diversity analysis. Due to the underrepresentation of hydrocarbon-impacted sites and soils, comparison of diversities was restricted to the marine and aqueous non-marine habitats and their subcategories. Within all habitats, the levels of diversity varied widely, but marine habitats showed higher diversity than freshwater habitats (Student t-test p-value = 0.037), with most of the marine environments (72%) showing medium to very high *Aminicenantes* diversity (Table 3). Within marine habitats, a higher average diversity rank was observed in coastal samples, and a lower average diversity was observed in hydrothermal vent samples. Indeed, coastal samples *Aminicenantes* diversities were significantly higher than those in all other marine environments (p-value ranging from 0.0004 to 0.041). Hydrothermal vent samples *Aminicenantes* diversities were significantly lower than those in coastal, and deep marine sediment samples (p-values 0.0004, and 0.04, respectively).

Aminicenantes diversity within aquatic non-marine environments varied, with high diversities observed in spring/groundwater samples and the single sample from a salt marsh. Significantly lower diversities were observed in samples from freshwater temperate environments (p-value = 0.002).

We also correlated diversity rankings to environmental conditions including temperature, salinity, and oxygen tension (Table 3). Interestingly, while no clear correlation was identified between temperature, or salinity and diversity levels of *Aminicenantes* at $OTU_{0.03}$, a positive highly significant correlation existed between the dataset diversity rank and the environment's oxygen tension (Spearman rank correlation coefficient = 0.4, p-value = 5.3E-9).

Discussion

In this study, we utilized *in silico* database mining approaches to provide an updated and expanded taxonomic outline of the candidate phylum *Aminicenantes* using near full-length 16S rRNA gene sequences, as well as to examine the global patterns of *Aminicenantes* distribution using high throughput (Pyrosequencing and Illumina) generated 16S rRNA gene datasets. We report that: 1. Members of the *Aminicenantes* are present in a substantial fraction (918 out of 3,141) of high throughput-generated datasets examined, where they represent a minor/rare fraction of the community, with very few exceptions. 2. Members of the

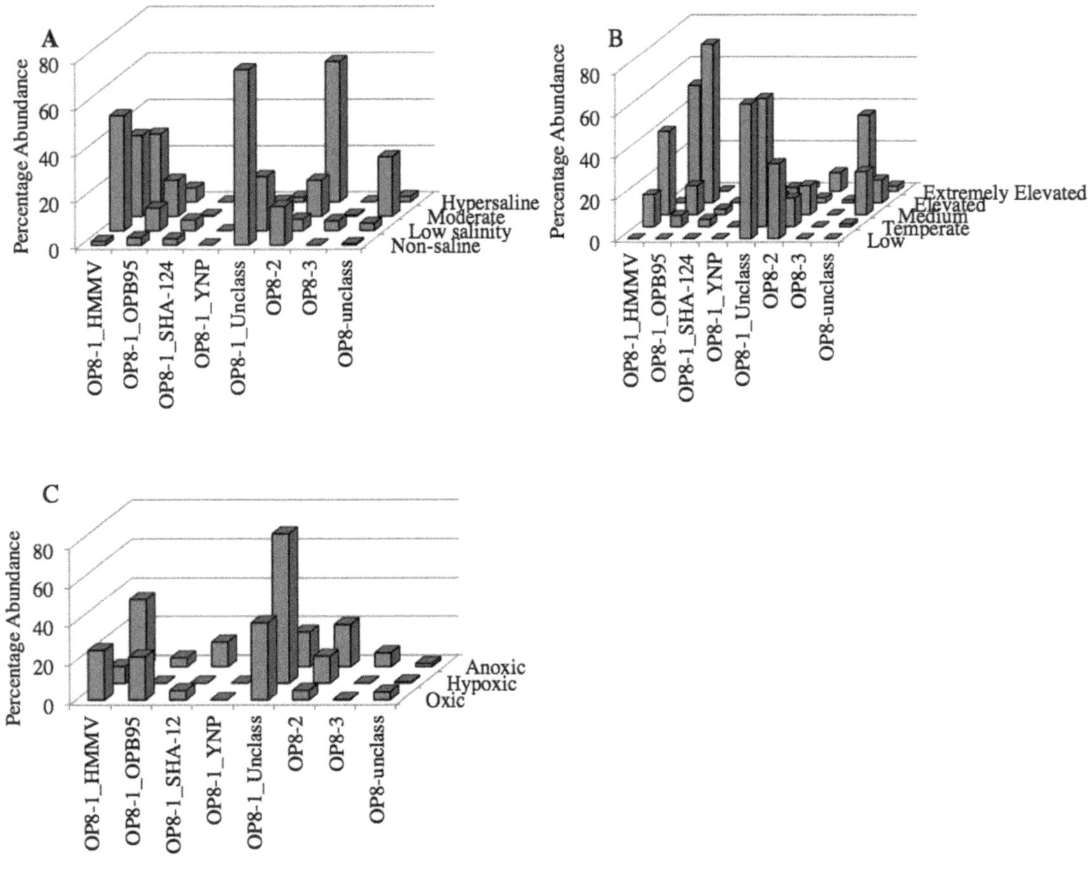

Figure 3. Relative abundance of *Aminicenantes*-affiliated sequences in different environments sub-classified according to (A) temperature, (B) oxygen tension, and (C) salinity.

Table 3. Diversity rankings of all datasets classified according to habitat and prevalent environmental conditions.

Habitat/ Environmental parameter	Average diversity rank±SD	Number of samples belonging to this diversity rank				
		Very low	Low	Medium	High	Very high
Marine	115±68.5	11	1	6	10	12
Pelagic	89.9±63.3	1	0	2	1	0
Coastal	169.1±39.6	0	1	0	1	7
Coral	115.4±57.3	1	0	1	3	0
Deep_sed	106.4±65.9	5	0	3	4	5
Hyd_vent	34.3±59.1	4	0	0	1	0
Non-marine	96±52.8	26	38	34	31	21
Freshwater	93.7±51.6	26	38	34	30	18
Spring/GW	187.5±8.8	0	0	0	0	3
Salt marsh	148	0	0	0	1	0
Hydrocarbon-Impacted Soil	96.5±83.4	3	0	0	0	4
Soil	174.5	0	0	0	0	1
Salinity						
Non-saline	95.8±52.4	29	38	34	29	24
Low-salinity	115.8±69.6	11	0	6	10	14
Hypersaline	115.3±47.4	0	1	0	2	0
Temperature						
Temperate	102.6±56.1	33	39	40	40	38
Elevated	24.3±47	7	0	0	1	0
O2 tension						
Anoxic	88.3±74.5	12	1	1	6	7
Hypoxic	84.6±44.1	26	36	37	27	5
Oxic	157.4±49.5	2	2	2	7	26

Aminicenantes are ubiquitous, being encountered in all different types of habitats and across all spectra of environmental parameters (temperature, salinity, and oxygen tension) examined. 3. Distinct differences exist between the relative abundance of the *Aminicenantes* across different habitats and environmental conditions. 4. Members of the *Aminicenantes* exhibit a distinct community structure patterns across various datasets, and these patterns appear to be, mostly, driven by habitat variations rather than prevalent environmental parameters.

Utilizing high throughput-generated datasets of partial 16S rRNA gene sequences in dedicated sequence repositories (VAMPS, MG-RAST, and GenBank SRA) for analyzing patterns of prokaryotic diversity represents an extremely valuable, yet largely overlooked, resource. Next generation sequencing datasets are often deposited with a single accession number per dataset, often with inadequate metadata, and, unlike Sanger-generated sequences, these datasets are not readily amendable to online search queries. Nevertheless, when properly exploited, these datasets represent an excellent resource for testing specific ecological hypothesis. Examining *Aminicenantes* diversity in 3,141 distinct datasets, comprising a total of ~1.8 billion partial sequences clearly demonstrates the presence of members of this candidate phylum in a large number (29.2% of datasets examined) of habitats. However, the *Aminicenantes* always represented a minor fraction of the overall community and often exhibited an extremely rare distribution: The relative abundance of the *Aminicenantes* was less than 0.01% of the total community in 70.1% of datasets examined, 0.01–0.1% in 16.1% of datasets examined, 0.1–1% in 12.9% of datasets, and more than 1% in only 0.9% of datasets examined (Table S1 in File S1). The reason for the occurrence, survival, and retention of various lineages as members of the rare biosphere (e.g. less than 0.1%) in various environments is an issue that has previously been thoroughly debated [6,53,54]. Possible reasons explaining this phenomenon vary and range between filling very specialized niches, acting as a backup system that readily responds to seasonal variations encountered in various ecosystem, exhibiting extremely slow growth or dormancy, introduction to the ecosystem through recent immigration of these rare phylotypes to the sampling site, or introduction to the dataset through contamination during sampling, DNA extraction, or amplification. Indeed, several of these explanations are plausible to elucidate the role of extremely rare members of the *Aminicenantes* in their respective ecosystems. Regardless, it is reasonable to assume that the detection of the *Aminicenantes* above a certain empirical threshold (e.g. 1%, equivalent to 10^5 cells/gram or ml in a community with a cell count of 10^7) reflects its successful colonization and propagation in a specific habitat, and suggests its importance in fulfilling vital ecosystem services that justifies its retention in that habitat. Therefore, examination of the few datasets in which the *Aminicenantes* are present in relatively higher abundances could offer a window on what factors are conducive for *Aminicenantes* survival and propagation *in-situ*. Datasets with more than 1% *Aminicenantes* relative abundance (Table S1, Figure S1 in File S1)

(0.9% of the total number of datasets) were not restricted to one habitat type or one environmental condition, but occurred within the majority of the five habitats examined, and across a wide range of environmental conditions. Therefore, it is improbable that a single, specific, environmental condition e.g. hypersalinity or extreme temperature represents the only scenario for eliciting a competitive niche for the *Aminicenantes*. Rather, we argue that conditions at which *Aminicenantes* propagates appear to be induced by other types of natural or anthropogenic stressors, which effectively preclude a large fraction of the population, opening the window for *Aminicenantes* to propagate. This is apparent from the fact that many of the datasets with >1% *Aminicenantes* relative abundance came from environments with variable types of environmental stressors e.g. high levels of hydrocarbons (e.g. Alberta oil sands tailing ponds, Petroleum reservoirs in Huabei, China, and north slope oil facility) [55–57], or high levels of metal (arsenic) contamination in Araihazar, Bangladesh [47].

Overall relative abundance of the *Aminicenantes* appeared to vary widely across various habitats, as well as across specific environmental conditions. The *Aminicenantes* appear to be most abundant in hydrocarbon-impacted environments, being encountered in 71.4% of the datasets (10/14), with an average abundance of 0.321%. The association of specific lineages and phylotypes with hydrocarbon-impacted environments regardless of its origin (natural or anthropogenic), or chemical composition (natural gas, petroleum, enrichments on a single substrate) has previously been noted ([58,59]. This prevalence in hydrocarbon-impacted settings is in agreement with the notion that success and propagation of members of the *Aminicenantes* in a specific environment is contingent on the occurrence of specific environmental stressors (hydrocarbon contamination and possibly associated anaerobiasis and high sulfide levels in such habitats) that partially alleviates competition, allowing for successful propagation of members of the *Aminicenantes*. The *Aminicenantes* were also identified in a considerable fraction of marine and aquatic non-marine habitats (Table 1, Table S1 in File S1). However, the complexity and variability of geochemical parameters encountered in these heterogeneous ecosystems prevents us from deciphering what exact environmental characteristics, or combination thereof, within these habitats favored *Aminicenantes* propagation. Correlating *Aminicenantes* abundance to environmental conditions (temperature, salinity, and oxygen tension) revealed that while members of the *Aminicenantes* could be encountered in a wide range of environmental conditions, it appears to exhibit significantly higher abundances in anoxic (compared to oxic and microoxic) habitats and a significantly lower abundance in low temperature (compared to temperate and elevated temperature) habitats. The relatively higher abundance of the *Aminicenantes* in anoxic environments suggests a prevalent anaerobic/facultative mode of metabolism within the *Aminicenantes*. Indeed, the majority of studies where the *Aminicenantes* represented more than 1% the total bacterial community originated from seemingly anaerobic habitats (e.g. arsenic contaminated ground water from Bangladesh, Guayamas methane seeps, and hypoliminion sites in Lake Mendota).

Analysis of the *Aminicenantes* community structure was conducted by: 1. Utilizing the classification of all (47,351) next generation sequences identified to examine the *Aminicenantes* community structure in various types of habitats and across various environmental conditions, and 2. PCA analysis of the *Aminicenantes* community structure in datasets where they exhibited relatively higher abundances (n>50). Overall, it appears that factors impacting *Aminicenantes* community structure are mostly habitat-driven (i.e. similar community structure observed in similar habitats), rather than driven by prevalent environmental conditions (temperature, salinity, oxygen tension) within an ecosystem. For example, class OP8-3 was exclusively identified in hydrothermal vent habitats; order OP8-1_ HMMV represented the majority of *Aminicenantes* sequences encountered in coral associated, pelagic, and deep marine habitats; OP8-1_unclassified represented the majority of sequences in aqueous non-marine habitats; and OP8-2 represented the majority of sequence in hydrocarbon-impacted habitats. The role of prevalent environmental condition in shaping the *Aminicenantes* microbial community is less certain, mostly due to the inadequate representation of special categories e.g. normal (body) temperature, elevated temperature, and hypersaline environments. However, one notable exception in which an environmental parameter appears to play a clear role in shaping the *Aminicenantes* microbial community is the distinct prevalence of order OP8-1_ HMMV in multiple low salinity datasets regardless of their habitat.

Finally, it is interesting to note that *Aminicenantes* sequences were identified across all ranges of salinity and temperatures including those conducive to the growth of obligate halophiles and hyperthermophiles, respectively. Collectively, the detection of *Aminicenantes*-affiliated sequences across environmental extremes, coupled to their observed ubiquitous distribution on a global scale and the distinct patterns of community structure exhibited argues for a high level of intraphylum metabolic and adaptive diversity within the *Aminicenantes*. Therefore it is probable that *Aminicenantes* cells in nature exhibit multiple distinct metabolic capabilities, wide array of survival weapons, and various adaptive strategies. This, in turn, highlights the importance of obtaining multiple genomic assemblies that adequately represents the broad phylogenetic diversity of this phylum, as well as its wide environmental distribution to truly gauge the pangenomic diversity within the *Aminicenantes*. The recently acquired genomic information from single cell-based efforts from Sakinaw Lake represents admirable effort to investigate this understudied and yet-uncultured lineage. However, information from such assemblies should not be extrapolated to describe all members of the *Aminicenantes*. Indeed, the discovery of novel capabilities within well-establish lineages e.g. phototrophy amongst *Acidobacteria* [60], methane oxidation amongst the *Verrucomicrobia* [61,62], anaerobic oxidation of ammonia amongst the *Planctomycetes* [63] highlights the importance of continued efforts to decipher and expand genomic diversity within various bacterial phyla.

Supporting Information

File S1 Contains the files: Figure S1 *Aminicenantes* relative abundance in different habitat types. **Figure S2** *Aminicenantes* relative abundance in response to various geochemical conditions. **Table S1** Summary of all high throughput-generated datasets analyzed in this study. **Table S2** List of all near full-length 16S rRNA sequences belonging to the *Aminicenantes* and their class/order level phylogenetic affiliations to *Aminicenantes*.

Acknowledgments

We would like to thank Dana Brunson at the OSU high performing computer center for technical assistance.

Author Contributions

Conceived and designed the experiments: MSE NHY. Performed the experiments: IFF JPD. Analyzed the data: IFF JPD NHY. Wrote the paper: MSE IFF.

References

1. Dojka MA, Hugenholtz P, Haack SK, Pace NR (1998) Microbial diversity in a hydrocarbon- and chlorinated-solvent-contaminated aquifer undergoing intrinsic bioremediation. Appl Environ Microbiol 64: 3869–3877.
2. Hugenholtz P, Pitulle C, Hershberger KL, Pace NR (1998) Novel division level bacterial diversity in a Yellowstone hot spring. J Bacteriol 180: 366–376.
3. Teske A, Hinrichs KU, Edgcomb V, de Vera Gomez A, Kysela D, et al. (2002) Microbial diversity of hydrothermal sediments in the Guaymas Basin: evidence for anaerobic methanotrophic communities. Appl Environ Microbiol 68: 1994–2007.
4. Roesch LF, Fulthorpe RR, Riva A, Casella G, Hadwin AK, et al. (2007) Pyrosequencing enumerates and contrasts soil microbial diversity. ISME J 1: 283–290.
5. Wang Y, Sheng HF, He Y, Wu JY, Jiang YX, et al. (2012) Comparison of the levels of bacterial diversity in freshwater, intertidal wetland, and marine sediments by using millions of illumina tags. Appl Environ Microbiol 78: 8264–8271.
6. Sogin ML, Morrison HG, Huber JA, Mark Welch D, Huse SM, et al. (2006) Microbial diversity in the deep sea and the underexplored "rare biosphere". Proc Natl Acad Sci U S A 103: 12115–12120.
7. Hugenholtz P (2002) Exploring prokaryotic diversity in the genomic era. Genome Biol 3: REVIEWS0003.
8. Vergin KL, Urbach E, Stein JL, DeLong EF, Lanoil BD, et al. (1998) Screening of a fosmid library of marine environmental genomic DNA fragments reveals four clones related to members of the order Planctomycetales. Appl Environ Microbiol 64: 3075–3078.
9. Treusch AH, Kletzin A, Raddatz G, Ochsenreiter T, Quaiser A, et al. (2004) Characterization of large-insert DNA libraries from soil for environmental genomic studies of Archaea. Environ Microbiol 6: 970–980.
10. Elshahed MS, Najar FZ, Aycock M, Qu C, Roe BA, et al. (2005) Metagenomic analysis of the microbial community at Zodletone Spring (Oklahoma): insights into the genome of a member of the novel candidate division OD1. Appl Environ Microbiol 71: 7598–7602.
11. Kielak AM, van Veen JA, Kowalchuk GA (2010) Comparative analysis of acidobacterial genomic fragments from terrestrial and aquatic metagenomic libraries, with emphasis on acidobacteria subdivision 6. Appl Environ Microbiol 76: 6769–6777.
12. Narasingarao P, Podell S, Ugalde JA, Brochier-Armanet C, Emerson JB, et al. (2012) De novo metagenomic assembly reveals abundant novel major lineage of Archaea in hypersaline microbial communities. ISME J 6: 81–93.
13. Albertsen M, Hugenholtz P, Skarshewski A, Nielsen KL, Tyson GW, et al. (2013) Genome sequences of rare, uncultured bacteria obtained by differential coverage binning of multiple metagenomes. Nat Biotechnol 31: 533–538.
14. Wrighton KC, Thomas BC, Sharon I, Miller CS, Castelle CJ, et al. (2012) Fermentation, hydrogen, and sulfur metabolism in multiple uncultivated bacterial phyla. Science 337: 1661–1665.
15. Pelletier E, Kreimeyer A, Bocs S, Rouy Z, Gyapay G, et al. (2008) "Candidatus Cloacamonas acidaminovorans": genome sequence reconstruction provides a first glimpse of a new bacterial division. J Bacteriol 190: 2572–2579.
16. Rinke C, Schwientek P, Sczyrba A, Ivanova NN, Anderson IJ, et al. (2013) Insights into the phylogeny and coding potential of microbial dark matter. Nature 499: 431–437.
17. Youssef NH, Blainey PC, Quake SR, Elshahed MS (2011) Partial genome assembly for a candidate division OP11 single cell from an anoxic spring (Zodletone Spring, Oklahoma). Appl Environ Microbiol 77: 7804–7814.
18. Campbell JH, O'Donoghue P, Campbell AG, Schwientek P, Sczyrba A, et al. (2013) UGA is an additional glycine codon in uncultured SR1 bacteria from the human microbiota. Proc Natl Acad Sci U S A 110: 5540–5545.
19. McLean JS, Lombardo MJ, Badger JH, Edlund A, Novotny M, et al. (2013) Candidate phylum TM6 genome recovered from a hospital sink biofilm provides genomic insights into this uncultivated phylum. Proc Natl Acad Sci U S A 110: E2390–2399.
20. Tourna M, Stieglmeier M, Spang A, Konneke M, Schintlmeister A, et al. (2011) Nitrososphaera viennensis, an ammonia oxidizing archaeon from soil. Proc Natl Acad Sci U S A 108: 8420–8425.
21. Konneke M, Bernhard AE, de la Torre JR, Walker CB, Waterbury JB, et al. (2005) Isolation of an autotrophic ammonia-oxidizing marine archaeon. Nature 437: 543–546.
22. Girguis PR, Cozen AE, DeLong EF (2005) Growth and population dynamics of anaerobic methane-oxidizing archaea and sulfate-reducing bacteria in a continuous-flow bioreactor. Appl Environ Microbiol 71: 3725–3733.
23. Bates ST, Clemente JC, Flores GE, Walters WA, Parfrey LW, et al. (2013) Global biogeography of highly diverse protistan communities in soil. ISME J 7: 652–659.
24. Freitas S, Hatosy S, Fuhrman JA, Huse SM, Welch DB, et al. (2012) Global distribution and diversity of marine Verrucomicrobia. ISME J 6: 1499–1505.
25. Bergmann GT, Bates ST, Eilers KG, Lauber CL, Caporaso JG, et al. (2011) The under-recognized dominance of Verrucomicrobia in soil bacterial communities. Soil Biol Biochem 43: 1450–1455.
26. Jones RT, Robeson MS, Lauber CL, Hamady M, Knight R, et al. (2009) A comprehensive survey of soil acidobacterial diversity using pyrosequencing and clone library analyses. ISME J 3: 442–453.
27. Buckley DH, Huangyutitham V, Nelson TA, Rumberger A, Thies JE (2006) Diversity of Planctomycetes in soil in relation to soil history and environmental heterogeneity. Appl Environ Microbiol 72: 4522–4531.
28. Gilbert JA, Meyer F, Jansson J, Gordon J, Pace N, et al. (2010) The Earth Microbiome Project: Meeting report of the "1 EMP meeting on sample selection and acquisition" at Argonne National Laboratory October 6 2010. Stand Genomic Sci 3: 249–253.
29. Huse SM, Ye Y, Zhou Y, Fodor AA (2012) A core human microbiome as viewed through 16S rRNA sequence clusters. PLoS One 7: e34242.
30. Knight R, Jansson J, Field D, Fierer N, Desai N, et al. (2012) Unlocking the potential of metagenomics through replicated experimental design. Nat Biotechnol 30: 513–520.
31. Chouari R, Le Paslier D, Daegelen P, Ginestet P, Weissenbach J, et al. (2005) Novel predominant archaeal and bacterial groups revealed by molecular analysis of an anaerobic sludge digester. Environ Microbiol 7: 1104–1115.
32. Joynt J, Bischoff M, Turco R, Konopka A, Nakatsu CH (2006) Microbial community analysis of soils contaminated with lead, chromium and petroleum hydrocarbons. Microb Ecol 51: 209–219.
33. Losekann T, Knittel K, Nadalig T, Fuchs B, Niemann H, et al. (2007) Diversity and abundance of aerobic and anaerobic methane oxidizers at the Haakon Mosby Mud Volcano, Barents Sea. Appl Environ Microbiol 73: 3348–3362.
34. Dhillon A, Teske A, Dillon J, Stahl DA, Sogin ML (2003) Molecular characterization of sulfate-reducing bacteria in the Guaymas Basin. Appl Environ Microbiol 69: 2765–2772.
35. McDonald D, Price MN, Goodrich J, Nawrocki EP, DeSantis TZ, et al. (2012) An improved Greengenes taxonomy with explicit ranks for ecological and evolutionary analyses of bacteria and archaea. ISME J 6: 610–618.
36. Quast C, Pruesse E, Yilmaz P, Gerken J, Schweer T, et al. (2013) The SILVA ribosomal RNA gene database project: improved data processing and web-based tools. Nucleic Acids Res 41: D590–596.
37. Johnson M, Zaretskaya I, Raytselis Y, Merezhuk Y, McGinnis S, et al. (2008) NCBI BLAST: a better web interface. Nucleic Acids Res 36: W5–9.
38. Goecks J, Nekrutenko A, Taylor J, Galaxy T (2010) Galaxy: a comprehensive approach for supporting accessible, reproducible, and transparent computational research in the life sciences. Genome Biol 11: R86.
39. Larkin MA, Blackshields G, Brown NP, Chenna R, McGettigan PA, et al. (2007) Clustal W and Clustal X version 2.0. Bioinformatics 23: 2947–2948.
40. Youssef N, Steidley BL, Elshahed MS (2012) Novel high-rank phylogenetic lineages within a sulfur spring (Zodletone Spring, Oklahoma), revealed using a combined pyrosequencing-sanger approach. Appl Environ Microbiol 78: 2677–2688.
41. Dalevi D, Hugenholtz P, Blackall LL (2001) A multiple-outgroup approach to resolving division-level phylogenetic relationships using 16S rDNA data. Int J Syst Evol Microbiol 51: 385–391.
42. Meyer F, Paarmann D, D'Souza M, Olson R, Glass EM, et al. (2008) The metagenomics RAST server - a public resource for the automatic phylogenetic and functional analysis of metagenomes. BMC Bioinformatics 9: 386.
43. Benson DA, Cavanaugh M, Clark K, Karsch-Mizrachi I, Lipman DJ, et al. (2013) GenBank. Nucleic Acids Res 41: D36–42.
44. Schloss PD, Westcott SL, Ryabin T, Hall JR, Hartmann M, et al. (2009) Introducing mothur: open-source, platform-independent, community-supported software for describing and comparing microbial communities. Appl Environ Microbiol 75: 7537–7541.
45. Youssef NH, Elshahed MS (2009) Diversity rankings among bacterial lineages in soil. ISME J 3: 305–313.
46. Team RDC (2011) R: A Language and Environment for Statistical Computing. Reference Index. Vienna, Austria: R Foundation for Statistical Computing.
47. Legg TM, Zheng Y, Simone B, Radloff KA, Mladenov N, et al. (2012) Carbon, metals, and grain size correlate with bacterial community structure in sediments of a high arsenic aquifer. Front Microbiol 3: 82.
48. Flores GE, Campbell JH, Kirshtein JD, Meneghin J, Podar M, et al. (2011) Microbial community structure of hydrothermal deposits from geochemically different vent fields along the Mid-Atlantic Ridge. Environ Microbiol 13: 2158–2171.
49. Madrid VM, Taylor GT, Scranton MI, Chistoserdov AY (2001) Phylogenetic diversity of bacterial and archaeal communities in the anoxic zone of the Cariaco Basin. Appl Environ Microbiol 67: 1663–1674.
50. Zinger L, Amaral-Zettler LA, Fuhrman JA, Horner-Devine MC, Huse SM, et al. (2011) Global patterns of bacterial beta-diversity in seafloor and seawater ecosystems. PLoS One 6: e24570.
51. Hollister EB, Engledow AS, Hammett AJ, Provin TL, Wilkinson HH, et al. (2010) Shifts in microbial community structure along an ecological gradient of hypersaline soils and sediments. ISME J 4: 829–838.
52. Chu H, Fierer N, Lauber CL, Caporaso JG, Knight R, et al. (2010) Soil bacterial diversity in the Arctic is not fundamentally different from that found in other biomes. Environ Microbiol 12: 2998–3006.
53. Elshahed MS, Youssef NH, Spain AM, Sheik C, Najar FZ, et al. (2008) Novelty and uniqueness patterns of rare members of the soil biosphere. Appl Environ Microbiol 74: 5422–5428.
54. Pedros-Alio C (2012) The rare bacterial biosphere. Ann Rev Mar Sci 4: 449–466.

55. Stevenson BS, Drilling HS, Lawson PA, Duncan KE, Parisi VA, et al. (2011) Microbial communities in bulk fluids and biofilms of an oil facility have similar composition but different structure. Environ Microbiol 13: 1078–1090.
56. Saidi-Mehrabad A, He Z, Tamas I, Sharp CE, Brady AL, et al. (2013) Methanotrophic bacteria in oilsands tailings ponds of northern Alberta. ISME J 7: 908–921.
57. Li H, Yang SZ, Mu BZ, Rong ZF, Zhang J (2006) Molecular analysis of the bacterial community in a continental high-temperature and water-flooded petroleum reservoir. FEMS Microbiol Lett 257: 92–98.
58. Elshahed MS, Senko JM, Najar FZ, Kenton SM, Roe BA, et al. (2003) Bacterial diversity and sulfur cycling in a mesophilic sulfide-rich spring. Appl Environ Microbiol 69: 5609–5621.
59. Davis JP, Struchtemeyer CG, Elshahed MS (2012) Bacterial communities associated with production facilities of two newly drilled thermogenic natural gas wells in the Barnett Shale (Texas, USA). Microb Ecol 64: 942–954.
60. Bryant DA, Costas AM, Maresca JA, Chew AG, Klatt CG, et al. (2007) Candidatus Chloracidobacterium thermophilum: an aerobic phototrophic Acidobacterium. Science 317: 523–526.
61. Pol A, Heijmans K, Harhangi HR, Tedesco D, Jetten MS, et al. (2007) Methanotrophy below pH 1 by a new Verrucomicrobia species. Nature 450: 874–878.
62. Dunfield PF, Yuryev A, Senin P, Smirnova AV, Stott MB, et al. (2007) Methane oxidation by an extremely acidophilic bacterium of the phylum Verrucomicrobia. Nature 450: 879–882.
63. Strous M, Fuerst JA, Kramer EH, Logemann S, Muyzer G, et al. (1999) Missing lithotroph identified as new planctomycete. Nature 400: 446–449.
64. Tamura K, Peterson D, Peterson N, Stecher G, Nei M, et al. (2011) MEGA5: molecular evolutionary genetics analysis using maximum likelihood, evolutionary distance, and maximum parsimony methods. Mol Biol Evol 28: 2731–2739.

20

Complete Nucleotide Sequence and Analysis of Two Conjugative Broad Host Range Plasmids from a Marine Microbial Biofilm

Peter Norberg[1], Maria Bergström[2], Malte Hermansson[2]*

1 Department of Infectious Diseases, University of Gothenburg, Göteborg, Sweden, 2 Department of Chemistry and Molecular Biology, Microbiology, University of Gothenburg, Göteborg, Sweden

Abstract

The complete nucleotide sequence of plasmids pMCBF1 and pMCBF6 was determined and analyzed. pMCBF1 and pMCBF6 form a novel clade within the IncP-1 plasmid family designated IncP-1 ς. The plasmids were exogenously isolated earlier from a marine biofilm. pMCBF1 (62 689 base pairs; bp) and pMCBF6 (66 729 bp) have identical backbones, but differ in their mercury resistance transposons. pMCBF1 carries Tn*5053* and pMCBF6 carries Tn*5058*. Both are flanked by 5 bp direct repeats, typical of replicative transposition. Both insertions are in the vicinity of a resolvase gene in the backbone, supporting the idea that both transposons are "res-site hunters" that preferably insert close to and use external resolvase functions. The similarity of the backbones indicates recent insertion of the two transposons and the ongoing dynamics of plasmid evolution in marine biofilms. Both plasmids also carry the insertion sequence IS*Pst1*, albeit without flanking repeats. IS*Ps1*is located in an unusual site within the control region of the plasmid. In contrast to most known IncP-1 plasmids the pMCBF1/pMCBF6 backbone has no insert between the replication initiation gene (*trfA*) and the vegetative replication origin (*oriV*). One pMCBF1/pMCBF6 block of about 2.5 kilo bases (kb) has no similarity with known sequences in the databases. Furthermore, insertion of three genes with similarity to the multidrug efflux pump operon *mexEF* and a gene from the NodT family of the tripartite multi-drug resistance-nodulation-division (RND) system in *Pseudomonas aeruginosa* was found. They do not seem to confer antibiotic resistance to the hosts of pMCBF1/pMCBF6, but the presence of RND on promiscuous plasmids may have serious implications for the spread of antibiotic multi-resistance.

Editor: Finbarr Hayes, University of Manchester, United Kingdom

Funding: Financial support for this study was provided by The Swedish Research Council (contract no. 2002-214), to M.H. and by the University of Gothenburg. The funders had no role in study design, data collection and analysis, decision to publish, or preparation of the manuscript.

Competing Interests: The authors have declared that no competing interests exist.

* E-mail: malte.hermansson@cmb.gu.se

Introduction

Plasmids are independent from the bacterial host chromosome and conjugative broad-host range (BHR) plasmids can transfer between many different species in a bacterial community. Thus, plasmids are a particularly fluid part of the bacterial genome and one that may provide the cell with new traits that allow adaptation to selection forces, especially to transient environmental changes and challenges. Plasmids are likely to evolve differently in different bacterial backgrounds [1] and might also be influenced by environmental conditions [2]. Thus analysis of plasmids from different environments is important for our understanding of these important mobile elements (MGE). There are a number of studies about plasmids in marine environments [3,4,5,6,7,8,9,10,11], as well as reports about the role of marine plasmids in antibiotic resistance [12,13,14,15] and population dynamics [16,17], for excellent reviews on this topic see [18,19]. Still, compared with many other environments, plasmids from marine habitats are not well studied and fully sequenced plasmids with marine origins are underrepresented in the databases.

Using an exogenous isolation strategy, we have earlier isolated plasmids from marine bulk water, biofilms and the surface microlayer [20]. One of these plasmids, pMCBF1 isolated from a biofilm, was shown to have a BHR and transferred to many Gram-negative bacteria, including *Planctomyces maris* [21]. We also measured comparatively high transfer rates of pMCBF1 from a *P. putida* donor to indigenous bacteria, directly in seawater [4]. pMCBF1 was therefore an interesting BHR marine biofilm plasmid to characterize further. We determined the complete nucleotide sequences of pMCBF1 and the related pMCBF6, and a brief summary of the plasmids was presented earlier [1]. It was also shown that they form a novel clade within the promiscuous IncP-1 plasmid incompatibility group [1].

Since pMCBF1 and pMCBF6 are the only representatives of their IncP-1 clade, and except for pMLUA1, pMLU3 and pMUA4 [22], so far the only sequenced IncP-1 plasmids isolated from marine environments, it is interesting to analyze both their backbone genes and their accessory genes, and to compare these with other IncP-1 plasmids. Here we present a detailed analysis of pMCBF1 and pMCBF6.

Materials and Methods

Bacterial strains, plasmids and growth conditions

P. putida UWC1 [23] and *P. putida* KT2440 [24] with pMCBF1 and pMCBF6 were grown overnight at 26°C in Luria-Bertani media [25] with 10 g NaCl per liter and supplemented with $HgCl_2$ at 17 mg per liter. *Escherichia coli* with pMCBF1 and pMCBF6 were grown overnight at 37°C in the same medium supplemented with ampicillin (50 mg/liter).

Mercury and antibiotic resistance tests

To test for phenyl mercury resistance, *P. putida* KT2440 with either pMCBF1 or pMCBF6 were grown overnight and spread as lawns on LB plates. Discs that were dipped in solutions of phenyl mercury chloride (saturated solution in ether) were dried and placed on the lawns. Plates were inoculated overnight at 30°C after which the diameters of clearing zones around the discs were recorded. *P. putida* KT2440 without plasmids was used as controls. For the antibiotic resistance tests susceptibility discs (OXOID, UK) were placed on bacterial lawns of *E. coli* HB101 [26] and CAG [27] with and without pMCBF1 and pMCBF6 and incubated in 37°C overnight and clearing zones recorded as above.

Sequencing and sequence analysis

Sequencing of pMCBF1 and pMCBF6 was performed earlier in our laboratory using standard techniques [1]. Sequence alignments presented here were created using the eBioX program, the phylogenetic network was created using the SplitsTree program [28], and the similarity analysis was carried out using the SimPlot program [29]. The content of the genomes of the IncP-1 plasmids vary and all genes are not present in all plasmids [1]. The network constructed here was based on the largest segment that we could identify in which all genes were present in all analyzed plasmids (TraC to TraM).

Nucleotide sequence accession numbers

The complete sequences from pMCBF1 and pMCBF6 are deposited with GeneBank CoreNucleotide (accession # AY950444 and EF107516, respectively).

Results and Discussion

pMCBF1 and pMCBF6 represent the first fully sequenced IncP-1 plasmids from marine environments. Generally the difference between the IncP-1 plasmids sequenced so far lies in the various insertions of resistance and catabolic genes, often as part of MGEs, onto a more or less conserved backbone structure. pMCBF1 and pMCBF6 are obvious examples of this since they differ only in their mercury resistance (Hg^R) transposons that have inserted on otherwise identical backbones (Fig. 1A). Furthermore, the similarity of pMCBF1 and pMCBF6 strongly indicates that the insertions of Tn*5053* in pMCBF1 and Tn*5058* in pMCBF6 are recent events.

Plasmid backbone

Although the focus of most previous studies has been on accessory genes and MGEs, it has been shown that also the IncP-1 plasmid backbones vary and can be divided into different evolutionary clades. Hitherto, twelve phylogenetic clades have been described for the IncP-1 plasmid group, which are designated α, β-1, β-2, γ, δ, ε, ς, κ, η, θ [1,30,31,32,33,34,35] and two unnamed groups (Fig. 2). The pMCBF1/pMCBF6 plasmids described here are so far the only described plasmids from the IncP-1 ς-clade.

As shown in Fig 2, the members of the IncP-1 plasmid backbone family form a star-phylogeny. Such topology is typical when there has been extensive recombination [36,37], exponential population growth [38,39], or if members of each clade have evolved in different distinct separated populations (i.e., long time of evolution since a common ancestor result in long external branches). It has previously been demonstrated that homologous recombination has contributed to the evolution of the IncP-1 backbones [1]. IncP-1 plasmids have also been found in different environments, such as soil, wastewater treatment plants, and in human pathogens and fish pathogens [35], and as described here in a marine biofilm. Furthermore, it was recently demonstrated that the IncP-1 plasmid backbone has evolved in, and genetically adapted to, vastly different host bacterial species [1]. It is thus likely that the IncP-1 backbone at least periodically has evolved in different separated populations, which may explain the diversity and the star-shaped phylogeny of this BHR plasmid family.

We have identified 44 open reading frames (ORFs) in the pMCBF1/pMCBF6 backbone. pMCBF1 has 24 ORFs and pMCBF6 28 ORFs that are related to accessory genes or functions such as transposons, various resistance functions, IS elements and multi-drug efflux systems. There are various genetic distances between the pMCBF1/pMCBF6 backbone genes and the corresponding genes in plasmids from the other clades, which is probably explained by the history of recombination [1]. As an example, most of the backbone genes such as replication, conjugation and stable maintenance genes share 58–95% similarity with genes from plasmids pADP-1 [40], pB4 [41], R751 [31] and pTSA [42] from the β clade, which is the most studied IncP-1 clade.

The origin of replication (*oriV*) of pMCBF1/pMCBF6 resembles other IncP-1 plasmids. The nine TrfA DNA binding sites (iterons) in the vicinity of the pMCBF1/pMCBF6 *oriV* have the 17 bp consensus sequence N(T/C/G)GCCCCTCA(A/T)(A/G/C)T(A/G)T(C/T)A, and have been conserved to some degree compared to both RK2 and R751. The R751 iterons by comparison have the sequence (A/C/G)NGCCCC(T/C)C A(A/T)(G/C)(T/G)GTCA. The pMCBF1/pMCBF6 iterons are arranged as eight direct repeats and one inverted repeat, similar to the iterons close to *oriV* in R751.

The TrfA protein is necessary for the replication of iteron containing theta replication plasmids (e.g. [43]) and *trfA* has been used as a base for plasmid classification [44]. We suggested earlier that pMCBF1 and pMCBF6 had replication functions that were different from known plasmid incompatibility groups based on lack of hybridization with the *inc/rep* Couturier probes [20]. However, the general lack of similarity of the Couturier probes with many environmentally isolated plasmids has been noted in several investigations [9,18,45,46,47]. We now know that the IncP-1 Couturier probe only detects the IncP-1 α clade [48]. In fact, the nucleotide similarity between pMCBF1/pMCBF6 and the Couturier IncP-1 probe is only about 60% and these plasmids could therefore not be targeted by the probe.

The pMCBF1/pMCBF6 backbone has no insert between the replication initiation gene *trfA* and the *oriV*. Except for the IncP-1 β2 plasmids pBP136 and pA1, most other IncP-1 plasmids have an insert, usually several kilobases (kb), that separates *trfA* and *oriV*.

Just as for the IncP-1 β plasmids, pMCBF1/pMCBF6 lack the *parABCDE* genes that play a key role in stability of IncP-1 α plasmids. Although the *parABCDE* genes may be absent from pMCBF1/pMCBF6, they apparently possess the *korA-incC-korB* locus that mediates accurate plasmid segregation. Surface exclusion systems also protect the resident plasmid from competition, by preventing new plasmids to enter the cell. pMCBF1/pMCBF6 carry the *trbK* homologue, which confers surface exclusion in plasmid RK2. We know from recombination analyses that pMCBF1/pMCBF6 have been involved in recombination events [1], which strongly suggest that these plasmids have been in the

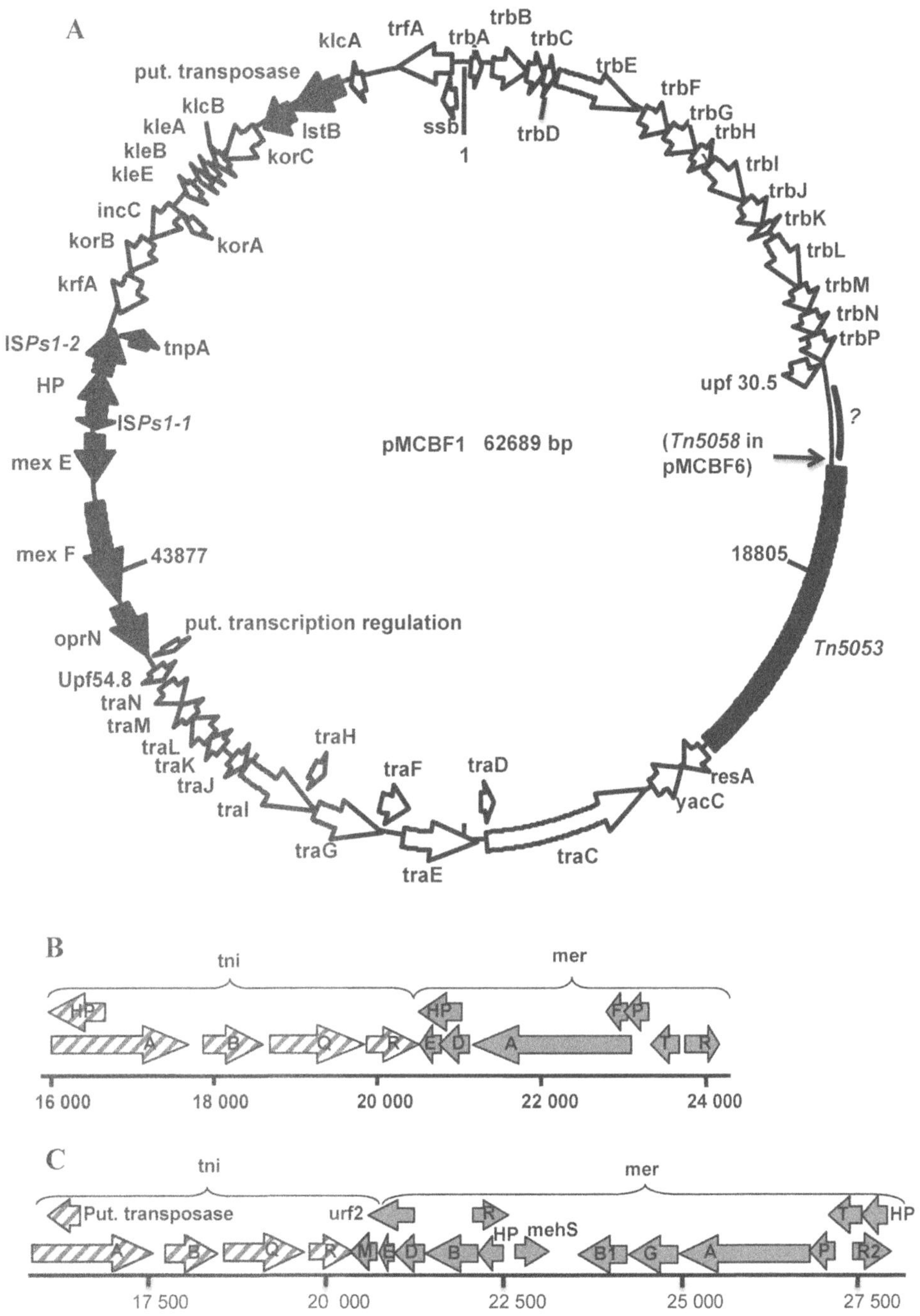

Figure. 1. Genetic map of the IncP-1 ς plasmids pMCBF1 and pMCBF6. The backbones of the two plasmids are identical and differ only with regard to inserted transposons. Panel (A) shows pMCBF1 with Tn*5053*. Coding regions are shown by arrows, indicating the direction of transcription. Unfilled arrows denote plasmid backbone genes, black arrows denote accessory genes. An arrow points to the insertion site of Tn*5058* in the pMCBF6 backbone. The section of the pMCBF1/pMCBF6 backbone that has a low similarity to other sequences in the databases is marked with a question mark. Panels B and C shows transposon Tn*5053* on pMCBF1 and Tn*5058* on pMCBF6, respectively. Striped arrows denote genes in the transposition module (tni) and gray arrows denote mercury resistance (mer) genes. HP means hypothetical proteins.

same host cell as other IncP-1 plasmids [1]. Thus, it seems that the surface exclusion systems of pMCBF1/pMCBF6 have a "leakiness" that allow recombination to occur.

One block of about 2.5 kb, between the inserted Hg^R transposons and the *trb* section (marked with a question mark in Fig.1) has a low similarity (73% identity) to *Methylophaga* sp. JAM7 [49].

Accessory genes, transposons and insertion sequences

The DNA segment bearing the mercury resistance determinant on pMCBF1 (Fig. 1B) shows 99% nucleotide similarity to Tn*5053* [50]. There are five base pair direct repeats in the flanking plasmid DNA, which are typically formed after Tn*5053* transposition [50]. Interestingly, Tn*5053* in pMCBF1 is inserted close to the resolvase gene (*resA*) in the backbone (Fig. 1A) which supports the suggestion that Tn*5053* is a "resolvase-site-hunter" that insert in "hot-spots"

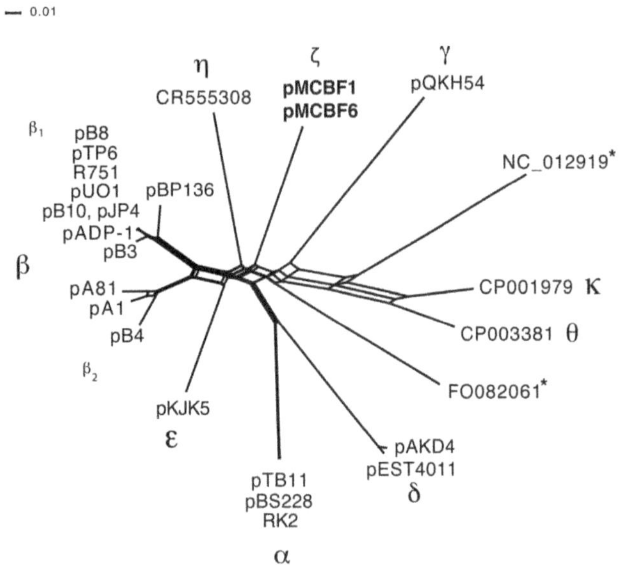

Figure. 2. Phylogenetic network of clades of the IncP-1 plasmid family. The network is based on representative plasmids from each of the twelve previously described phylogenetic clades of the IncP-1 plasmid family. The network is based on the genetic segment harboring the *traC* - *traM* genes for all analyzed plasmids. The two plasmids pMCBF1 and pMCBF6 described in this study are highlighted in bold. Clades without designated names are marked with *. Previously described intra-clade recombinants [1] were not included in the analysis. The figure is updated from [1].

associated with *res* sites and resolvase genes [51]. Minakhina *et al.* suggested that the plasmid resolvase may participate in the transposition event, but the exact mechanism is not clear [51]. If so, this is an interesting case in which the transposon is dependent on an external site-specific recombination system. The two inverted repeats (IR) that form the ends of Tn*5053* in pMCBF1 have 2 and 4 mismatches compared with other published sequences [52]. There are also 4 mismatches between the two IR of Tn*5053* in pMCBF1. However, this apparently did not stop transposition of Tn*5053* into pMCBF1. Tn*5053* belongs to a superfamily of diverse elements including retroviruses, phage Mu, insertion sequences of the IS*3* family, and transposons Tn*552*, Tn*7* and Tn*5058* [53]. They are found on plasmids within the IncP-1 group (pPUO1, pB11 and pMCBF1/pMCBF6) as well as on other plasmid groups, such as the *Enterobacter cloacae* plasmid pELC_A [54].

The DNA segment bearing the mercury resistance determinant on pMCBF6 (Fig. 1C) is highly similar to Tn*5058* in *Pseudomonas* sp. ED23-33 [55]. The length of both sequences are 12, 373 bp and the nucleotide similarity varies from 96 to 99% except for the first 257 bp where similarity is 84%, but the IR at the start of the element again has a perfect match with Tn*5058*. The element has 25 bp inverted repeats with 2 mismatches. The complex Tn*5058* is part of the Tn*5053* family (see above) [53,56]. Judging from comparisons with the closest relatives of Tn*5058*, such as Tn*21*, Tn*501*, Tn*5053*, Tn*5041D*, Tn*5718* and others, it was suggested that Tn*5058* was formed by a complicated series of recombination events involving these elements [56]. Tn*5058* is also carried by other IncP-1 plasmids such as pIJB1, pWEC911 and pTP6. Interestingly, Tn*5058* has been detected in bacteria preserved for 15.000–40.000 years in permafrost grounds [56]. When compared to several Tn*5058* collected in modern days, there were high sequence similarities suggesting that these transposons are genetically stable over time. We compared the Tn*5058* in plasmid pMCBF6 with the Tn*5058* isolated from permafrost, using the sliding window protocol implemented in the SimPlot program. Our results support previous suggestions that the Tn*5058* is indeed well conserved and there is a high nucleotide identity between the transposon Tn5058 isolated here and the transposon Tn5058 isolated from permafrost in most regions. Least conserved was the gene *tniA* with a 200 bp region of only than 80% identity and *merB* and *merR* with regions of less than 95% identity between the Tn*5058* in pMCBF6 and the Tn*5058* isolated from permafrost. Tn*5058* has two copies of *merR*, *merB* and *merD*. Interestingly, it was suggested that the duplications in some of the *mer* genes in Tn*5041D* were the result of an integration, via homologous recombination, of a *mer* containing circular DNA structure [55]. The circular cassette was speculated to originate from an ancestral donor in which the *mer* genes were flanked by IS elements. The duplication of *mer* genes in Tn*5058* might also have been the result of a similar event [55]. Just as for Tn*5053* in pMCBF1, Tn*5058* is flanked by 5 base pair direct repeats in pMCBF6, which strongly indicate a transposition event [50,52]. These flanking direct repeats are also seen when Tn*5058* is inserted in IncP-1 plasmid pIJB1. The insertion of Tn*5058* in pMCBF6 is only 63 bp from the site where Tn*5053* inserted in pMCBF1, which is close to the resolvase gene (*resA*) in the plasmid backbone. The insertion site further confirms that Tn*5058* is a "res-site hunter" [51].

The sequence data suggest that Tn*5053* confers mercury resistance towards inorganic mercury while Tn*5058* also carries resistance towards organo-mercury compounds. When the two plasmids were tested for phenyl mercury chloride (PhHg) resistance the diameter of the clearing zones around PhHg discs was only 6 mm for *P. putida* 2440(pMCBF6) but 13 mm for *P. putida* 2440(pMCBF1) and the plasmid-free *P. putida* 2440, confirming the PhHg resistance of pMCBF6 but not of pMCBF1.

A 2982 bp sequence between *kfrA* and *mexE* in both pMCBF1 and pMCBF6 shows 99% (nucleotide) similarity with insertion sequence IS*Ps1* (Fig. 1A). The 24 bp IR regions with 4 mismatches characteristic of IS*Ps1* [57] are found in pMCBF6/pMCBF1. IS*Ps1* is related to the IS*L3* family, but is larger than the usual IS*L3* elements which are 1300–1550 bp. IS*Ps1* has been found in several copies in *Pseudomonas stutzeri*, where it has inserted into, and inactivated, catabolic genes [57], and in *Yersinia ruckeri* plasmid pYR1 [58]. Interestingly, IS*Ps1* was also recently found on plasmid pAMEC615 from the marine *Alteromonas macleodii* isolated from the English Channel [59]. In *Pseudomonas stutzeri*, IS*Ps1* was flanked by eight bp direct repeats (DR), indicating a replicative transposition event [57], but no such flanking DR was found in any of pMCBF1/pMCBF6, pAMEC615 or pYR1. If DR were formed during a replicative transposition in these plasmids, they may have degenerated with time, perhaps because these sequences are not under selective pressure. Alternatively, the IS*Ps1*was inserted by some other recombination mechanism than replicative transposition. The insertion point of IS*Ps1* in pMCBF1/6, within the plasmid control region between *kfrA* and *upf54.8*, seems to be unique among IncP-1 plasmids, indicating that insertion points other than the ones between *tra* and *trb* regions and between the replication region and *oriV*, are possible.

A region containing a putative remnant of an insertion event is located between *klcA* and *klcB*. This region contains an ORF with closest similarity (74%) to a putative transposase, as well as an IstB homolog. IstB is associated with IS*21* family insertion sequences. The function of IstB is unknown, but it may assist in transposition [60]. We find no DR or IR as signs of a recent insertion. Insertions of transposons in the *klcA*/*klcB* region are seen also in IncP-1 plasmid RP1/RK2 [61].

Adjacent to IS*Ps1* is an inserted region that shows a high similarity to part of the recently sequenced plasmid pALIDE201 from *Alicycliphilus denitrificans* isolated from anaerobic sewer sludge [62]. Furthermore, three genes within this region show similarity (73–67%) to *mexEF* and a gene from the NodT family, which are part of tripartite multi-drug resistance-nodulation-division (RND) families. Such efflux systems are common on chromosomes of many Gram-negative bacteria [63], e.g. the *mexEF-oprN* in *Pseudomonas aeruginosa*. The IncP-1 β plasmid pB4 has a region with about 80% similarity to a RND efflux system in *P. denitrificans*, this was the first plasmid that was shown to carry a RND efflux system [41]. Adjacent to the RND region in pB4 a putative transposon terminus was found close to a Tn*3*-like transposase, and the authors speculated that the RND efflux system might have been transferred to pB4 from the chromosome of an unknown gram-negative bacterium by transpositional cointegrate formation [41]. Although IS*Ps1* is found adjacent to the *mex*-region in pMCBF6/pMCBF1, we do not know if this IS element was involved in the *mex* insertion. The MexEF-OprN system in *P. aeruginosa* confers resistance to aromatic hydrocarbons, fluoroquinolones, chloramphenicol, triclosan and trimethoprim [63]. Our experimental data shows that when tested in disk diffusion tests for antibiotic sensitivity on agar plates, *E. coli* with pMCBF1 or pMCBF6 did not confer a higher resistance to chloramphenicol, nalidixic acid or trimethoprim than plasmid free cells. Further studies are needed to define the possible function of MexEF-OprN in pMCBF1/pMCBF6. Our finding that plasmids from a marine biofilm carry a multi-drug efflux system shows that pB4 is not unique, and that many different efflux systems may potentially be borne by promiscuous plasmids. The possibly serious implications of multi-drug efflux systems on such plasmids and the effects this may have on the spread of multi-resistance to human pathogens have been discussed [41].

The majority of the well-known IncP-1 plasmids, and plasmids in general, originate from soil environments or from a clinical environment and other man-made systems, such as wastewater treatment plants [35]. pMCBF1/pMCBF6 are among the first IncP-1 plasmids that are sequenced from marine environments. Three IncP-1ε plasmids (pMLUA1, pMLUA3, pMLUA4) were recently isolated from the air-water interface from an estuary in Portugal [22,64]. Analysis of these, and other plasmids from marine bacteria, has often revealed new features [7,10,11,22]. Analysis of the pMCBF1/pMCBF6 plasmids also showed, among other things, that they form a new clade within the IncP-1 plasmid group.

Acknowledgments

We would like to thank Robert Almstrand for skillful technical assistance.

Author Contributions

Conceived and designed the experiments: MB MH. Performed the experiments: MB. Analyzed the data: PN MB MH. Contributed reagents/materials/analysis tools: PN MB. Wrote the paper: PN MB MH.

References

1. Norberg P, Bergström M, Jethava V, Dubhashi D, Hermansson M (2011) The IncP-1 plasmid backbone adapts to different host bacterial species and evolves through homologous recombination. Nat Comm 2: 268.
2. Kirzhner V, Paz A, Volkovich Z, Nevo E, Korol A (2007) Different clustering of genomes across life using the A-T-C-G and degenerate R-Y alphabets: early and late signaling on genome evolution? J Mol Evol 64: 448–456.
3. Cook M, Osborn A, Bettandorff J, Sobecky P (2001) Endogenous isolation of replicon probes for assessing plasmid ecology of marine sediment microbial communities. Microbiology 147: 2089.
4. Dahlberg C, Bergström M, Hermansson M (1998) In situ detection of high levels of horizontal plasmid transfer in marine bacterial communities. Appl Environ Microbiol 64: 2670–2675.
5. Hermansson M, Jones GW, Kjelleberg S (1987) Frequency of antibiotic and heavy metal resistance, pigmentation, and plasmids in bacteria in the marine air-water interface. Appl Environ Microbiol 53: 2338–2342.
6. Ma Y, Paulsen IT, Palenik B (2011) Analysis of two marine metagenomes reveals the diversity of plasmids in oceanic environments. Environ Microbiol 14: 453–466.
7. Petersen J, Frank O, Göker M, Pradella S (2013) Extrachromosomal, extraordinary and essential—the plasmids of the Roseobacter clade. Appl Microbiol Biotech 97: 2805–2815.
8. Sobecky P, Mincer T, Chang M, Toukdarian A, Helinski D (1998) Isolation of broad-host-range replicons from marine sediment bacteria. Appl Environ Microbiol 64: 2822–2830.
9. Sobecky PA, Mincer TJ, Chang MC, Helinski DR (1997) Plasmids isolated from marine sediment microbial communities contain replication and incompatibility regions unrelated to those of known plasmid groups. Appl Environ Microbiol 63: 888–895.
10. Zhong Z, Caspi R, Helinski D, Knauf V, Sykes S, et al. (2003) Nucleotide sequence based characterizations of two cryptic plasmids from the marine bacterium Ruegeria isolate PR1b. Plasmid 49: 233–252.
11. Zhong Z, Toukdarian A, Helinski D, Knauf V, Sykes S, et al. (2001) Sequence analysis of a 101-Kilobase plasmid required for agar degradation by a Microscilla isolate. Appl Environ Microbiol 67: 5771–5779.
12. Yang J, Wang C, Shu C, Liu L, Geng J, et al. (2013) Marine sediment bacteria harbor antibiotic resistance genes highly similar to those found in human pathogens. Microb Ecol 65: 975–981.
13. Phelan RW, Clarke C, Morrissey JP, Dobson ADW, O'Gara F, et al. (2011) Tetracycline resistance-encoding plasmid from Bacillus sp. strain #24, isolated from the marine sponge haliclona simulans. Appl Environ Microbiol 77: 327–329.
14. Baya AM, Brayton PR, Brown VL, Grimses DJ, Russek-Cohen E, et al. (1986) Coincident plasmids and antimicrobial resistance in marine bacteria isolated from polluted and unpolluted Atlantic Ocean samples. Appl Environ Microbiol 51: 1285–1292.
15. Sizemore RK, Colwell RR (1977) Plasmids carried by antibiotic-resistant marine bacteria. Antimicrob Ag Chemoth 12: 373–382.
16. Palenik B, Ren Q, Tai V, Paulsen IT (2009) Coastal Synechococcus metagenome reveals major roles for horizontal gene transfer and plasmids in population diversity. Environ Microbiol 11: 349–359.
17. Pradella S, Päuker O, Petersen J (2010) Genome organisation of the marine Roseobacter clade member Marinovum algicola. Arch Microbiol 192: 115–126.
18. Smalla K, Sobecky PA (2002) The prevalence and diversity of mobile genetic elements in bacterial communities of different environmental habitats: insights gained from different methodological approaches. FEMS Microbiol Ecol 42: 165–175.
19. Sobecky PA, Hazen TH (2009) Horizontal gene transfer and mobile genetic elements in marine systems. Methods in molecular biology (Clifton, NJ) 532: 435–453.
20. Dahlberg C, Linberg C, Torsvik VL, Hermansson M (1997) Conjugative plasmids isolated from bacteria in marine environments show various degrees of homology to each other and are not closely realted to well characterized plasmids. Appl Environ Microbiol 63: 4692–4697.
21. Dahlberg C, Bergström M, Andreason M, Christensen BB, Molin S, et al. (1998) Interspecies bacterial conjugation by plasmids from marine environmnents visualized by *gfp* expression. Mol Biol Evol 15: 385–390.
22. Oliveira CS, Moura A, Henriques I, Brown CJ, Rogers LM, et al. (2013) Comparative genomics of IncP-1ε plasmids from water environments reveals diverse and unique accessory genetic elements. Plasmid 70: 412–419.
23. McClure NC, Weightman AJ, Fry JC (1989) Survival of Pseudomonas putida UWC1 containing cloned catabolic genes in a model activated-sludge unit. Appl Environ Microbiol 55: 2627–2634.
24. Nelson KE, Weinel C, Paulsen IT, Dodson RJ, Hilbert H, et al. (2002) Complete genome sequence and comparative analysis of the metabolically versatile Pseudomonas putida KT2440. Environ Microbiol 4: 799–808.
25. Maniatis T, Fritsch EF, Sambrook J (1982) Molecular cloning: a laboratory manual. Cold Spring Harbor, New York: Cold Spring Harbour Laboratory.
26. Boyer HW, Roulland-Dussoix D (1969) A complementation analysis of the restriction and modification of DNA in Escherichia coli. J Mol Biol 41: 459–472.
27. Singer M, Baker TA, Schnitzler G, Deischel SM, Goel M, et al. (1989) A collection of strains containing genetically linked alternating antibiotic resistance elements for genetic mapping of *Escherichia coli*. Microbiol Rev 53: 1–24.
28. Huson DH (2005) Application of phylogenetic networks in evolutionary studies. Mol Biol Evol 23: 254–267.
29. Lole KS, Bollinger RC, Paranjape RS, Gadkari D, Kulkarni SS, et al. (1999) Full-length human immunodeficiency virus type 1 genomes from subtype C-infected seroconverters in India, with evidence of intersubtype recombination. J Virol 73: 152–160.

30. Pansegrau W, Lanka E, Barth PT, Figurski DH, Guiney DG, et al. (1994) Complete nucleotide sequence of Birmingham IncP alpha plasmids. Compilation and comparative analysis. J Mol Biol 239: 623–663.
31. Thorsted PB, Macartney DP, Akhtar P, Haines AS, Ali N, et al. (1998) Complete sequence of the IncP plasmid R751: Implications for evolution and organisation of the IncP backbone. J Mol Biol 282: 969–990.
32. Vedler E, Vahter M, Heinaru A (2004) The completely sequenced plasmid pEST4011 contains a novel IncP1 backbone and a catabolic transposon harboring tfd genes for 2,4-dichlorophenoxyacetic acid degradation. J Bacteriol 186: 7161–7174.
33. Haines AS, Akhtar P, Stephens ER, Jones K, Thomas CM, et al. (2006) Plasmids from freshwater environments capable of IncQ retrotransfer are diverse and include pQKH54, a new IncP-1 subgroup archetype. Microbiology (Reading, England) 152: 2689–2701.
34. Bahl MI, Hansen LH, Goesmann A, Sørensen SJ (2007) The multiple antibiotic resistance IncP-1 plasmid pKJK5 isolated from a soil environment is phylogenetically divergent from members of the previously established alpha, beta and delta sub-groups. Plasmid 58: 31–43.
35. del Castillo CS, Bin Jang H, Hikima J-i, Jung TS, Morii H, et al. (2013) Comparative analysis and distribution of pP9014, a novel drug resistance IncP-1 plasmid from Photobacterium damselae subsp. piscicida. Int J Antimicrob Agents 42: 10–18.
36. Schierup MH, Hein J (2000) Recombination and the molecuar clock. Mol Biol Evol 17: 1578–1579.
37. Schierup MH, Hein J (2000) Consequences of Recombination on Traditional Phylogenetic Analysis. Genetics 156: 879–891.
38. Holmes EC, Pybus OG, Harvey PH (1999) The molecular population dynamics of HIV-1. In: Crandall KA, editor. The Evolution of HIV. Baltimore: The Johns Hopkins University Press. pp. 177–207.
39. Slatkin M, Hudson RR (1991) Pairwise comparisons of mitochondrial-DNA sequences in stable and exponentially growing populations. Genetics 129: 555–562.
40. Martinez B, Tomkins J, Wackett LP, Wing R, Sadowsky MJ (2001) Complete nucleotide sequence and organization of the atrazine catabolic plasmid pADP-1 from Pseudomonas sp. strain ADP. J Bacteriol 183: 5684–5697.
41. Tauch A, Schlüter A, Bischo N, Goesmann A, Meyer F, et al. (2003) The 79,370-bp conjugative plasmid pB4 consists of an IncP-1b backbone loaded with a chromate resistance transposon, the strA- strBstreptomycin resistance gene pair, the oxacillinase gene blaNPS-1, and a tripartite antibiotic efflux system of the resistance-nodulation-division family. Mol Gen Genomics 268: 570–584.
42. Tralau T, Cook AM, Ruff J (2001) Map of the IncP1{beta} plasmid pTSA encoding the widespread genes (tsa) for p-toluenesulfonate degradation in Comamonas testosteroni T-2. Appl Environ Microbiol 67: 1508–1516.
43. Thomas CM, editor (2000) The Horizontal gene pool. Amsterdam, The Netherlands: Harwood Academic Publ. 419 p.
44. Couturier MF, Bex P, Bergquist L, Maas WK (1988) Identification and classification of bacterial plasmids. Microbiol Rev 52: 375–395.
45. Kobayashi N, Bailey M (1994) Plasmids isolated from the sugar beet phyllosphere show little or no homology to molecular probes currently available for plasmid typing. Microbiology 140: 289–296.
46. Osborn AM, Pickup RW, Saunders JR (2000) Development and application of molecular tools in the study of IncN-related plasmids from lakewater sediments. FEMS Microbiol Lett 186: 203–208.
47. van Elsas JD, McSpadden Gardener BB, Wolters AC, Smit E (1998) Isolation, characterization, and transfer of cryptic gene-mobilizing plasmids in the wheat rhizosphere. Appl Environ Microbiol 64: 880–889.
48. Heuer H, Smalla K (2012) Plasmids foster diversification and adaptation of bacterial populations in soil. FEMS Microbiol Rev 36: 1083–1104.
49. Villeneuve C, Martineau C, Mauffrey F, Villemur R (2012) Complete genome sequences of Methylophaga sp. strain JAM1 and Methylophaga sp. strain JAM7. J Bacteriol 194: 4126–4127.
50. Kholodii GY, Yurieva OV, Lomovskaya OL, Gorlenko ZM, Mindlin SZ, et al. (1993) Tn5053, a mercury resistance transposon with integron's ends. J Mol Biol 230: 1103–1107.
51. Minakhina S, Kholodii G, Mindlin S, Yurieva O, Nikiforov V (1999) Tn5053 family transposons are res unters sensing plasmidal res sites occupied by cognate resolvases. Mol Microbiol 33: 1059–1068.
52. Partridge SR (2011) Analysis of antibiotic resistance regions in Gram-negative bacteria. FEMS Microbiol Rev 35: 820–855.
53. Kholodii GY, Mindlin SZ, Bass IA, Yurieva OV, Minakhina SV, et al. (1995) Four genes, two ends, and a res region are involved in transposition of Tn5053: a paradigm for a novel family of transposons carrying either a mer operon or an integron. Mol Microbiol 17: 1189–1200.
54. Ren Y, Ren Y, Zhou Z, Guo X, Li Y, et al. (2010) Complete Genome Sequence of Enterobacter cloacae subsp. cloacae Type Strain ATCC 13047. J Bacteriol 192: 2463–2464.
55. Kholodii G, Gorlenko Z, Mindlin S, Hobman J, Nikiforov V (2002) Tn5041-like transposons: molecular diversity, evolutionary relationships and distribution of distinct variants in environmental bacteria. Microbiology 148: 3569–3582.
56. Mindlin S, Minakhin L, Petrova M, Kholodii G, Minakhina S, et al. (2005) Present-day mercury resistance transposons are common in bacteria preserved in permafrost grounds since the Upper Pleistocene. Res Microbiol 156: 994–1004.
57. Bolognese F, di Lecce C, Galli E, Barbieri P (1999) Activation and inactivation of Pseudomonas stutzeri methylbenzene catabolism pathways mediated by a transposable element. Appl Environ Microbiol 65: 1876–1882.
58. Welch TJ, Fricke WF, McDermott PF, White DG, Rosso M-L, et al. (2007) Multiple antimicrobial resistance in plague: An emerging public health risk. PLoS ONE 2: e309.
59. Lopez-Perez M, Gonzaga A, Rodriguez-Valera F (2013) Genomic diversity of deep ecotype Alteromonas macleodii isolates: evidence for pan-Mediterranean clonal frames. Gen Biol Evol 5: 1220–1232.
60. Smith CA, Pinkney M, Guiney DG, Thomas CM (1993) The ansestral IncP replication system consisted of contigous *oriV* and *trfA* segments as deduced from a comparison of the nucleotide sequences of diverse IncP plasmids. J Gen Microbiol 139: 1761–1766.
61. Pinyon JL, Hall RM (2011) Evolution of IncP-1α Plasmids by Acquisition of Antibiotic and Mercuric Ion Resistance Transposons. Microb Drug Resist 17: 339–343.
62. Oosterkamp MJ, Veuskens T, Plugge CM, Langenhoff AAM, Gerritse J, et al. (2011) Genome sequences of Alicycliphilus denitrificans strains BC and K601T. J Bacteriol 193: 5028–5029.
63. Poole K (2001) Multidrug resistance in Gram-negative bacteria. Curr Opin Microbiol 4: 500–508.
64. Oliveira CS, Lázaro B, Azevedo JSN, Henriques I, Almeida A, et al. (2012) New molecular variants of epsilon and beta IncP-1 plasmids are present in estuarine waters. Plasmid 67: 252–258.

Hypernatremia in Dice Snakes (*Natrix tessellata*) from a Coastal Population: Implications for Osmoregulation in Marine Snake Prototypes

François Brischoux[1]*, Yurii V. Kornilev[2]

1 Centre d'Etudes Biologiques de Chizé, CEBC UMR 7372 CNRS-ULR, Villiers en Bois, France, 2 Bulgarian Society for the Protection of Birds, NCC "Poda", Burgas, Bulgaria

Abstract

The widespread relationship between salt excreting structures (e.g., salt glands) and marine life strongly suggests that the ability to regulate salt balance has been crucial during the transition to marine life in tetrapods. Elevated natremia (plasma sodium) recorded in several marine snakes species suggests that the development of a tolerance toward hypernatremia, in addition to salt gland development, has been a critical feature in the evolution of marine snakes. However, data from intermediate stage (species lacking salt glands but occasionally using salty environments) are lacking to draw a comprehensive picture of the evolution of an euryhaline physiology in these organisms. In this study, we assessed natremia of free-ranging Dice snakes (*Natrix tessellata*, a predominantly fresh water natricine lacking salt glands) from a coastal population in Bulgaria. Our results show that coastal *N. tessellata* can display hypernatremia (up to 195.5 mmol.l^{-1}) without any apparent effect on several physiological and behavioural traits (e.g., hematocrit, body condition, foraging). More generally, a review of natremia in species situated along a continuum of habitat use between fresh- and seawater shows that snake species display a concomitant tolerance toward hypernatremia, even in species lacking salt glands. Collectively, these data suggest that a physiological tolerance toward hypernatremia has been critical during the evolution of an euryhaline physiology, and may well have preceded the evolution of salt glands.

Editor: Ulrich Joger, State Natural History Museum, Germany

Funding: Funding was provided by the CNRS (France). The funders had no role in study design, data collection and analysis, decision to publish, or preparation of the manuscript.

Competing Interests: The authors have declared that no competing interests exist.

* E-mail: francois.brischoux@gmail.com

Introduction

Living in seawater entails physiological consequences such as water loss and salt gain, and coping with these constraints represents one of the principal challenges of secondarily marine vertebrates [1]. Accordingly, marine tetrapods (i.e., mammals, birds, turtles, snakes, lizards and crocodiles) display specific adaptations related to the maintenance of osmotic balance. For instance, marine mammals have specialized nephrons which allow highly concentrated urine [2]. Although marine reptiles lack the ability to excrete excess salt in urine, they have evolved salt glands that secrete concentrated salt solution [3,4].

The widespread relationship between marine life and presence of specific salt-excretory structures, found across very different taxa, strongly suggests that the ability to excrete excess salt has been critical during the invasion of marine environments by tetrapods. However, as with most evolutionary processes, transitional steps are missing and are seldom represented by fossil remains [5]. In addition, crucial characteristics such as physiology and/or behaviour do not print well within the fossil records [5]. Yet, some research works have proposed scenarios of the evolution of an euryhaline physiology during the transition to marine life [6,7,8]. For instance, Dunson and Mazzotti [6] have proposed four successive steps that should ultimately lead to an efficient maintenance of the osmotic balance. The first step consists of a primary reliance on behavioural osmoregulation (e.g., frequent obligate fresh water drinking [9]). The second step involves a reduction in salt gain and water loss through permeable surfaces [10,11,12]. The third and fourth steps include the evolution of rudimentary salt secreting features and their subsequent development [4], which would ultimately allow exploiting more saline, thus larger, oceanic areas [13].

Clearly, these successive stages would ultimately allow organisms to progressively become emancipated from regular access to fresh water, and thus to thrive in saline environments. Likely, this has led to the conclusion that marine tetrapods could maintain their water balance without consuming fresh water [14,15]. However, recent investigations have challenged this paradigm. Specifically, the most detailed studies performed on marine snakes have shown that species having a functional salt gland cannot equilibrate their hydromineral balance without access to fresh water [12,16]. Dehydration in seawater has been shown to occur in amphibious sea snakes (Laticaudine sea kraits) as well as in fully marine species (Hydrophine sea snakes) [12,16,17]. In addition, elevated plasmatic sodium concentrations have been measured in various marine snake species [18–25]. These studies have led to the hypothesis that the development of a physiological tolerance to hypernatremia may have been an important feature of the evolution of marine snakes [25]. However, data gathered under experimental conditions show that fresh water species lacking salt glands (including coastal presumably salt tolerant species) rapidly accumulate large salt loads when acclimated in brackish and salt water [24,26]. In most of these cases, the resulting hypernatremia was lethal [24,26]. Taken together, these elements highlight the lack of information from groups thought to resemble transitional forms between the land and the sea (species lacking salt glands but occasionally using salty environments) to draw a comprehensive picture of the evolution of an euryhaline physiology in these organisms.

Snakes provide a suitable study system with which to clarify the steps that allowed coping with osmotic challenge [13]. Indeed, this lineage displays a unique gradient of habitat use which allows studying groups thought to resemble transitional forms between the land and the sea [13,27]. Importantly, these transitional steps can be investigated at several phylogenetic scales along a gradient of habitat use [27]. Species that are salt tolerant, but imperfectly marine (i.e., lacking salt glands) represent a powerful opportunity to investigate the early steps of the evolution of an euryhaline physiology. In the current investigation, we examine such a study system. The European Dice snake (*Natrix tessellata*) is a typical semi-aquatic fresh water natricine species that occurs over Eurasia (broadly from Italy to China [28]). Although this species relies primarily on fresh water bodies to forage for fish and amphibians, some populations are known to use, more or less extensively, brackish or saline habitats, thereby offering the possibility to investigate an intermediate step during the evolution to marine life. In this study we report natremia (plasma sodium concentration, an indicator of the osmotic challenge linked to marine life in snakes [25]) measured in free-ranging Dice snakes inhabiting a coastal ecotone between freshwater and the Black Sea in Bulgaria. In combination with a review of plasmatic sodium concentration of snakes, these results are discussed in the light of the secondary transition to marine life in tetrapods.

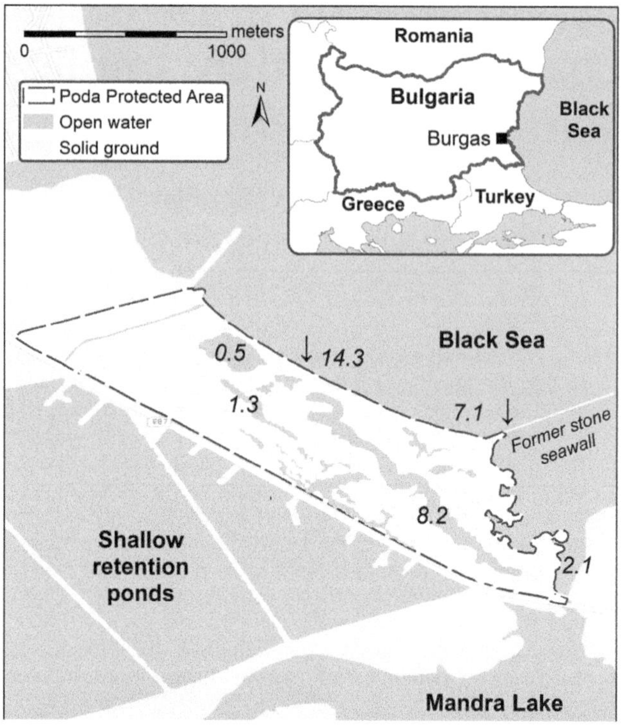

Figure 1. Map of the study area. The upper panel gives the location of the Poda Protected Areas in the vicinity of Bourgas, Bulgaria. Emergent lands are indicated in white, water is indicated in light grey. The lower panel shows the Poda Protected Area (dashed area). Emergent lands are indicated in white and water is indicated in light grey. Numbers designate salinity (‰) recorded for three ponds, two locations on the shore of the Black Sea, and one location at the mouth of the Mandra Lake. The two arrows show sites where tracks from snakes commuting between the land and the Black Sea were observed.

Materials and Methods

Ethics statement

All procedures were approved by French and Bulgarian regulations (Comité d'éthique Poitou-Charentes approval number CE2013-5 to FB; Ministry of Environment and water of Bulgaria permit to YVK: 298/09.03.2011).

Study species

Natrix tessellata is a medium-sized (up to 130 cm [29]) species with an extended Palearctic distribution: from central Europe to northern Egypt and east as far as north-western China [30,31]. It is a typical semi-aquatic natricine foraging mainly for fish, and to a lesser extent for amphibians in streams, rivers, and lakes [28].

Although the vast majority of *N. tessellata* populations rely on fresh water bodies, few do occur in saline environments along the coasts of the Adriatic Sea [28,32–35], the Ionian and Aegean Seas [36], the Black Sea ([37–40], this study) and the Caspian Sea [41,42]. In most of these cases, *N. tessellata* occurs in brackish waters of lagoons, salt marshes and river mouths.

Study site

We surveyed a population of Dice snakes on the southern Bulgarian Black Sea coast, in the "Poda" Protected Area (Fig. 1). The Poda wetland (1 km^2) consists of a coastal ecotone inserted between a large predominantly freshwater reservoir (Mandra Lake) and the Black Sea (Fig. 1). Poda is mainly composed of an alternation of shallow pools of water (usually <1 m deep) intersected by embankment lands and lower, temporarily flooded areas. The proximity of the Mandra Lake and the Black Sea and seasonal climatic fluctuations create a wide variety of aquatic habitats spanning from fresh (<1‰) and brackish (1–10‰) to salt water (>10‰), including some hyperhaline waters during the summer (32–33 ‰, [43], Fig. 1).

Field procedures

In April 2012, a total number of 19 snakes were captured by hand. Snakes were typically found while basking in the sun. Individuals were measured (snout-vent length [SVL] and total length [TL], ± 0.5 cm), weighted (± 1 g), and sexed by eversion of the hemipenis. Feeding and reproductive status were assessed by gentle palpation.

Only large adult females (>140 g, $N = 13$) were blood-sampled to avoid putative detrimental effects of the procedure on smaller individuals, and to avoid sex effects on plasmatic parameters. Blood (~400 µl) was sampled through cardiocentesis using 30 G-needles. A small fraction (10 µl) of the blood was collected in a micro-capillary tube and centrifuged on site in a minihaematocrit Compur M1101 (Bayer) for 3 min to record haematocrit (packed blood cell volume, %). The remaining blood was centrifuged (3 min at 8,000 G) and the plasma was separated and stored at -25°C until assays were processed. Plasma sodium concentrations were assessed with an ISE module on a Pentra C 200 (Horiba Medical Ltd) compact chemistry analyzer.

At the end of the procedures (usually <30 min), snakes were released at the location of capture.

Water samples were collected from water bodies where *N. tessellata* were observed foraging and/or in the vicinity of which we

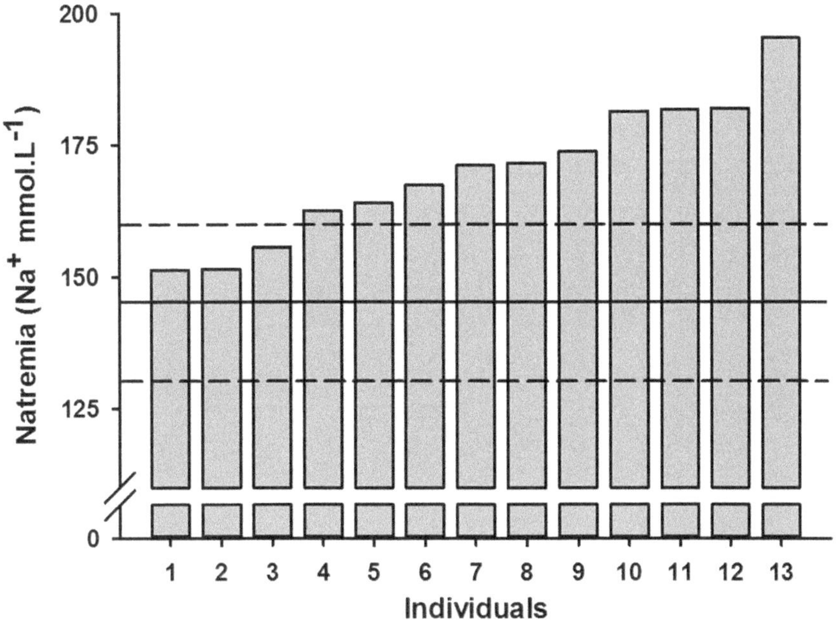

Figure 2. Natremia (plasma sodium concentration) of thirteen free-ranging individual *N. tessellata* captured at Poda Protected Area, Bulgaria. The dashed lines indicate the range of normonatremia (130–160 mmol.l^{-1} [45]) and the horizontal black line indicates mean normonatremia (145 mmol.l^{-1}). For clarity, individuals are ranked by ascending order of natremia.

collected snakes. These sampled stations included 3 ponds within Poda, 2 sites along the coast of the Black Sea as well as 1 site situated at the mouth of the Mandra Lake (Fig. 1). Salinity (‰) was assessed with a Pocket Salt Meter (PAL-ES2, Atago).

Analyses

We quantified a body condition index (BCI) using residual scores from the linear regression between body size (SVL) and body mass (both variables were log transformed for linearity [44]). We excluded individuals with prey in the stomach from the BCI calculations.

Relationships between natremia and possible correlates (BCI, Hct) were investigated using Spearman rank correlations.

Results and Discussion

Free-ranging *N. tessellata* display highly variable plasma sodium concentrations (mean 169.9±13.2 mmol.l^{-1}) ranging from normonatremia (which range from 130 to 160 mmol.l^{-1} in non-mammalian tetrapods [45]) to hypernatremia (up to 195.5 mmol.l^{-1}, Fig. 2). Most individuals (N = 10, 77%) displayed hypernatremia, and only three snakes had values within the range of normonatremia (130–160 mmol.l^{-1} [45]). Classically, deviations of the osmotic balance trigger several behavioural and physiological adjustments in snakes. For instance, dehydrated and/or hypernatremic individuals tend to seclude themselves in well-buffered shelters in order to reduce additional water loss [8,9]. Such behaviour is usually accompanied by a thermal depression and decreased metabolism which result in a strong reduction in activity levels [46,47]. Apparently, the high natremia we recorded did not trigger such adjustments in *N. tessellata*. Indeed, all the individuals we captured were basking in the sun, or actively moving in the open, and tried actively to evade capture. Remnants of food were palpated in three individuals that did not display particularly low plasma sodium (155.6, 167.4 and 173.8 mmol.l^{-1} Na^+ respectively). In addition, BCI was not related to plasma sodium (Spearman rank correlation, $r_s = -0.47$, $p>0.05$), suggesting an absence of long-term effect of hypernatremia on foraging.

The elevated natremia we found in some individuals could be the result of mere dehydration (i.e., the more dehydrated an individual, the more concentrated its body fluids). However, this hypothesis seems unlikely as we did not find any relationship between plasma sodium and BCI (see above) or haematocrit (Spearman rank correlation, $r_s = 0.20$, $p>0.05$), two parameters known to correlate with hydration state [17,48]. Our results rather suggest that free-ranging *N. tessellata* gained salt during their day-to-day activities. Indeed, measurements of environmental salinity in Poda showed that most potential foraging areas were saline (up to 14.3‰, Fig. 1, see also [43]). Likely, Dice snakes foraged in water bodies that were brackish or saline, or indeed at sea (as witnessed by tracks of snakes commuting to the Black Sea, Fig. 1); and gained salt passively through permeable surfaces. Marine snakes display a significant reduction in salt gain and water loss through permeable surfaces [10–12,49], and future studies should assess skin permeability to water and sodium in *N. tessellata* and compare coastal *versus* inland populations.

Voluntary or incidental (e.g., during prey capture) salt water drinking is an additional process that leads to salt gain [15]. Accordingly, marine forms usually display an increased ability to discriminate water salt content and to avoid salt water drinking [12,16,26,50,51]. We do not know whether *N. tessellata* is able to discriminate water salt content and/or to avoid salt water drinking and such issues need to be clarified. In addition to salt water drinking avoidance, many marine taxa can rely on behavioural osmoregulation such as fresh water drinking. Indeed, dehydrated and hypernatremic marine snakes are known to drink large amounts of fresh water when available to restore osmotic balance [9,12,16,25]. Interestingly, two individuals (151.3 and 162.6 mmol.l^{-1} Na^+ respectively) regurgitated copious amounts of fresh water upon capture, suggesting that these individuals have drank shortly before. The variety of aquatic habitats found in Poda (fresh, brackish and salt water, Fig 1) over a small spatial scale,

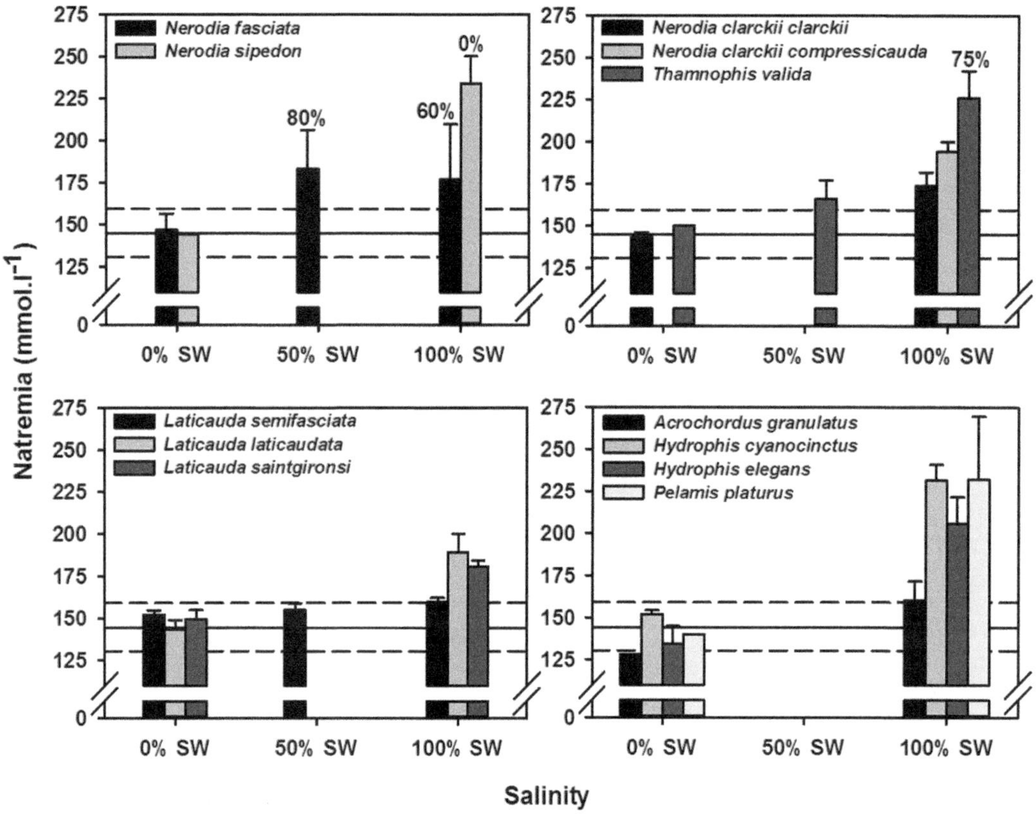

Figure 3. Published data on snake natremia. These data were available from strictly fresh water species (*Nerodia fasciata* and *N. sipedon* [24,26]), salt tolerant species lacking salt glands (*N. clarckii clarckii, N. clarckii compressicauda, Thamnophis valida* [24,26]), amphibious sea kraits with functional salt glands (*Laticauda saintgironsi, L. laticaudata, L. semifasciata* [24,25]) and fully marine sea snakes with functional salt glands (*Acrochordus granulatus, Hydrophis elegans, H. cyanocinctus, Pelamis platurus* [18,19,21–23]. The dashed lines indicate the range of normonatremia (130–160 mmol.l^{-1} [45]) and the horizontal black line indicates mean normonatremia (145 mmol.l^{-1}). Numbers above the bars indicate survival rates (no number = 100%). Data are mean values per species ± SD.

may allow hypernatremic Dice snakes to easily access fresh water, and thus to periodically restore osmotic balance. Accordingly, specific environments characterized by low and/or variable salinity may have facilitated evolutionary transitions to marine life in snakes by allowing regular access to relatively fresh water over short time-scales and decreasing the cost of osmotic maintenance [13].

More generally, a review of plasmatic sodium concentration of snakes experimentally maintained in fresh water or seawater gives additional insights to our results (Fig. 3 and references therein). These data suggest that when kept in fresh water, irrespective of their primary habitat (i.e., fresh- *versus* seawater) all species shared similar normonatremia (~140–150 mmol.l^{-1}, Fig. 3). Similarly, in full-strength seawater, plasma sodium increased in all species regardless from their osmoregulatory attributes (i.e., presence/absence of salt glands, Fig. 3, [25]). Importantly, survival differed among species and treatments, with strictly fresh water species (*Nerodia fasciata* and *N. sipedon*) having decreased short-term (i.e., hours) survival in osmotically challenging treatments (Fig. 3). Survival decreased also for salt tolerant species, although to a lesser extent, with only one species (*Thamnophis valida*) out of three having its short-term survival decreased in full-strength seawater (Fig. 3). The other salt tolerant species survived for several days in full-strength seawater. Survival stayed high (100%) in both groups of marine adapted snakes (Fig. 3). Overall, these patterns seem to indicate that along a continuum of habitats use between fresh- and seawater, snake species display a concomitant physiological tolerance toward high plasma sodium, even in species lacking salt glands (Figs. 2 and 3). In turn, this physiological flexibility would allow reducing detrimental effects of salt gain such as decreased activity levels and decreased short-term survival [26]. Such resistance would allow individuals to periodically access fresh water, and hence to occasionally restore their osmotic balance. Ultimately, in marine species having salt glands, such flexibility would allow excreting excess salt when natremia exceeds high thresholds [22], which would substantially decrease energetic costs linked to salt gland functioning [25].

In conclusion, the combination of these data strongly suggest that the development of a physiological tolerance toward deviations of the osmotic balance (e.g., increased plasma sodium) might have been a critical innovation in the evolution of an euryhaline physiology and may well have preceded the evolution of salt glands. Although only few populations of *N. tessellata* are found in saline environments, our results show that these populations may be salt tolerant, and use saline water bodies despite lacking salt glands. In this respect, *N. tessellata* seems a promising study model (i.e., a marine snake prototype) of the secondary transition to marine life in vertebrates.

Acknowledgments

We thank Bruno Michaud for natremia assays and Leslie Babonis for sharing data on sea krait natremia. Andéaz Dupoué, Frédéric Angelier,

Olivier Lourdais and an anonymous referee provided insightful comments on an earlier draft of the MS.

Author Contributions

Conceived and designed the experiments: FB YVK. Performed the experiments: FB YVK. Analyzed the data: FB YVK. Contributed reagents/materials/analysis tools: FB YVK. Wrote the paper: FB YVK.

References

1. Schmidt-Nielsen K (1983) Animal physiology: adaptations and environments. Cambridge Univ. Press.
2. Ortiz RM (2001) Osmoregulation in marine mammals. J Exp Biol 204: 1831–1844.
3. Peaker M, Linzell J (1975) Salt glands in birds and reptiles. Cambridge Univ. Press.
4. Babonis LS, Brischoux F (2012) Perspectives on the convergent evolution of tetrapod salt glands. Integr Comp Biol 52: 245–256.
5. Mazin J.-M., de Buffrénil V (2001) Secondary adaptation of tetrapods to life in water. Verlag Dr. Friedrich Pfeil.
6. Dunson WA, Mazzotti FJ (1989) Salinity as a limiting factor in the distribution of retiles in Florida Bay: a theory for the estuarine origin of marine snakes and turtles. Bull Mar Sci 44: 229–244.
7. Roe LJ, Thewissen JGM, Quade J, O'Neil JR, Bajpai S, et al.(1998) Isotopic approaches to understanding the terrestrial-to-marine transition of the earliest cetaceans. pp 399–422 in Thewissen JGM, ed. The Emergence of Whales: Evolutionary Patterns in the Origin of Cetacea. New York: Plenum Press.
8. Lillywhite HB, Sheehy III CM, Zaidan III F (2008a) Pitviper scavenging at the intertidal zone: an evolutionary scenario for invasion of the sea. BioScience 58: 947–955.
9. Bonnet X, Brischoux F (2008) Thirsty sea snakes forsake their shelter during rainfall. Austral Ecol 33: 911–921.
10. Dunson WA, Robinson GD (1976) Sea snake skin: permeable to water but not to sodium. J Comp Physiol 108: 303–311.
11. Dunson WA, Stokes GD (1983) Asymmetrical diffusion of sodium and water through the skin of sea snakes. Physiol Zool 56: 106–111.
12. Lillywhite HB, Babonis LS, Sheehy III CM, Tu M-C (2008b) Sea snakes (*Laticauda* spp.) require fresh drinking water: implication for the distribution and persistence of populations. Physiol Biochem Zool 81: 785–796.
13. Brischoux F, Tingley R, Shine R, Lillywhite HB (2012a) Salinity influences the distribution of marine snakes: implications for evolutionary transitions to marine life. Ecography 35: 994–1003.
14. Randall D, Burggren W, French K (2002) Eckert Animal Physiology. W.H. Freeman, New York, USA.
15. Houser DS, Crocker DE, Costa DP (2005) Ecology of water relations and thermoregulation. eLS published online (doi:10.1038/npg.els.0003216).
16. Lillywhite HB, Brischoux F, Sheehy III CM, Pfaller JB (2012) Dehydration and drinking responses in a pelagic sea snake. Integr Comp Biol 52: 227–234.
17. Brischoux F, Rolland V, Bonnet X, Caillaud M, Shine R (2012b) Effects of oceanic salinity on body condition in sea snakes. Integr Comp Biol 52: 235–244.
18. Dunson WA (1968) Salt gland secretion in the pelagic sea snake *Pelamis*. Am J Physiol 215: 1512–1517.
19. Dunson WA, Dunson MK (1973) Convergent evolution of sublingual salt glands in the marine file snake and the true sea snakes. J Comp Physiol 86: 193–208.
20. Dunson WA, Dunson MK (1974) Interspecific differences in fluid concentration and secretion rate of sea snake salt glands. Am J Physiol 227: 430–438.
21. Dunson WA, Dunson MK (1975) The relation between plasma Na concentration and salt gland Na-K ATPase content in the Diamondback terrapin and the Yellow-bellied sea snake. J Comp Physiol 101: 89–97.
22. Dunson WA, Packer RK, Dunson MK (1971) Sea snakes: an unusual salt gland under the tongue. Science 173: 437–441.
23. Duggan RT, Lofts B (1978) Adaptation to freshwater in the sea snake *Hydrophis cyanocinctus*: Tissue electrolytes and peripheral corticosteroids. Gen Comp Endo 36: 510–520.
24. Babonis LS, Miller SN, Evans DH (2011) Renal responses to salinity change in snakes with and without salt glands. J Exp Biol 214: 2140–2156.
25. Brischoux F, Briand MJ, Billy G, Bonnet X (2013) Variations of natremia in sea kraits (*Laticauda* spp.) kept in seawater and fresh water. Comp Biochem Physiol A 166: 333–337.
26. Dunson WA (1980) The relation of sodium and water balance to survival in sea water of estuarine and freshwater races of the snakes *Nerodia fasciata*, *N. sipedon*, and *N. valida*. Copeia 2: 268–280.
27. Murphy JC (2012) Marine invasions by non-sea snakes, with thoughts on terrestrial–aquatic–marine transitions. Integr Comp Biol 52: 217–227.
28. Mebert K (Ed.) (2011) The Dice Snake, *Natrix tessellata*: Biology, Distribution and Conservation of a Palaearctic Species. Mertensiella 18, DGHT, Rheinbach, Germany.
29. Stoyanov A, Tzankov N, Naumov B (2011) Die Amphibien und Reptilien Bulgariens. Edition Chimaira, Frankfurt am Main, Germany.
30. Gruschwitz M, Lenz S, Mebert K, Lanka V (1999) *Natrix tessellata* (Laurenti, 1768) – Würfelnatter. – In: Böhme W (Ed.): Handbuch der Amphibien und Reptilien Europas Band 3/II Schlangen (Serpentes) II:. – AULA-Verlag, Wiesbaden, Germany: 581–644.
31. Baha El Din S (2006) A Guide to the Reptiles and Amphibians of Egypt. American University in Cairo Press, Egypt.
32. Mršić N (1997) Reptiles (Reptilia) of Slovenia. – Institution of Republic of Slovenia for Matters of School, Ljubljana (in Slovene).
33. Tome S (2002) Snakes why do we fear them?!? – Prirodoslovni muzej Slovenije, Ljubljana (in Slovene).
34. Jelić D, Lelo S (2011) Distribution and status quo of *Natrix tessellata* in Croatia, and Bosnia and Herzegovina. Mertensiella 18: 217–224.
35. Žagar A, Krofel M, Govedič M, Mebert K (2011) Distribution and habitat use of dice snakes, *Natrix tessellata*, in Slovenia. Mertensiella 18: 207–216.
36. Ioannidis Y, Mebert K (2011) Habitat preferences of *Natrix tessellata* at Strofylia, northwestern Peloponnese, and comparison to syntopic *N. natrix*. Mertensiella 18: 302–310.
37. Nöllert A, Nöllert C, Ritter A (1986) Einige Beobachtungen zur Herpetofauna der bulgarischen Schwarzmeerküste und Südwestbulgariens (Teil 2 – Die Reptilien). Herpetofauna 8: 30–341.
38. Thieme U (1986) Die Amphibien- und Reptilienfauna der südost- bulgarischen Küste. Teil IV Reptilien III. Sauria 8(3): 3–6.
39. Carlsson M, Kärvemo S, Tudor M, Sloboda M, Mihalca AD, et al. (2011) Monitoring a large population of dice snakes at Lake Sinoe in Dobrogea, Romania. Mertensiella 18: 237–244.
40. Naumov B, Tzankov N, Popgeorgiev G, Stojanov A, Kornilev Y (2011) The dice snake (*Natrix tessellata*) in Bulgaria: distribution and morphology. Mertensiella 18: 288–297.
41. Tuniyev B, Tuniyev S, Kirschey T,Mebert K (2011) Notes on the dice snake (*Natrix tessellata*) from the Caucasian Isthmus. Mertensiella 18: 343–356.
42. Mebert K, Trapp B, Kreiner G, Billing H, Speybroeck J, et al. (2011) Nocturnal activity of *Natrix tessellata*, a neglected aspect of its behavioral repertoire. Mertensiella 18: 234–236.
43. Poda Protected Area Management Plan (2002) Poda Protected Area Management Plan 2002–2010. Bulgarian Society for the Protection of Birds/ Ministry of the Environment and Waters, Sofia, Bulgaria [in Bulgarian].
44. Bonnet X, Naulleau G (1995) Estimation of body reserves in living snakes using a body condition index (B.C.I.). In: Llorente G, Santos X, Carretero MA, editors. Scientia Herpetologica. Barcelona: Asociacion Herpetológica Española. p. 237–240.
45. Campbell TW (2004) Blood Biochemistry of Lower Vertebrates. In: 55th Annual Meeting of the American College of Veterinary Pathologists (ACVP) & 39th Annual Meeting of the American Society of Clinical Pathology (ASVCP), ACVP and ASVCP (Eds.). American College of Veterinary Pathologists & American Society for Veterinary Clinical Pathology, Middleton WI, USA.
46. Ladyman M, Bradshaw D (2003) The influence of dehydration on the thermal preferences of the Western tiger snake, *Notechis scutatus*. J Comp Physiol B 173: 239–246.
47. Ladyman M, Bradshaw D, Bradshaw F (2006) Physiological and hormonal control of thermal depression in the tiger snake, *Notechis scutatus*. J Comp Physiol B 176: 547–557.
48. Peterson CC (2002) Temporal, population, and sexual variation in hematocrit of free-living desert tortoises: correlational tests of causal hypotheses. Can J Zool 80: 461–470.
49. Lillywhite HB, Menon JG, Menon GK, Sheehy III CM, Tu M-C (2009) Water exchange and permeability properties of the skin in three species of amphibious sea snakes (*Laticauda* spp.). J. Exp. Biol. 212: 1921–1929.
50. Dunson WA (1986) Estuarine populations of the Snapping turtle (*Chelydra*) as a model for the evolution of marine adaptation in reptiles. Copeia 3: 741–756.
51. Kidera N, Mori A, Tu M-C (2013) Comparison of freshwater discrimination ability in three species of sea kraits (*Laticauda semifasciata*, *L. laticaudata* and *L. colubrina*). J Comp Physiol A 199: 191–195.

Changes in Carbon Pool and Stand Structure of a Native Subtropical Mangrove Forest after Inter-Planting with Exotic Species *Sonneratia apetala*

Weizhi Lu[1,2], Shengchang Yang[1], Luzhen Chen[1], Wenqing Wang[1], Xiaona Du[1], Canmou Wang[1], Yan Ma[1], Guangxuan Lin[3], Guanghui Lin[2,4]*

1 Key Laboratory of the Ministry of Education for Coastal and Wetland Ecosystems, School of Life Sciences, Xiamen University, Xiamen, Fujian, China, **2** Division of Marine Sciences and Technology, Graduate School at Shenzhen, Tsinghua University, Shenzhen, Guangdong, China, **3** Administrative Bureau of Zhanjiang National Mangrove Nature Reserve, Zhanjiang, Guangdong, China, **4** Ministry of Education Key Laboratory for Earth System Modeling, Center for Earth System Science, Tsinghua University, Beijing, China

Abstract

In this study, we compared stand structure, biomass and soil carbon pools, and litterfall production between a mixed mangrove forest consisting of *Aegiceras corniculatum* inter-planted with the exotic *Sonneratia apetala* and a native monospecific forest dominated by *A. corniculatum* in the intertidal area of Zhanjiang, Guangdong Province, southeast China. The goal of this study was to test the hypothesis that inter-planting fast growing exotic mangrove *S. apetala* into subtropical native mangrove forests will significantly increase C sequestration. Although the tree heights and basal diameters of *S. apetala* were significantly higher than those of *A. corniculatum*, the density of the 12-year-old *S. apetala* trees in the mixed forest was much smaller than that of *A. corniculatum* in the monospecific forest. In contrast to several previous studies on *S. apetala* forests planted directly on mangrove-free mudflats, the mixed mangrove forest showed no significant difference in either standing biomass or soil carbon pools from the native monospecific mangrove forest ($p = 0.294$ and 0.073, respectively) twelve years after inter-planting with *S. apetala*. Moreover, carbon cycling was likely speeded up after inter-planting *S. apetala* due to higher litterfall input and lower C/N ratio. Thus, inter-planting fast-growing *S. apetala* into native mangrove forest is not an effective way to increase carbon sequestration in this subtropical mangrove forest. Given that exotic plant species may exert negative impact on native mangrove species and related epifauna, this fast-growing mangrove species is not suitable for mangrove plantation projects aiming mainly at enhancing carbon sequestration.

Editor: John F. Valentine, Dauphin Island Sea Lab, United States of America

Funding: This study was supported financially by National Natural Science Foundation of China (30930017) and by National Key Basic Research Program (or 973 program) of Ministry of Science and Techology, China (2013CB556601). The funders had no role in study design, data collection and analysis, decision to publish, or preparation of the manuscript.

Competing Interests: The authors have declared that no competing interests exist.

* E-mail: lingh@mail.tsinghua.edu.cn

Introduction

Mangrove wetlands have great ecological and economic value, including high primary productivity, effective carbon (C) storage, high epifaunal diversity and great benefits for aquaculture [1–5]. However, there is increasing concern about the continued loss and degradation of mangroves, as 35% to 86% of the global mangrove area has been lost during the last several decades [6–9]. In addition, as international climate agreements emphasize Reduced Emissions from Deforestation and Degradation (REDD+) as a key option for mitigating climate change [10], attention has been drawn to C sequestration potentials of mangrove forests, salt marshes and seagrass beds, as well as measures to enhance these C sinks [10–12].

Afforestation and reforestation can be effective methods for increasing forest ecosystem C sequestration [13,14]. Large-scale efforts have been made to restore degraded mangroves and create new mangrove forests around the world [15,16]. In the tropical and subtropical areas of mainland China, plantation is a common method for restoring mangrove forests [17]. Native to Bangladesh, *Sonneratia apetala* Buch. Ham, a fast-growing mangrove tree species of Sonneratiaceae family, has been widely used for mangrove restoration projects in many locations along the southeastern coasts of China during last three decades. It was estimated that the total area of *S. apetala* plantations in China reached 3800 ha, which accounts for more than 50% of total replanted mangrove area [17]. However, there were still debates on whether *S. apetala* in China is an invasive species or a great restoration species [18].

In most cases, *S. apetala* afforestation projects were implemented on mudflats with low salinity, where native mangrove forests had been removed or destroyed [18]. In Guangdong and Fujian provinces, however, *S. apetala* trees were often inter-planted inside native mangrove forests for landscape renovation to make the appearance prettier because of *S. apetala* tall and "willow-like" canopy. Moreover, this exotic mangrove species has been naturally spread into many native mangrove forests in these regions since it was introduced to China in late 1970's [18]. Previous researches have studied carbon accumulation potentials of *S. apetala* plantations in destroyed mangrove forests. For example, Ren et al. demonstrated that *S. apetala* plantations sequestered large amounts of C in both biomass and sediments [19]. Other studies also showed that *S. apetala* has higher rates of C accumulation in biomass and sediments than many native mangrove species [19–22]. However, these previous studies focused only on monoculture plantations of *S. apetala* on mudflats. Little was known about whether inter-planting *S. apetala* into native mangrove forests can also increase C accumulation. In this study, we hypothesized that planting this exotic fast-growing mangrove species into native mangrove forests should have greater potentials to increase C accumulation than native mangroves.

In the present study, we compared stand structure, C pools in standing biomass and sediments, and litterfall production between a subtropical mangrove forest dominated by native *Aegiceras corniculatum* (L.) Blanco and a mixed mangrove forest of *A. corniculatum* inter-planted with the exotic *S. apetala*. The objective of this study was to test the hypothesis that inter-planting fast growing exotic mangrove species such as *S. apetala* into subtropical mangrove forests will significantly increase C accumulation. *A. corniculatum* of the Myrsine family, commonly known as black mangrove, is a species of mangrove shrub with multiple stems, which has a large distribution area in southeast China.

Materials and Methods

Study site description

The study site (21.5676°N, 109.7562°E) is located at the confluence of Liangguang River and Gaoqiao River, northeast of the Beibu Gulf, in Guangdong Province. The Gaoqiao mangrove forest is at the core zone of Zhanjiang Mangrove National Nature Reserve (Fig. 1), the largest mangrove nature reserve in China with a total mangrove area of approximately 1200 ha.

The local climate is characterized by strong seasonal variation with high precipitation. According to the data from our eddy flux tower (about1.0 km apart from the study site) (Fig. 1), the annual precipitation at our study site was 1168 mm in 2010, and the air temperature ranged from 8.7°C to 37.3°C, with an annual mean of 23.2°C (Fig. 2). The study site experiences regular diurnal tides, with an inundation once during each lunar day, and the average tide range is 1.3 m but the highest astronomical tides can reach up to 1.9 m above the soil surface.

The dominant mangrove species at the study site was *A.corniculatum*, accompanying by *Bruguiera gymnorrhiza* (L.) Lamk, *Rhizophora stylosa* Griff and *Kandelia obovata* Sheue, Liu et Yong, forming separate pure patches in the mixed forest. Most *A. corniculatum* trees were in dwarf form, with a mean canopy height of about 2.2 m. The age of these mangrove trees was estimated to be greater than 60 years. In 1999, the seedlings of *S. apetala* were planted into a monospecific *A. corniculatum* tree stand of about 1 ha and then developed gradually into a mixed forest. The initial seedling density of *S. apetala* was set at 1,667 seedlings ha^{-1} and the average row width was approximately 1 m. At the time of inter-planting, the mean height of *S. apetala* seedlings was about 40 cm (personal observation by Guangxuan Lin, one of the co-authors of this paper and the staff at Zhanjiang National Mangrove Nature Reserve).

Forest structure surveys and biomass measurements

We investigated two mangrove forests for this study after applying for the permission of the Zhanjiang Mangrove National Nature Reserve. The first forest was a native mangrove forest (hereafter 'monospecific forest'), composed of only ~60 year old *A. corniculatum* trees. The other was a mixed forest, composed of 12 year old *S. apetala* trees inter-planted into~60 year old *A. corniculatum* forest (hereafter 'mixed forest'). Five 10 m ×10 m survey plots (as replicates for each forest type) were selected along the middle line of each forest at 50 m intervals (Fig. 1).

In January 2011, we recorded canopy height (*H*) and diameter at breast height (*DBH* or *D* in this paper) of each *S. apetala* individual tree in all five quadrats as mentioned above. Because of their extensive root systems, excavating trees of *S. apetala* would have been very destructive to surrounding trees. Instead, the relative growth equations obtained from a similar *S. apetala* forest in northwest Leizhou Bay, about 75 km south the study site [19], were used to estimate the aboveground biomass (W_{AGB}), belowground biomass (W_{BGB}) and total biomass (W_{Total}) using their *D* and *H* values:

$$W_{AGB} = 0.280(D^2H)^{0.693} \quad (1)$$

$$W_{BGB} = 0.038(D^2H)^{0.759} \quad (2)$$

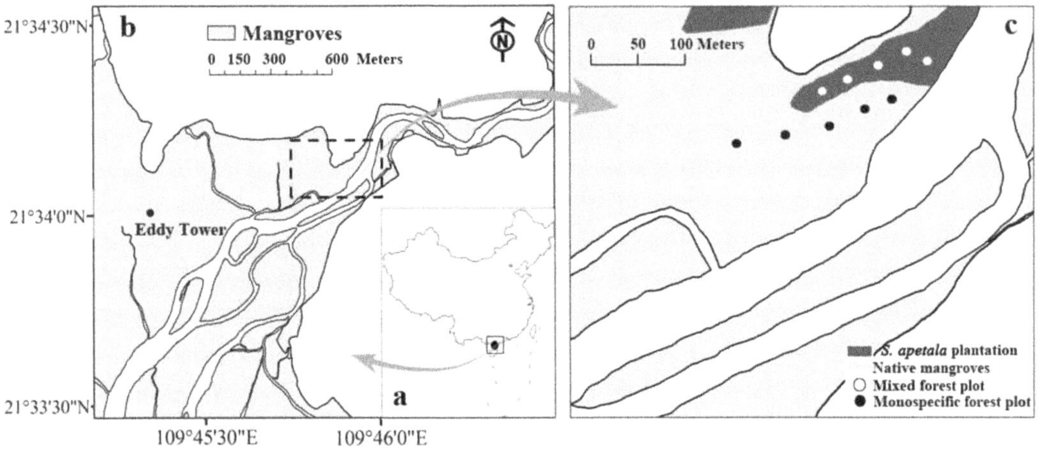

Figure 1. The location of Gaoqiao mangrove research station (a and b) and the experimental plots in the mixed and monospecific mangrove forests (c).

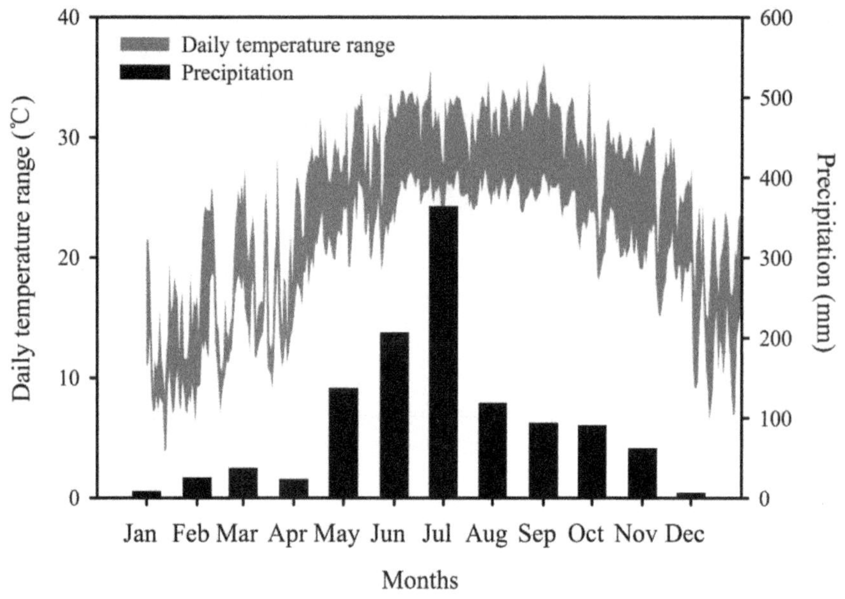

Figure 2. Daily air temperature range and monthly precipitation in 2010 for the Gaoqiao mangrove research station.

$$W_{\text{Total}} = 0.312(D^2H)^{0.705} \quad (3)$$

One shortcoming of this approach for W_{BGB} measurements was that the aerial roots were not separated from normal roots [19], even though they functioned differently and might have very different carbon content.

We also surveyed all trees of *A. corniculatum* in both the monospecific native mangrove forest and the mixed mangrove forest, and recorded the tree heights. Although the stems of *A. corniculatum* trees were usually higher than 1.3 m, most of them had more than one branch at the height of 1.3 m. Thus, we measured the basal diameters and basal areas at the height of 5 cm above the soil surface for *A. corniculatum* forest structure survey. We didn't use relative growth equations to estimate the biomass of *A. corniculatum* because we had no suitable allometric growth equations for shrub mangrove species with multiple stems. Instead, three1 m ×1 m quadrats were selected in the center of each forest for the measurements of *A. corniculatum's* aboveground biomass (AGB) and belowground biomass (BGB). The AGB included the total weight of leaves, branches, stems and trunk, but the biomass of the propagules (semi-viviparous fruits) was excluded since it was only a very small proportion of total biomass, although the propagules could be an important component of litterfall during the reproductive months (see below). For the BGB measurements, the soils within a soil cube of 1 m^3 were excavated, and then carefully washed over a fine screen to collect all the roots. For the mixed mangrove forest, the roots of *S. apetala* were removed by hands because their biomass was already estimated by the allometric growth equations mentioned above. These roots had more spongy cortex than those of *A. corniculatum*, facilitating their identification between each other. The fresh weights of AGB and BGB components from each quadrat were measured in the field using a balance (MS24KLIPE, Metteler-Toledo, Greifensee, Switzerland). The biomass of each component was then calculated using the above data.

Biomass C/N ratio and carbon pool measurements

About 100–200 g fresh weight of each biomass component (leaves, branches and roots) from five trees of each species in the two forest-types were sub-sampled for water, carbon (C) and nitrogen (N) content analyses. We used the mean value of five trees in the same survey plot for each forest type as one replicate, so there were only five replicates for each species or forest type. These samples were oven dry at 60°C for at least 72hours to a constant weight, and then their dry weights were measured for water content calculations. The dried samples were tested for C and N contents using an elemental analyzer (Vario EL III Elemental Analyzer, Hanau, Germany). The C storages in all biomass components of two mangrove species were calculated separately using their biomass and C content data, and then summed up for both the monospecific and mixed mangrove forest.

Litterfall collection

Litterfall was collected using 1 m^2 round baskets covered with 1.5 mm fiberglass mesh. One basket was set up in each plot, with a total of 10 baskets for the two mangrove forests. The baskets were placed 1.5 m above the soil to avoid immersion by tidal water, and litterfall was collected twice a month from January 1, 2010 to December 31, 2010. Leaves, branches, twigs and fruits (or propagules, i.e. semi-viviparous fruits, for *A. corniculatum*) were separated manually and dry weight was determined using an electronic balance (ML2001, Metteler-Toledo, Greifensee, Switzerland) after oven-dry at 60°C for at least 72 hours to a constant weight. The components of litterfall were not separated between two species in the mixed mangrove forest due to the labor shortage in the field. Monthly litterfall production was calculated using the accumulated all litter during each month.

Soil sampling and property measurements

Soil samples down to 60 cm depth were collected using a soil core sword (C040903, Eijkelkamp, Holland) in July 2010. Five cores were randomly taken in each plot, for a total of 50 cores collected for the two mangrove forests. Each core was cut into three sections: 0–20 cm, 20–40 cm and 40–60 cm. Fixed-volume cores were also taken in each section with a soil-cutting ring to measure bulk density. These soil samples were dried at 60°C for at least 72 hours to a constant weight and the dry weights were taken using the Metteler-Toledo's ML 2001 electronic balance. Soil bulk density was calculated as the ratio of soil dry weight to the volume of the soil-cutting ring. The dried samples were ground with a ball grinding mill (Retsch Mixer Mill MM400, Verder, Germany) and passed through an 80-mesh sieve. A 10-mg subsample was taken from each soil sample and then measured for its C and N contents as described above.

Statistical analyses

For each parameter, we calculated the mean and standard deviation (SD) of the five replicates for each forest. The differences in tree density, basal area, height, biomass, C/N ratio among forest types were evaluated using one-way ANOVA test. The assumptions for the one-way ANOVA test, including normal distribution and homogenous variance of the data, were first tested. Since the tree density, basal area, height and leaf C/N ratio had an inhomogeneous variance, we log-transformed these values before the ANOVA tests. Repeated measure ANOVA was used to examine the seasonal change in litterfall production over time in two mangrove forests. The differences in the biomass and soil C pools between the two forest types were evaluated by *t*-test after a power analysis was conducted. All statistical analyses were performed using SPSS version 13.0 (SPSS Inc., Chicago, USA). Power analysis was performed using PASS software version 2008 (Kaysville, Utah, USA).

Results

Forest structure and biomass

The tree height of the 12 year-old *S. apetala* was significantly higher ($p<0.001$) than *A. corniculatum* (Table 1). Both basal area and tree height of *A. corniculatum* in the mixed forest were significantly lower (both $p<0.001$) than those in the monospecific forest (Table 1). Moreover, the basal area and tree height of *S. apetala* were significantly higher ($p<0.001$) than those of *A. corniculatum* in the same forest (Table 1). There was no significant difference ($p=0.425$) in the tree density of *A. corniculatum* between the mixed and monospecific forests, which was about 30 times greater than that of *S. apetala* in the mixed forest (Table 1).

The total amounts of biomass in the monospecific and mixed forests were comparable at 102.54±9.83 and 126.33±32.19 Mg ha^{-1}, respectively (*t*-test, $p=0.280$)(Table 2). However, the total biomass of *A. corniculatum* in the monospecific forest was four times higher than that in the mixed forest (Table 2). The 12-year-old *S. apetala* comprised the majority of the biomass in the mixed forest (about 77.13%) (Table 2). The biomass components of *S. apetala* had significantly lower C/N ratios than their counterparts of *A. corniculatum* except the branches ($p=0.084$, 0.030 and 0.020 for the branches, leaves and roots, respectively) (Table 3). However, the C/N ratios of *A. corniculatum* biomass were not significantly different between the mixed and monospecific mangrove forests ($p=0.689$, 0.439 and 0.379 for the branches, leaves and roots, respectively).

Litterfall production

Annual litterfall production during 2010 was 1322.5 and 574.7 g m^{-2} for the mixed and monospecific forest, respectively (Fig. 3). Litterfall from the mixed forest showed a significant seasonal dynamic ($p<0.001$), with a gradual increase from January, reaching a peak in May and then maintaining relatively high production from May to October (Fig. 3). The two peak litterfall production periods were May and October (Fig. 3). However, litterfall production of *A. corniculatum* did not show obvious seasonal pattern except for a peak production in July-August (Fig. 3). Moreover, there were significant differences in the litterfall production between two forests over months of the year (repeated measure ANOVA, $p<0.001$).

Soil properties

There was no significant difference in bulk density differ between two forests at each depth (Fig. 4a). However, the soil C concentration in the mixed forest was significantly higher than in the monospecific forest at both 0–20 cm and 40–60 cm layers (Fig. 4b). There was no significant difference in C density between the two forest types at both 0–20 cm and 20–40 cm layers (Fig. 4c).The C/N ratio of top layer soils in the mixed mangrove forest was significantly lower ($p<0.001$) than that in the monospecific forest (Table 3).

Carbon storage in biomass and soil

Total C storage in biomass was 45.73±4.43 Mg C ha^{-1} and 53.03±9.49 Mg C ha^{-1} for the monospecific forest and mixed forests, respectively, and the total C storage in soil was 68.15±6.81 and 78.76±5.13 Mg C ha^{-1}, respectively (Fig. 5). However, there was no significant difference in biomass C storage between the two mangrove forests (*t*-test, $p=0.294$; with a result of 0.805 from the power analysis) (Fig. 5). Furthermore, there was no significant difference in soil C storage between the two mangrove forests (*t*-test, $p=0.073$; with a result of 0.761 from the power analysis) (Fig. 5).

Discussion

C sequestration potential of *S. apetala* forests

Compared with native mangrove species such as *A. corniculatum*, *Avicennia marina*, *K. obovata* and *B. gymnorrhiza*, the exotic mangrove species *S. apetala* has much greater biomass amount and C

Table 1. Selected properties of soils and tree canopy structure for two different mangrove forests (mean±SD).

Forest type	Soil texture	Species	Age (yrs)	Height (m)	Basal diameter (cm)	Basal area (m^2 ha^{-1})	Density (ind ha^{-1})
Mixed forest	Silty clay	*S. apetala*	12	13.64±0.88^c	15.86±3.31^c	73.15±9.80^b	1866.67±568.62^a
		A. corniculatum	~60	1.74±0.08^a	2.94±0.16^a	38.48±5.43^a	56846.67±10826.27^b
Monospecific forest	Silty clay	*A. corniculatum*	~60	2.02±0.05^b	4.02±1.25^b	80.82±12.47^b	63200±6053.10^b
		$F_{2,6}$		1813.27	58.01	16.29	66.40
		p value		<0.001	<0.001	<0.001	<0.001

Different lower case letters indicate significant differences at $p<0.05$ between two mangrove species or the same species between two forest types.

Table 2. Aboveground biomass (AGB) and belowground biomass (BGB) of two different mangrove forests (mean±SD) (Unit: Mg ha^{-1})*.

Forest type	Species	Stem or Trunk	Branch	Leaf	AGB	BGB	Total
Mixed forest	*S. apetala*	49.27 ± 15.54^b	21.58 ± 4.87^b	4.39 ± 0.54^a	79.28 ± 21.39^b	18.47 ± 5.42^b	97.45 ± 26.72^b
		(50.12±2.39%)	(22.36±1.25%)	(4.57±0.32%)	(81.42±0.41%)	(18.87±0.43%)	
	A. corniculatum	11.67 ± 4.28^a	2.84 ± 0.86^a	2.56 ± 0.37^a	17.07 ± 5.47^a	11.81 ± 0.65^a	28.88 ± 5.47^a
		(39.37±8.27%)	(9.68±1.25%)	(8.94±0.60%)	(57.99±8.83%)	(42.04±9.28%)	
Monospecific forest	*A. corniculatum*	58.75 ± 5.94^b	14.85 ± 2.71^b	9.58 ± 1.20^b	83.17 ± 9.83^b	19.37 ± 2.55^b	102.54 ± 9.83^b
		(57.27±0.41%)	(14.40±1.32%)	(9.31±0.30%)	(80.99±1.86%)	(19.17±4.37%)	
$F_{2,6}$		18.91	25.54	48.50	21.19	4.232	18.17
p value		0.003	0.001	<0.001	0.002	0.071	0.003

*Here AGB did not include the biomass of *A. corniculatum* propagules (semi-viviparous fruits), while the BGB of *S. apetala* included all aerial roots. Different lowercase letters indicate significant differences at $p<0.05$ between two mangrove species or the same species between two forest types. The values in the parentheses were the percentage of each component biomass to the total biomass for a given species.

accumulation capacity [19–23]. For example, Ren et al. observed that the total C stock in a 10-year-old *S. apetala* plantation in Leizhou, Guangdong was 49.0 Mg C ha^{-1} [19], while Liao et al. reported that the total aboveground biomass in a 10-year-old *S. apetala* plantation in Haikou, Hainan reached 82.1 Mg DW ha^{-1} [20]. Chen et al. also demonstrated experimentally that at least during the first 25 months the total C stock in *S. apetala* planting in Shenzhen, Guangdong was up to 18.74 Mg C ha^{-1} [24]. Because of such high C sequestration capacity, this fast-growing exotic mangrove species from Bangladesh is often recommended for coastal afforestation or restoration projects aiming at increasing C sink [19,22], even though it potentially becomes an invasive species in the southern coast region of China and posses serious threats to native mangrove forests [18].

In contrast to these above studies where the *S. apetala* trees or seedlings were planted directly on mangrove-free mudflats, the C storage did not increase 12 years after the *S. apetala* seedlings was

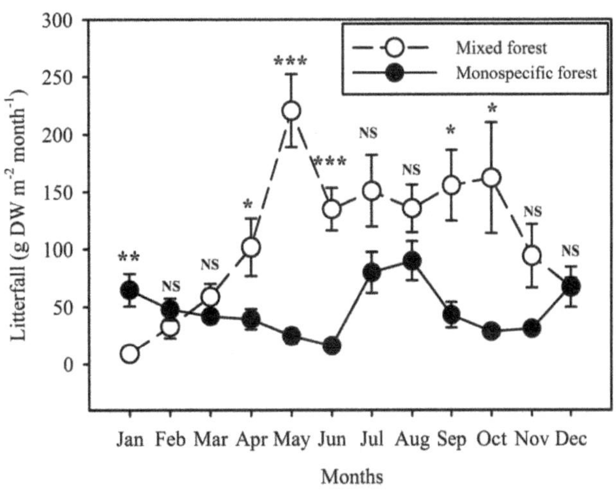

Figure 3. Monthly litterfall production for the *A. corniculatum* monospecific mangrove forest and the mixed mangrove forest of *S. apetala+A. corniculatum* from January to December 2010 (mean±SD; n = 5). Significant differences at each month between two forests are indicated by *$p<0.05$, **$p<0.01$ and ***$p<0.05$, while NS indicates not significant difference at $p>0.05$. Note that in the mix forest the litter from two mangrove species were not separated.

inter-planted into the native mangrove forest of *A. corniculatum* forest in the present study. Although the *S. apetala* trees were significantly taller and bigger than the native mangrove species *A. corniculatum*, the C storage in the biomass of the mixed forest was only slightly increased but not significant when compared to the native monospecific mangrove forest of *A. corniculatum*. We found that the biomass of the native mangrove *A. corniculatum* in the mixed forest was just a quarter of the same species' biomass in the monospecific forest due to significantly smaller basal diameter, lower basal area and lower tree height. Moreover, the density of the 12-year-old *S. apetala* trees was just about a thirtieth of that for the *A. corniculatum* trees in the monospecific forest. The total C storage in the mixed forest did not show significant increase after inter-planting *S. apetala* trees because the increase in biomass C accumulation by *S. apetala* was almost compensated by the decrease in plant biomass of *A. corniculatum* trees underneath them.

Most previous studies focused on C accumulation in biomass following restoration, neglecting C accumulation in soils, which could be very important in determining long-term C sequestration [5,25]. Our results showed that C accumulation in the soil in these subtropical mangrove forests was greater than that in the biomass, which was consistent with a previous study [10]. A previous study showed that *S. apetala* monoculture plantations could sequestrate significant amounts of C in sediments probably due to the increasing input of dead roots and litterfall [19]. However, Chen et al. found no significant increase in soil carbon-storage in *S. apetala* monoculture forests [24]. The study of Chen et al. only measured the C accumulation in the top 20 cm soils and the experiment lasted less than 3 years after the seedlings were planted, which could substantially underestimated the total C storage in the sediments. In the present study, we investigated the C pools in the sediments down to 60 cm deep and focused on *S. apetala* trees of 12 years old, but we still did not find significant increase in C storage in the sediment after inter-planting *S. apetala* into existing native mangrove forest. Thus, it will take much longer time for the exotic mangrove plants to increase C accumulation in sediments through the increasing input of dead roots and litterfall.

Change in C turnover following introduction of exotic mangrove species

Previous studies demonstrated that, following the invasion of alien species, ecosystems may become more fragile and unstable due to unstable C pools and fluxes [26,27]. In most cases, litterfall

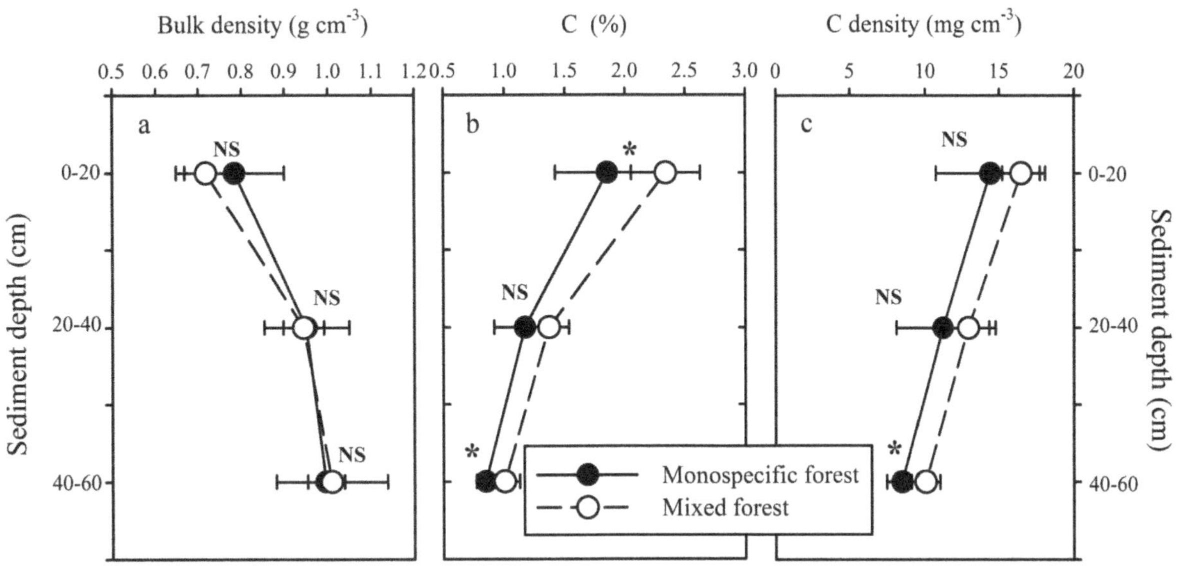

Figure 4. Bulk density (a), carbon concentration (b), and carbon density (c) at different soil depths for a mixed mangrove forest and a monospecific mangrove forest (mean±SD; n = 5). Significant differences between two forests are indicated by *$p<$ 0.05, **$p<0.01$ and ***$p<0.05$, while NS indicates no significant difference at $p>0.05$.

production was significantly higher in the alien forest than native forests [28–30]. This phenomenon was confirmed by the comparison in the litterfall production between alien *S. apetala* and native mangrove species in China [31]. Zan et al. observed an annual litterfall production of 1660 g DW m^{-2} for a *S. apetala* forest in Shenzhen [31], while Han et al. reported even higher litterfall production (1895 g DW m^{-2}) in a six- to seven-year-old forest of *S. apetala* in Leizhou, Guangdong [32]. The results for the litterfall production of *S. apetala* forest from the present study were comparable with these values, but much greater than those of many native mangrove forests in the same region (average annual litterfall production was only 1049±295 g DW m^{-2}) [33–38]. Thus, our results indicated that inter-planting *S. apetala* seedlings into native mangrove forest could supply significantly higher amount of litterfall to mangrove soils, which would speed up the C cycling. In addition, C and N interactions play a key role in determining the speed of C decomposition and thus C loss, and lower C/N ratio is often associated with faster litter decomposition [39,40]. In the present study, we found that *S. apetala* tissues had significantly lower C/N ratio than those of *A. corniculatum*, with the exception of the branch.

From significantly higher litterfall production and lower biomass C/N ratio associated with the exotic mangrove species *S. apetala*, we can conclude that inter-planting *S. apetala* trees into native mangrove forest can substantially increase the speed of C cycling, which may lead to C loss from the mangrove forests.

It is worthy pointing out that the leaf biomass in the monospecific stand was higher than in the mixed stand (native and non-native together). There are three reasons for this phenomenon. First, the litterfall in the mixed forest composed of more branches and fruits mainly from *S. apetala* than the monospecific forest. Secondly, the life-span of *A. corniculatum* leaves was more than 24 months, much longer than that of *S. apetala* leaves (about 6.5 months) [41,42].Thirdly, the biomass allocation in *A. corniculatum* changed significantly after the *S. apetala* inter-planting, with 80.99±1.86% of total biomass allocated to aboveground biomass in the monospecific native mangrove forest but only 57.99±8.83% to the same biomass component in the mixed forest (Table 2). The reduced allocation to AGB after *S. apetala* inter-planting came from all biomass components, with the stem (trunk) biomass decreased the most (Table 2).

Table 3. C/N ratios of soils and plant biomass for two types of mangrove forests (mean±SD).

Forest type	Soil	Species	Branch	Leaf	Root
Mixed forest	17.04±1.23[a]	*S. apetala*	96.47±14.37[a]	31.15±2.58[a]	163.3±23.71[a]
		A. corniculatum	197.06±22.75[a]	53.01±7.73[b]	395.06±65.70[b]
Monospecific forest	20.71±2.75[b]	*A. corniculatum*	177.13±76.80[a]	49.15±0.83[b]	324.86±104.19[ab]
		$F_{2,6}$	3.86	26.53	8.08
t_4	-4.76	*p* value	0.084	0.030	0.020
p value	0.001				

Different lowercase letters indicate significant differences at $p<0.05$ between two mangrove species or the same mangrove species between forest types.

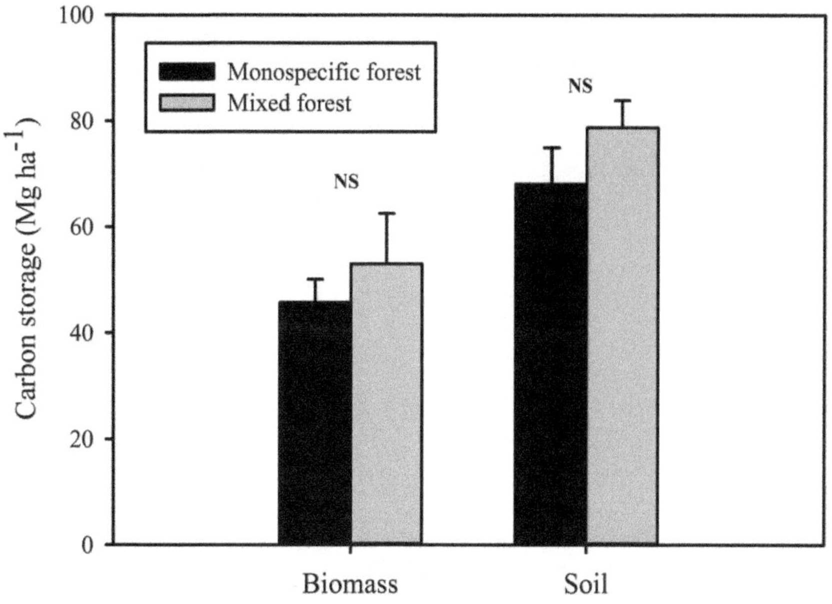

Figure 5. Comparison in the carbon storage of biomass and soils between a mixed mangrove forest and a monospecific mangrove forest. Significant differences between two forests are indicated by $^{*}p< 0.05$, $^{**}p<0.01$, $^{***}p<0.05$, while NS indicates not significant difference at $p>0.05$.

Implications for managements of mangrove restoration projects

Mangrove forests are recognized among the most productive and C rich ecosystems in the world, having critical ecological resources and providing great ecosystem services to human beings [5,11,43]. However, mangrove forests are being destroyed at an average annual rate of 1–2% [1], and the situation is much worse in China, with almost half of national mangrove forests being destroyed or degraded [17]. Since the 1990s, the Chinese government has invested greatly to reforest and afforest mangroves to increase the coastal vegetation cover, to protect the shoreline from tidal surges, and to conserve biodiversity along the south-eastern coastlines of China [17,18].

The planting of *S. apetala* is a core approach to restore mangrove wetlands and thus increase coastal C sequestration in China [17,19]. However, when *S. apetala* trees are inter-planted into native mangrove forests, interspecific competition is not taken into consideration in most of the reforestation project [24]. Our study found that *S. apetala* had a largely negative effect on native mangroves such as *A. corniculatum* possibly because of shading effects. After the exotic *Sonneratia* trees were planted in the native mangrove forest, lots of pre-existing trees died off because of shading effects. The fast-growing *Sonneratia* trees quickly grew over the existing *Aegiceras* tree canopy shading the pre-existing trees, and then new seedlings of *Aegiceras* grew under the canopy of *Sonneratia* trees, which might have caused them having smaller basal diameters than the trees in the nearby intact native mangrove forest. Unfortunately there was no data collection before the *Sonneratia* planting to verify this process, but it was quite possible according to the field observations by Guangxuan Lin. In a separate study on the same sites of this study, Cai et al. found that there were quite different species composition, animal density and secondary productivity between the *S. apetala* and *A. corniculatum* forests [44]. Thus, more studies are needed to understand the overall impact of alien species on native plant and animal communities of mangrove forests. Given that exotic *S. apetala* may exert negative impact on native mangrove species with only marginal increases in C sequestration, we would not recommend this fast-growing exotic mangrove species for mangrove reforestation or afforestation projects aiming at enhancing C sequestration.

Acknowledgments

The authors would like to thank Yihui Zhang and Jianxiang Feng for their assistance throughout the study. We also thank Abby Lunstrum, Yvonne Rogers and Guozhong Wu for editing the English of our manuscript.

Author Contributions

Conceived and designed the experiments: WL SY LC WW Guanghui Lin. Performed the experiments: WL XD CM YM Guangxuan Lin Guanghui Lin. Analyzed the data: WL SY Guanghui Lin. Contributed reagents/materials/analysis tools: WL Guanghui Lin. Wrote the paper: WL Guanghui Lin. Language proofread: SY LC WW.

References

1. Alongi DM (2002) Present state and future of the world's mangrove forests. Environmental Conservation 29: 331–349.
2. Bouillon S, Borges AV, Castaneda-Moya E, Diele K, Dittmar T, et al. (2008) Mangrove production and carbon sinks: a revision of global budget estimates. Global Biogeochemical Cycles 22: GB003052.
3. Sathirathai S, Barbier EB (2001) Valuing mangrove conservation in southern Thailand. Contemporary Economic Policy 19: 109–122.
4. Walters BB, Ronnback P, Kovacs JM, Crona B, Hussain SA, et al. (2008) Ethnobiology, socio-economics and management of mangrove forests: a review. Aquatic Botany 89: 220–236.
5. Donato DC, Kauffman JB, Murdiyarso D, Kurnianto S, Stidham M, et al. (2011) Mangroves among the most carbon-rich forests in the tropics. Nature Geoscience 4: 293–297.
6. Duke NC, Meynecke JO, Dittmann S, Ellison AM, Anger K, et al. (2007) A world without mangroves? Science 317: 41.

7. Giri C, Ochieng E, Tieszen LL, Zhu Z, Singh A, et al. (2011) Status and distribution of mangrove forests of the world using earth observation satellite data. Global Ecology and Biogeography 20: 154–159.
8. Gong P, Niu ZG, Cheng XA, Zhao KY, Zhou DM, et al. (2010) China's wetland change (1990–2000) determined by remote sensing. Science In China Series D-earth Sciences 53: 1036–1042.
9. Irving AD, Connell SD, Russell BD (2011) Restoring coastal plants to improve global carbon storage: reaping what we sow. PLoS Biology 6: e18311.
10. Yee SM (2010) REDD and BLUE carbon: carbon payments for mangrove conservation. Capstone advisory committee final capstone project signature form: MAS Marine Biodiversity and Conservation Capstone Project.
11. Alongi DM (2011) Carbon payments for mangrove conservation: ecosystem constraints and uncertainties of sequestration potential. Environmental Science and Policy 14: 462–470.
12. McLeod E, Chmura GL, Bouillon S, Salm R, Bjork M, et al. (2011) A blueprint for blue carbon: toward an improved understanding of the role of vegetated coastal habitats in sequestering CO_2. Frontiers in Ecology and the Environment 9: 552–560.
13. Li DJ, Niu SL, Luo YQ (2012) Global patterns of the dynamics of soil carbon and nitrogen stocks following afforestation: a meta-analysis. New Phytologist 195: 172–181.
14. Metz B (2007) Climate change 2007: mitigation of climate change: contribution of Working Group III to the Fourth Assessment Report of the Intergovernmental Panel on Climate Change. London: Cambridge University Press.
15. Barbier EB (2006) Natural barriers to natural disasters: replanting mangroves after the tsunami. Frontiers in Ecology and the Environment 4: 124–131.
16. Chang SE, Adams BJ, Alder J, Berke PR, Chuenpagdee R, et al. (2006) Coastal ecosystems and tsunami protection after the December 2004 Indian Ocean tsunami. Earthquake Spectra 22: 863–887.
17. Chen LZ, Wang WQ, Zhang YH, Lin GH (2009) Recent progresses in mangrove conservation, restoration and research in China. Journal of Plant Ecology-Uk 2: 45–54.
18. Ren H, Lu HF, Shen WJ, Huang CL, Guo QF, et al. (2009) *Sonneratia apetala* Buch. Ham in the mangrove ecosystems of China: an invasive species or restoration species? Ecological Engineering 35: 1243–1248.
19. Ren H, Chen H, Li ZA, Han WD (2010) Biomass accumulation and carbon storage of four different aged *Sonneratia apetala* plantations in Southern China. Plant and Soil 327: 279–291.
20. Liao WB, Zheng DZ, Zeng SF (1990) Study on the biomass of *Sonneratia carseolaria* community. Journal of Forest Research: 47–54.
21. Miao SY, Chen GZ, Chen ZT (1998) Biomasses and distribution patterns of mangrove populations in Zhanjiang Nature Reserve, Guangdong, China. Guihaia: 16–19.
22. Ren H, Jian SG, Lu HF, Zhang QM, Shen WJ, et al. (2008) Restoration of mangrove plantations and colonisation by native species in Leizhou bay, South China. Ecological Research 23: 401–407.
23. Bosire JO, Dahdouh-Guebas F, Walton M, Crona BI, Lewis RR, et al. (2008) Functionality of restored mangroves: a review. Aquatic Botany 89: 251–259.
24. Chen LZ, Zeng XQ, Tam NFY, Lu WZ, Luo ZK, et al. (2012) Comparing carbon sequestration and stand structure of monoculture and mixed mangrove plantations of *Sonneratia caseolaris* and *S. apetala* in Southern China. Forest Ecology and Management 284: 222–229.
25. Craft C, Megonigal P, Broome S, Stevenson J, Freese R, et al. (2003) The pace of ecosystem development of constructed *Spartina alterniflora* marshes. Ecological Applications 13: 1417–1432.
26. Didham RK, Tylianakis JM, Gemmell NJ, Rand TA, Ewers RM (2007) Interactive effects of habitat modification and species invasion on native species decline. Trends in Ecology and Evolution 22: 489–496.
27. Larson DL, Anderson PJ, Newton W (2001) Alien plant invasion in mixed-grass prairie: effects of vegetation type and anthropogenic disturbance. Ecological Applications 11: 128–141.
28. Dassonville N, Vanderhoeven S, Gruber W, Meerts P (2007) Invasion by *Fallopia japonica* increases topsoil mineral nutrient concentrations. Ecoscience 14: 230–240.
29. Lindsay EA, French K (2005) Litterfall and nitrogen cycling following invasion by *Chrysanthemoides monilifera* ssp. rotundata in coastal Australia. Journal of Applied Ecology 42: 556–566.
30. Ordonez A, Wright IJ, Olff H (2010) Functional differences between native and alien species: a global-scale comparison. Functional Ecology 24: 1353–1361.
31. Zan QJ, Wang YJ, Liao WB (2001) Biomass and net productivity of *Sonneratia carseolaria- Sonneratia apetala* mangrove forest. Journal of Wuhan Botanical Research 15: 391–397.
32. Han WD, Gao XM, Teunissen E (2001) Study on *Sonneratia apetala* productivity in restored forests in Leizhou Peninsula, China. Journal of Forest Research 12: 229–234.
33. Lin P, Lu CY, Lin GH, Chen RH, Su L (1985) Studies on mangrove ecosystem of Jiulongjiang river estuary in China. The biomass and productivity of *Kandelia candel* community. Journal of Xiamen University (Natural Science) 24: 508–514.
34. Lin P, Lu CY, Wang GL, Chen HX (1990) Biomass and productivity of *Bruguiera sexangula* mangrove forest in Hainan island, China. Journal of Xiamen University (Natural Science) 29: 209–213.
35. Lin P, Lu CY, Wang GL, Chen HX (1990) Study on dynamics of litter fall of *Bruguiera sexangula* mangrove in Hainan island, China. Acta Phytoecologica et Geobotanica Sinica 14: 69–74.
36. Lu CY, Zheng FZ, Lin P (1988) Study on litter fall production of *Kandelia candel* mangrove community in Estuary. Journal of Xiamen University (Natural Science) 27: 459–463.
37. Zhang QM, Chen YF (2003) Production and seasonal change pattern of litter fall of *Rhizophora apeculata* in Sanya River mangroves, Hainan island. Acta Ecologica Sinica 23: 1977–1983.
38. Zheng FZ, Lu CY, Zheng WJ, Lin P (2000) Seasonal dynamics of litter fall and energy flow through the leaf litter of *Kandelia candel* Mangrove in Jiulongjiang Estuary, Fujian Province, China. Journal of Xiamen University (Natural Science) 39: 693–698.
39. Finzi AC, Moore DJP, DeLucia EH, Lichter J, Hofmockel KS, et al. (2006) Progressive nitrogen limitation of ecosystem processes under elevated CO_2 in a warm-temperate forest. Ecology 87: 15–25.
40. Luo YQ, Field CB, Jackson RB (2006) Does nitrogen constrain carbon cycling, or does carbon input stimulate nitrogen cycling? Ecology 87: 3–4.
41. Moriya H, Komiyama A, Prawiroatmodjo S, Ogino K (1986) Specific characteristics of leaf dynamics. In: Ogino K, Chihara M, editors. Biological System of Mangroves: A Report of the East Indonesian Mangrove Expedition. pp. 123–136.
42. Clarke PJ (1994) Baseline studies of temperate mangrove growth and reproduction: demographic and litterfall measures of leafing and flowering. Australian Journal of Botany 42: 37–48.
43. Alongi DM (2012) Carbon sequestration in mangrove forests. Carbon Management 3: 313–322.
44. Cai LZ, Xu P, Fu SJ, Peng X, Cao J, et al. (2012) Secondary productivity of macrobenthos in mangrove and salt marsh in Gaoqiao of Zhanjiang, Guangdong Province of South China. Chinese Journal of Applied Ecology 23: 965–971.

23

Post-Glacial Expansion and Population Genetic Divergence of Mangrove Species *Avicennia germinans* (L.) Stearn and *Rhizophora mangle* L. along the Mexican Coast

Eduardo Sandoval-Castro[1,2], Richard S. Dodd[3], Rafael Riosmena-Rodríguez[4], Luis Manuel Enríquez-Paredes[2], Cristian Tovilla-Hernández[5], Juan Manuel López-Vivas[3], Bily Aguilar-May[6], Raquel Muñiz-Salazar[1]*

1 Escuela de Ciencias de la Salud, Universidad Autónoma de Baja California, Ensenada, Baja California, México, **2** Facultad de Ciencias Marinas, Universidad Autónoma de Baja California, Ensenada, Baja California, México, **3** Department of Environmental Science, Policy, and Management, University of California, Berkeley, Berkeley, California, United States of America, **4** Departamento Académico de Biología Marina, Universidad Autónoma de Baja California Sur, La Paz, Baja California Sur, México, **5** El Colegio de la Frontera Sur, Tapachula, Chiapas, México, **6** Instituto Tecnológico Superior de Villa La Venta, Huamanguillo, Tabasco, México

Abstract

Mangrove forests in the Gulf of California, Mexico represent the northernmost populations along the Pacific coast and thus they are likely to be source populations for colonization at higher latitudes as climate becomes more favorable. Today, these populations are relatively small and fragmented and prior research has indicated that they are poor in genetic diversity. Here we set out to investigate whether the low diversity in this region was a result of recent colonization, or fragmentation and genetic drift of once more extensive mangroves due to climatic changes in the recent past. By sampling the two major mangrove species, *Rhizophora mangle* and *Avicennia germinans*, along the Pacific and Atlantic coasts of Mexico, we set out to test whether concordant genetic signals could elucidate recent evolution of the ecosystem. Genetic diversity of both mangrove species showed a decreasing trend toward northern latitudes along the Pacific coast. The lowest levels of genetic diversity were found at the range limits around the Gulf of California and the outer Baja California peninsula. Lack of a strong spatial genetic structure in this area and recent northern gene flow in *A. germinans* suggest recent colonization by this species. On the other hand, lack of a signal of recent northern dispersal in *R. mangle,* despite the higher dispersal capability of this species, indicates a longer presence of populations, at least in the southern Gulf of California. We suggest that the longer history, together with higher genetic diversity of *R. mangle* at the range limits, likely provides a gene pool better able to colonize northwards under climate change than *A. germinans*.

Editor: João Pinto, Instituto de Higiene e Medicina Tropical, Portugal

Funding: International Community Foundation Grant No.: 20030900; 14th Internal Call for Support to Investigation Projects of the UABC, Grant No. 0843; CONACYT-SEMARNAT-INE 2009, Grant No. 107696. The funders had no role in study design, data collection and analysis, decision to publish, or preparation of the manuscript.

Competing Interests: The authors have declared that no competing interests exist.

* E-mail: ramusal@uabc.edu.mx

Introduction

Cyclical climatic oscillations during the last few million years have had a great effect on biota, causing isolation, genetic subdivision and speciation in some cases and/or hybridization and homogenization in others [1]. In many cases, the range of a species can be considered to be highly dynamic, with periods of directional or isotropic growth of range expansions, and of contractions followed by re-expansions [2]. Given that populations of temperate species suffered extirpations from glacial ice sheets, most studies have focused on them and the effects of past climate change on tropical species are still poorly understood.

Populations are dynamic in space and time, and peripheral populations are often of special conservation concern as they are predicted to experience local extinctions at a greater rate, and to have reduced recolonization potential due to smaller population sizes and more limited dispersal [3]. For tropical species, future climate change could provide suitable habitat at higher latitudes providing the opportunity for them to expand their range limits [4,5,6,7]. Population expansions can leave detectable signatures in the distribution of genetic diversity [8] that help us to trace their evolution through time. Therefore, a retrospective view of demographic changes in populations can provide insights into their colonization potential under future climate change.

Mangroves are highly productive tropical ecosystems that support numerous food chains in the coastal zone and neighboring ecosystems [9]. The presence of mangroves has a strong effect on fishery yields [10], since they provide food to offshore systems as a source of carbon [11]. Mangroves also provide protection against erosion of coastlines by reducing wave energy, locking sediment in place with their roots and promoting sedimentation [12]. Therefore, understanding the response of mangrove species to climate change is of utmost importance for the management of coastal resources.

The global distribution of mangrove marsh is mainly influenced by temperature, restricting species to warm tropical and subtropical latitudes [13]. Extreme cold events have been hypothesized to explain range transformations and severe regional extinctions and latitudinal limits of mangrove distribution [8,14,15]. Because salt marsh ecosystems are mostly linear along coastal areas, they are likely to respond in more predictable ways to climate change than more dispersed terrestrial systems. Indeed, a recent study suggests that *Rhizophora mangle* L. experienced range limit oscillations during glacial – interglacial cycles, to the extent that modern ranges are still constrained by post-glacial re-colonization along the Brazilian coast [8].

Mexican mangroves cover 770,057 ha [16] of which 66% is located along the Atlantic coast (11% in the Gulf of Mexico and 55% on the Yucatan peninsula). Along the Pacific coast, the largest mangrove forests are located in "Marismas Nacionales", followed by the mangrove forests along the coast of southern Mexico in Oaxaca and Chiapas. Mangroves around the Gulf of California are less extensive and they represent the northern natural limit of mangrove species along the Pacific coast [17]. Since mangrove forests of Mexico include populations at their northern range limit, one of the research questions driving our study was whether low levels of genetic diversity reported earlier by Sandoval-Castro et al. [18] for *R. mangle* in northern Mexico are a result of recent colonization or fragmentation and genetic drift of once more extensive mangroves due to climatic changes in the recent past. Recently colonized populations are expected to harbor decreased levels of genetic diversity [19,20,21], although exceptions to this may result when long-distance dispersal (LDD) events are increasingly frequent [22]. For temperate species that invaded new habitats following receding ice sheets, long distance dispersal events can help explain the speed of species' advance and unexpectedly high genetic diversity in populations distant from the putative refugia [23]. However, for tropical species, climate-related advances and retreats were most likely over relatively short distances and not into vacant habitat. This is particularly true of mangroves along the east Pacific coast that are more or less linearly distributed. Although mangrove distributions are (and likely were in the past) dependent on suitable substrate and therefore are not continuous, climatic-related advances were most likely to have followed a slow diffusion model rather than LDD. Theoretical expectations of such a model are low genetic diversity and low divergence in the newly formed populations [23].

Recent colonization, or fragmentation of older populations could have important consequences as to the likelihood of these northernmost populations providing well-adapted seed sources for colonization at higher latitudes as climate changes in the future.

Therefore, we hypothesize that Gulf of California populations of Mexican mangroves are recent colonizations following climatic restriction southwards during the last glacial. Propagule sources for northward advance along the northern Pacific coast would most likely be dominated by leading edge populations in this narrow linear system. Assuming a contiguous dispersal model (local diffusion) over non-contiguous long distance dispersal, founder populations should exhibit low genetic diversity and in this stepping stone system low genetic divergence [22]; that would lead to the following expectations that we test here: 1) All mangrove species of the ecosystem should show a concordant pattern of genetic diversity; low genetic diversity in Gulf of California populations and little genetic differentiation among populations compared to those further south along the Mexican coast, 2) This pattern would be more marked along the Pacific than the Atlantic coast, because the former includes the latitudinal limit of mangroves and source populations are more linearly distributed along the Pacific than the Caribbean coast In order to elucidate these hypotheses, we analyzed and compared patterns of genetic variation of the two major mangrove species in Mexico (*A. germinans* and *R. mangle*) to interpret demographic and biogeographic processes along the Pacific and Atlantic coasts of Mexico.

Materials and Methods

Ethics Statement

Permission to collect samples in each location was obtained from General Directorate for Wildlife (Dirección General de Vida Silvestre de la Secretaría de Medio Ambiente y Recursos Naturales, SEMARNAT).

Plant Material and DNA Isolation

Leaf tissue from 448 individuals of *A. germinans* and 600 of *R. mangle* was collected from mangrove forests along the Pacific and the Atlantic coasts of Mexico. Sampling data of *R. mangle* from the Baja California peninsula and the Gulf of California have already been reported in a previous work [18]. Remaining locations from the Central and Southern Pacific coast and from the Atlantic are new to this study (Table 1). Along the Pacific coast, sampling was performed from the Gulf of California to "La Encrucijada", near to the Mexican border with Guatemala. Along the Atlantic coast, the sampling localities were situated in the Gulf of Mexico and around the Yucatan Peninsula (Fig. 1). Samples were taken from specimens separated by at least 30 m to prevent consanguinity and to maximize the probability of collecting diverse genotypes. The samples were dehydrated and stored in silica gel until DNA extraction. Total genomic DNA was isolated from approximately 200 mg dry weight of leaf using a modified CTAB/PVP method [18].

Microsatellite Analysis

Individuals were genotyped at seven loci (AgT4, AgT7, AgT8, AgT9, AgD6, AgD13 and CA_002) previously designed for *A. germinans* [24,25,20] and six loci (Rm7, Rm11, Rm19, Rm21, Rm38, and Rm46) previously designed for *R. mangle* [26]. These repeat motifs were isolated from genomic DNA and are expected, *a priori*, to be neutral with respect to natural selection. The forward primers were fluorescent-labeled with FAM, VIC, PET and NED (Applied Biosystems Inc). All amplifications were performed on a MyCycler BIORAD thermal cycler in 20 μL PCR reactions containing 1×Buffer (10 mM Tris HCl, 50 mM KCl, pH 8.3, SIGMA), 2.0 mM $MgCl_2$, 200 μM of each dNTP, 0.15 μM of each primer, 1 unit *Taq* DNA polymerase (SIGMA) and 20 ng of genomic DNA. With the exception of some small variations on annealing temperatures, all microsatellite loci were amplified with a similar thermocycler profile. The profile consisted of an initial denaturation step at 95°C for 5 min, followed by 35 amplification cycles as follows: 95°C for 30 s, annealing temperature (50°C for all the loci in *R. mangle* and for the loci AgT4, AgT7, AgT8 and AgT9, 55°C for AgD13 and CA_002 and 59°C for AgD6 in *A. germinans*), for 30 s and 45 s at 72°C, ending with an extension cycle at 72°C for 30 min. To ensure reproducibility and consistency in PCR amplification, approximately 5% of samples were re-amplified. In addition, a negative control was run for each set of PCR reactions and genotyped to check for contamination. Amplified products were run on an ABI 310 automated DNA

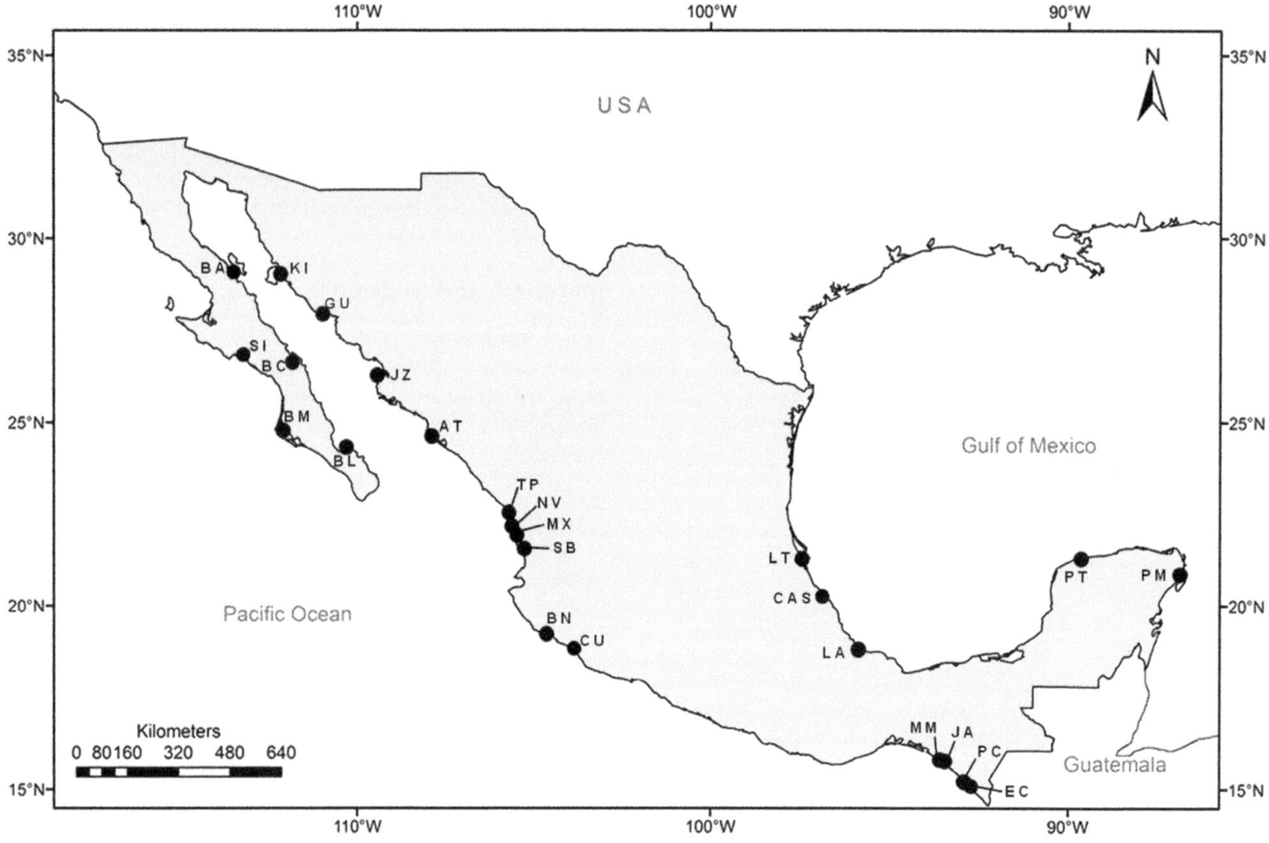

Figure 1. Geographic location of the sampling sites of mangrove species sampled along Mexican coasts.

sequencer, and microsatellite alleles were visualized and scored in the program GeneMarker 1.97 (Softgenetics).

Data Analysis

Genetic diversity. To determine spatial patterns in genetic diversity that might confirm latitudinal trends consistent with recent colonization northwards, we estimated the number of alleles (A), unbiased expected heterozygosity (H_E) and observed heterozygosity (H_O) for each locus across all localities using GDA 1.1 [27]. Allelic richness per locus (AR) and per population were calculated using FSTAT 2.9.3 [28]. Global tests for deviation from Hardy-Weinberg equilibrium were performed using a Markov chain algorithm and linkage between all pairs of loci was estimated using GENEPOP 4.0 [29,30] with significance levels determined using the Markov chain method. For all Markov chain tests, the default parameters in GENEPOP were used with 100 batches of 1000 iterations each. Null alleles, large allele dropout and stutter peaks were explored using Micro-Checker 2.2.3 [31].

Clustering analyses. We applied the Bayesian clustering algorithm implemented in STRUCTURE v2.2.3 [32] as an exploratory analysis to infer population genetic structure, assigning individuals (probabilistically) without *a priori* knowledge of population boundaries. STRUCTURE uses individual multilocus genotype data to cluster individuals into K groups while minimizing Hardy-Weinberg disequilibrium and gametic phase disequilibrium between loci within groups [32]. STRUCTURE runs were based on 500,000 iterations after a burn-in of length 500,000 and assumed correlated allele frequencies and an admixture model with an estimated proportion α of admixed individuals. To check for Markov chain Monte Carlo (MCMC) convergence, we performed 10 replicates for each K value and checked the consistency of results. The most likely number of clusters (K) was considered to be the K value with the highest $Pr(X|K)$ [32,33]. Also, the optimal K value was calculated after the ΔK method described by Evanno et al. [34].

Genetic structure. Genetic differentiation among populations was evaluated at each locus and over all loci by calculating the measures of relative genetic differentiation among populations defined under the infinite allele model (IAM; F_{ST}) [35] and the stepwise mutation model (SMM; R_{ST}) [36,37]. The presence of phylogeographic structure was assessed by using SPAGeDi 1.4 [38] through permutations of allele sizes among alleles within a single locus (pR_{ST}) (10,000 permutations). This analysis compares the observed R_{ST} value (before randomization) with the distribution of pR_{ST} values obtained for all possible configurations of allele size permutations. If observed R_{ST} is within the upper 5% of the distribution of pR_{ST} the contribution of mutations to population differentiation is non-negligible compared with genetic drift and migration [39]. Population genetic structure was also examined using hierarchical analysis of molecular variance (AMOVA) among the clusters determined by STRUCTURE v.2.3.3. At this level, genetic differentiation was quantified with F-statistics [35] using ARLEQUIN v.3.5. [40]. The distribution of genetic variation was assessed at four hierarchical levels: among groups, among populations within groups, among individuals within populations and within individuals [41]. Statistical significance of the variance was tested by 10,000 non-parametric permutations.

Table 1. Genetic diversity of *A. germinans* and *R. mangle* in terms of total number of alleles (A), Allelic richness (AR), expected (H_E) and observed heterozygosity (H_O), standard errors in parentheses. (N) Number of sampled individuals, (–) no sample.

Sampling site	Code	N	A	AR	H_E	H_O	N	A	AR	H_E	H_O
Pacific coast		***A. germinans***					***R. mangle***				
Bahía de los Ángeles	BA	–	–	–	–	–	†48	8	1.25	0.09	0.06
Laguna San Ignacio	SI	–	–	–	–	–	†30	9	1.43	0.13	0.09
Bahía Kino	KI	29	7	1.00	0.00	0.00	†27	9	1.33	0.10	0.08
Bahía Concepción	BC	18	10	1.29	0.08	0.07	†26	9	1.43	0.12	0.12
Bahía Magdalena	BM	34	9	1.08	0.01	0.01	†30	10	1.48	0.18	0.19
Balandra	BL	30	9	1.15	0.02	0.02	†30	11	1.60	0.18	0.26
Guaymas	GU	24	14	1.41	0.07	0.06	†28	13	1.76	0.15	0.13
El Jitzámuri	JZ	30	15	1.37	0.05	0.05	†28	12	1.87	0.26	0.26
Bahía de Altata	AT	30	19	1.94	0.21	0.21	†29	15	2.12	0.26	0.27
Teacapán	TP	30	22	2.12	0.24	0.24	†29	14	2.12	0.32	0.24
Novillero	NV	24	24	2.21	0.25	0.22	39	14	2.05	0.32	0.33
Mexcaltitán	MX	17	18	2.03	0.22	0.18	26	13	2.00	0.31	0.32
San Blas	SB	30	19	1.96	0.23	0.18	29	14	2.07	0.33	0.33
Barra Navidad	BN	10	12	1.56	0.10	0.10	10	13	2.17	0.36	0.30
Laguna Cuyutlán	CU	–	–	–	–	–	18	12	1.85	0.28	0.16
Mar Muerto	MM	16	31	3.32	0.42	0.36	–	–	–	–	–
Joaquín Amaro	JA	20	31	3.18	0.38	0.35	31	17	2.49	0.42	0.40
Laguna Panzancola	PC	10	28	3.51	0.51	0.51	30	22	2.80	0.47	0.46
La Encrucijada	EC	–	–	–	–	–	11	19	3.09	0.48	0.56
Mean Pacific				1.94(0.21)	0.19(0.04)	0.17(0.04)			1.94(0.12)	0.26(0.03)	0.25(0.03)
Atlantic coast											
Laguna Tamiahua	LT	28	41	3.52	0.43	0.36	30	13	1.92	0.26	0.20
Laguna Casitas	CA	–	–	–	–	–	16	12	1.99	0.25	0.27
Laguna Alvarado	LA	30	42	3.62	0.45	0.41	–	–	–	–	
Progreso-Telchac	PT	30	42	4.14	0.56	0.48	27	13	2.08	0.35	0.27
Puerto Morelos	PM	8	23	3.26	0.55	0.44	28	16	2.37	0.38	0.31
Mean Atlantic				3.64(0.18)	0.50(0.03)	0.42(0.03)			2.09(0.10)	0.31(0.03)	0.26(0.02)
Total		448	93	2.3(0.24)	0.26(0.04)	0.23(0.04)	600	34	1.97(0.1)	0.27(0.02)	0.25(0.03)
Mean Pacific/Mean Atlantic				0.53 (0.06)††	0.37 (0.08)††	0.41 (0.09)††			0.93 (0.05)††	0.85 (0.09)††	0.96 (0.09)††

†Data previously reported by Sandoval-Castro et al. [18].
††Standard errors estimated after 5,000 bootstraps.

Migration rates. On the basis of Structure and SPAGeDi analysis, we classified localities into populations. The magnitude and direction of gene flow were estimated among populations for which mutations had a non-significant effect on differentiation $R_{ST} \leq pR_{ST}$. Estimates of evolutionary patterns of gene flow were obtained according to the maximum likelihood approach implemented in MIGRATE v.3.2.7 [42,43]. MIGRATE uses a coalescence approach to estimate migration rates (Nm) among populations, assuming a constant per-locus mutation rate. This approach is judged to estimate gene flow more accurately than other F_{ST} methods, especially when multiple loci are employed [42]. We used 10 short-chain searches and three long-chain searches over the number of assayed microsatellite loci to obtain the magnitude and direction of gene flow according to Beerli and Felsenstein [42]. For each locus, the program was run for 10 consecutive exploratory chains with lengths of 5×10^6 genealogy visits to adjust the driving values for both the run and the 3 long chains; the last chain was used to generate the presented results. Each of the long chains visited 5×10^8 genealogies, sampling 5000 after an initial burn-in of 10000 steps. The program assumes discrete populations and generations, mutation-drift equilibrium, non-selective effects and the stepwise mutation model for microsatellite markers [44]. Furthermore, we estimate recent levels of gene flow between populations by using the program BAYESASS [45], which uses transient levels of linkage disequilibrium produced by recent migrants or their immediate descendants to infer levels of migration into populations. The program uses MCMC sampling in a Bayesian statistical framework to estimate gene flow in the recent past. The run involved 3×10^6 MCMC iterations, discarding the first 10^6. The program estimates the mean value for migration rate, and a 95% confidence interval for the estimate.

Isolation by distance. To test the significance of Isolation By Distance (IBD), the correlation between genetic and geographic distance matrices was tested using a Mantel test with 2000 permutations [46]. The geographical distances between samples were based on coastline distances and the genetic distance was expressed as $F_{ST}/(1-F_{ST})$ following Rousset [47]. IBD analysis was only performed on the mainland Pacific coast populations (KI, GU, JZ, AT, TP, NV, MX, SB, BN, CU, MM, JA, PC and EC). Given that mangroves from Atlantic (LT, CAS, LA, TP, PM) do not have a clear connection with the Pacific they were removed from this analysis. Because of potentially complex scenarios of dispersal into and around the Gulf of California, populations SI, BM, BL, BC, and BA were not included in the IBD analyses. Since we hypothesized a recent colonization of northern mangrove populations, we also performed an IBD analysis restricted to the northwestern coast (KI, GU, JZ, AT, TP, NV, MX, SB).

Results

Genetic Diversity

Microchecker analyses showed no evidence of null alleles nor allelic drop out that might interfere with posterior analysis on genetic diversity. We detected 93 alleles in *A. germinans*, of which 35.5% were exclusive to the Atlantic, 35.5% to the Pacific and 29% were common to both coasts. For *R. mangle*, only 34 alleles were detected, of which 11.8% were exclusive to the Atlantic, 38.2% to the Pacific and 50% were shared between both coasts.

Genotypic linkage disequilibrium was not detected among any of the pairwise loci comparisons across all populations. This suggests no evidence for selective sweeps and supports the expectation that the markers are neutral with respect to natural selection. Both mangrove species showed lower levels of genetic diversity around the Gulf of California, where mangrove forests are less extensive and more fragmented. Over all populations, genetic diversity was not significantly different between *A. germinans* and *R. mangle* (Table 1). For *A. germinans*, we detected higher levels of genetic diversity in the Atlantic populations, whereas for *R. mangle* greater diversity was detected among Pacific coast populations. Since direct comparisons of diversity at the different loci for the two species cannot be made, we standardized genetic diversity along the Pacific coast to that of the Atlantic coast by dividing mean Pacific diversity by mean Atlantic diversity (Table 1). Genetic diversity in both mangrove species showed a decreasing trend northward along the Pacific coast; lowest values were detected along the northern coasts of the Gulf of California, where mangrove species reach their natural range limits (Pearson correlation coefficients between observed heterozygosity and latitude $r = -0.84$ $P<0.0001$ for both *A. germinans* and *R. mangle* (Fig. 2A). For all three measures of genetic diversity, a ratio close to one for *R. mangle* indicated no significant difference in genetic diversity between the two coasts. However, for *A. germinans*, ratios ranged from about 0.4 to 0.5, indicating much lower diversity along the Pacific coast. This lower genetic diversity was supported by an ANCOVA analysis applied to the corrected variation (residuals) of genetic diversity (HO) versus latitude as a covariate ($P = 0.001$; Fig. 2B).

Clustering Analyses

Applying the Bayesian analysis in STRUCTURE and the approach of Evanno et al. [34], the most likely number of genetic clusters was six for *A. germinans* and five for *R. mangle*. Of the six clusters for *A. germinans*, five were detected along the Pacific coast and one along the Atlantic coast (Fig. 3). Although proportional assignment of individuals to clusters indicated admixture in most populations, the admixture proportions of the STRUCTURE clusters revealed five spatially distinct groups (shown in lower frame of Fig. 3). Gulf of California and peninsula populations formed, two groups; one more or less homogeneous group including populations BL, BM, KI, GU and JZ, and a group comprising a single sampling location (BC) shown in red in Fig. 3. The latter showed very little admixture of the STRUCTURE clusters and was most likely the result of a recent founder event by an individual with a genotype that was unrepresentative of the nearby populations that were likely the colonizing source. Sample locations (AT, TP, NV, MX) from "Marismas Nacionales" formed a third group that graded into the Central Pacific coast group (SB, BN). However, these sampling locations shared genotypes among themselves and with those from the Gulf of California. The fifth group included southern sampling locations along the Pacific coast (MM, JA, PC) and comprised a single STRUCTURE cluster, with individuals having minimal levels of admixture Although a spatial pattern was evident for *R. mangle*, only the Atlantic coast sampling sites formed a unique cluster (Fig. 4). Pacific coast populations formed four clusters with high levels of admixture at all but the most northerly sampling locations.

Genetic Structure

In general, overall pairwise values of F_{ST} were lower than R_{ST}, but significantly different from zero. AMOVA analysis showed that most of the genetic variation was among clusters defined by STRUCTURE (Table 2). According to SPAGeDi analyses, differentiation among all populations taking into account allele sizes ($R_{ST} = 0.60$ and 0.80 for *R. mangle* and *A. germinans* respectively) was significantly larger ($\alpha = 0.05$, $P<0.0001$) than differentiation based on allele identities ($F_{ST} = 0.47$ and 0.54), indicating that stepwise mutations contributed to overall among-

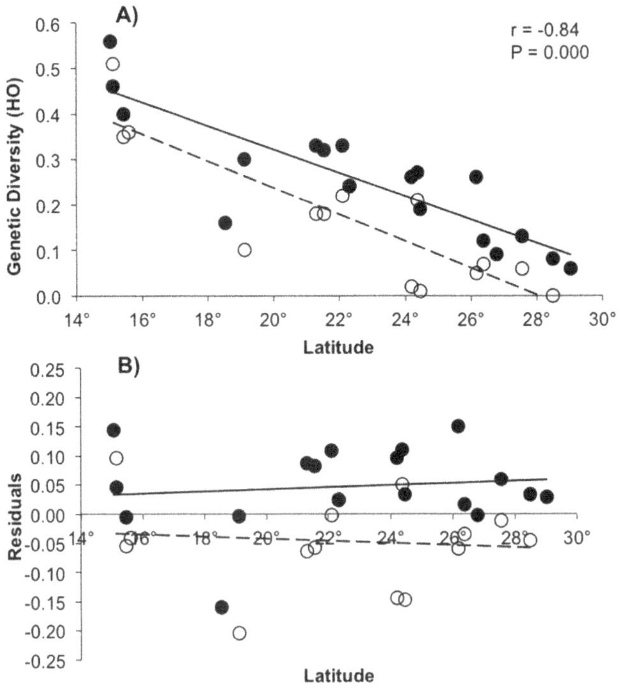

Figure 2. Lower genetic diversity was observed at northern margin limits. A) Genetic diversity decreases with latitude in mangrove species *A. germinans* (open circles, dashed line) and *R. mangle* (black circles, solid line) along Mexican Pacific coast. (r = Pearson's correlation). B) ANCOVA analysis showing lower genetic diversity of *A. germinans* at Pacific coast ($\alpha = 0.05$, $P = 0.001$).

population differentiation. However, when we only analyzed populations from the northwestern coast ("Marismas Nacionales" and "Gulf of California"), in both mangrove species R_{ST} was not significantly higher than F_{ST} ($\alpha = 0.05$, $P = 0.379$), suggesting that more recent evolutionary processes of migration and genetic drift were responsible for differentiation over this geographic range.

Isolation by Distance

For *A. germinans*, isolation by distance was significant over all sampling locations along the Pacific coast and over the more restricted geographic range of the northwestern coast (Fig. 5A, B). For *R. mangle*, isolation by distance was significant only over the more restricted northwester geographic range (Fig. 5C, D).

Migration Rates

In view of the minimal effect of mutations on genetic structure along the north-west coast, we focused on estimating directional gene flow between the Gulf of California (GC) and "Marismas Nacionales" (MN), including Central Pacific (CP) as a control (Fig. 6). Evolutionary-scale rates of migration (MIGRATE) showed a higher magnitude than contemporary-scale rates (BAYESASS). In *A. germinans* the estimated number of migrants per generation (Nm) ranged from 0.83 to 1.81 and from 0.00 to 0.26 (MIGRATE and BAYESASS respectively) (Fig. 6A), whereas in *R. mangle* Nm ranged from 0.44 to 1.94 and from 0.00 to 0.05 (Fig. 6B). Bayesian estimates based on MCMC simulations performed using BAYESASS suggest that for *A. germinans* contemporary gene flow is predominantly northward along the northwest Pacific coast, with higher rates from CP to MN and from MN to GC. On other hand, for *R. mangle* the BAYESASS analysis showed no significant evidence of recent gene flow, whereas on an evolutionary-scale, MIGRATE detected greater southward gene flow from GC to MN and CP. These inferred migration rates suggest that *R. mangle* populations in the Gulf of California are older than those of *A. germinans*. This is consistent with the present more northerly distribution of *R. mangle* along the Peninsula of Baja California today.

Discussion

Today, mangroves in the Gulf of California and along the Pacific coasts of the Baja California peninsula form the northern limit of this ecosystem in the eastern Pacific. Populations are relatively small and fragmented and prior research has indicated that they are poor in genetic diversity [18,48,49]. We set out to examine patterns of genetic diversity that would help determine whether low diversity in this region was a result of recent founder events, or of fragmentation and genetic drift of once more extensive mangroves due to climatic changes in the recent past. Here, we discuss our hypotheses for recent colonization in the Gulf of California in the light of genetic diversity along the Pacific and Atlantic coasts of Mexico and evaluate the roles of differential dispersal in the two major taxa of the mangrove ecosystem.

Are Patterns of Genetic Diversity Concordant in the Two Major Mangrove Species?

We found that broad patterns of genetic diversity along the Mexican Pacific coast were consistent for both species; a linear decrease in genetic diversity with latitude, becoming impoverished in the Gulf of California. Although patterns of genetic diversity were broadly similar, demographic processes inferred from population structure and directional migration rates revealed a more complex evolution of populations and the ecosystem. For both species, populations from "Marismas Nacionales" and the Central Pacific were highly admixed and levels of admixture declined into the Gulf of California (Fig. 3, 4). Based on cluster

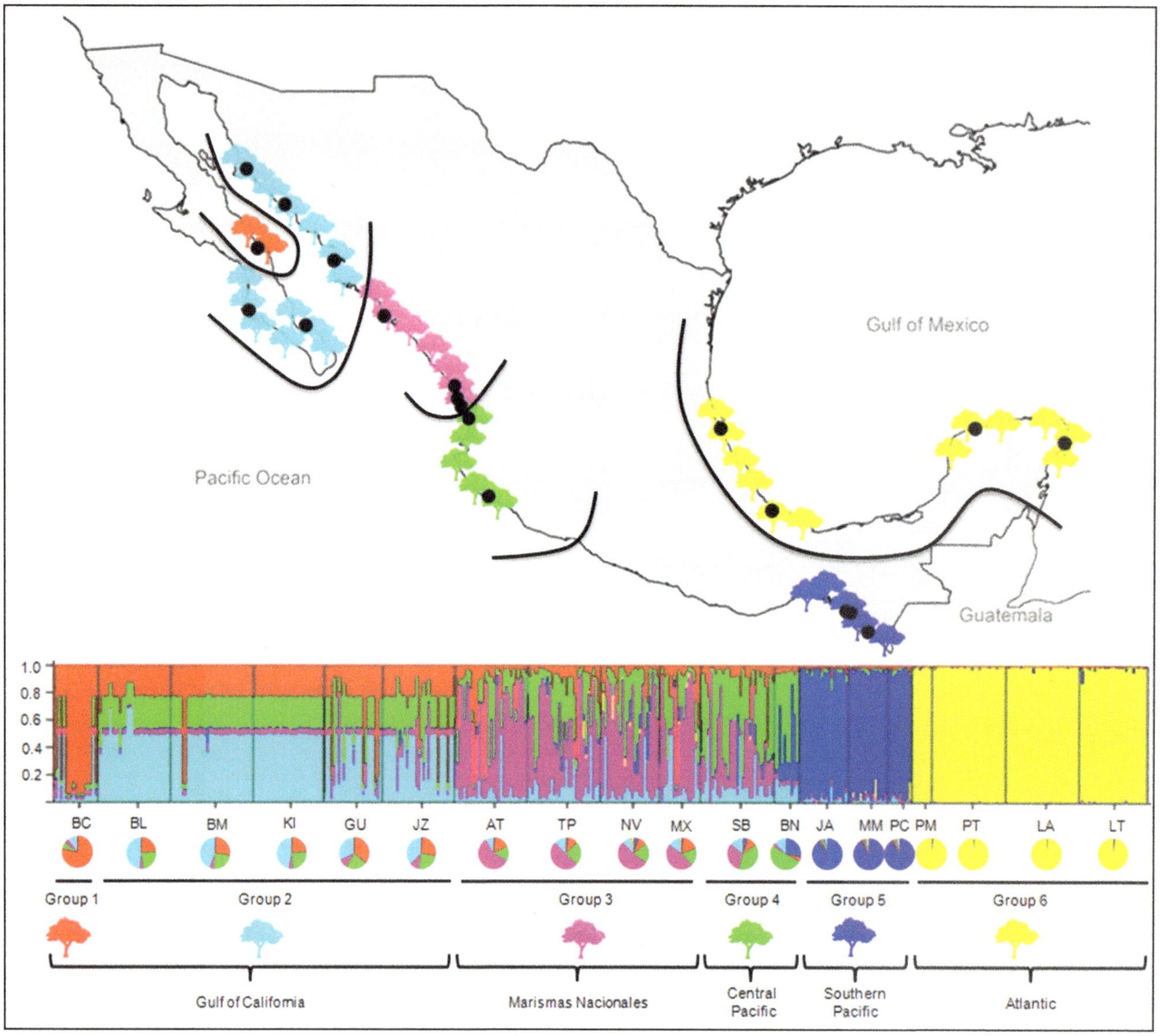

Figure 3. STRUCTURE plot of *Avicennia germinans* along Mexican coasts. Lines in the map indicate the spatial borders of the groups; vertical bars represent individuals whose genotypes have been apportioned into 6 clusters; colors in pie charts represent the percentage of assignment (Q values) of each group.

assignments from the program STRUCTURE, we detected shared ancestry between populations of *A. germinans* from the Gulf of California and those from "Marismas Nacionales" and Central Pacific, consistent with Gulf of California populations being the product of recent founder events, for example the BC population represented by the red color in the STRUCTURE analysis. For *A. germinans*, this hypothesis was supported by estimates of recent directional migration inferred using the program BAYESASS, which showed a predominantly northward migration from "Marismas Nacionales" to the Gulf of California.

Despite comparable latitudinal trends in genetic diversity, we found contrasting demographic signals in the two species. Gulf of California populations of *R. mangle* shared very little recent ancestry with those of "Marismas Nacionales" and Central Pacific and we found no detectable recent migration among these populations. This would seem to indicate: 1) A longer presence of *R. mangle* in the Gulf of California, 2) Confounding due to unsampled source populations, 3) Effects associated with enhanced dispersability of *R. mangle* propagules. We were careful to sample most populations along the Pacific mainland coast and, although extinct populations could have been the source of Gulf populations, it seems unlikely that they would have been genetically distinct from "Marismas Nacionales" and Central Pacific. Today "Marismas Nacionales" is one of the more extensive areas of mangrove forests along the Mexican Pacific coast. Therefore, we find it unlikely that unsampled source populations will have confounded our results. Our data do indicate that *R. mangle* propagules are more effectively dispersed than those of *A. germinans*. Whereas, southern Mexican Pacific populations of *A. germinans* formed a discrete cluster in the STRUCTURE analysis, those of *R. mangle* were admixed with "Marismas Nacionales" and Central Pacific populations. The intervening coastline between central and southern populations is not suitable for mangrove establishment, so long-distance dispersal appears to have been

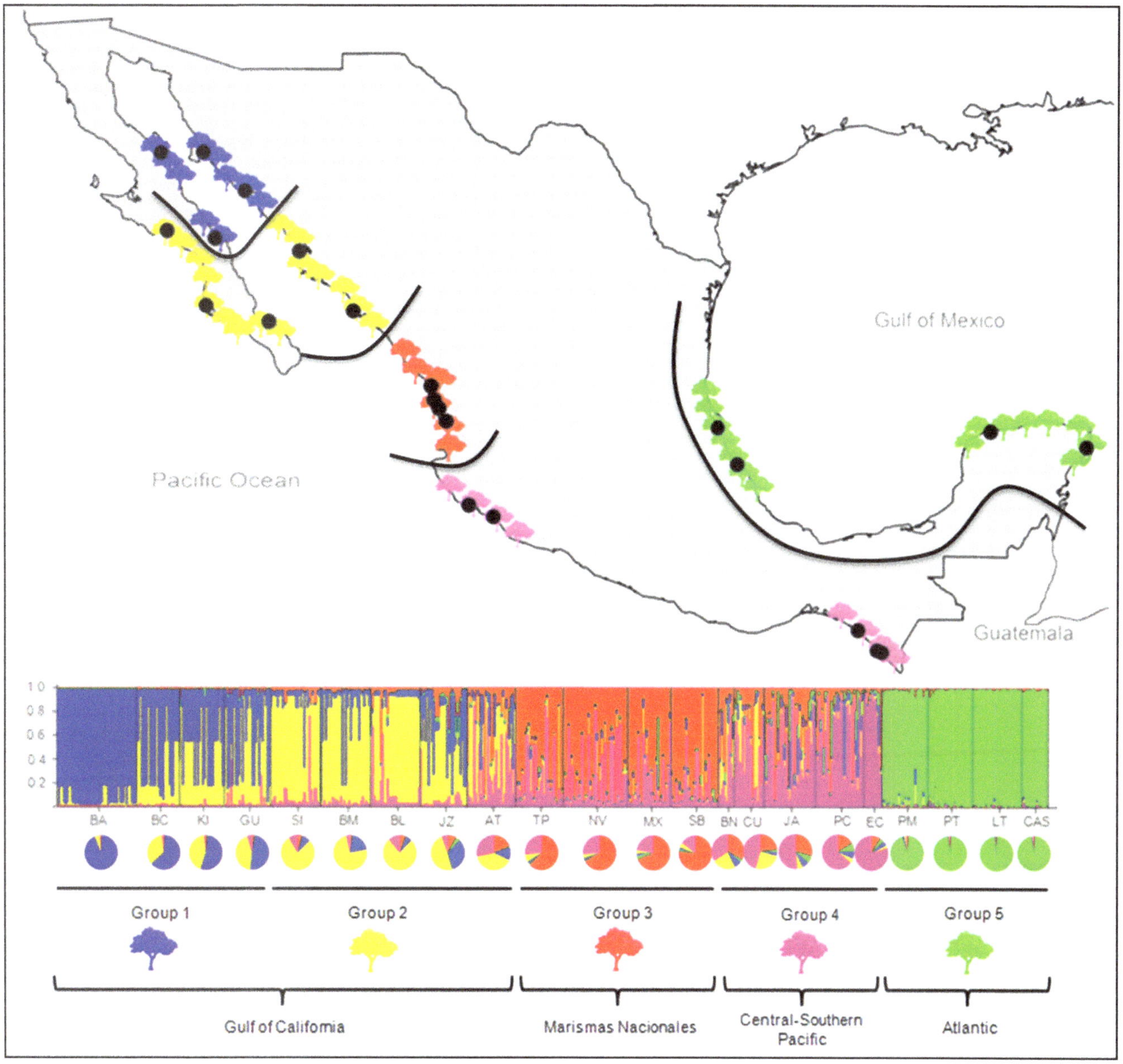

Figure 4. STRUCTURE plot of *Rhizophora mangle* along Mexican coasts. Lines in the map indicate the spatial borders of the groups; vertical bars represent individuals whose genotypes have been apportioned into 5 clusters; colors in pie charts represent the percentage of assignment (Q values) of each group.

more effective for *R. mangle*. Nevertheless, there is no evidence of shared ancestry between Gulf of California populations of *R. mangle* and those from the southern Mexican coast to suggest a source of founder propagules for Gulf populations. We find the most likely explanation of the contrasting demographic patterns between *A. germinans* and *R. mangle* is that populations of the latter species have been present somewhere in the Gulf of California longer than *A. germinans*. Interestingly, evolutionary rates of migration estimated from MIGRATE, also yielded contrasting results; no significant rates of migration between the Gulf of California and "Marismas Nacionales" or Central Pacific for *A. germinans*, but significant southward migration from the Gulf to "Marismas Nacionales" for *R. mangle*. Taken together, these data support a scenario of glacial restriction followed by Holocene latitudinal advance for *A. germinans*, and possible persistence of populations of *R. mangle* somewhere in the Gulf of California during the harsh conditions of the last glacial maximum.

Are Latitudinal Patterns of Genetic Diversity Comparable between Atlantic and Pacific Coasts in Mexico?

Whereas, we detected a clear linear trend of decreasing genetic diversity among populations of both species along the Pacific coast, no trend was detected along the Atlantic coast. Indeed, STRUCTURE assigned all individuals from the Atlantic to single clusters for each species. These opposing trends support our hypothesis that a stronger pattern of decreasing genetic diversity would be expected along the Pacific coast because of the more linear distribution of mangroves and inclusion of the populations from the range limit here. Sherrod and McMillan [50] argued that

Table 2. Population genetic structure of *A. germinans* and *R. mangle* in Mexico.

	A. germinans						*R. mangle*					
	IAM			**SMM**			**IAM**			**SMM**		
Source of variation	% var.	FST	P-value	% var.	RST	P-value	% Var.	FST	P-value	% Var.	RST	P-value
Among groups	52.95	0.52*	0.000	81.00	0.81*	0.000	44.82*	0.45*	0.000	58.50*	0.58*	0.000
Among populations within groups	5.67	0.12*	0.000	2.35	0.12*	0.000	6.65*	0.12*	0.000	5.46*	0.13*	0.000
Among individuals within populations	4.33	0.10*	0.000	2.79	0.16*	0.000	3.22*	0.06*	0.000	0.00	0.00	0.854
Among all individuals	37.05	0.62*	0.000	13.86	0.85*	0.000	45.31*	0.54*	0.000	33.37*	0.62*	0.000

Hierarchical AMOVA calculated by the Infinite Allele Model (IAM) and the Stepwise Mutation Model (SMM); percentage of molecular variation (% var.) explained by the hierarchical level. *Statistically significant ($p<0.05$). The group numbers and population assignment was according to STRUCTURE results.

populations of *A. germinans* along the mainland of the Gulf of Mexico advanced and went extinct during climatic cycles of the Pleistocene. Today, whereas both species are common components of Floridan mangroves, *A. germinans* extends further north than *R. mangle.* However, along the Pacific coast, *R. mangle* has the northernmost distribution [17]. We found that mean genetic diversity of *A. germinans* was about twice as great in the Atlantic compared with the Pacific coast, but that mean genetic diversity was about equal for *R. mangle.* This is in part, due to greater sharing of alleles between the two coasts in *R. mangle*; *R. mangle* shared almost 50% of alleles between Atlantic and Pacific populations, while *A. germinans* shared around 29% of alleles. This could be an effect of higher mutation rates at the studied loci in *A. germinans,* or of more extensive early dispersal of *R. mangle* before closure of the Central American Isthmus (CAI). Genetic structure analysis showed the highest genetic differentiation among populations from the Atlantic and the Pacific coasts, supporting previously reported molecular data [48,51,52,53,54,55,56] and highlighting the effect of geographical isolation following elevation of the CAI around 3.5 million years ago.

Is there Evidence of Earlier Colonization by R. Mangle, a Species with More Effective Propagule Dispersal?

Today, *R. mangle* extends further north than *A. germinans* along the Pacific coast. As discussed earlier, the patterns of genetic diversity that we observed suggest a longer presence of *R. mangle* in the Gulf of California and the Baja California peninsula. This could be explained by earlier colonization of *R. mangle* because of its more effective dispersal capabilities [57,58] and could explain the higher admixture in *R. mangle*, however, this would not explain the presence of genetic structure in both mangrove species. Therefore, we prefer an interpretation that populations of *R. mangle* persisted, perhaps in the lower Gulf, through the last glacial, which could explain the greater genetic diversity observed in *R. mangle* than *A. germinans* along the Gulf of California. However, it is difficult to explain why *R. mangle* has been more successful than *A. germinans* at the range limit along the north-western coast of Mexico. The answer may lie with shoreline topography, which may propitiate local extinction of mangroves on coastlines with no gradual gradient [59]. Climatic niche modeling for Pacific coast populations of *A. germinans* hindsighted onto last glacial maximum climate from the PMIP database, suggests that mangroves could have survived along the Sinaloa coastline and into the lower Gulf of California through the last glacial (Dodd unpublished data). This is also supported by fossil record of prop roots of A. germinans in BC population dating from the Pliocene [60]. Mangrove forests along the Gulf of California are discontinuously distributed as a narrow band along the littoral because of the persistence of rocky coastline with reduced tidal range. This condition could favor *R. mangle* over *A. germinans* because the former is more tolerant to changes in the hydroperiod, which have occurred periodically through time [59].

In conclusion, our data provide compelling evidence that the Gulf of California and outer Baja California were recently colonized by *A. germinans* through northward propagule dispersal, perhaps from relictual stands around present-day "Marismas Nacionales". On the other hand, *R. mangle* showed no signal of recent northern dispersal despite the greater dispersal capability of this species, suggesting the presence of populations at least in the southern Gulf of California for a longer time. Future climate change could favor colonization at higher latitudes, but the source propagules must be well adapted to the arid and highly saline conditions. Populations with a longer history at a site are likely to be better adapted to local conditions than founders from

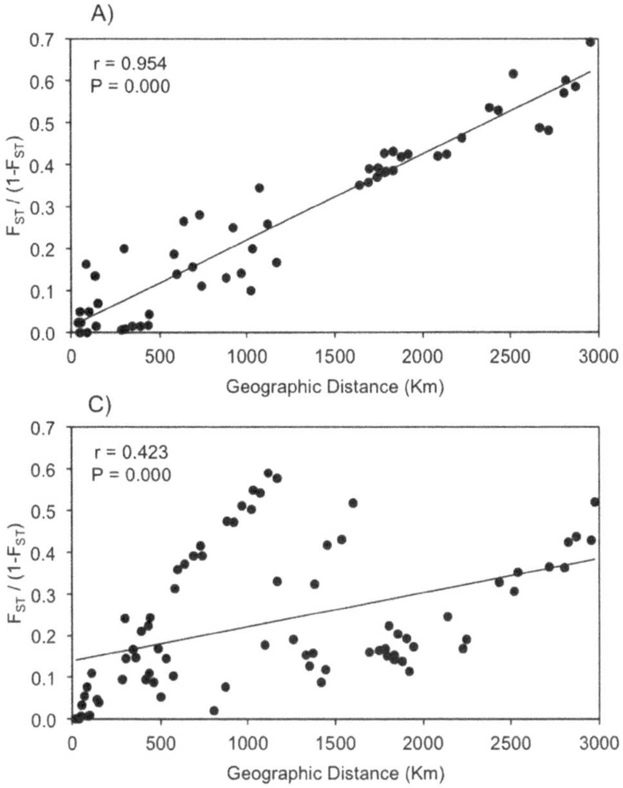

Figure 5. Isolation by distance. Correlations and probabilities were estimated from a Mantel test with 2000 repeats of bootstrap resampling. The y-axis is $F_{ST}/(1-F_{ST})$ following Rousset (1997). A, B) *A. germinans,* C, D) *R. mangle* on Pacific and northwestern coast, respectively.

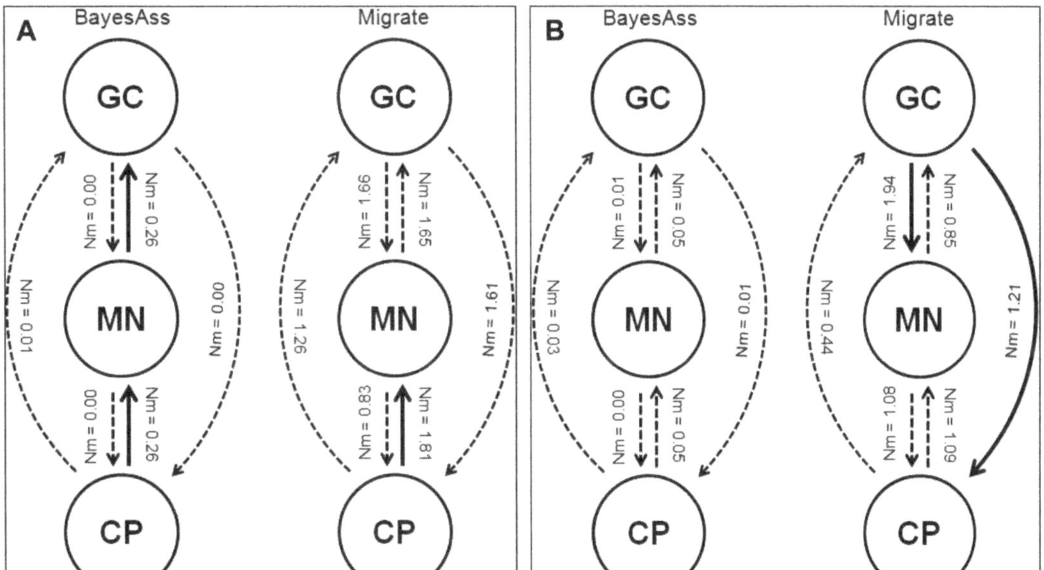

Figure 6. Estimates of the historical gene flow and migration directions among Gulf of California (GC), Marismas Nacionales (MN) and Central Pacific (CP). Numbers and arrows performed by MIGRATE; and estimates of recent migration rates and directions performed by BAYESASS for the northwestern Mexican mangrove populations of A) *A. germinans* and B) *R. mangle.* Solid arrows show significant gene flow.

populations under more benign conditions. Under this scenario, *R. mangle* would serve as a better adapted gene pool for latitudinal advances under climate change, even though *A. germinans* showed evidence of more successful northward migration in recent time.

Acknowledgments

The authors want to thank N.L. Victoria-Cota, S.C. Moreno-Medina, for their help in the laboratory work. This work represents part of Eduardo Sandoval-Castro's Ph.D. dissertation, financed by CONACYT scholarship 201092. This study was financed through grant funded by the CONACyT-SEMARNAT project (no. 107696).

Author Contributions

Conceived and designed the experiments: RMS. Performed the experiments: ESC. Analyzed the data: ESC RSD. Contributed reagents/materials/analysis tools: RMS ESC. Wrote the paper: ESC RSD. Sampling collection: ESC RRR CTH JMLV BAM. Made important contribution in the discussion section: RSD RMS LMEP CTH RRR.

References

1. Hewitt GM (2004) A climate for colonization. Heredity 92: 1–2.
2. Excoffier L, Foll M, Petit RJ (2009) Genetic Consequences of Range Expansions. Annu Rev Ecol Evol S 40: 481–501.
3. Peterman WE, Feist SM, Semlitsch RD, Eggert LS (2013) Conservation and management of peripheral populations: Spatial and temporal influences on the genetic structure of wood frog (Rana sylvatica) populations. Biol Conserv 158: 351–358.
4. Perry C, Mendelssohn I (2009) Ecosystem effects of expanding populations of Avicennia germinans in a Louisiana salt marsh. Wetlands 29: 396–406.
5. Jacobs DK, Haney TA, Louie KD (2004) Genes, Diversity, andD Geologic Process on the Pacifc Coast. Annu Rev Earth Pl Sc 32: 601–652.
6. Milá B, McCormack JE, Castañeda G, Wayne RK, Smith TB (2007) Recent postglacial range expansion drives the rapid diversification of a songbird lineage in the genus Junco. P Roy Soc B-Biol Sci 274: 2653–2660.
7. Yamano H, Sugihara K, Nomura K (2011) Rapid poleward range expansion of tropical reef corals in response to rising sea surface temperatures. Geophys Res Lett 38: L04601.
8. Pil MW, Boeger MRT, Muschner VC, Pie MR, Ostrensky A, et al. (2011) Postglacial north-south expansion of populations of *Rhizophora mangle* (*Rhizophoraceae*) along the Brazilian coast revealed by microsatellite analysis. Am J Bot 98: 1031–1039.
9. Tomlinson PB (1994) The Botany of Mangroves. Cambridge, UK: Cambridge University Press 419 p.
10. Aburto-Oropeza O, Ezcurra E, Danemann G, Valdez V, Murray J, et al. (2008) Mangroves in the Gulf of California increase fishery yields. P Natl Acad Sci USA 105: 10456–10459.
11. Ong JE (1993) Mangroves - a carbon source and sink. Chemosfere 27: 1097–1107.
12. Thampanya U, Vermaat JE, Sinsakul S, Panapitukkul N (2006) Coastal erosion and mangrove progradation of Southern Thailand. Estuar Coast Shelf S 68: 75–85.
13. Duke NC, Ball MC, Ellison JC (1998) Factors influencing biodiversity and distributional gradients in mangroves. Global Ecol Biogeogr 7: 27–47.
14. Rull V (1998) Evolución de los manglares Neotropicales: La crisis del Eoceno. Interciencia 23: 255–262.
15. Graham A (2006) Paleobotanical evidence and molecular data in reconstructing the historical phytogeography of *Rhizophoraceae*. Ann Mo Bot Gard. Missouri. 325–334.
16. Acosta-Velázquez J, Rodríguez-Zuñiga T, Díaz-Gallegos JR, Cerdeira-Estrada S, Troche-Souza C, et al. (2009) Assessing a nationwide spatial distribution of mangrove forest for Mexico: an analysis with high resolution images. 33rd International Symposium on Remote Sensing of Environment. Stressa, Italy. 1–4.
17. Pacheco-Ruiz I, Zertuche-Gonzalez JA, Meling-Lopez AE, Riosmena-Rodriguez R, Orduña-Rojas J (2006) El límite geográfico de *Rhizophora mangle* para el Golfo de California. Ciencia y Mar 28: 19–22.
18. Sandoval-Castro E, Muñiz-Salazar R, Enríquez-Paredes LM, Riosmena-Rodríguez R, Dodd RS, et al. (2012) Genetic population structure of red mangrove (*Rhizophora mangle L.*) along the northwestern coast of Mexico. Aquat Bot 99: 20–26.
19. McCauley DE (1991) Genetic consequences of local population extinction and recolonization. Trends Ecol Evol 6: 5–8.
20. Hewitt G (2000) The genetic legacy of the Quaternary ice ages. Nature 405: 907–913.
21. Dutech C, Maggia L, Tardy C, Joly H, Jarne P (2003) Tracking a genetic signal of extinction-recolonization events in a neo-tropical tree species: *Vouacapoua americana* Aublet in French Guiana. Evolution 57: 2753–2764.
22. Bialozyt R, Ziegenhagen B, Petit RJ (2006) Contrasting effects of long distance seed dispersal on genetic diversity during range expansion. J Evol Biol 19: 12–20.
23. Petit RJ, Aguinagalde I, de Beaulieu J -L, Bittkau C, Brewer S, et al. (2003) Glacial refugia: hotspots but not melting pots of genetic diversity. Nature 300: 1563–1565.
24. Nettel A, Rafii F, Dodd RS (2005) Characterization of microsatellite markers for the mangrove tree *Avicennia germinans L. (Avicenniaceae)*. Mol Ecol Notes 5: 103–105.
25. Ceron-Souza I, Rivera-Ocasio E, Funk SM, McMillan WO (2006) Development of six microsatellite loci for black mangrove (Avicennia germinans). Mol Ecol Notes 6: 692–694.
26. Rosero-Galindo C, Gaitan-Solis E, Cardenas-Henao H, Tohme J, Toro-Perea N (2002) Polymorphic microsatellites in a mangrove species, *Rhizophora mangle* L (Rhizophoraceae). Mol Ecol Notes 2: 281–283.
27. Lewis P, Zaykin D (2001) Genetic Data Analysis: Computer program for the analysis of allelic data.
28. Goudet J (2001) FSTAT, a program to estimate and test gene diversities and fixation indices. 2.9.3 ed. Switzerland: Institut d'Ecologie, Bâtiment de Biologie Université de Lausanne.
29. Raymond M, Rousset F (1995) GENEPOP Version 1.2.: population genetic software for exact test and ecumenicism. J Hered 86: 248–249.
30. Rousset F (2008) GENEPOP 007: a complete re-implementation of the GENEPOP software for Windows and Linux. Mol Ecol Resour 8: 281–283.
31. Van Oosterhout C, Hutchinson WF, Willis DPM, Shipley P (2004) MICRO-CHECKER: software for identifying and correcting genotyping errors in microsatellite data. Mol Ecol Notes 4: 535–538.
32. Pritchard JK, Stephens M, Donnelly P (2000) Inference of population structure using multilocus genotype data. Genetics 155: 945–959.
33. Falush D, Stephens M, Pritchard JK (2003) Inference of population structure using multilocus genotype data: Linked loci and correlated allele frequencies. Genetics 164: 1567–1587.
34. Evanno G, Regnaut S, Goudet J (2005) Detecting the number of clusters of individuals using the software structure: a simulation study. Mol Ecol 14: 2611–2620.
35. Weir BS, Cockerham CC (1984) Estimating F-statistics for the analysis of population structure. Evolution 38: 1358–1370.
36. Michalakis Y, Excoffier L (1996) A Generic Estimation of Population Subdivision Using Distances Between Alleles With Special Reference for Microsatellite Loci. Genetics 142: 1061–1064.
37. Slatkin M (1995) A measure of population subdivision based on microsatellite allele frequencies. Genetics 139: 457–462.
38. Hardy OJ, Vekemans X (2002) SPAGeDi: A versatile computer program to analyse spatial genetic structure at the individual or population levels. Mol Ecol Notes 2: 618–620.
39. Hardy O, Charbonnel N, Fréville HMH (2003) Microsatellite allele size: a simple test to asses their significance on genetic differentiation. Genet Soc Am 163: 1467–1482.
40. Excoffier L, Lischer HEL (2010) Arlequin suite ver 3.5: a new series of programs to perform population genetics analyses under Linux and Windows. Mol Ecol Resour 10: 564–567.
41. Excoffier L, Smouse P, Quattro J (1992) Analysis of molecular variance inferred from metric distances among DNA haplotypes: application to human mitochondrial DNA restriction data. Genetics 131: 479–491.
42. Beerli P, Felsenstein J (1999) Maximum-Likelihood Estimation of Migration Rates and Effective Population Numbers in Two Populations Using a Coalescent Approach. Genetics 152: 763–773.
43. Beerli P, Palczewski M (2010) Unified Framework to Evaluate Panmixia and Migration Direction Among Multiple Sampling Locations. Genetics 185: 313–326.
44. Otha J, Kimura M (1973) A model of mutation appropriate to estimate the number of electrophoretically detectable alleles in a finite population. Genet Res 22: 201–204.
45. Wilson GA, Rannala B (2003) Bayesian Inference of Recent Migration Rates Using Multilocus Genotypes. Genetics 163: 1177–1191.
46. Jensen J, Bohonak A, Kelley S (2005) Isolation by distance, web service. BMC Genet 6: 13.
47. Rousset F (1997) Genetic Differentiation and Estimation of Gene Flow from F-Statistics Under Isolation by Distance. Genetics 145: 1219–1228.
48. Nettel A, Dodd RS (2007) Drifting propagules and receding swamps: Genetic footprints of mangrove recolonization and dispersal along tropical coasts. Evolution 61: 958–971.
49. Nettel A, Dodd RS, Afzal-Rafii Z, Tovilla-Hernandez C (2008) Genetic diversity enhanced by ancient introgression and secondary contact in East Pacific black mangroves. Mol Ecol 17: 2680–2690.
50. Sherrod CL, McMillan C (1985) The distributional history and ecology of mangrove vegetation along the northern Gulf of Mexico coastal region. Contrib Mar Sci 28: 129–140.

51. Dodd RS, Afzal-Rafii Z, Kashani N, Budrick J (2002) Land barriers and open oceans: effects on genetic diversity and population structure in *Avicennia germinans L. (Avicenniaceae)*. Mol Ecol 11: 1327–1338.
52. Dodd RS, Raffi ZA, Bousquet Melou A (2000) Evolutionary divergence in the pan-Atlantic mangrove *Avicennia germinans*. New Phytol 145: 115–125.
53. Castillo-Cárdenas MF, Toro-Perea N, Cárdenas-Henao H (2005) Population genetic structure of neotropical mangrove species on the colombian Pacific coast: *Pelliciera rhizophorae (Pellicieraceae)*. Biotropica 37 (2): 266–273.
54. Salas-Leiva DE, Mayor-Durán VM, Toro-Perea N (2009) Genetic diversity of the black mangrove (*Avicennia germinans* L.) in Colombia. Aquat Bot 91: 187–193.
55. Ceron-Souza I, Bermingham E, McMillan W, Jones F (2012) Comparative genetic structure of two mangrove species in Caribbean and Pacific estuaries of Panama. BMC Evol Biol 12: 205.
56. Takayama K, Tamura M, Tateishi Y, Webb EL, Kajita T (2013) Strong genetic structure over the American continents and transoceanic dispersal in the mangrove genus *Rhizophora (Rhizophoraceae)* revealed by broad-scale nuclear and chloroplast DNA analysis. Am J Bot 100: 1191–1201.
57. Rabinowitz D (1978) Dispersal Properties Of Mangrove Propagules. Biotropica 10: 45–57.
58. Sousa WP, Kennedy PG, Mitchell BJ, Ordóñez BM (2007) Supply-side ecology in mangroves: do propagule dispersal and seedling establishment explain forest structure? Ecol Monogr 77 (1): 53–76.
59. Woodroffe CD, Grindrod J (1991) Mangrove Biogeography: The Role of Quaternary Environmental and Sea-Level Change. J Biogeogr 18: 479–492.
60. Ledesma-Vázquez J, Berry RW, Johnson ME, Gutiérrez-Sanchez S (1997) El Mono chert: A shallow-water chert from the Pliocene Infierno Formation, Baja California Sur, Mexico. Geol Soc Am Special Papers 318: 73–81.

24

Integration at the Round Table: *Marine Spatial Planning in Multi-Stakeholder Settings*

Erik Olsen[1]*, David Fluharty[2], Alf Håkon Hoel[3], Kristian Hostens[4], Frank Maes[5], Ellen Pecceu[4]

1 Institute of Marine Research, Bergen, Norway, **2** School of Marine and Environmental Affairs, University of Washington, Seattle, Washington, United States of America, **3** Institute of Marine Research, Tromsø, Norway, **4** Animal Sciences, Aquatic Environment and Quality, Bio-environmental Research, Institute for Agricultural and Fisheries Research (ILVO), Oostende, Belgium, **5** Maritime Institute, Ghent University, Ghent, Belgium

Abstract

Marine spatial planning (MSP) is often considered as a pragmatic approach to implement an ecosystem based management in order to manage marine space in a sustainable way. This requires the involvement of multiple actors and stakeholders at various governmental and societal levels. Several factors affect how well the integrated management of marine waters will be achieved, such as different governance settings (division of power between central and local governments), economic activities (and related priorities), external drivers, spatial scales, incentives and objectives, varying approaches to legislation and political will. We compared MSP in Belgium, Norway and the US to illustrate how the integration of stakeholders and governmental levels differs among these countries along the factors mentioned above. Horizontal integration (between sectors) is successful in all three countries, achieved through the use of neutral 'round-table' meeting places for all actors. Vertical integration between government levels varies, with Belgium and Norway having achieved full integration while the US lacks integration of the legislature due to sharp disagreements among stakeholders and unsuccessful partisan leadership. Success factors include political will and leadership, process transparency and stakeholder participation, and should be considered in all MSP development processes.

Editor: Judi Hewitt, University of Waikato (National Institute of Water and Atmospheric Research), New Zealand

Funding: The project was jointly funded by the EU FP-7 research project "Monitoring and evaluation of spatially managed areas (MESMA)" www.mesma.org, and by internal funding from the authors institutions: the Institute of Marine Research, University of Washington, Ghent University and the Institute for Agricultural and Fisheries Research (ILVO). The funders had no role in study design, data collection and analysis, decision to publish, or preparation of the manuscript.

Competing Interests: The authors have declared that no competing interests exist.

* Email: eriko@imr.no

Introduction

The management of marine ecosystems and their human activities underwent dramatic changes since the 1992 Rio Declaration on Environment and Development. Today sustainable development is still a fundamental principle, but following the 2002 Johannesburg Declaration, sustainable management has been expanded to encompass not only single species or sectors, but the whole ecosystems through the "ecosystem approach to management" (EBM) [1,2]. EBM is a powerful concept based on analysing and managing the ecosystems from a holistic approach, taking into account all components, pressures and impacts [3,4].

Because ecosystems are spatially explicit, area-based management approaches offer a suitable and efficient way of implementing EBM into practice [5,6]. One such approach is marine spatial planning (MSP), which offers an effective perspective to deal with the challenging issues of multiple use and multiple (cumulative) impacts in EBM [7]. Even though a spatial perspective has been successful in terms of leading to novel and forward-looking ecosystem-based management plans, like in Norway and Australia [8–11], the development of marine spatial plans is a complex process at the borders between science, management and politics. Several approaches, like the UNESCO 10-step approach [12], have been developed to lead the practitioners safely from start to finish during planning. The establishment of effective governance of the planning, implementation and review processes, is a fundamental step to develop sound MSP [12,13].

Because MSP by definition is multi-sectoral, a potentially large number of managers, stakeholders and policy-makers are involved, each accustomed to operate on his own (*i.e.* within specific sectors). Successful MSP means getting all these actors to communicate and work together in an integrated way. Therefore, integration means crossing boundaries at professional, physical, institutional or administrative level [14]. To develop the appropriate measures in an integrated (ecosystem) MSP setting, the integration of concerns and interests across sectors (horizontal integration) and between governmental levels or between government and stakeholders (vertical integration) is required. Integration is fundamental to MSP and especially important to pro-actively resolve spatial conflicts [14].

In this paper we explore how such integration has been dealt with in three markedly different MSP processes: the Belgian MSP covering the whole Belgian EEZ in the North Sea, the Norwegian Integrated Management plans – three plans covering the Norwegian EEZ, and the current US National Ocean Policy (Figure 1) regional planning process now being implemented. We illuminate how the integration of concerns and interests varies with

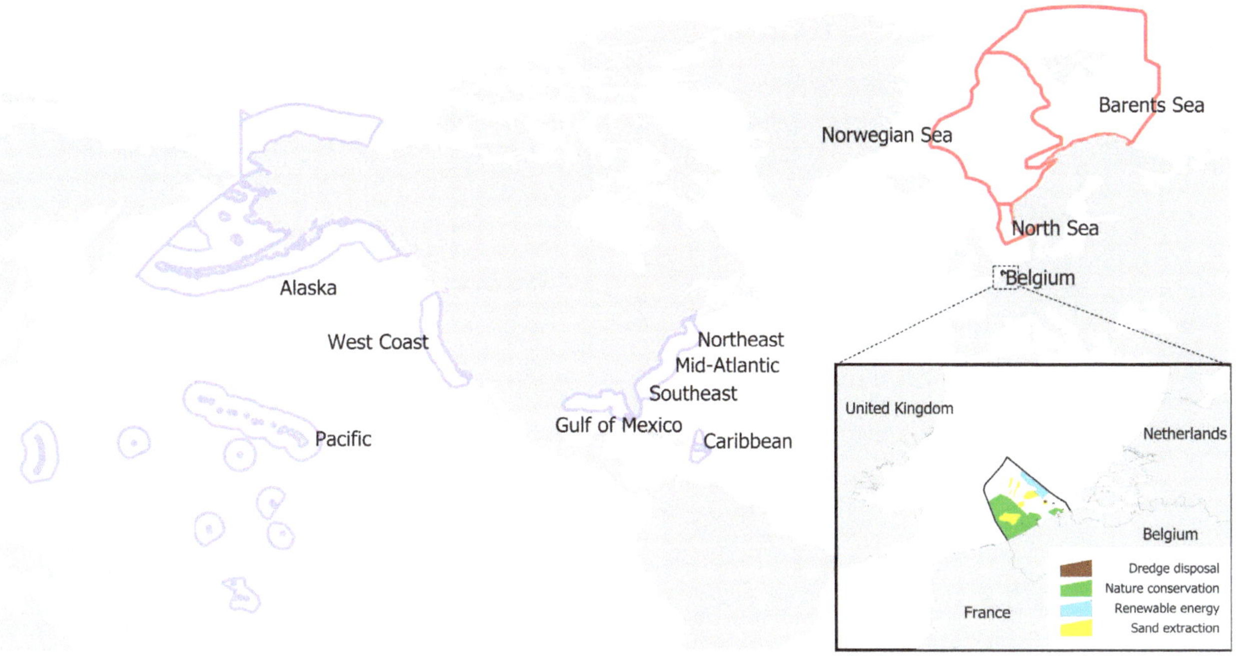

Figure 1. Marine spatial planning (MSP) study areas around the North Atlantic. United States (blue), Norway (red) and Belgium (Green, detail in inset). The US ocean policy area is divided into 9 planning areas (named in figure) that span the EEZ, but the borders between each of these areas have not been officially defined.

the context and commonalities that allow for a successful integration of different viewpoints and hence successful governance of MSP.

The Three Case Studies

Belgium: from Masterplan to MSP

In March 2014, Belgium approved a legally binding Marine Spatial Plan (Royal Decree of 20 March 2014 on the adoption of the MSP) for the Belgian part of the North Sea (BPNS), a small area of ca. 3500 km^2 covering the territorial sea, the continental shelf and the entire Belgian Exclusive Economic Zone (EEZ). MSP in Belgium is the result of a long process, going back to initial attempts to implement the 1999 Marine Environmental Protection Act (MEPA), followed by the so-called "Masterplan for the BPNS" in 2003–2005. Since 2012 an explicit legal basis for MSP in Belgium is embedded in the MEPA, which clearly indicates the environmental roots of MSP. Therefore, governments and users of the sea in Belgium have to take into account several legal principles, namely the principle of prevention, the precautionary principle, the polluter pays principle and the restoration principle (art. 5 MEPA).

Belgium functions as a multi-level government with authorities divided among local, regional, federal (and European) levels. The marine competences are divided between the federal state and the Flemish region, and within each level, competences are fragmented over several departments [15]. No less than 17 governmental institutions (both Flemish and federal) have some form of competence at sea. Both governments are legally equal in adopting their legislation within their fields of competence (Figure 2). Consultation between the two governments can be held at ministerial level, between their cabinets or between their administrations. Matters in which both governments have different exclusive competences can be jointly dealt with in formal cooperation agreements.

It took until 2003, with the appointment of a federal Minister for the North Sea, to set a major step forward in the Belgian MSP process. This Minister was mandated to spatially co-ordinate all activities and competences at sea, except for fisheries which is a regional (Flemish) competence. He initiated a more strategic approach to the (potential) conflicting claims between users of the BPNS which emerged between 1999 and 2003. He developed the BPNS Masterplan, which actually combined several political decisions concerning marine matters that were taken by the federal Council of Ministers and implemented by a number of Royal decrees. The plan was further adopted in two phases: in 2003 with an agreement on the delimitation of several zones for two economic activities- aggregate extraction and offshore renewable energy - and later in 2005 with the delimitation of nature conservation areas for birds and habitat protection. However, it remained essentially a zoning plan, without a legal basis for a planning process or an integrated policy approach, nor a clear and transparent process for stakeholder involvement and public participation [16].

In Belgium, there is strong vertical integration between the highest levels (Ministers) and their public administrations. Moreover, each governmental level has its own advisory bodies (Figure 2), representing the major stakeholders, e.g. the Federal Council for Sustainable Development (FRDO), the Flemish Council for Environment and Nature (MiNa Council) and the Flemish Strategic Advisory body for Agriculture and Fisheries (SALV). It is important to recognize that despite the existence of sectoral cooperation agreements and advisory bodies, the concerned parties only have the mandate and responsibility for their own competences.

Stakeholders have been involved in the MSP process since the Masterplan was presented in 2003, but this happened only on a sectoral basis and mostly through bilateral consultations behind closed doors [17]. This was in line with an old Belgium practice of informal consultations within and between the governments and

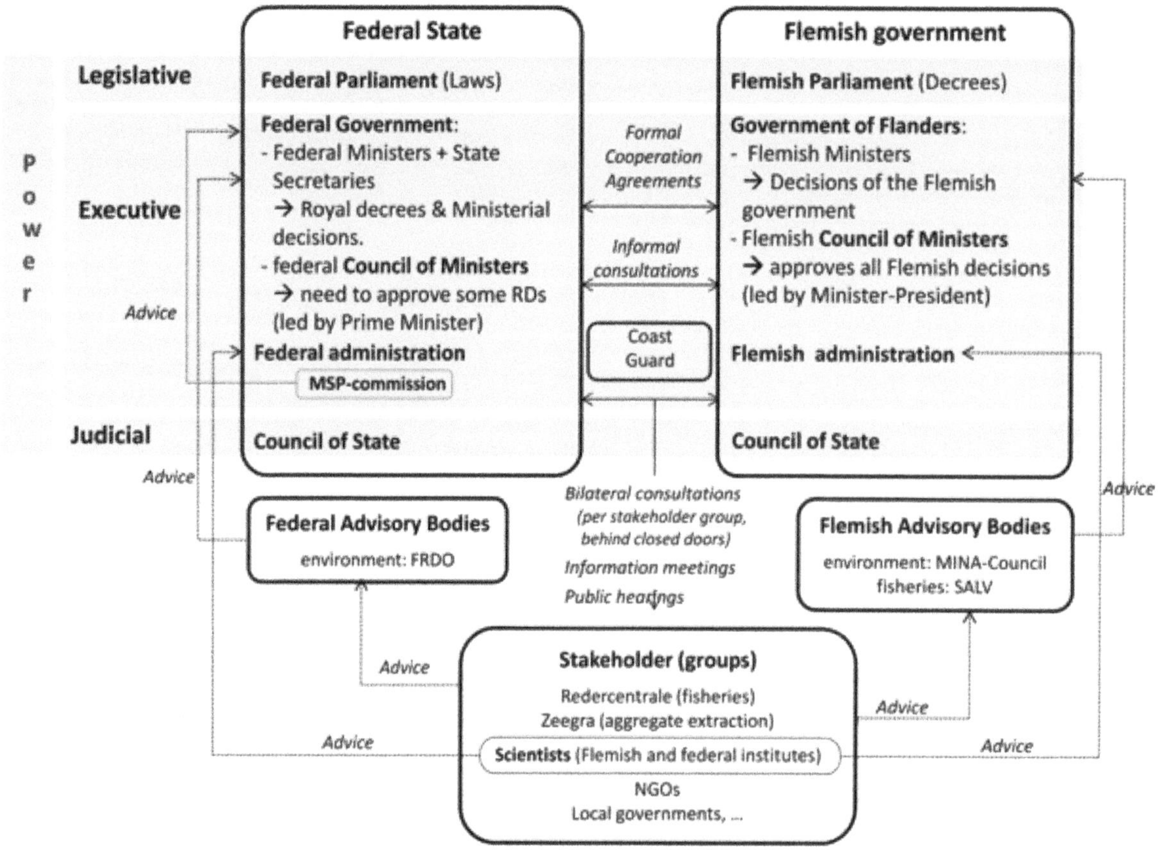

Figure 2. Institutional and stakeholder integration in Belgium. Institutional bodies, stakeholder participation and integration, related to marine and maritime governance in the Belgian part of the North Sea.

administrations, as well as between the governments and stakeholders. As such the Masterplan was drawn up on an *'ad hoc'* basis, taking into account the individual and selective demands of sectoral organizations and individual stakeholders. As a consequence, consultation with other sea users than the ones in the focus of the Masterplan was not deemed necessary. After 2006, the Masterplan has been adjusted a few times, in particular to take into account shipping interests and to designate a new marine protected area (2008–2010). However, these developments were once again dealt with at sectoral level.

In 2012, the newly elected federal government re-appointed the same Minister for the North Sea, and a second phase of MSP started. The new marine spatial plan departed from the BPNS Masterplan, but with a clear aim to establish a 'legally binding' marine spatial plan and process. The draft spatial plan, which was again based on informal and formal consultations with stakeholders, was subjected to a series of administrative consultations and advice from a new Advisory Commission on MSP. The latter was formally established by a Royal Decree in November 2012 and involves all federal authorities with competences at sea, next to 'invited' Flemish authorities. As such, an integration between all federal governmental departments concerning MSP became formally embedded in the new MSP legislation.

Next to the MSP Advisory Commission, experts were consulted, stakeholder participation was organized and a strategic environmental assessment (SEA) of the spatial plan was conducted. In July 2013, an updated plan was approved by the Council of Ministers, followed by consultation with the public and neighboring countries and advice from advisory bodies and regional authorities. The Ministry of Environment took into account 140 remarks and finally the Council of Ministers approved the MSP by Royal Decree on 20 March 2014.

The new Belgian marine spatial plan comprises the coordinates of all delimitated zones for the activities that are allowed and limited or prohibited in the BPNS, including some new zones for future activities (energy atolls, sustainable aquaculture, etc.). The Royal Decree also contains several annexes outlining the spatial context and explaining the policy choices that have been made. The long term vision to implement MSP in Belgium foresees concrete objectives for the period 2014–2020 and a revision of the spatial plan every six years, although in between adaptation of the plan is not excluded.

A major shortcoming of the Belgian MSP might be that fisheries is only partly included in the MSP, as this is a not a competence of the federal government (Flemish competence). Nevertheless, consultations took place at the highest political (ministerial) level, and certain limitations for fisheries are included in the new MSP, such as the prohibition of fisheries in the offshore renewable energy concession zones, and a limitation for certain fishing techniques (mainly classic bottom disturbing trawling techniques) in specific parts of the Natura 2000 area 'Vlaamse Banken' in order to achieve a good environmental status for the BPNS.

Norway: Integrated Management Plans

As a response to the international [1] and regional drivers (North Sea ministerial conference in 1997) for EBM as well as national pressures from the petroleum industry to get access to new areas further north and along the coast, the new government heralded marine Integrated Management Plans (IMPs) in its coming-to-power declaration in 2001 [18]. The development of these IMPs started in 2002, with the first plan, for the Lofoten–Barents Sea area implemented in 2006 [8,18], the Norwegian Sea plan in 2009 [9], and the last plan for the North Sea and Skagerrak area in 2013 [19]. These IMPs are regional plans integrating all human uses of the area in a spatial context through sectoral zoning and geographical analysis. The Lofoten-Barents Sea plan has been revised once in 2011, the Norwegian Sea plan is due for revision in 2014.

An inter-ministerial steering group led by the Ministry of Environment was set up to coordinate the development and implementation of these IMPs. The steering group had members from all relevant sector-ministries (e.g., Foreign Affairs, Fisheries and Coastal Affairs, Petroleum and Energy). This strong political (top-down) steering group tasked the different institutions and directorates under each ministry to contribute to the development and implementation of the plan. Each plan was customized for the particular ecosystem, but with a similar overarching strategic objective: "*to allow for sustainable use while ensuring continued ecosystem health*" [20]. The duality in this main strategic objective reflects the underlying conflict to integrate two widely different objectives, namely the push for increased industrial (petroleum) developments and the protection of the ecosystems (to ensure the health and survival of species and habitats) [21].

Integration of different interests and concerns is central to the Norwegian MSP process, and already commenced at the start of the planning process in 2002. The various sectors and levels (of government) have been integrated by jointly developing the knowledge base and management measures across sectors and between levels of government (Figure 3). Joint groups, forums and meeting places across sectors have been set up to achieve horizontal – cross-sectoral integration. Hearings, open public meetings and sectoral meetings with the government help to foster cooperation between the various governmental levels and with stakeholders in particular.

The Norwegian planning process is centralized, where decisions are made by the government and subjected to parliamentary approval. As such, all IMPs have been passed with broad political approval in the Parliament. Involvement of local government levels (e.g. counties) has been poor, mainly because ocean management is a national issue in Norway, with only limited local control over marine resources or management of marine space.

The overarching conflict of industrial use versus nature protection has been a central issue for debate at all levels, in particular in the management and risk forums (Figure 3). In these multi-sector groups the different views allowed for cross-sectors insight and discussion, and although this has not resolved the conflict, it has increased mutual understanding and respect, based on the personal, professional and institutional relationships that have been established during the planning and implementation process. The networking has increased trust among the participants, which made it possible to discuss difficult and challenging issues.

Imposing consensus between the sectoral government institutions was instrumental to develop and implement the Lofoten-Barents Sea plan in a 4-year time frame. This first IMP spearheaded the development of the other plans and created much of the structure and methods to allow the two later plans to be created in 3-year processes. However, seeking consensus could also lead to suppression of substantial and value-based differences. It can be questioned whether the decision-makers would be better served if the differing views and options in the central conflict (industrial use vs. nature conservation) would have been clearly presented as different choices rather than hidden in a compromise. A consensus approach limits the options for the decision-makers, who normally prefer a range of options which allows them more room to govern.

The existing sectoral legislation and management structures have been the main barrier for integration. Using soft-law in processes like the development of IMP's, is a tradition in Norway, and is fast, pragmatic and efficient. Also, by implementing through existing legislation and institutional structures, the policy is rapidly and effectively translated into practice without spending much time on developing new legislation or governance structures. However, this approach does little to integrate concerns across the existing sectoral legislation. Nevertheless, many new sectoral laws passed since 2002, took the integrated management approach into consideration and specifically required integration between sectors.

US: Ocean Policy Executive Order

Influenced by international MSP developments and strongly promoted by major environmental organizations, the US started a process towards developing an Ocean Policy [22] including MSP. The first steps were already taken by the Bush administration, but development efforts increased in 2008 with the new Obama administration. After 18 months of intensive planning, the National Ocean Policy was effectuated through President Obama's Executive Order 13547 (July 19, 2010). This Executive Order established the National Ocean Council, consisting of 27 federal agencies, departments and offices. Additionally, the US marine and Great Lakes areas are divided into nine regions, which can form regional planning bodies on a voluntary basis to develop the planning process in their respective regions (Figure 4). The National Ocean Council published an Implementation Plan [23], that further gives guidance to the administration and partner institutions on how the Ocean Policy should be implemented [24]. The Obama Ocean Policy approach relies on existing legal mechanisms [25].

The Ocean Policy can be seen as a MSP process, although the term 'marine spatial planning' is not used in the most recent documents as it has become politicized to the point that the US Congress in 2012 and 2013 refused to fund any MSP related activities. The majority in the House of Representatives see MSP as an unnecessary layer of government, hampering the development of business activities at sea. Also industrial (oil, fishermen, etc.) lobby groups (e.g., National Ocean Policy Coalition, see http://oceanpolicy.com) advocated against MSP. Therefore, MSP implementation has ended up in the middle of the partisan battlefield, even though the first steps were taken by the Bush Administration with bipartisan support. As a result, the federal support for the Ocean Policy has decreased compared to what was anticipated in 2010. Currently, the Ocean Policy implementation is totally dependent on support from the State and regional levels, who are leading the way while the federal level lags behind. Substantial financial support has however been provided by private donors to MSP efforts in the Northeast and Mid-Atlantic regions. It is critically important that top-down and bottom-up processes are linked in the implementation of the US Ocean Policy [26], however the situation today is charitably defined as experimental and otherwise defined as chaotic when looking across all nine regions.

Some States have taken a leadership positions in MSP for their coastal environments, e.g., Rhode Island, Massachusetts, Oregon, New York, Hawaii and Florida, with Washington State not far behind [27]. Also, California has taken major steps to develop

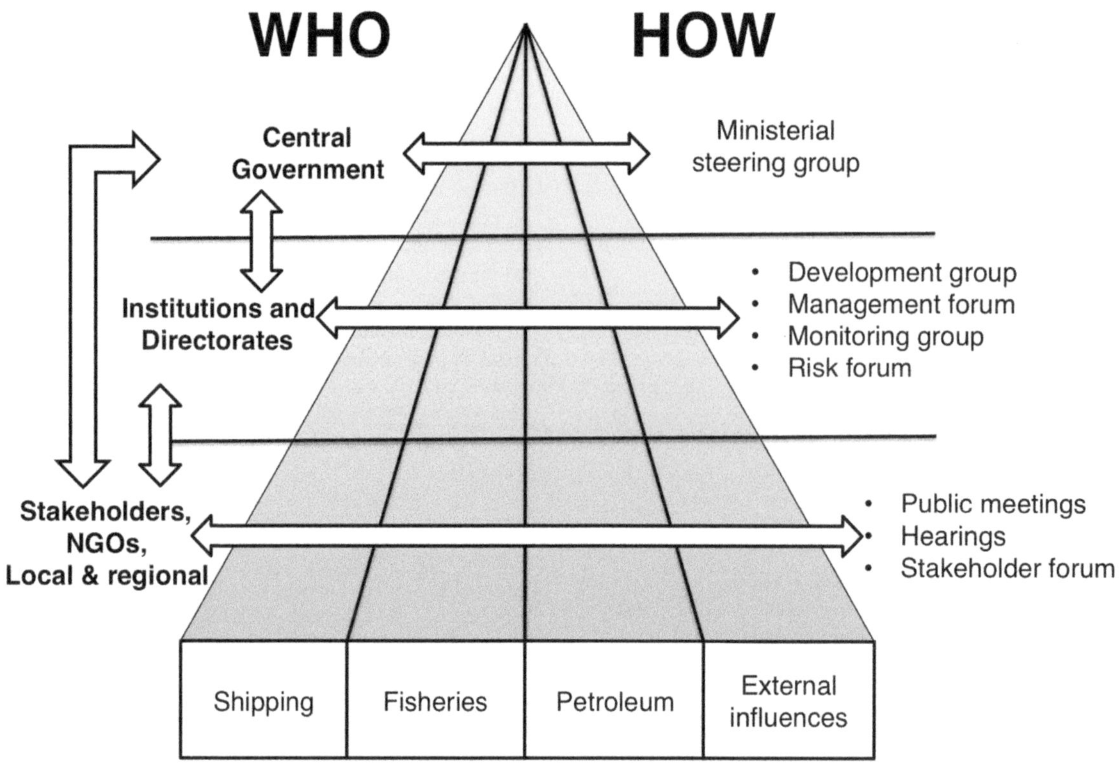

Figure 3. Institutional, government, and stakeholder integration in Norway. Institutional bodies, stakeholder participation and integration, related the development and implementation of the Integrated Management plan for the Lofoten – Barents sea area.

marine spatial planning through a marine reserve network in state waters. In addition, Maine, New Hampshire, Massachusetts, Rhode Island and Connecticut together with federal partners formed the Northeast Regional Planning body (RPB) as envisioned by the Ocean Policy (Figure 4). The Northeast RPB constituted itself, meets regularly, and is on its way to develop a regional plan in 2015. Also, the Mid-Atlantic and Southeast Atlantic have formed RPBs, with the Mid-Atlantic having started a planning process in 2014, while the Southeast is a bit delayed by the limited participation from Florida. These RPBs are expected to produce regional MSPs by 2015 for review and approval by the National Ocean Council but this timeline is ambitious.

Other regions, except Alaska, have taken up on the Ocean Policy and have or are in the process of forming RPBs. Hawaii (part of the Pacific region) had a state plan under development when the Ocean Policy was being put forward. The greatest challenge in the Pacific region is distance, as it includes Hawaii and all US Pacific Islands, spanning seven time-zones and the international date-line. Finding time, place and funding to bring regional representatives together, is logistically very demanding. The Pacific RPB may ultimately adopt a more archipelagic approach as has been done for fisheries management [28].

The National Ocean Policy envisions a regional approach to develop MSPs for large management regions, *in casu* ecosystems, based on the regional administrative capacity. However, it is recognized that the desire for standard approaches across the nation may not be accommodated at the regional level. Therefore, a flexible structure of Federal agency engagement with States and local governments is required. All current RPBs have wide participation of federal, state and tribal governmental institutions that have formal mandates for ocean management (Figure 4). The RPB work and meetings are open to the public and public interaction is actively encouraged through specific sessions where the public is allowed to ask questions or make comments to the full RPB. These occasions are actively used by self-selected and motivated NGOs, interest groups and the public to voice their perspectives.

Discussion - Comparing Integration in the Three Cases

Spatial scale of the managed area

Belgium, Norway and the US represent three different approaches of implementing MSP, stemming from both geography and governance [26]. Whereas the Belgium plan spans the whole EEZ and Belgian national waters, an area of a few thousand square kilometres, both the Norwegian and US plans span millions of square kilometres under different climate and biogeographic regions, making them much more complex than the Belgium plan area. To better handle the varied climatological and geographic patterns, both the US and Norway chose to divide their marine areas into (eco)regions and are developing separate plans for each area. Regardless of the size of the spatially managed area, it remains a complex issue to manage all types of human activities under one umbrella, requiring integration between different sectors, stakeholders and governmental levels. Reducing the size of the planning area will reduce complexities, but only to a certain level, as the multi-sectoral cross-political aspects remain, irrespective of spatial scale.

Legislation, political will and leadership

The three countries differ markedly in terms of governance: Belgium is an EU member and marine governance has to adhere to EU policies (e.g. Common Fisheries Policy) and directives (e.g.,

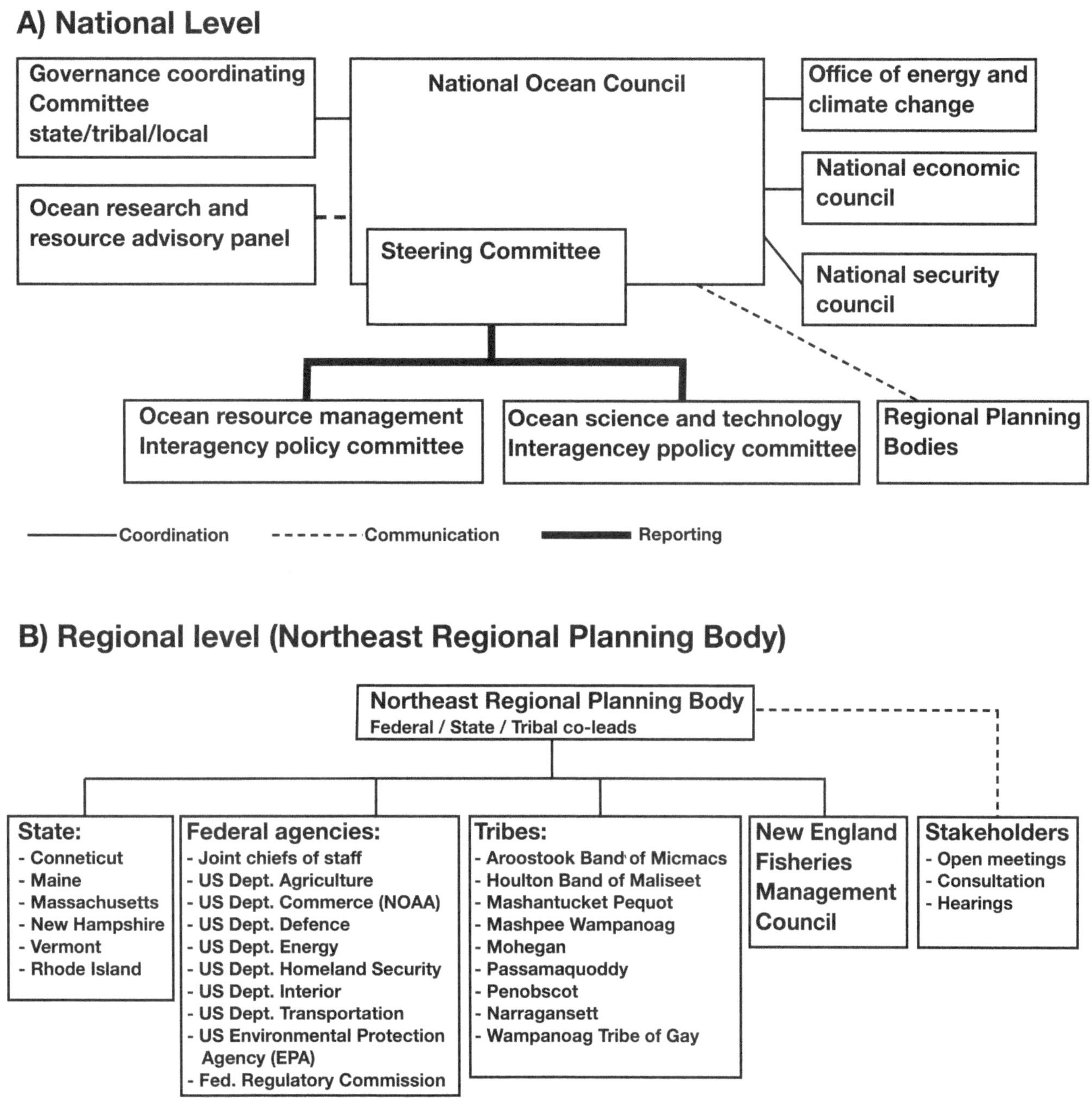

Figure 4. Governance structure and integration in the US Ocean Policy development. A) At a national level leading the Ocean Policy development. B) Regional implementation in regional planning bodies (RPB), exemplified by the Northeast RPB.

Habitats and Bird, Marine Strategy Framework, Water Framework). Belgium has a strong formal division of power between the federal and regional levels, although integration between both governmental levels is assured through formal agreements and informal consultations. This is also reflected in the protracted Belgian MSP process: after more than 10 years, the MSP plan and process are firmly embedded in environmentally driven legislation.

Norway is not a member of the EU and has an ethnically homogenous population. The government is well integrated across sectors both at central and regional levels. The Norwegian Integrated Management Plans (IMPs) were passed through the Parliament, but are not embedded in Norwegian legislation. They are rather government reports with MSP objectives and a vision on how they will achieve these objectives through a range of management options, including spatial zoning.

In the US the independency of the states led to a geographic division in marine management at the border between federal and state waters, mostly three nautical miles offshore. The US Ocean Policy process, based on a Presidential Executive Order, does not have backing in the Congress, which led to funding restrictions to develop or continue the MSP process. The US Ocean Policy became heavily politicized in 2010 and has since been drawn into a struggle among stakeholders largely along commercial users of ocean space and biodiversity conservation, spilling over into the partisan debates in Congress. Even though embedding MSP in legislation creates a more stable solution in heterogeneous societies like Belgium and the US, achieving overall trust and buy-in is instrumental for a successful implementation (and funding) of the MSP process and plan.

In contrast to Norway and Belgium, broad political support for implementation of MSP has not been reached in the US, due to the lack of strong stakeholder support for MSP by commercial and non-commercial users. Strong political leadership was instrumental to achieve that bipartisan support for MSP in both Belgium and Norway. In Belgium the personal charisma, tenacity and clout of the Minister of the North Sea was pivotal to lead the planning process and persuade the different parties of the necessity of MSP. In Norway it was the tenacity and clout of the Ministry of Environment, strongly supported by the Ministries of Fisheries and Foreign Affairs that pushed the MSP process. Moreover, achieving broad political buy-in was relatively easy in Norway, as the MSP process was partly driven by Norway's foreign policy. Foreign priorities are very stable in Norway and do not change with changing governments. In the US leadership has certainly been shown at the Executive level but it is not sufficient to carry through to the legislative leadership.

Drivers and incentives

Leaving aside political and ethnic differences between the countries, socio-economics, i.e. the importance of certain marine industries in the national economy of each country, help to explain some of the differences in their governance approach to MSP. In Belgium, maritime infrastructure (shipping and ports) is very important, next to aggregate extraction and to a lesser extent fisheries. Since 2003 offshore renewable energy (wind-farms) developments have been the main driver for MSP in Belgium. However, adherence to EU environmental directives (Habitats, Bird, Marine Strategy Framework Directive) has created a second driver in MSP in order to identify areas of high biological value (Natura2000 sites) and to achieve a "good environmental status" of the ecosystem. In Norway, the marine natural resources (oil and fish) form the basis for the country's welfare, and sea-based industries (oil/gas, fisheries and aquaculture) are of fundamental socio-economic importance. It is a high priority for all governmental levels to create (sustainable) marine value in Norway. The health of the marine ecosystem is understood as an important prerequisite for sustainable value creation, but this is not seen as an absolute hindrance to human uses. Therefore, protecting areas by excluding economic activities is not common in Norway.

In the US, sea-based industries, shipping and fishing are of great socio-economic importance, but the country is not as dependent on the exploitation of marine resources (oil/gas and fish) as Norway. The importance and political power of the different sectors vary regionally, e.g., in the Gulf of Mexico the petroleum industry is given priority, while in the Northeast region offshore renewable wind energy has been a driver for MSP. The US has a strong terrestrial tradition of using protection (national parks and forests) as a management measure. This has been transferred to the marine environment with the establishment of several MPAs with a high protection level like the Papahānaumokuākea Marine National Monument (MNM) in the north-western Hawaiian islands, the Marianas Trench MNM and the Pacific Remote Islands MNM.

The economic incentives have led to different objectives in the respective MSP processes. Although all three countries try to combine ecosystem health with economic use, they differ in their approach. While Norway has a clearly stated dual goal of sustainable use while ensuring a healthy ecosystem, Belgium is focusing on protection of valuable habitats and ecosystem components while finding room for new uses and minimizing conflicts. The US objectives are complex and not prioritized in terms of trade-offs between economic use and nature conservation.

Stakeholder participation and transparency

Integrating stakeholders in the planning process has been important, but done in different ways in the three countries. In Norway and Belgium, stakeholders were brought early on board, which led to acceptance of the MSP process, although with much debate about specific details of the plan. In the US, stakeholders were invited to comment during the 18 month process leading up to the federal Ocean Policy Executive Order. This process built on many years of responding to the US Commission on Ocean Policy and the non-governmental Pew Oceans Commission processes. However, there was considerable surprise that MSP was the most tangible action required by the Executive Order. This shift in approach resulted in several industrial stakeholder groups forming lobby-groups against MSP implementation, which blocked necessary incremental funding of the MSP process. The resultant stakeholder skepticism conveniently fed into the partisan political battle in Washington D.C. In the meantime, stakeholder involvement and transparency is much more effective at the state and regional level in the US. Threats and opportunities are tackled depending on one's perspective from existing spatial conflicts, and through new drivers primarily associated with both non-renewable (import/export) and renewable (offshore) energy. A credible threat to established interest adds to the acceptance and continuation of the MSP process in most US regions. Full transparency, open meetings, public documents and active solicitation of stakeholder opinions, make the US stand out compared to Belgium and Norway, where transparency was more limited and controlled.

Conclusion – Integration at the MSP Round Table

Ecosystem-based management in general and MSP in particular is a multifaceted management approach that combines the management of multiple sectors and goals under one roof through integration. Integration can be measured both in relation to geographic scale and governance scope [14] according to which all three cases (Belgian, Norwegian and the US) can be classified as "highly integrated". However, there are some marked differences between the cases in their horizontal and vertical integration (Table 1).

Horizontally, all three processes are similarly well-integrated, but it has been achieved through different governance approaches. In Belgium a "Minister for the North Sea" was appointed to lead the process, Norway set one ministry (Environment) in charge to lead a multi-ministerial steering group in which the existing ministries were tasked to cooperate, while the US set down an Executive (Ocean Policy) Task Force, directly reporting to the Council on Environmental Quality in the White House and operating through government agencies on a (eco)regional basis. Both the Belgian and Norwegian processes are integrated on a national level, while the US proposed to integrate on a regional level (which is sensible as US regions are much larger than most European states). Horizontal integration across sectors relies strongly on each sector perceiving it to be treated in a fair and equitable manner. Therefore, meeting places and forums should be neutral – amalgamating the Arthurian 'round table' of Camelot that created peace among the noble but imperious knights closest to the king (in our examples the relevant ministers or president).

Vertically, Belgium and Norway are well-integrated from parliament (through the executive government levels) to stakeholders, while the US is well integrated from the Executive level

Table 1. Horizontal, vertical and stakeholder integration in the MSP processes in Belgium, Norway and the US.

	Horizontal integration	Vertical integration	Stakeholder integration
Belgium	Good integration. Minister for the North Sea ensured a high government focus and top-level integration. Important federal and partly regional integration achieved	All levels of government well integrated	Consultation and formal roles in planning group
Norway	Good integration both at top government levels and at government institutional levels	All levels of government integrated	Consultation and participation in some meetings
US	Good integration between sectors at state and regional levels where direct threats or opportunities are perceived. Poor integration at national level due to lack of prioritization of objectives and failure to produce adequate resources to support on-going efforts	Lacking integration with lawmakers (congress).	Open public meetings of regional planning bodies. Consultation at national level. Strong stakeholder lobbying

downwards, but lacks integration (acceptance) from the lawmakers (Congress).

Next to informal consultations with sectoral stakeholder groups, stakeholders also have a formalized role in some planning forums in Belgium. In Norway and the US, stakeholder participation remains informal and voluntary or self-selected (through consultations), although in the US stakeholder involvement is more comprehensive, reflected in the strong lobbying work of stakeholder groups to the government and the ability to file lawsuits based on administrative procedures and environmental compliance. Planning meetings are open to the public in the US, in contrast to Norway where meetings in governmental steering groups are held behind closed doors. The latter was also true for Belgium in the first MSP (2003–2005), but changed with the adoption process of the second MSP (2013–2014) when stakeholder participation and public consultation were included in the process. Although the current political situations may be regarded as challenging, the three cases illustrate the benefit of having a wide political backing to efficiently implement broad management processes like MSP. Where the MSP process tends to be top-down, formally defined processes and leadership are deemed necessary, while also bottom-up processes are highly valuable when it comes to integrating across vertical and horizontal scales.

In Belgium and Norway the process has led to effective and practical implementation of MSP, while in the US implementation is still pending. Our analysis shows that all three planning efforts can be regarded as well-integrated in intent, and that integration is important for a successful MSP. In addition to seeking and supporting political leadership, planners should keep in mind that processes and structures that facilitate effective integration are needed. Transparency and openness of access to stakeholders to make their perspectives be heard and analysed in the total process is a large part of what is required to ensure fair and equitable treatment, in relation to both horizontal and vertical integration.

Acknowledgments

This work was carried out as part of the EU FP-7 project "Managing and evaluation of spatially managed areas" (MESMA).

Author Contributions

Conceived and designed the experiments: EO DLF AHH FM EP. Performed the experiments: EO DLF AHH KH FM EP. Analyzed the data: EO DLF AHH KH FM EP. Contributed reagents/materials/analysis tools: EO DLF AHH KH FM EP. Contributed to the writing of the manuscript: EO DLF AHH KH FM EP.

References

1. United Nations (2002) Report of the World Summit on Sustainable Development. Johannesburg. 173pp.
2. Barange M (2003) Ecosystem science and the sustainable management of marine resources: from Rio to Johannesburg. Frontiers in Ecology and the Environment 1: 190–196.
3. Link JS, Browman HI (2014) Integrating what? Levels of marine ecosystem-based assessment and management. ICES Journal of Marine Science. doi:10.1093/icesjms/fsu026.
4. Browman HI, Stergiou KI (2004) Perspectives on ecosystem-based approaches to the management of marine resources. Marine Ecology Progress Serie 274: 269–303.
5. Ogden JC (2010) Marine spatial planning (MSP): A first step to ecosystem-based management (EBM) in the Wider Caribbean. Revista de biología tropical 58(3): 71–79.
6. Douvere F (2008) The importance of marine spatial planning in advancing ecosystem-based sea use management. Marine Policy 32: 762–771.
7. Katsanevakis S, Stelzenmüller V, South A, Sørensen TK, Jones PJS, et al. (2011) Ecosystem-based marine spatial management: Review of concepts, policies, tools, and critical issues. Ocean & Coastal Management 54: 807–820.
8. Olsen E, Gjøsæter H, Røttingen I, Dommasnes A, Fossum P, et al. (2007) The Norwegian ecosystem-based management plan for the Barents Sea. ICES Journal of Marine Science 64: 599–602.
9. Ottersen G, Olsen E, van der Meeren GI, Dommasnes A, Loeng H (2011) The Norwegian plan for integrated ecosystem-based management of the marine environment in the Norwegian Sea. Marine Policy 35: 389–398.
10. Day V, Paxinos R, Emmett J, Wright A, Goecker M (2008) The marine planning framework for South Australia: A new ecosystem-based zoning policy for marine management. Marine Policy 32: 535–543.
11. Kenchington RA, Day JC (2011) Zoning, a fundamental cornerstone of effective Marine Spatial Planning: lessons learnt from the Great Barrier Reef, Australia. J Coast Conserv 15: 271–278.
12. Ehler C, Douvere F (2009) Marine Spatial Planning: a step-by-step approach toward ecosystem-based management. Paris: Intergovernmental Oceanographic Commission and Man and the Biosphere Programme. IOC Manual and Guides No. 53, ICAM Dossier No. 6. 99pp.
13. Qiu W, Jones PJS (2013) The emerging policy landscape for marine spatial planning in Europe. Marine Policy 39: 182–190.
14. Portman ME (2011) Marine spatial planning: achieving and evaluating integration. ICES Journal of Marine Science 68: 2191–2200.
15. Pecceu E, Hostens K, Maes F (2013) The evolution of marine protected areas in the Belgian Part of the North Sea. A case study report for Work Package 6 of the MESMA project (www.mesma.org). MESMA report. 49pp.
16. Maes F, Vanhulle A, Lescrauwaet AK (2013) Marine spatial planning. In: Lescrauwaet, AK, Pirlet, H, Vereleye, T, Mees, J, Herman, R, editors. Compendium for Coast and Sea: Integrating Knowledge on the Socio-economic, Environmental and Institutional Aspects of the Coast and Sea in Flanders and Belgium. Ostend. pp. 261–273.
17. Bogaert D, Cliquet A, De Waen D, Maes F (2008) The designation of marine protected areas in Belgium. An analysis of the decision making process. In: Bogaert D, Maes F, editors. Who rules the coast?. Antwerp. pp. 59–105.
18. Anon (2001) Politisk grunnlag for en samarbeidsregjering. Sem-erklæringen. Sem. 52pp. [in Norwegian]

19. Hoel AH, Olsen E (2012) Integrated Ocean Management as a Strategy to Meet Rapid Climate Change: The Norwegian Case. AMBIO 41: 85–95.
20. Anon (2006) Helhetlig forvaltning av det marine miljø i Barentshavet og havområdene utenfor Lofoten (forvaltningsplan). Oslo: Det kongelige miljøverndepartement. Oslo. 141pp. [in Norwegian]
21. Misund OA, Olsen E (2013) Lofoten–Vesterålen: for cod and cod fisheries, but not for oil? ICES Journal of Marine Science 70: 722–725.
22. Lubchenco J, Sutley N (2010) Proposed U.S. Policy for Ocean, Coast, and Great Lakes Stewardship. Science 328: 1485–1486.
23. Anon (2013) National Ocean Policy Implementation Plan. Executive Office of the President - National Ocean Council. Washington DC. 36pp.
24. National Ocean Council (2013) Marine planning handbook. Executive Office of the President. Washington DC. 22pp.
25. Environmental Law Institute (2009) Marine Spatial Planning in US Waters: An Assessment of Existing Legal Mechanisms, Anticipated Barriers, and Future Opportunities. Washington DC. 94pp.
26. Sievanen L, Leslie HM, Wondolleck JM, Yaffee SL, McLeod KL, et al. (2011) Linking top-down and bottom-up processes through the new U.S. National Ocean Policy. Conservation Letters 4: 298–303.
27. Collie JS, Vic Adamowicz WL, Beck MW, Craig B, Essington TE, et al. (2013) Marine spatial planning in practice. Estuarine, Coastal and Shelf Science 117: 1–11.
28. Glazier E (2011) Ecosystem Based Fisheries Management in the Western Pacific. Blackwell-Wiley Publishers, Hoboken, New Jersey. 400pp.

PERMISSIONS

All chapters in this book were first published in PLOS ONE, by The Public Library of Science; hereby published with permission under the Creative Commons Attribution License or equivalent. Every chapter published in this book has been scrutinized by our experts. Their significance has been extensively debated. The topics covered herein carry significant findings which will fuel the growth of the discipline. They may even be implemented as practical applications or may be referred to as a beginning point for another development.

The contributors of this book come from diverse backgrounds, making this book a truly international effort. This book will bring forth new frontiers with its revolutionizing research information and detailed analysis of the nascent developments around the world.

We would like to thank all the contributing authors for lending their expertise to make the book truly unique. They have played a crucial role in the development of this book. Without their invaluable contributions this book wouldn't have been possible. They have made vital efforts to compile up to date information on the varied aspects of this subject to make this book a valuable addition to the collection of many professionals and students.

This book was conceptualized with the vision of imparting up-to-date information and advanced data in this field. To ensure the same, a matchless editorial board was set up. Every individual on the board went through rigorous rounds of assessment to prove their worth. After which they invested a large part of their time researching and compiling the most relevant data for our readers.

The editorial board has been involved in producing this book since its inception. They have spent rigorous hours researching and exploring the diverse topics which have resulted in the successful publishing of this book. They have passed on their knowledge of decades through this book. To expedite this challenging task, the publisher supported the team at every step. A small team of assistant editors was also appointed to further simplify the editing procedure and attain best results for the readers.

Apart from the editorial board, the designing team has also invested a significant amount of their time in understanding the subject and creating the most relevant covers. They scrutinized every image to scout for the most suitable representation of the subject and create an appropriate cover for the book.

The publishing team has been an ardent support to the editorial, designing and production team. Their endless efforts to recruit the best for this project, has resulted in the accomplishment of this book. They are a veteran in the field of academics and their pool of knowledge is as vast as their experience in printing. Their expertise and guidance has proved useful at every step. Their uncompromising quality standards have made this book an exceptional effort. Their encouragement from time to time has been an inspiration for everyone.

The publisher and the editorial board hope that this book will prove to be a valuable piece of knowledge for researchers, students, practitioners and scholars across the globe.

LIST OF CONTRIBUTORS

Kentaro Honda and Masahiro Nakaoka
Akkeshi Marine Station, Field Science Center for Northern Biosphere, Hokkaido University, Aikappu, Akkeshi, Hokkaido, Japan

Yohei Nakamura
Graduate School of Kuroshio Science, Kochi University, Nankoku, Kochi, Japan

Wilfredo H. Uy
Institute of Fisheries Research and Development, Mindanao State University at Naawan, Misamis Oriental, The Philippines

Miguel D. Fortes
The Marine Science Institute, CS, University of the Philippines-Diliman; Diliman, Quezon City, The Philippines

Thi Hong Hanh Bui and Shing Yip Lee
Australian Rivers Institute and School of Environment, Griffith University Gold Coast Campus, Southport, Qld, Australia

Stephen D. A. Smith and Ana Markic
National Marine Science Centre, Southern Cross University, Coffs Harbour, New South Wales, Australia

Delphine Rocklin
UMR M101, Aménagement des Usages, des Ressources et des Espaces marins et littoraux, Observatoire des Sciences de l'Univers - Institut Universitaire Européen de la Mer, Universitéde Brest, Brest, France

Harold Levrel and Johanna Herfaut
UMR M101, Aménagement des Usages, des Ressources et des Espaces marins et littoraux, Unitéd'Economie Maritime, Ifremer, Plouzané, France

Mickaël Drogou and Gérard Veron
Département des Ressources Biologiques et Environnement, UnitéSciences et Techniques Halieutiques, Ifremer, Plouzané, France

Sherie C. Harding, Barbara P. Nash, Erich U. Petersen, A. A. Ekdale and Christopher D. Bradbury
Department of Geology and Geophysics, University of Utah, Salt Lake City, Utah, United States of America

M.Darby Dyar
Department of Astronomy, Mount Holyoke College, South Hadley, Massachusetts, United States of America

Ingrid Tulp, Marieke Keller, Hendrik V. Winter and Martin de Graaf
Institute for Marine Resources and Ecosystem Studies (IMARES), Wageningen University, IJmuiden, The Netherlands

Jacques Navez and Willy Baeyens
Department of Analytical and Environmental Chemistry, Vrije Universiteit Brussel, Brussels, Belgium
Section de Pétrographie-Minéralogie-Géochimie, Koninklijk Museum voor Midden-Afrika, Musée Royal de l'Afrique Centrale, Tervuren, Belgium

Matthew J. Slater and Selina M. Stead
School of Marine Science and Technology, Newcastle University, Newcastle upon Tyne, Tyne and Wear, United Kingdom

Yunus D. Mgaya
Department of Aquatic Sciences and Fisheries, University of Dar es Salaam, Dares Salaam, Tanzania

Siobhan C. Watkins, Paul K. Hayes and Joy E. M. Watts
School of Biological Sciences, University of Portsmouth, Portsmouth, United Kingdom

James R. Smith
School of Pharmacy and Biomedical Sciences, University of Portsmouth, Portsmouth, United Kingdom

Chris L. Dupont, John P. McCrow, Lisa Zeigler Allen, Andrew E. Allen, Robert Friedman, J. Craig Venter and Jeff Hoffman
Microbial and Environmental Genomics, J. Craig Venter Institute, San Diego, California, United States of America

John Larsson, Karolina Ininbergs, Johannes Asplund-Samuelsson, Narin Celepli, Martin Ekman, Björn Brindefalk and Birgitta Bergman
Department of Ecology, Environment and Plant Sciences, Stockholm University, Stockholm, Sweden

Shibu Yooseph, Alexander R. Richter, Aaron Tenney, Daniel Brami and Jonathan H. Badger
Informatics Group, J. Craig Venter Institute, San Diego, California, United States of America

Johannes Goll, Mathangi Thiagarajan, Andrey Tovchigrechko and Douglas B. Rusch
Informatics Group, J. Craig Venter Institute, Rockville, Maryland, United States of America

Andrew J. Lucas
Marine Physical Laboratory, Scripps Institution of Oceanography, University of California San Diego, San Diego, California, United States of America

Åke Hagström
Swedish Institute for the Marine Environment (SIME), University of Gothenburg, Gothenburg, Sweden

Anders F. Andersson and Daniel Lundin
KTH Royal Institute of Technology, Science for Life Laboratory, School of Biotechnology, Solna, Sweden

Johan A. A. Nylander
Department of Biodiversity Informatics, Swedish Museum of Natural History, Stockholm, Sweden

Erling Norrby
Center for History of Science, The Royal Swedish Academy of Sciences, Stockholm, Sweden

Jarone Pinhassi
Centre for Ecology and Evolution in Microbial Model Systems, Linnaeus University, Kalmar, Sweden

Lalit Kumar
University of New England (UNE), School of Rural and Environmental Sciences, Armidale, New South Wales, Australia

Jaqueline Thorner
University of New England (UNE), School of Rural and Environmental Sciences, Armidale, New South Wales, Australia

National Marine Science Centre, Southern Cross University (SCU), Coffs Harbour, New South Wales, Australia

Stephen D. A. Smith
National Marine Science Centre, Southern Cross University (SCU), Coffs Harbour, New South Wales, Australia

Katherine L. Yates and David S. Schoeman
Environmental Science Research Institute, University of Ulster, Coleraine, United Kingdom

Faculty of Science, Health, Education and Engineering, University of the Sunshine Coast, Maroochydore DC, Queensland, Australia

Emie Durighello, Joseph Alexander Christie-Oleza, Jean Armengaud
CEA, DSV, IBEB, Lab Biochim System Perturb, Bagnols-sur-Céze, France

Gilles Havik and Milena Holmgren
Resource Ecology Group, Wageningen University, Wageningen, The Netherlands

Alessandro Catenazzi
Department of Zoology, Southern Illinois University Carbondale, Carbondale, Illinois, United States of America

Juan Chen, Duan-Ye Xiong, Wen-Hua Wang, Wen-Jun Hu, Martin Simon, Juan Chen, Xiang Liu and Hai-Lei Zheng
Key Laboratory for Subtropical Wetland Ecosystem Research of MOE, College of the Environment and Ecology, Xiamen University, Xiamen, Fujian, China

Qiang Xiao
Laboratory of Biological Resources Protection and Utilization of Hubei Province, Hubei Institutes for Nationalities, Enshi, Hubei, China

Ting-Wu Liu
Department of Biology, Huaiyin Normal University, Huaian, Jiangsu, P.R. China

Andrea K. Townsend
Department of Wildlife, Fish, and Conservation Biology, University of California Davis, Davis, California, United States of America

Christopher M. Barker
Department of Pathology, Microbiology, and Immunology, School of Veterinary Medicine, University of California Davis, Davis, California, United States of America

MohamedSalah Romdhane and Frida Ben Rais Lasram
UR 03AGRO1 Ecosystémes et Ressources Aquatiques, INAT (Institut National Agronomique de Tunisie), Tunis, Tunisia

Tarek Hattab
UR 03AGRO1 Ecosystémes et Ressources Aquatiques, INAT (Institut National Agronomique de Tunisie), Tunis, Tunisia
UMR 212 Ecosystémes Marins Exploités, IRD (Institut de Recherche pour le Développement), Séte, France

Philippe Cury and François Le Loćh
UMR 212 Ecosystémes Marins Exploités, IRD (Institut de Recherche pour le Développement), Séte, France

Camille Albouy
Département de biologie, Chimie et Géographie, UQAR (Universitédu Québec áRimouski), Québec Canada

Chérif Sammari
Laboratoire du milieu marin, INSTM (Institut National des Sciences et Technologies de la Mer), Salammbô, Tunisia

Fabien Leprieur
Laboratoire Ecologie des Systémes Marins Côtiers UMR 5119, UM2 (Universitéde Montpellier 2), Montpellier, France

Antonio Pusceddu, Cristina Gambi and Cinzia Corinaldesi
Dipartimento di Scienze della Vita e dell'Ambiente, Universitá Politecnica delle Marche, Ancona, Italy

Mariaspina Scopa
Dipartimento di Scienze della Vita e dell'Ambiente, Universitá Politecnica delle Marche, Ancona, Italy
Istituto Zooprofilattico Sperimentale dell'Abruzzo e del MoliséG. Caporalé, Experimental Regional Centre for Fishery and Acquacolture, Termoli, Italy

Roberto Danovaro
Dipartimento di Scienze della Vita e dell'Ambiente, Universitá Politecnica delle Marche, Ancona, Italy
Stazione Zoologica Anton Dohrn, Napoli, Italy

Yaping Chen and Yong Ye
Key Laboratory of the Ministry of Education for Coastal and Wetland Ecosystem, College of the Environment and Ecology, Xiamen University, Xiamen, Fujian, China

Ibrahim F. Farag, James P. Davis, Noha H. Youssef and Mostafa S. Elshahed
Department of Microbiology and Molecular Genetics, Oklahoma State University, Stillwater, Oklahoma, United States of America

Peter Norberg
Department of Infectious Diseases, University of Gothenburg, Göteborg, Sweden

Maria Bergström and Malte Hermansson
Department of Chemistry and Molecular Biology, Microbiology, University of Gothenburg, Göteborg, Sweden

François Brischoux
Centre d'Etudes Biologiques de Chizé, CEBC UMR 7372 CNRS-ULR, Villiers en Bois, France

Yurii V. Kornilev
Bulgarian Society for the Protection of Birds, NCC "Poda", Burgas, Bulgaria

Shengchang Yang, Luzhen Chen, Wenqing Wang, Xiaona Du Canmou Wang and Yan Ma
Key Laboratory of the Ministry of Education for Coastal and Wetland Ecosystems, School of Life Sciences, Xiamen University, Xiamen, Fujian, China

Weizhi Lu
Key Laboratory of the Ministry of Education for Coastal and Wetland Ecosystems, School of Life Sciences, Xiamen University, Xiamen, Fujian, China
Division of Marine Sciences and Technology, Graduate School at Shenzhen, Tsinghua University, Shenzhen, Guangdong, China

Guangxuan Lin
Administrative Bureau of Zhanjiang National Mangrove Nature Reserve, Zhanjiang, Guangdong, China

Guanghui Lin
Division of Marine Sciences and Technology, Graduate School at Shenzhen, Tsinghua University, Shenzhen, Guangdong, China
Ministry of Education Key Laboratory for Earth System Modeling, Center for Earth System Science, Tsinghua University, Beijing, China

Raquel Muñiz-Salazar
Escuela de Ciencias de la Salud, Universidad Autónoma de Baja California, Ensenada, Baja California, México

Eduardo Sandoval-Castro
Escuela de Ciencias de la Salud, Universidad Autónoma de Baja California, Ensenada, Baja California, México
Facultad de Ciencias Marinas, Universidad Autoónoma de Baja California, Ensenada, Baja California, México

Luis Manuel Enríquez- Paredes
Facultad de Ciencias Marinas, Universidad Autónoma de Baja California, Ensenada, Baja California, México

Richard S. Dodd, Juan Manuel López-Vivas
Department of Environmental Science, Policy, and Management, University of California, Berkeley, Berkeley, California, United States of America

Rafael Riosmena-Rodríguez
Departamento Académico de Biología Marina, Universidad Autónoma de Baja California Sur, La Paz, Baja California Sur, México

Cristian Tovilla-Hernández
El Colegio de la Frontera Sur, Tapachula, Chiapas, México

Bily Aguilar-May
Instituto Tecnológico Superior de Villa La Venta, Huamanguillo, Tabasco, México

Erik Olsen
Institute of Marine Research, Bergen, Norway

David Fluharty
School of Marine and Environmental Affairs, University of Washington, Seattle, Washington, United States of America

Alf Håkon Hoel
Institute of Marine Research, Tromsø, Norway

Kristian Hostens and Ellen Pecceu
Animal Sciences, Aquatic Environment and Quality, Bio-environmental Research, Institute for Agricultural and Fisheries Research (ILVO), Oostende, Belgium

Frank Maes
Maritime Institute, Ghent University, Ghent, Belgium

Index